Food Safety
Implementation

from farm to fork

Food Safety
Implementation

from farm to fork

Puja Dudeja PhD
Post Graduate Institute of Medical Education and Research
(PGIMER)
Chandigarh

Amarjeet Singh MD
Professor
Department of Community Medicine
PGIMER
Chandigarh

Sukhpal Kaur
Lecturer
National Institute of Nursing Education
PGIMER
Chandigarh

CBS Publishers & Distributors Pvt Ltd

New Delhi • Bengaluru • Chennai • Kochi • Kolkata • Mumbai
Hyderabad • Nagpur • Patna • Pune • Vijayawada

Food Safety
Implementation
from farm to fork

ISBN: 978-93-85915-25-3

Copyright © Authors and Publisher

First Edition: 2016

Published by Satish Kumar Jain and produced by Varun Jain for

CBS Publishers & Distributors Pvt Ltd

4819/XI Prahlad Street, 24 Ansari Road, Daryaganj, New Delhi 110 002, India.

Ph: 23289259, 23266861, 23266867

Fax: 011-23243014

Website: www.cbspd.com

e-mail: delhi@cbspd.com; cbspubs@airtelmail.in.

Corporate Office: 204 FIE, Industrial Area, Patparganj, Delhi 110 092

Ph: 4934 4934   Fax: 4934 4935

e-mail: publishing@cbspd.com; publicity@cbspd.com

Branches

- **Bengaluru:** Seema House 2975, 17th Cross, K.R. Road,
  Banasankari 2nd Stage, Bengaluru 560 070, Karnataka
  Ph: +91-80-26771678/79   Fax: +91-80-26771680   e-mail: bangalore@cbspd.com
- **Chennai:** 7, Subbaraya Street, Shenoy Nagar, Chennai 600 030, Tamil Nadu
  Ph: +91-44-26680620, 26681266   Fax: +91-44-42032115   e-mail: chennai@cbspd.com
- **Kochi:** Ashana House, No. 39/1904, AM Thomas Road, Valanjambalam,
  Ernakulam 682 018, Kochi, Kerala
  Ph: +91-484-4059061-62-64-65   Fax: +91-484-4059065   e-mail: kochi@cbspd.com
- **Kolkata:** 6/B, Ground Floor, Rameswar Shaw Road, Kolkata-700 014, West Bengal
  Ph: +91-33-22891126, 22891127, 22891128   e-mail: kolkata@cbspd.com
- **Mumbai:** 83-C, Dr E Moses Road, Worli, Mumbai-400018, Maharashtra
  Ph: +91-22-24902340/41   Fax: +91-22-24902342   e-mail: mumbai@cbspd.com

Representatives

- **Hyderabad** 0-9885175004
- **Nagpur** 0-9021734563
- **Patna** 0-9334159340
- **Pune** 0-9623451994
- **Vijayawada** 0-9000660880

Printed at: India Binding House, Noida, UP

Foreword

Food is one item that every human being or animal consumes every day. The Indian human population being more than 1.3 billion, every morsel of the 3 billion plus meals per day needs to be safe to ensure the safety of health of people and to reduce the national burden of expenditure on public health. It, therefore, becomes the collective responsibility of every citizen to remain aware about all possible hazards emerging from inappropriate practices.

It is imperative to remember that safety of the food sold in the supermarkets or smaller stores comes from proper operations at each stage of the production chain starting right from the field until it reaches the plate. Even the animal needs to consume safe feed to ensure the safety of the food consumed by human beings. The new food safety standards under the Food Safety and Standards Act will only prescribe the safety limits of different contaminants like residues of pesticides and veterinary drugs, heavy metals, mycotoxins, NOTS and microbiological contaminants. Truly speaking, the safety of food will come from compliance with best practices and training along the food supply chain. And regular surveillance, data analysis and dissemination will be the key to operators taking appropriate steps for ensuring safe food and prevention from food or waterborne diseases. Greater attention is desirable in respect of the street food vendors and school canteen supplies as well as the vulnerable consumer groups like infants, young children, pregnant women and senior citizens.

This book presents a good collection of case studies, research work and information on the legal aspects, and should serve as a useful practical guide to all stakeholders, be it the consumers, food manufacturers, canteens, caterers, transporters or even medical practitioners and hospitals. The age-old methods followed in India reflect on the best practices applied by all present the logic for food safety when modern technology was not available to us. A thorough reading of this book should encourage the introduction of a food safety program in the course curriculum in our medical and other colleges and also at the school level.

S. Dave

Preface

Food safety is a vital issue in India with frequent reports of outbreaks of foodborne illness (FBI) resulting in substantial costs to individuals, health care system and the country. Mishandling of food has been implicated as the cause of 97% of all foodborne outbreaks. However, most of these are preventable through proper implementation of food safety measures. The kitchen has been described as the frontline in the battle against foodborne disease.

Human behavior is intimately linked with violation or infringement of food safety principles when we discuss quality of food in eating establishments, i.e. behavior of suppliers, producers, food business operators (FBOs) or food handlers (cooks or waiters). As per the health promotion concept in public health, there are two approaches to promote compliance of stakeholders with recommended food safety practices—educative and legislative. Both are important and need to be used together. Government has done its job by enacting Food Safety and Standards Regulations (FSSR) 2011. Simultaneously we need to promote awareness levels of community and their skills through on various aspects of food safety. For this standard text material is needed. This book will serve this purpose.

Over the past two decades, there has been an increasing consumer demand for safe and high quality foods. Paralleling this demand, there has been a rise in food safety hazards also due to increased processing and lengthening of food safety chain 'from farm to fork'. This book is an attempt to provide a comprehensive approach to ensure food safety. All the issues in the food chain at the farm, during storage, transportation, retail, cooking and serving have been covered comprehensively. These have been discussed taking practical and real-life food safety related examples which will provide an invaluable source of information. It will provide readers with a completely new outlook for appreciation of food safety issues. The editors of the book are public health specialists (medical doctors, faculty in apex medical institutes) with an in-depth understanding of the concepts of food safety and their application.

It is envisioned that the book will support the needs of all those working in the area of food safety. The book has been written in simple language, easy to understand and comprehensible by the layman. Nevertheless, it will serve as a useful guide for others too. The book has been conceptualized keeping in mind the needs of all the community members on food safety. The books available in the market on this subject have followed a syllabus-based approach for the students of home science, food technology, etc. They are full of technical jargon. However, our idea is to demystify the science of food safety so that the practical steps to ensure food safety can be understood and followed by the community. The comprehensiveness, originality and practical utility with universal appeal will make it stand out among other books in the market. This book provides key insight to the need of systematic and effective measures to be taken to avoid outbreaks of FBI.

The contents of the book have been designed to cover all the ignored, yet important, areas in the field of food safety. It will be an excellent reference for anyone involved in, studying or considering entering the food industry. The knowledge and information contained provides realistic, practical and very usable information.

Puja Dudeja
Amarjeet Singh
Sukhpal Kaur

Contributors

Alka Ahuja PhD Scholar
Punjab University, Chandigarh
dralka48@yahoo.com

Arun Gupta MD, DNB (Community Medicine)
drgupta.arun@yahoo.com

Dalbir Singh MD (Forensic Medicine)
PGIMER, Chandigarh

Gunjan Grover Research Scholar
PGIMER, Chandigarh
ggrover243@gmail.com

H Ravi Rammamurthy MD (Pediatrics)
drhravi@gmail.com

Ishwarpreet Kaur PhD Scholar
PGIMER, Chandigarh
ishwarpreet814@yahoo.co.in

Jaideep MPH, PhD
deep.1050@gmail.com

Mamta Bansal MSc, PhD
drmamta1503@gmail.com

Maninder Kaur PhD Scholar
Punjab University, Chandigarh
maninder2911@yahoo.in

Meenakshi Sharma MPH, PhD
mnxmph@gmail.com

Nancy Sahni PhD
PGIMER, Chandigarh
drnancysahni@yahoo.com

Neha Chanana PhD Scholar
PGIMER, Chandigarh
drneha.mph@gmail.com

Nidhi Bhatnagar MD, DNB (Community Medicine)
Army College of Medical Sciences, New Delhi
nidhibhatnagar20@gmail.com

Rohit Tewari MD (Pathology)
rohittewariaa@gmail.com

Ruchi Sharma PhD Scholar
PGIMER, Chandigarh
ruchi_sharma158@yahoo.com

Satinder Pal Singh MD (Forensic Medicine)
GMC Sec32, Chandigarh
spsingh9988@yahoo.in

Shalini Dwivwdi MSc, PhD
shalinidwivedi0904@gmail.com

Sonia Puri MD (Community Medicine)
GMC Sec32, Chandigarh
soniagpuri@gmail.com

Sonika Raj MPH, PhD
sonikagoel007@yahoo.com

Sukhbir Singh PhD
Punjab University, Chandigarh
buntysen2k@gmail.com

Sumeet Kaur
GMC Sec32, Chandigarh

Suninder Kaur PhD
suninderk@hotmail.com

Surjinder Singh PhD Scholar
PGIMER, Chandigarh
sukhwinder_dr@yahoo.co.in

Tavleen Kaur BDS, MPH
drtavleen95@gmail.com

Contents

Section V: Food Handlers: An Important Link in Food Safety

Section VI: Government Efforts to Ensure Food Safety

Section VII: Food Safety in Special Conditions of life

Section VIII: Newer Issues in Food Safety

Section IX: Miscellaneous

Section I

Introduction to Food Safety

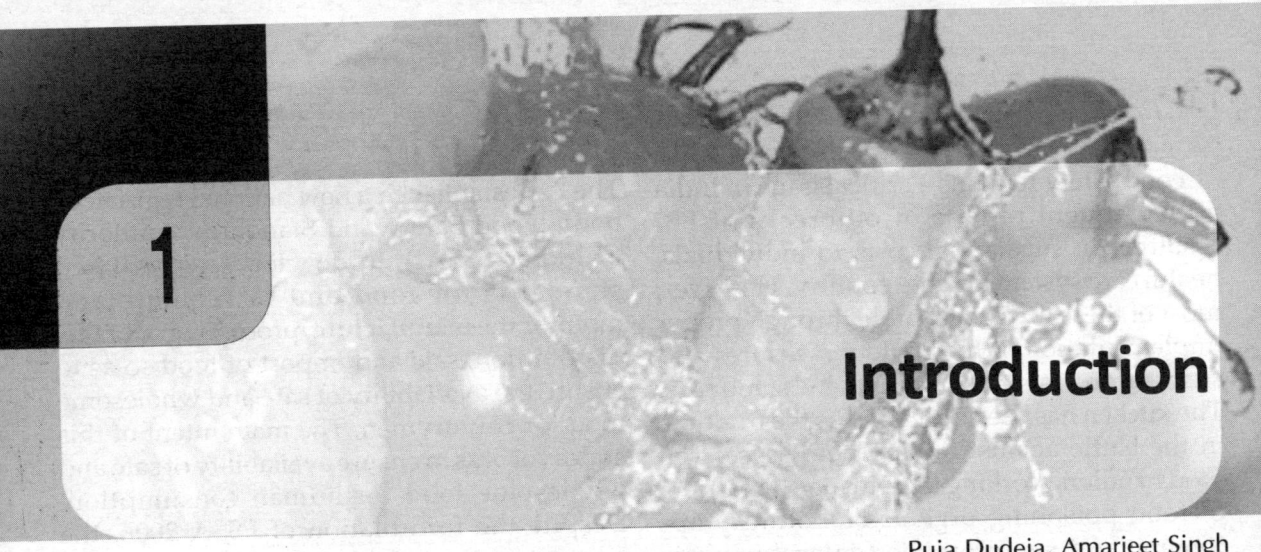

1

Introduction

Puja Dudeja, Amarjeet Singh

Food provides us with energy and nutrients for daily activities of our life. We expect our food to be not only nourishing but also safe alongside. Until few decades back, food safety was not a much discussed issue in our society, as most of us would eat home cooked food. Nevertheless, food safety can be compromised both in home available food as well as in outside food. Our ancestors would gather/hunt/cook food and consume as soon as possible after cooking. However, with industrialization and advancements in food storage, transport and processing technologies there has been a lengthening in the food chain from farm to fork. The food industry in the current scenario does mass production followed by marketing and distribution of food items for more profit which was not there earlier. This can be ascribed to fast changes in our life style and eating pattern. The growing market for foods provides both challenges and opportunities to people involved in food business. Food safety which was 'one off' local/personal phenomenon has become global now, e.g. Incidents of Bovine Spongiform encephalopathy, adulteration of meat with horse meat, rejection of Indian Alphanso mangoes exported to Europe, etc. These are issues concerned with safety of food at international level.

In the past two decades, there has been an exponential rise in growth of food industry leading to an increase in international trade of foodstuffs. The variety of food items available in the market has also increased tremendously. Shelves of supermarkets are flooded with variety of such food products. Most of the fruits and vegetables are now available throughout the year. Usual season of growing and location are now irrelevant for the availability of a food item. Increase in travel both for work and holidays have also brought many people in contact with different cultures and cuisines. This has resulted in greater demand for more exotic foods leading to new opportunities for food trade. It is difficult to assure the safety of a food item which has travelled miles before reaching our table or has been handled by people in eating establishments. It may be contaminated with harmful chemicals, pesticides, antibiotic residues, pathogenic microorganisms, etc and may cause foodborne illness (FBI) on consumption.

Today the consumers are much more aware, sensible and cognizant of the choice they have in terms of quality of food products. Despite all the rhetorics, presently in India consumer have to accept food products at the face value even if raw material used is of low quality.

Food safety remains a critical issue in India with frequent reports of outbreaks of FBI resulting in substantial costs to individuals, health care system and the country. However, most of these are preventable through proper implementation of food safety measures and strict enforcement of food hygiene standards. The kitchen has been described as the frontline in the battle against foodborne disease. The food handlers working therein may be carriers of food poisoning organisms both asymptomatic and symptomatic (enterotoxigenic staphylococci). The main tools to fight this battle are implementation of legislative measures in commercial EEs, training of food handlers and managers and sanitary inspections that may result in improved food handling practices.

At an international level Hazard Analysis and Critical Control Point (HACCP), Safe Quality Food, ISO certifications, Food Safety Management System, etc. are recognized as part of food safety assurance system. These concentrate on preventive strategies to ensure safety of food products.

In India, the concept of food safety is now being looked into seriously. A new era in food safety has been initiated by formulation of the Food Safety and Standards Act (FSSA) 2006.

The Act established a new national regulatory body, Food Safety and Standards Authority of India(FSSAI), to develop science based standards for food and to regulate and monitor the manufacture, processing, storage, distribution, sale and import of food so as to ensure the availability of safe and wholesome food for countrymen. The main intent of this endeavor was to ensure availability of safe and wholesome food for human consumption. Despite the formulation of FSSA 2006, the present status of food hygiene in eating establishments in India is dismal. Food safety is essentially an issue of prevention. For ensuring food safety, basic hygiene and cleanliness in manufacturing units or kitchens can immensely contribute towards food safety. Good infrastructure with adequate lighting, ventilation, hand washing and toilet facilities for food handlers, adequate and clean storage facilities, clean and maintained equipment utensils, etc. are the most important factors that facilitate food safety.

The current law is comprehensive and has given due importance to the principles of health promotion like clean environment, safe water and proper waste disposal facilities. It is expected that this will improve the quality of food for consumers and ensure safety of food.

2

Food Safety Concepts in Charaka Samhita

Jaideep Kumar

Ayurveda is one of the great gifts of the sages of ancient India to mankind. Ayurveda indicates the science by which life in its totality is understood. Ayurveda, regarded as a Holistic manual of Life and Age, describes a lifestyle thats in harmony with nature. Ayurveda considers the individual as whole and seeks to re-establish harmony between all the constituents of the body and a perfect balance of the tripod—Mind, Body and Spirit. Basically Ayurveda is health promotive–preventive–curative and nutritive—all self contained.

Ayurveda is grounded in a metaphysics of the *panchmahabhuta* ('five great elements'; earth, water, fire, air and ether)—all of which compose the universe, including the human body. *Rasa dhatu* (chyle), *rakta dhatu* (blood), *mamsa dhatu* (flesh), *medha dhatu* (fat), *asthi dhatu* (bone), *majja dhatu* (marrow) and *shukra dhatu* (semen or female reproductive tissue) are held to be the *saptadhatus*, i.e. seven primary constituent elements of the body. Ayurveda stresses a balance of three humors or energies: *vata* (wind/air), *pitta* (bile) and *kapha* (phlegm). According to Ayurveda, these three regulatory principles—*dosas* (*tridosa*)—are important for health, because when they are in balanced state, the body is healthy, and when imbalanced, the body has diseases.

Ayurveda, the age old science of life, has always emphasized to maintain the health and prevent the diseases by following proper diet and lifestyle regimen rather than treatment and cure of the diseases. Thousands of years ago, Ayurveda pointed out the importance of preventive over curative approach. The two principle objectives of Ayurveda are:

1. '*Swasthasya Swasthya Rakshanam* (स्वस्थस्य स्वास्थ्यरक्षणं)'—To prolong life and promote perfect health (add years to life and life to years).

2. '*Aturasya Vikar Prashamanamcha* (अतुरस्य विकार प्रशमनाम्चा)'—To completely eradicate the disease and dysfunction of the body.

First objective of Ayurveda gives stress on prolonging health life and to remain healthy instructions about daily routine (*dinacharya*) and seasonal regimes (*ritucharya*) are described in it. In addition to this, food/dietary schedules for different times of the day, different seasons, according to one's age and most importantly, to suit one's individual constitution (*prakriti*) are elaborated in the classical texts of Ayurveda. *Ahara* (food) and dietary practices are emphasized in Ayurveda as essential component for healthy living as well as for the treatment of diseases.

5

In *Charaka Samhita*, in context to the origin of *Purusa* (man) and his diseases the concluding remark of *Punarvasu Atreya* to accept the 'Ahara' as a causative factor for both, carries the historical importance of dietetics. Among three *Upastambhas* (supporting factors) of life, the *Ahara* (diet), *Nidra* (sleep) and *Brahmacharya*, the diet is an essential factor for maintenance of healthy life. Being supported by these three well regulated factors of life, the body is endowed with strength, complexion and growth and continues up till the full span of life.

Acharya *Charaka* has mentioned that, *Anna* (food) is the best sustainer of life. Taking in account the utility of food not only at individual but also at universal level, it has been mentioned that whatever beneficial for worldly happiness, whatever pertains to the Vedic sacrifices leading to heaven, and whatever action leads to spiritual salvation are all said to be established in food. *Charaka* says that it is the *Ahara* (food) which maintains the equilibrium of bodily *dhatus* and helps in promotion of health and prevention of diseases. He classified the food articles in different ways.

Charaka Samhita explained intensely about health, hygiene, diet, lifestyle and medicine. According to this *Samhita*, the objects of the Ayurvedic science are two-fold, viz. the treatment of patients suffering from the disease and maintenance of positive health. Of all the factors for the maintenance of positive health, food taken in proper quantity occupies the most important position.

The quantity of food to be taken again depends upon the power of digestion and metabolism. The amount of food which, without disturbing the equilibrium of *dhatus* and *dosas* of the body, gets digested as well as metabolized in proper time is mentioned as the proper quantity. This power of digestion

and metabolism again varies according to season as well as of the individuals. Measurement of food is, in fact, of two types, viz. food as a whole and its different ingredient having different tastes like sweet, sour, etc.

The concept of taste is very important in context to food safety according to Ayurvedic perception. According to this perception, if food as a whole is taken according to the prescribed measurement but its ingredients having different tastes are not in prescribed ratio, the equilibrium of *dhatus* and *dosas* gets definitely disturbed due to imbalance in the composing *rasas* (tastes). When employed properly, *rasas* maintained the body and their incorrect utilization results in the vitiation of *dosas*.

Because of this importance, priority is given to description of *rasas* in *Charaka Samhita*. *Charaka* has mentioned the six *rasas*: Sweet, astringent, bitter, sour, salty and pungent are that humans experience. Just like the *dosas*, each taste is composed of two elements. For instance, the sweet taste is composed of earth and water, the sour taste of fire and water, the salty of earth and fire, pungent of fire and air. The qualities are then expanded. Each taste then has qualities and also actions it has on substance. Substance like our bodies and the tastes having effects because of the qualities it holds. Each taste has a specific effect on the human body and has its specific characteristics.

The concept of wholesome diet is very nicely explained by the *Charaka* by correlating this with the *rasas* theory. According to this, a substance conductive to individual is called 'satmya' and use of such substance results in the well-being of that individual. This is of three types, viz. superior, inferior and mediocre. Use of all the *rasas* in the diet is superior type of *satmya* whereas use of only one *rasa* is of inferior type. In between the superior and inferior types is the mediocre type of *satmya*. He recommended that the inferior and mediocre types of *satmya* should be slowly changed over to the superior *satmya* for the good health.

Charaka has given the description of the eight factors which determine the utility of various types of food. These includes: *prakriti* (nature of food articles); *karana* (methods of their processing); *samyoga* (combination); *rasi* (quantity); *desa* (habitat); *kala* (time, i.e. stage of the disease or the stages of the individual); *upayogasamstha* (rules governing the intake of food) and *upayokin* (wholesomeness to the individuals who takes it). These eight factors are associated specifically with useful and harmful effects and they are conditioned by the one another.

Rules for taking food are pointed out in *Charaka Samhita*. According to this, healthy individuals as well as some of the patients should eat only that food in proper quantity which is hot, unctuous and not contradictory in potency and that too, after the digestion of previous food. Food should be taken in proper place equipped with all the accessories, without talking and laughing, with concentration of mind and paying due regards to oneself. Thus, the food taken in prescribed manners helps in bringing about the strength, complexion, happiness and longevity.

Quality of food with reference to their heaviness and lightness is elaborated by *Charaka*. He gives emphasis to *prakriti* or body constitution in consideration of food intake. Three basic *dosas*—*Vata*, *Pitta* and *Kapha* form seven types of *prakriti*. In this world each person is a unique being of unique heredity, unique environment, unique biochemical structure and unique mental status. For this reason all the natural and good food items cannot be effective with all the individuals to the same extent. Every individual should take a diet suitable to his predominant constitutional *dosas*, to balance them in different seasons.

Charaka explains the seasonal regimen in detail. It is called as *Tasyashiteeya Adhyaya*. It literally means—qualitative dietetics explained based on seasons. Suitable diets for every season are described in *Charaka Samhita*. This is divided into six parts according to six seasons. These seasons includes: *sisira* (late winter), *vasanta* (spring), *grisma* (summer), *varsa* (rainy season), *sarat* (autumn) and *Hemanta* (early winter). These six seasons are grouped under two *ayana*, i.e. *uttarayana* (*adana kala* or the period of dehydration) and *dakshinayana* (*visarga kala* or the period of hydration).

To provide the knowledge of suitable diet for different seasons, *Charaka* has added the knowledge of seasons themselves. The year according to Ayurveda is divided into two periods *Ayana* (solstice) depending on the direction of movement of sun that is *Uttarayana* (northern solstice) and *Dakshinayana* (southern solstice). Each is formed of three *Ritus* (seasons). The word *Ritu* means "to go." It is the form in which the nature expresses itself in a sequence in particular and specific in present forms in short, the seasons. *Adana Kala* (hot and dry)—The north movement of Sun brings about water loss in the body beginning from late winter to summer. Sun is dominant. *Visarga Kala* (cold and wet)—the southward movements of sun is coolant and forms the other seasons beginning with the rainy to early winter. Moon is dominant. Hence one has to adjust his diet based on these variations.

According to him, in the period of *visarga* winds are not very dry as they are during the periods of *adana*. The period of *visarga*, predominantly shares the qualities of moon and during this period, the moon, with the undertrained cooling property, continuously delight the world with its soothing rays. The period of *adana*, on the other hand is dominated by the quality of *agni* (fire). These two periods constitutes the time, season, taste and vitiation of *dosas* and bodily strength.

According to him "One's diet leads to promotion of strength and complexion only if he knows the wholesomeness according to different seasons dependent on behavior and diet". One should consciously resort to food and lifestyle habits that are opposite to the nature of the season and of the disease. Seasonal dietary recommendations are given by *Charaka*. He mentioned that *"Tasya Shitadiya Ahaarbalam Varnascha Vardhate. Tasyartusatmayam Vaditam Chestaharvyapasrayam,"* which means 'the strength and complexion of the person knowing the suitable diet and regimen for every season and practicing accordingly are enhanced.

Acharya Charaka has described various food articles as *Hitatama* (wholesome) and *Ahitatama* (unwholesome) *Ahara* by nature. Apart from elemental constitution of food various dietary rules and other factors like *matra* (quantity), *kala* (time or season), *kriya* (mode of preparation), *bhumi* (habitat or climate), *deha* (constitution of person), *desha* (body humour and environment), etc. also play a significant role in the acceptability of wholesome diet.

Description of indicated and contraindicated foods is also given in *Charaka Samhita*. It is mentioned that one should not regularly take heavy articles such as *vallura* (dried meat), dry vegetables. One should never take meat of a diseased animal. Moreover, one should not regularly take *kurchika* (boiled buttermilk), *Kilata* (inspissiated milk), pork, beaf, meat of buffalo, fish, curd, etc. On the other hand, regularly intake of *swastika* (a kind of rice harvested in sixty days), *sali, mudga,*

rock salts, rain water, *amalaka*, ghee, meat of animal dwelling in arid climate and honey is recommended in *Charaka Samhita*.

Incompatible diets are responsible for various disorders in human beings. Charaka has described eighteen factors responsible for food incompatibility as:

1. *Desha* (climate)
2. *Kala* (season)
3. *Agni* (digestive power)
4. *Matra* (quantity)
5. *Satmya* (accustom)
6. *Doshas* (tridosha)
7. *Samskara* (mode of processing)
8. *Ahara virya* (potency of food)
9. *Kostha* (bowel habits)
10. *Avastha* (state of health)
11. *Krama* (order of food intake)
12. *Parihara* (restriction)
13. *Upachara* (prescription)
14. *Paka* (cooking)
15. *Sanyoga* (combination)
16. *Hridya* (palatability)
17. *Sampad* (richness of quality)
18. *Vidhi* (rules of eating).

In addition to this, *Charaka* had illustrated the concept of poisoning very nicely. According to *Charaka*, the king is exposed to danger of being poisoned through food and regimens by the attendants secretly employed in his palace by another king having enmity, and also from his own wives. Therefore, the attendants should be carefully examined. Sometimes, women because of their ignorance, give undesirable material along with food to their husbands in order to be bestowed with auspiciousness. Similarly, cooks, etc. secretly employed by the enemy may give poison through food and regimens. Therefore, these retinues should be examined carefully. *Charaka* had mentioned the features poison giver as a person who behaves in an extremely suspicious manner, who is garrulous or who speaks very little, who has lost lustre of his face and who exhibits changes in his characteristic features.

Charaka described the characteristics of poisonous food. He mentioned that when a person exhibiting the characteristic features of a poison-giver is located, then the food, etc. served by him should not be taken immediately, but a part of it should be thrown over fire. If the food is poisoned, then the flame of the fire exhibits abnormal characteristics like variegated color of pea-cock feather. The smoke which comes out of such fire is sharp, intolerable and ununctuous, and it smells like a dead body. The flame which comes out makes a cracking noise; it moves spirally or it gets extinguished. If the water of wells and ponds are poisoned, then the water becomes foul-smelling, dirty and discolored.

According to *Charaka*, if the poison is added to drinks like alcohol, then blue lines appear over its surface or it becomes discolored. A person's own shadow is not reflected through such drinks or the shadow is reflected in a distorted manner. The poisoned food when kept in a pot gets discolored, and flies sitting on it succumb to death. When this poisoned food is seen by crows, their voice becomes feeble, and when the *cakor* bird sees it, its eyes become discolored.

Symptoms of poisoned person are also illustrated in *Charaka Samhita*. The smell of poisoned food and drinks causes headache, pain in the cardiac region and fainting. If touched, such poisoned food and drinks cause oedema and numbness in the hands, burning sensation and pinching pain in the fingers, and cracking of the nails. When put into the mouth, these poisoned food and drinks cause tingling sensation in the lips, swelling, numbness and discoloration of the tongue, tingling sensation in the teeth, stiffness of the jaw-bones (mandibular joints), burning sensation in the face, salivation and morbidity in the throat.

If the poisoned food and drinks have entered into the stomach, then the patient suffers from discoloration, sweating, asthenia, nausea, impairment of the vision, arrest of cardiac functions and appearance of drop like pimples all over the limbs. If the poisoned food and drinks enter into the colon, then the patient suffers from fainting, intoxication, unconsciousness, burning sensation, weakness, drowsiness and emaciation. The patient suffers from anemia when the poisoned food and drinks get localized in the abdomen. Intake of poisoned water causes edema, urticaria and pimples, and even death. If the tooth brushing twig is poisoned, when the brush-like tip of it gets withered, and the patient suffers from edema of the teeth, lips and muscles of the mouth. If the oil for application over the head is poisoned, then the patient suffers from hair-fall, headache and tumors in the head.

Intake of poisoned food vitiates the *kostha* (gastrointestinal tract) and external application of poisoned material afflicts the skin in the beginning. If the poison has reached the stomach, then the physician in the beginning should administer emetic therapy. If the poisonous material is located in the skin, then ointments and fomentation therapy, etc. should be administered. These therapeutic measures should be administered, keeping in view the nature of *dosas* and the strength of the patient.

Due to its great significance, the food has attained the highest position in this universe. Charaka has given much detail of the religious practices; those increase devotion towards diet and have good impact on the mind and body. Hence, before taking the food, for the betterment of life, it should be worshipped in its full religious and ethical discipline.

The concept of food safety is described in *Charaka Samhita* in a much illustrated way in comparison to current issues taken care by the selected ones. *Charaka* had not only described the general food safety concerns, he illustrated the topic with individualized focus according to the nature, body type of person. The concept of taste and analogism of food in correlation to foods safety are quoted by *Charaka* very nicely, which is on other hand is not much explored by the modern researchers. Serious attention is required to these unexplored issues to accommodate them in current food safety instruction both at community as well as individualized level.

Charaka Samhita, Sushruta Samhita and *Ashtanga Hridaya* are considered as the major triad of Ayurveda. In all these classical texts of Ayurveda, concepts of food along with its safety issues are described in detailed. In this chapter we have elaborated the food safety concept as discussed in *Charaka Samhita*, one of the three major classics of Ayurveda.

BIBLIOGRAPHY

1. Datta, N.; Chowdhury, K.; and Badal, J. Diet in Elderly: An Ayurvedic Perspective. *Scholars Journal of Applied Medical Sciences (SJAMS)*, 660–663. Retrieved June 10, 2014, from http://saspublisher.com/wp-content/uploads/2014/03/SJAMS-22B660-663.pdf

2. Kumar, J.; Roy, JD.; and Minhas, AS. *Suitability of Ayurvedic Graduates in Public Health Workforce of India-An Exploratory Study*. Germany: VDM Verlag Dr. Mullar GmbH and Co.KG, 2010.

3. Nathani, N. An Appraisal of the Concept of Diet and Dietetics in Ayurveda. *Asian Journal of Modern and Ayurvedic Medical Science*, 2(1). Retrieved from http://ajmams.com/viewpaper.aspx?pcode=ef5cc 006-6583-4806-b293-7d170f36d956, Jan 2013.

4. Sharma, RK.; Dash, B. (2011). *Charaka Samhita* (Vols 1–5). Varanasi, Uttar Pardesh, India: Chowkhamba Sanskrit Series Office, 2011.

5. Shukla, V.; and Tripathi, RD. (Eds.). (2003). *Charak Samhita of Agnivesa* (Vol. II). Delhi, India: Chaukhamba Sanskrit Pratishthan 2003.

6. Thakkar, J.; Chaudhari, S.; and Sarkar, PK. Ritucharya: Answer to the Lifestyle Disorders. *Ayu*, (2011, Oct-Dec) 32(4), 466–471. Retrieved June 11, 2014, from http://www.ncbi.nlm.nih.gov/pmc/articles/PMC3361919/

7. V, S. G. (n.d.). *Ayurvedic Concepts of Nutrition and Dietary Guidelines for Promoting/Preserving Health and Longevity*. Retrieved June 11, 2014, from http://nutritionfoundationofindia.res.in: http://nutritionfoundationofindia.res.in/FetchScriptpdf/festschrift%20%20 for %20%20Dr%20Gopalan/Section%201-scientific%20papers/Satyavati%204.11.pdf

3

Food Safety Concepts in Sushruta Samhita

Ayurveda, the age old science of life, has always emphasized to maintain the health and prevent the diseases by following proper diet and lifestyle regimen rather than treatment and cure of the diseases. The basic principle followed in the Ayurvedic system of medicine is 'Swasthasya Swasthya Rakshanam (स्वस्थस्य स्वास्थ्यरक्षणं)' which means to maintain the health of the healthy, rather than 'Aturasya Vikar Prashamanamcha (अतुरस्य विकार प्रष्शमनाम्चा)' means to cure the diseases of the diseased.

Thousands of years ago, Ayurveda pointed out the importance of preventive over curative approach. Ayurveda, regarded as a holistic manual of life and age, describes a lifestyle that's in harmony with nature. Ayurveda gives elaborate guidelines for achieving perfect health and remaining healthy in its "Swasthavritta" (literally meaning "on being healthy") through Dinacharya (daily routine) and Ritucharya (seasonal regimens). Comprehensive instructions are given on specific food/dietary schedules (for different times of the day, different seasons, according to one's age and most importantly, to suit one's individual constitution or "Prakriti"). The Ayurvedic description of health is:

"Samadosha, Samadhatu, Samagnischa malkriyah (समदोष, समधातु, समाग्निश्च मलक्रिय)

Prasannatmendriyamanah, Swastha itya-bhidhiyate (प्रसन्नात्मेंद्रियामन, स्वस्थ इत्याभिधियतेऋ).

That is Only he, whose dosas (Vata, Pitta, Kapha) dhatus (physical components—Rasa, Rakta, Mansa, Meda, Asthi, Majja and Shukra, i.e. Plasma, Blood, Flesh, Fats, Bones, Bone marrows and Semen respectively) and agni (digestive fire) is balanced, appetite is good, all tissues of the body and all natural urges are functioning properly, and whose mind, body and spirit (self) are cheerful or full of bliss, is a perfectly healthy person.

Ayurveda has views on how individuals should be involved in their own health and health care. The essentials of health care, viz., health education, personal hygiene and habits, exercise, dietary practices, food, sanitation, environmental sanitation, code of conduct and self-discipline, civic and spiritual values, treatment of minor ailments and injuries, etc. were emphasized and advocated in Ayurveda.

In the treatment of diseases, according to the Ayurvedic system of medicine, obser-vance of a diet regime is a very essential part of treatment. Ayurveda has given stress on Ahara (food) for good health. Diet has been scientifically and extensively linked to disease.

According to Ayurveda food is medicine and medicine is food. Eating correctly is the most important aspect of Ayurvedic life-style in both the short term and the long term. What is so-called 'correct' or 'compatible' depends on the individual and as the saying goes *'one man's meat is another man's poison'*. Different foods suit different people. Thus, the health of a community depends upon the adequate availability of the food and intelligent consumption of food.

Ayurveda included the food in the three support of life. Classical literature of Ayurveda contains ample knowledge about the food safety in the form of information about nature of food materials, proper quantity of food and protection of food. This is also supported by the mentioned information about the concept of importance of the site of food, characteristics of the person who put poison (suspected) along with the features of poisonous food. In addition to this, information about the testing techniques of poisoned food, diseases produced by poisoned food, purification therapy for person who consumed the poisoned food are also point out. Texts have also considered incompatible food similar to poison and artificial poison. Ayurvedic classical texts had included the food safety information very judicious. To attain good health, the need is only to explore and apply the things mentioned in the texts.

Sushruta Samhita is considered as the one of the major classical text of Ayurveda, in which concepts of food safety is described in very appropriate way. Some of the major recommendations of *Sushruta* about food are depicted here. *Sushruta* has mentioned the importance of food—as growth, strength, good health, complexion and the alertness of sense are traceable to food. According to him, ill health too can be traced to the dis-equilibrium in food which belongs to four categories. These categories are soft food

(*asita*), drinks (*pita*), lickables (*lidha*) and chewables (*khadita*).

Sushruta insisted that the food should be prepared by the experienced cook in the kitchen having the mentioned characteristics. After that, it should be kept at clean and pure spot, concealed from the view of public. Food should be made safe by admixture of anti-toxin medicine, and freed from all poison by reciting mantra over it. *Sushruta* mentioned the various recommendations for serving of food. Some of them are described here.

Containers for serving of various food items: *Sushruta* has also described the methods of serving the various items in respective containers. Some of them are mentioned here as-clarified butter should be served out in a vessel of steel (*kanta-loha*); liquid (*peya*), in a silver blow; and all kinds of fruit and confectionary (such as the *laddukas*), on leaves. The preparations of meat known as the *parisuska* and *pradigdha mansa* should be served on golden plates; fluid edibles and meat essences in silver bowls; *katvaras* and *kharas* in stone utensils; and cool boiled milk in copper vessels. Other drinks, wines and cordials should be given in earthen pots.

Arrangement of food items: The cook should place the bowls containing preparations of pulse and boiled rice on clean, spacious trays of fanciful design, and spread them out in front (of the king). All kinds of desserts, confectionary and dry fruits should be served on his right, while all soups, etc. meat-essences, drinks, cordials, milk and *peya* should be placed on his left. Bowls containing preparations of treacle should occupy a place midway between the two sets of bowls described above.

Place of food intake: *Sushruta* recommended that the intelligent physician, well conversant with the rules of serving dishes as above laid down, should attend upon the kind at his table, and spread out on the purified level floor of a solitary, beautiful, spacious, blissful, perfumed and flower-decorated chamber, and the kind should partake of those sacred and pleasant dishes, served neither hot nor cold, and cooked and seasoned in the desired mode, and possessed of their specific tastes.

Code for dining: *Sushruta* mentioned that the diner should sit on a comfortable raised seat, assumed a relaxed position and take interest in the meal. The wise person eats when hungry and prefers light, emollient, hot, largely liquid and agreeable food at the appropriate time and in appropriate quantity. He should not linger over the meal which should be enjoyed at a reasonable pace.

Order of food intake: According to *Sushruta,* the general convention for meal is to serve the sweet items to begin with, move on to sour and salty in middle and reserve the other *rasas* (tastes) to the last part of meal. Fruits such as tubers, pomegranate, etc. are often prescribed as starter. *Amalaka* fruit is appropriate at any stage of the meal.

Other recommendation for dining: During the cold months, when the nights are longer, substances, which tend to subdue the bodily humours which are naturally deranged during that season, should be eaten in the morning, while during the seasons, when the days are inordinately long, things which are congenial in those seasons should be eaten in the afternoon (in spring and autumn), when days and night are equal, the meal should be taken just at the middle part of the day and night.

A meal should not be eaten before the appointed time, nor before the appetite has fully come. Similarly, over or insufficient eating should be equally refrained from. Eating at an improper time and before the system feels light and free brings on a large number of diseases, and may ultimately lead to death. A meal eaten at an hour long after the appointed time tends to aggravate the bodily *Vayu*, which affects the digestive fire, and offers serious obstacles in the way of its digestion. The food thus digested with difficulty in the stomach creates discomforts and destroys all desire for a second meal.

Insufficient diet gives but in adequate satisfaction, and tends to weaken the body. Over eating, on the contrary, is attended with such distressing symptoms, as languor, heaviness of the body, disinclination for movements, and distension of the stomach, accompanied by rumbling in the intestines, etc. Hence it is recommended to take only as much food as he can easily, digest, which should be well cooked and made to possess all the commendable (adequately nutritive) properties. Moderation in diet is the golden rule, besides taking into consideration the demerits of a particular food before partaking thereof and the nature of the time (day or night) it is eaten.

Food eaten with a good appetite tastes pleasant and relishing. The food which is congenial to ones temperament begets no discomfort after the eating. Light food is soon digested. Emollient food gives tone and vigour to the system. Warm food improves the appetite. Food eaten neighter too slowly nor too hurriedly is uniformly digested. Food abounding in fluid components is not imperfectly digested, nor is attended by any acid reaction. Moderation in food leads to a happy and perfect digestion and tends to maintain the fundamental principles of the body in their normal state. More and more palatable dishes should be successively taken in the course of a meal.

During the course of a meal, the mouth should be frequently rinsed or gargled in as much as the palate thus constantly being cleansed becomes more susceptible to taste, and anything eaten thereafter is relished the better and gives all the pleasures of a first morsel. The palate affected with a sweet taste at the outset fails to appreciate the tastes of the successive dishes. Hence, the mouth should be washed at intervals during the meal. Sweet food eaten with a relish pleasurably affects the mind, brings joy, energy, strength, and happiness in its train, and contributes to the growth of the body; whereas the one of a contrary character is attended with opposite effects. The food, which does not satiate a man even after repeated eating should be considered as agreeable to him.

After finishing a meal water should be drunk in a quantity which would be beneficial. Food particles adhering to the teeth should be gently drawn out by means of a tooth-pick, in as much as if not removed a kind of fetor is produced in the mouth.

Precaution which should be taken for food intake: *Sushruta* mentioned that boiled rice food which is impure and dirty, infested with poison, or out of which another has eaten a portion as well as that which is full of weeds, pebbles, dust, etc. which the mind instinctively repels, or cooked on the previous day or which has been kept standing over-night, as well as that which is insipid or emits a fetid smell, should be similarly rejected. Also food which has been cooked long ago, or has become cold and hard, and has been re-warmed or which has been imperfectly strained, or is burnt and insipid should also not be served as food.

A diet which abounds in fluid courses should be refrained from. Only a single taste should not be enjoyed in the course of a meal. Cooked potherbs and course of diet abounding in acid taste should be avoided. Articles of one taste should not be eaten in large quantities at a time, nor should articles of various tastes be constantly indulged in.

A second meal should not be eaten on the same day in the event of the appetite having become dulled by a previous meal. Eating with a previous meal only partially digested seriously impairs the digestive functions. A man of dull or impaired appetite should refrain from eating heavy articles of food, as well as from partaking of large quantities of light substances. Cakes should never be eaten, and a double quantity of water should be taken if they are eaten at all out of hunger, by which their safe digestion would be ensured. Of drinks, lambatives and confectionary (solid food), each succeeding one is heavier than the one immediately preceding it in the order of enumeration. Heavy articles of food should be taken in half measures only, while the lighter ones may be eaten till satiety.

Liquid food or that which abounds in liquid substances should not be taken in large quantities. Dry articles of food taken in combination with a large number of other substances fail to do any injury to the stomach. Dry food (*anna*) taken alone cannot be completely digested. It is transformed into lumps in the stomach, is irregularly chymed, and produces deficient gastric digestion followed by a reactionary acidity. The ingested food, whether of a character that stamps it as belonging to the *vidathi* group or not, is but incompletely digested and gives rise to a reactionary acidity in the event of the *Pitta* being confined in the stomach, or in the intestines. Dry food (cakes, etc.), incompatible food combinations (milk with fish and so on), and those, which are long retained in the stomach in an undigested state, tend to impair the digestive functions (*agni*).

The eating together of both wholesome and unwholesome articles of food is called promiscuous eating (*samsanam*). Over or insufficient eating at intervals and at improper seasons goes by the denomination of irregular eating (*visamasana*). Eating before a former meal is thoroughly digested in the stomach is called *adhyasana*. These three kinds of eating are injurious, and speedily give rise to a variety of diseases, or may be ultimately attended with fatal consequences. The drin-+king of cold water helps the speedy digestion of a partially digested food, which has already been attended with a reactionary acidity, in as much as the coldness of the imbibed water tends to subdue the deranged *Pitta*, and the food thus moistened by the water naturally gravitates into the intestines.

Conduct after meal: According to Sushruta, the *Vayu* is increased after the completion of digestion, the Pitta, during the continuance of the process, while the *Kapha* is increased immediately after the act of eating. Hence, the *Kapha* is to be subdued after the close of a meal, and the intelligent eater should attain that end by partaking fruit of an astringent, pungent, or bitter taste, or by chewing a betel leaf prepared with broken areca nut, camphor, nutmeg, clove, etc. or by smoking, or by means of anything that instantaneously removes the viscidity in the cavity of the mouth, and permeates it with its own essence.

Then the eater should take rest, like a king, till the sense of drowsiness incidental to eating is removed. After this he should walk a hundred paces and lie down in a bed on his left side. After eating, a man should enjoy soft sounds, pleasant sights and tastes, sweet perfumes, soft and velvety touch, in short anything that ravishes the soul and enwraps the mind with raptures of joy, since such pleasurable sensations greatly help the process of digestion. Sounds, which are harsh and grating, sights, which are abominable, touches, that are hard and unpleasant, smells, which are fetid and disagreeable, encountered after a meal, or the eating of impure and execrable boiled rice, or a loud side-splitting laugh after a meal is followed by vomiting.

The after-meal siesta should not be long and continuous; basking before a fire, exposure to the sun, travelling, driving in a carriage, swimming, bathing, etc. should be avoided just after the close of a full and hearty meat.

Drinking of an abnormal quantity of water, irregular eating, voluntary suppression of any natural urging of the body, sleep in the day, keeping of late hours in the night, are the factors which interfere with the proper digestion of food and develop symptoms of indigestion even individual takes light and advantages diet at proper time. The food taken by a person under the influence of envy, passion, greed, or anger, etc. or by a man suffering from a chronic distemper, is not properly digested.

Types of indigestion: Types of indigestion are mentioned in *Sushruta Samhita*. According to this *Samhita*, a case of indigestion in which the undigested food matter acquires a sweet taste is called chymous (or mucous) indigestion (*amajirna*) that in which the undigested food acquires an acid taste in the stomach is called *vidagdha* indigestion. The form in which the food matter brought down into the stomach is partially or irregularly digested (one portion being digested, the other being not) followed by a pricking or piercing pain in the stomach and entire suppression of the flatus, is called *vistabdha* indigestion. The type known as indigestion of unassimilated chyle is characterized by the absence of any acid or sour eructations, but the patient feels no inclination for food in spite of the normal character of the eructations, if any. The type is further characterized by pain about the region of the heart, and water-brash. The unfavorable symptoms of indigestion are sudden loss of consciousness, delirium, vomiting, water-brash, languor with a gone-feeling in the limbs, and vertigo, etc. which may end in death. *Sushruta* also mentioned the treatment of all types of indigestion.

Fasting is beneficial in a case of *amajirna* indigestion. Ejection of the contents of the stomach gives relief in a case of *vidagdha* indigestion. Fomentation will alleviate a case of *vistabdha* indigestion, while in indigestion of unassimilated chyle, the patient should be confined to bed and fomentations, and digestive medicines should be administered as well. In a case of *vidagdha* indigestion the patient should be made to vomit the contents of his stomach with the help of warm water saturated with salt, while in a case of chymous indigestion the patient should forego all food till he is restored to his natural condition. A patient suffering from indigestion whose system has been cleansed and lightened with the above said appliances should go fasting till he is restored to his natural condition as regards the strength and humours of the body.

Poisonous food and guarding the important person like king from it: According to *Sushruta Samhita*, poisonous food may be given to the king by servants or relatives and sometimes by the ladies of the royal families to kill him. In this regards Sursruta mentioned the characteristics of royal kitchen, head of the kitchen, poisonous foods, etc. as described here.

Characteristics of royal kitchen: According to *Sushruta Samhita*, the royal kitchen should be a spacious chamber occupying an auspicious (south-east) corner of the royal mansion and built on a commendable site. The vessels and utensils (to be used in a royal kitchen) should be kept scrupulously clean. The kitchen should be kept clean, well lighted by means of a large number of windows and guarded with nets and fret works (against the intrusion of crows, etc.) Highly inflammable articles (such as hay, straw, etc.) should not be stocked in the royal kitchen whose ceiling should be covered with a canopy. The fire-god should be (daily) worshipped therein. Every person entering the kitchen must be watched and examined.

Characteristics of the head of the kitchen: *Sushruta* has pointed out that none but the trusted and proved friends and relatives should have access to the royal kitchen, or hold any appointment therein. He pointed out the importance of superintendent of royal kitchen. *Sushruta* insisted that a king should appoint a physician for the royal kitchen (to superintend the preparations of the royal fare). He pointed out that a physician of the royal kitchen should be very cautious and circumspect in the discharge of his duties, since food is the main stay of life, and the sole contributor to the safe continuance of the body.

According to him, the head or manager of the royal kitchen should generally possess the same qualifications as of a physician. The necessary qualifications of a superintendent of the royal kitchen are described by him as-he should come of a respectable family, should be virtuous in conduct, fondly attach to the person of his sovereign, and always watchful of the health of the king. He should be greedless, straight-forward, god-fearing, grateful, of handsome features, and devoid of irascibility, roughness, vanity, arrogance and laziness. He should be forbearing, self-controlled, cleanly, compassionate, well-behaved, intelligent, capable of bearing fatigue, well-meaning, devoted of good address, clever, skilful, smart, artless, energetic and marked with all the necessary qualifications (of a physician) as described before. He should be fully provided with all kinds of medicine and be highly esteemed by the members of his profession. In addition to this, *Sushruta* also recommended that he should be well-paid.

Sushruta has also mentioned indications for other kitchen staffs. He described that the every one employed in a royal kitchen such as, bearers, servers, cooks, soup makers, cake-makers (confectioners), should be placed under the direct control and supervision of the physician of the kitchen. The bearers and cooks in the royal kitchen should have their nails and hair clipped off and should bear turbans. They should be cleanly, civil, clever, obedient, good-looking, each charged with separate duties good-tempered, composed in their behaviour, well-bathed, greedless, determined, and prompt in executing the orders of their superiors.

Characteristic features of a poisoner: *Sushruta* has pointed out that head of kitchen or other authorised person should be there to spot a would be poisoner on the basis of signs/ characteristics such as unusual speech, action and facial expression; who become confused or silent on questioning ; speak irrationally or excessively; cracks knuckles or digs the earth with fingers; laughs, shakes and appears pale and fearful; breaks things with nails or scratches the head frequently and aimlessly; keeps an eye on the back door again and again; and behave as good sense has left him. *Sushruta* also mentioned that sometimes a good attendant may behave abnormally out of the fear of king, so the authorised person should inspect the food properly.

Indications of poisoned food and drink: According to *Sushruta*, all liquid substances such as wine, milk, water, etc. if anywise poisoned, are found to be marked with variegated stripes on their' surface and become covered over with froth and bubbles. Shadows are not reflected in such (poisoned) liquids and if they ever are, they look doubled, net-like (porous) thin and distorted. Preparations of potherbs, soups, boiled rice and cooked meat are instantaneously decomposed and become putrid, tasteless and emit little odour when in contact with poison. All kinds of food become tasteless, smelless and colourless when in contact with poison. Ripe fruit, under such conditions, is speedily decomposed and the unripe ones are found to get prematurely ripe.

Methods to detect the poisonous food: *Sushruta* has mentioned the methods to detect the poisonous food. According to him, a portion of the food prepared for the royal use should be first given to crows and flies and its poisonous character should be presumed if they instantaneously die on partaking of the same. Poisoned food bums making loud cracks, and when cast into the fire it assumes the colour of a peacock's throat, becomes unbearable, bums in severed and disjointed flames and emits irritating fumes and it cannot be speedily extinguished.

Sushruta has pointed out that specific change in the behavior of birds and animal occurs when they come in the contact of poisonous food. Some of them include-the eyes of a *Cakora* bird are instantaneously affected by looking at such poisoned food; the cooing of the cuckoo becomes hoarse and a *Kraunca* (heron) becomes excited; a peacock moves about and becomes sprightly; a swan cackles and a *Prsata* (a species of spotted deer) sheds tears and a monkey passes stool.

Characteristic features of poisoned water: A sheet of poisoned water becomes slimy, strong-smelling, frothy and marked with (black-coloured) lines on the surface. Frogs and fish living in the water die without any apparent cause. Birds and beasts that live (in the water) on its shores roam about wildly in confusion (from the effects of poison), and a man, a horse or an elephant, by bathing in this (poisoned) water is afflicted with vomiting, fainting, fever, a burning sensation and swelling of the limbs.

Symptoms of affected person: The vapours arising from poisoned food when served for use give rise to a pain in the cardiac region and produce headache arid restlessness of the eyes. A poison affecting the palms of the hands produces a burning sensation in them and leads to the falling off the finger-nails. Poisoned food pat-taken of through ignorance or folly, produces a stone-like swelling and numbness of the tongue, a loss of the faculty of taste and a pricking burning pain in that organ attended with copious mucous salivation.

Food mixed with poison, when it reaches the stomach gives rise to epileptic fits, vomiting, dysenteric stools, distension of the abdomen, a burning sensation, shivering and a derangement of the sense-organs. Food mixed with poisons if it reaches the intestines, gives rise to a burning sensation (in the body), epileptic fits, dysenteric stools, derangements of the organs of sense-perception, rumbling sounds in the abdomen and emaciation, and makes the complexion (of the sufferer) yellow.

Affects of poison on the human organism according to type of poisonous food: Root-poisons or poisonous roots produce a twisting pain in the limbs, delirium and loss of consciousness. A leaf-poison or poisonous leaf gives rise to yawning, difficult breathing and a twisting pain in the limbs. A fruit poison is attended with a swelling of the scrotum, a burning sensation in the body and an aversion to food. A flower-poison gives rise to vomiting, distension of the abdomen and loss of consciousness. A bark-poison, or pith-poison, or gum-poison is marked by a fetid in the mouth, roughness of the body, headache and a secretion of *Kapha* (mucus from the mouth). The effects of the poisonous milky exudations (of a tree, plant or creeper) are foaming from the mouth, loose stools (diar-rhoea) and heaviness of the tongue, whereas a mineral poison gives rise to pain in the heart, fainting and burning sensation in the region of the palate. All these are slow poisons

proving fatal only after a considerable length of time.

Treatment for poisoned person: *Sushruta* has described the treatment for a poisoned person very adequately according to type of poison as well as to the stages.

In addition to these topics *Sushruta* has pointed out the importance of wholesome and unwholesome foods. *Sushruta* described the things which are unwholesome through combination, incompatible preparation of food and objectionable proportions. Concept of rasas (tastes) and seasonal dietary recom-mendations (*ritu charyas*) are described in *Sushruta Samhita*.

In this way *Sushruta* describes the food as a source of life, strength, complexion and ojas of living being. He further mentioned that food is the primary agent in the genesis, existence and dissolution of even the divine Brahma.

BIBLIOGRAPHY

1. Bhishagratna, K. Sushruta Samhita (3 ed., Vols. 1–3). (J. Mitra, Ed.) Varanasi, Uttar Pardesh, India: Chowkhamba Sanskrit Series Office, 2005.

2. Department of AYUSH, MOHFW,GOI. RGGA HOSP Paprola. Retrieved April 18, 2012, from www.indianmedicine.nic.in: http://indian-medicine.nic.in/index4.asp? ssslid=431&sub-subsublinkid=103&lang=1.

3. Kumar, J.; Roy, JD.; Minhas, AS. (2010). Suitability of Ayurvedic Graduates in Public Health Workforce of India—*An Exploratory Study*. Germany: VDM Verlag Dr. Mullar GmbH & Co.KG, 2010.

4. Tripathi, AD. Sushruta Samhita. Varanasi: Chaukambha Sanskrit Sansthan, 1999.

5. Valiathan, MS. The Legacy of Sushruta (Vol. First). Chennai, Tamil Nadu, India: Orient Longman Private Limited. Retrieved March 20, 2014

4

The Quest for Food Safety from Past to Present

Puja Dudeja, Amarjeet Singh

Food is vital for survival. Safe food is the essence of life. It is essential for existence of humankind. But if we get contaminated food it can cause illness and mortality also. Foodborne diseases have been a constant concern of every society throughout the history of mankind. Though, fatality only happens in a minority of cases, the morbidity associated with the cases of food related illness has significant social and economic consequences. In ancient times, our ancestors would hunt, gather food and consume it on the same day to satisfy the basic need of hunger. Though they would not hesitate to eat the left over of large carnivores. It was not uncommon to eat reptiles, amphibians, insects, large and small mammals along with plants, nuts, fruits and roots. Early humans, probably by trial and error, also started to develop the art of recognition and avoidance of foods that were naturally toxic. There is a faint evidence of avoidance of berries and mushrooms during specific seasons. They would often hunt the weakest and sick prey as it was easy. In case the hunted animal was large they would also eat meat over many days. With the discovery of fire man started to cook food. This made the food tastier and also would eliminate threat from many foodborne disease causing micro-organisms.

Man started making efforts to keep food safe long time back time even before Biblical times. Various civilizations ensured food safety both through their own methods of preservation and food handling. They started the basic forms of food preservation, which possibly also made food safer, e.g. drying, salting, fermentation. Ancient Egyptians developed a storage tank designed to hold grain harvested from the fields called the *Silo*. Storing grain in a silo kept it cool, dry and able to last into the non-harvest months or longer. The Bible describes of the Hebrews receiving manna from heaven every morning as they get spoiled in a day. Romans first recognized the importance of freshness in fruits and other foods. They also salted their foods for preservation—a practice which still holds true. Later, it was found that if foods are kept cold they last longer. That is how some people started keeping meat and fish in the water fall to keep it fresh. Snow and ice were also identified as natural refrigerants.

It is known today that Egyptian Mummies were contaminated with foodborne biological agents like beef and pork tapeworm, liver flukes, whipworms, guinea worm, etc. Knowledge about the harms of unsafe food was present in our ancestors also. The link between bad food and sickness was well

19

understood by them. They were worried about the hygiene and safety of food. The ancient Greeks and Romans have also described the term food poisoning. The Greeks have mentioned that food poisoning was so common that tasters were employed to check the food before being served to the royal people. They have also described about food adulteration since food became an item of trade. Nutmeg was used to hide the taste and smell of decomposed meat before serving.

In 2000 BC Moses introduced laws to protect his people from food related disease, such as, the washing of clothes and bathing after the sacrificial slaughter of animals. Egyptians, Greeks and Romans also expressed similar concerns. Numerous taboos existed in the past that indirectly ensured food safety. For example, in 1800 BC Judea prohibited consumption of pork. In 500 BC Confucius, a Chinese spiritual leader warned against eating "sour rice." In India, list of unclean foods was developed for the first time in 500 BC, which included food items like meat cut with a sword, dog meat, etc.

Bible text refers to food consumption, prohibition of certain foods and pathological conditions which could result from improper consumption. A lot of evidence is available depicting that food safety was given due importance in ancient India (Dealt in detail in separate chapters)

The middle ages saw developments in food regulations. Ergot (Claviceps purpura) was commonly known as St Anthonys Fire and a number of outbreaks of ergotism were reported. There were 40,000 deaths reported to it in a single incidence in France in 944. King John developed the first food law in 1202.

During the period between 1300 and 1750 mold poisoning killed many people in England and other parts of Europe. This continued till nineteenth century. Ergotism

and Alimentary Toc Aleukia (ATA) were responsible for significant illnesses in Eurpoe. During World war II also thousands of Russians died of ATA as they were forced to eat infected grain. Food related parasites like Ascaris, Trichura, Taenia, Fasciola were also discovered during this time.

The period of industrial revolution between 1750–1900 witnessed migration of lot of people to cities. The demand for food for people in cities grew. These led to the concept of mass production of food. There was a change in agricultural practices, processing and trade patterns. There were many episodes of mass food poisoning. This led to an increase in demand for food which would not spoil or cause illness. Napoleon Bonaparte in early 1800s felt the need to prevent the food from spoilage for his soldiers. Nicolas Appert offered a solution to put the food in jars with lids and boil. People during this time were still struggling to find out what actually spoiled the food and was the cause of sickness after taking such food. Canning came as the solution till the time there were episodes of botulism and lead poisoning. At this stage, the scientific basis of food poisoning was not fully understood. Nevertheless, the advances in microbiology during 19th century contributed to a great extent in understanding the concept of food spoilage.

Father of microbiology, Antonie van Leeuwenhoek, reported the existence of microorganisms. The term food hygiene was first heard in 19th century. The issue was not thoroughly debated till the latter part of nineteenth century when Pasteur, Ferdinand Julius Cohn, and August Gartneu began to demonstrate that, although people couldn't see them, there were organisms in the air, soil, animals and water that can make us sick. Pasteur's work on pasteurization and fermentation had an mammoth impact on the science of food safety. The advances in microbiology complemented those in food technology.

August Gartner in 1888 diagnosed Bacillus enteritidis as cause of foodborne illness in fifty seven people who had eaten beef of a sick cow. Major milestones in food hygiene and safety in 19th century are given in Table 4.1.

Table 4.1: Milestones of food safety in 19th century

Year	Scientist	Advances
1810	Nicholas Appert	Basis of commercial heat processing
1850	John Tyndall	Tyndallization process for germ reduction in heat-sensitive foods
1855	Friedrich Küchenmeister	Relationship between pork tapeworms in humans and the parasitic infection cysticercus cellulosa
1860	Friedrich Albert Zenker	Proved the infectiveness of parasitic trichinae
1864	Louis Pasteur	Pasteurization process for the preservation of food
1880	Gartner	First isolation of Salmonella from foodpoisoning outbreak
1890	Multiple Scientists	Work in field of dairy bacteriology
1895	Carl von Linde	Cooling process to preserve food

Early in the 20th century, when food safety was a major concern to the public, two technologies, milk pasteurization and retort canning, were developed, which prevented foodborne diseases. Social changes during this era included rapid increase in women work force. The food processing industry started growing and brought convenience to those cooking and consuming home cooked food. This led to decline in consumption of home cooked food. Along with this the culture of eating out started.

Russel worked in area of food Processing and Microbiology from 1895–1928. He gave the first textbook of Dairy Bacteriology in 1928. One of the breakthrough achievements of the era was by MA Barber in 1914 that purposefully got food poisoning by spoiling milk and then consuming it and clarified the link between spoiled food and illness.

Explosions of technological advancements brought a revolution in food safety in twentieth century. After having understood the concept of germs, low temperature maintaining food safe, the world was looking for ways to keep foods cold. Simultaneously, with the change in society from agricultural to industrial, the desire for freshness in shipped beef and meat was felt. This gave birth to refrigeration technology.

The first home coolers which came in use were metal lined, insulated boxes with ice (icebox). The parallel advances in medical sciences benefited food safety technology. John gorrie, a physician, came up with the idea to treat patients with respiratory illnesses by putting them in rooms that had been artificially made cold. This idea became the first work of mechanical refrigeration. Later refrigerators became an important equipment of all kitchens.

The story of Minmata disease in Japan by consuming fish contaminated with methyl mercury was unfolded in 1970s. In 1981, the tragedy of tainted oil in Spain made many children sick by consuming rapeseed oil in unlabeled plastic containers as a cheap substitute for olive oil. The superbug *E. coli* was also discovered when cases of bloody diarrhea were analyzed in detail. Newer issues like bovine spongiform encephalopathy also emerged.

These episodes of foodborne illnesses were paralleled by advances in food science and technology and formulation of food laws. The history of modern food safety legislation can be traced back to Victorian England, when widespread adulteration of food was a serious problem. This was not only fraudulent, but was often dangerous. For example, toxic salts of lead and mercury were sometimes used to provide additional colour in sugar confectionery intended for children. The urgent need to curb these practices lead to the introduction of the first Food Adulteration Act in 1860. In India the first law against adulteration was passed in 1954 'Prevention of Food Adulteration Act'. World Health Organization (WHO) gave guidelines on investigation and control of foodborne disease outbreaks.

Even in 21st century, foodborne disease remains a major threat to public health, as new foodborne pathogens have emerged. With rapid globalization forces spearheading the social changes food science scenario has also changed. More and more people are moving to cities. There is a rapid increase in food trade. Globalization of the food supply has led to the rapid and widespread international distribution of foods. Changes in micro-organisms lead to the constant evolution of new pathogens, development of antibiotic resistance, and changes in virulence of known pathogens. In many countries, as people increasingly consume food prepared outside the home, growing numbers are potentially exposed to the risks of poor hygiene in commercial foodservice settings. All of these emerging challenges require that public health specialists continue to adapt to methods to changing environment with improved methods to combat these threats. Many developments have been there in epidemiological studies of foodborne illnesses (FBI).

New ways to process the food and deliver it to the stores with increased shelf life are being developed (discussed in detail in a separate Chapter 51). Food habits and food culture of people have significantly changed with changing time. Simultaneously newer food safety issues have emerged. Apart from advances in microbiology and food processing techniques there have been developments in area of public health also. Preventive approaches in the form Hazards Analysis and Critical Control Point (HACCP), training of food handlers, ISO Certifications, Safe Quality Food, Food Safety Management Systems, Risk Assessments, Risk Analysis and Risk Communication, etc. are successfully employed to ensure safety of food (discussed in detail in a separate Chapters 23, 24, 25).

Conclusion

With thousands of years of experience of food safety combined with over 150 years of food microbiology and the advances in microbiology and food preservation techniques the problems of food safety faced by mankind should have been resolved. However, the opposite is true with increased reports of foodborne illnesses (FBI). Rather this has become an important public health problem. Woolen in 1999 wrote, "Millions of words of advice and millions of pounds spent but the problem is getting worse". Problems of food safety are both faced by developed and developing countries. Mortlock et al., 2000 documented that food safety is a problem, which can be handled through a multidisciplinary approach like HACCP, Human Error Analysis, Social Marketing, Cost Benefit Analysis, Training Needs Analysis, etc. Application of multiple preventive actions at same time has an important role to play in safety of food.

Today, however the situation is that we still have to worry about food safety. While we know enough about why and how of safe food but the journey does not end here. There are more and more challenges to face in keeping the fuel for our bodies from being our "last meal."

BIBLIOGRAPHY

1. Asha. Emerging Sectors of Indian Economy. Global Journal of Management and Business Studies 2013; 3(5): 491–196.

2. Caballero, B; Trugo, L; Finglas (eds). Encyclopedia of food sciences and nutrition. Amastradam. Elsevierpublications, 2003.

3. Foodborne disease outbreaks: guidelines for investigation and control. World Health Organization. Switzerland WHO Press; 2008.

4. Griffith JC. Food safety: where from and where to? British Food Journal. 2006; 108(1): 6–15.

5. Motarjemi Y. (ed). Encyclopedia of Food Safety. Elsevier publications; 2014.

Section II

Epidemiology of Foodborne Illnesses

5

Burden of Foodborne Illnesses

Ruchi Sharma, Puja Dudeja, Amarjeet Singh

This morning when I (Ruchi) reached my OPD, my assistant told me that there was a case in the emergency. Without even wasting a second, I rushed to examine the patient. My subconscious mind was expecting a case of chest pain with an underlying heart attack. To my surprise, it was a young boy with history of pain in the stomach, loose stools, vomiting and fever. It was a clear cut case of food poisoning. I had a sigh of relief. After examining the patient, I comforted the relatives that there was nothing to worry. I prescribed some antibiotics and IV fluids to the boy. On taking a detailed history, I came to know that my patient had consumed mayonnaise sandwiches from a street vendor last evening.

Such episodes resulting from compromised food safety are quite common in our daily life. Most often, with easy availability of over the counter drugs; we tend to manage these things through self-medication. Even if there is a need to visit a doctor these incidents are soon erased from our memories. Most of the episodes are not reported. Have you ever wondered what is the exact burden of such episodes? At our homes, we can roughly estimate one episode per household in three months or one episode per household in a month. Since the occurrence depends upon the

virulence, type and number of pathogen, and the immunity of the individual; even after eating the same dish members of the family may be differently affected. One of them may fall sick while others may not develop any symptom.

Foodborne disease outbreaks make the news daily. We can assume that billions of people fall ill every year, and that many die, because they ate food contaminated with bacteria, viruses, parasites or chemicals. But no-one has ever quantified the problem comprehensively. Indeed, we have only a sketchy idea of how many people suffer from foodborne diseases every year.

The recent reports of melamine-contaminated milk powder in China remind us that foodborne illnesses can hit at anytime and anywhere. Over 50,000 children in China suffered kidney problems and four died from drinking the contaminated milk powder, which was also exported to dozens of countries. There will be no surprise if more such victims are reported in future also.

Foodborne Illness: Definition and Nature

For human beings, food is essential both for growth and for the maintenance of life. It supplies the energy and materials required to build and replace tissues, to carry out work and to maintain the body's defences against disease. But food can also be responsible for ill-health. Food, we eat, is equally relished by bacteria and other pathogens. This has been noted particularly with fatty foods such as salami, cheese, chocolate, milk, meat and ice creams. Such foods serve as good culture medium for bacteria to grow and multiply rapidly. Depending on the food's composition and conditions of storage, a pathogen present at low and possibly harmless levels may grow to numbers sufficient to produce illness. It can then lead to group of diseases termed as food-borne illness (sometimes called "foodborne disease," "foodborne infection," or "food poisoning"). It is a common, costly—yet preventable—public health problem. The term "food poisoning" has often been used in some countries, but it is an expression that can sometimes be restrictive or misleading. *Foodborne illness* (FBI) or *foodborne disease* are now the generally preferred terms. Foodborne disease can be defined as:

"Any disease of an infectious or toxic nature caused by or thought to be caused by the consumption of food or water".

Foodborne illnesses result (FBI) from consumption of food containing pathogens such as bacteria, viruses, parasites or the food contaminated by poisonous chemicals or bio-toxins.

FBI comprise a broad spectrum of diseases. These are responsible for substantial morbidity and mortality worldwide. Microbial pathogens, biotoxins, and chemical contaminants in food represent serious threats to the health of thousands of millions of people. In most cases the illnesses caused by foodborne micro-organisms, principally bacteria, are associated with gastrointestinal symptoms of nausea, vomiting, stomach pains and diarrhoea. Since diarrhoea is a common clinical symptom in FBI, many of these diseases are referred to as "diarrhoeal diseases". Estimates vary, but it is generally believed that in developed countries less than 10%, or even only 1%, of cases of FBI ever reaches official statistics. In countries with fewer resources, under-reporting must be even greater.

For taking any worthwhile public health action for this problem, it is important to understand the epidemiology (extent and pattern) of FBI. It will help in streamlining prevention and control efforts. Proper data will help administrators in appropriately allocating resources to control them. It will also facilitate monitoring and evaluation of food safety measures. Even development of new food safety standards and assessment of the cost-effectiveness of interventions will be benefitted by data on FBI.

Food safety has emerged as an important public health issue in 21st century. Increasing urbanization has led to greater requirements for transport, storage and preparation of food. In developing countries street vendors cater to bulk of the snacks/meal needs of the masses. In developed countries also, up to 50% of the food budget may be spent on food prepared outside the home. In the era of globalization, food is prepared in bulk at one outlet and then is distributed across the country or even throughout the globe. All these changes lead to situations in which a single source of contamination can have widespread consequences.

The population in developing countries is more prone to suffer from FBI because of multiple reasons. For example, in poor countries it is common to see lack of access to clean water for food preparation; inappropriate transportation and storage of foods; high ambient temperature/humidity and lack of awareness regarding safe and hygienic food practices. Moreover, majority of the developing countries have limited capacity to implement rules and regulations regarding food safety. In addition, there is lack of effective surveillance and monitoring systems for FBI. Inspection systems for food safety are weak. Also, educational programs regarding awareness of food hygiene are more or less non-existent.

It is worthwhile to note that no individual (age group, gender, community) is immune to FBI. However, susceptibility to infection can be affected by a range of factors such as age, general health, nutrition, immune status and whether a person is undergoing medical treatment. Foodborne infection can be mild or even asymptomatic in some individuals. But it can be severe and often life-threatening in others. In people with low gastric acidity, increased survival of ingested pathogens can reduce the required infective dose, thereby increasing the risk of infection. This is often found in the elderly and may help explain their increased susceptibility to foodborne infections.

In some cases organisms attark susceptible individuals with a vengeance. For example, infection by Listeria monocytogenes can turn out to be serious in pregnant women; where the mother may experience relatively mild symptoms but infection of the foetus can result in abortion, stillbirth or premature labour. Similarly, Listeriosis is also more than 300 times more common in AIDS patients than in the general population. Cancer patients and other immunocompromised individuals can suffer from bacteremia as a result of mild initial infection. Vero toxin producing *E. coli* generally results in bloody diarrhoea but can cause the Hemolytic Uremic Syndrome (HUS), characterized by thrombocytopenia, hemolytic anemia and acute kidney failure, particularly in children.

Children/infants are more susceptible to FBI for several reasons. Their immune systems are not yet fully developed so their ability to fight infection is reduced. Children have a lower body weight than adults which means a smaller dose of a pathogen can sicken them. They have limited control over their diet and related food safety risks. Finally, they have reduced stomach acid production which protects adults by killing harmful bacteria in stomach.

Foodborne Illness Outbreaks

An outbreak of FBI occurs when a group of people consume the same contaminated food and two or more of them come down with the same illness.

- It may be a group that ate a meal together somewhere, or it may be a group of people who do not know each other at all, but who all happened to buy and eat the same contaminated item from a grocery store or restaurant. For an outbreak to occur, something must have happened to contaminate a batch of food that was eaten by a group of people.

- Often, a combination of events contributes to the outbreak. A contaminated food may be left out at room temperature for many hours, allowing the bacteria to multiply to high numbers, and then be insufficiently cooked so that all the bacteria are not killed.

Many outbreaks are local in nature. They are recognized when a group of people realize that they all became ill after a common meal, and someone calls the local health department. This classic local outbreak might follow a catered meal at a reception or eating a meal at an understaffed restaurant on a particularly busy day. For example, 400 persons may fell sick together with diarrhea after consuming lunch served at a wedding. The cause identified was the bacteria Salmonella in ice-cream.

Other examples from India include.

- 126 cases of food poisoning were reported from Navsari. They had history of consuming food in a marriage ceremony which was possibly contaminated as it was found to be fermented.
- 153 cases of food poisoning were reported from Rajkot. They had history of consumption of food in a marriage ceremony and the probable source of infection was milk products.
- 86 cases of food poisoning were reported from Mandsaur. They had consumed food in a local ceremony.

These are a few examples of FBI out breaks noted from IDSP site. In most cases foods are not contaminated intentionally. Rather from carelessness or insufficient education or training of food handler in food safety waitres the main culprit. In some cases, contamination may also be deliberate as, for example, in the misuse of food additives such as prohibited colouring agents. In one serious case in 1981 in Spain, contaminated industrial rapeseed oil was sold for human consumption, killing more than 500 people and crippling more than 20,000.

In the United States, where people eat outside home frequently, 58% of FBI cases originate from commercial food facilities. The globalization of the food industry has resulted in occurrence of international viral foodborne outbreaks. These days FBI outbreaks are increasingly being reported that are more widespread, that affect persons in many different places, and that are spread out over several weeks.

Causes for increase in reported foodborne illness incidences—The last two decades have been characterized by a number of developments which can help to explain the increase in the number of cases in a number of countries.

1. **Changes in animal husbandry:** Modern intensive animal husbandry practices introduced to maximize production seem to have led to the emergence and increased prevalence of Salmonella serovars and/or Campylobacter in herds of all the most important production animals (poultry, cattle, pig).

2. **Changes in agronomic process**: The use of manure and chemical fertilizers, as well as the use of untreated sewage or irrigation water containing pathogens/toxic material undoubtedly contributes to the increased risk associated with fresh fruit and vegetables.

3. **Increase in international trade:** This has three main consequences: (i) the rapid transfer of microorganisms from one country to another, (ii) the time between processing and consumption of food is increasing, leading to increased opportunity for contamination and time/temperature abuse of the products and hence the risk of FBI, and (iii) the population is more likely to be exposed to a higher number of different strains/types of foodborne pathogens.

4. **Changes in food technology:** Advances in processing, preservation, packaging, shipping and storage technologies on a global scale have enabled the food industry to supply a greater variety of foods, especially ready-to-eat foods. The increased use of refrigeration to prolong shelf-life has contributed to the emergence of FBI.

5. **Increase in susceptible populations:** Advances in medical treatment have resulted in an increasing number of the elderly and immune-compromised people. In many industrialized countries, the absolute number of the elderly is rapidly increasing.

Similarly, the population of patients with AIDS is rapidly increasing. These patients show a clear increase in susceptibility to Salmonella (relative risk of infection increased by 20–100) and to Campylobacter (35-fold increase in relative risk), as well as an increased risk of more severe clinical, manifestations.

6. **Increase in travel:** Globalization of FBI results also result from increased travel. As a result of travelling, a person can be exposed to a FBI in one country and expose others to the infection in a location thousands of miles from the original source of infection. Depending on their destination, travelers are estimated to run 20 to 50% risk of contracting FBI.

7. **Change in lifestyle and consumer demands:** Previously unrecognized microbial hazards have emerged as a result of changes in food consumption, like the increasing consumption of fresh fruit and vegetables in a number of countries. While dining in restaurants and salad bars was relatively rare 50 years ago, they are today a major source of food consumption in a number of countries. As a result, an increasing number of outbreaks are associated with food prepared outside the home.

These characteristics, associated with changes in food production and distribution have generated a new outbreak scenario. Traditional outbreaks were characterized by an acute and locally limited number of cases, with a high inoculum dose and a high attack rate sometimes because of a food handler error in a small kitchen shortly before consumption, often after a social event. In contrast, new outbreaks are often spreading over a wide geographic area involving different parts of a country or even internationally with a potentially high number of patients involved. The originating event can be a low-level contamination of a widely distributed food, often industrially processed. In these cases food contamination is not the result of a terminal food handling error but the consequence of an event in the early stages of the food-chain.

Burden of Foodborne Illness (FBI)

FBI causes loss of productivity, expensive medical care along with potential serious complications and/or chronic disabilities. Accurate burden-of-illness estimates for foodborne diseases are useful to characterize and prioritize resources dedicated to addressing the problem of these diseases. Here 'Burden of disease' has been defined as the incidence and/or prevalence of morbidity, disability, and mortality associated with acute and chronic manifestations of disease.

The overall burden of disease is estimated using various composite measures of population health status such as the Disability-Adjusted Life Year (DALY), which is a time-based measure that combines years of life lost due to premature mortality and years of life lost due to time lived in disability or states of less than full health. Diarrheal diseases alone—a considerable proportion of which is foodborne-kills 2.2 million people globally every year, but the burden arising from all foodborne diseases is clearly larger. **The full extent of the burden and cost of foodborne diseases is still unknown but is thought to be substantial.** To date, however, no precise and consistent global information exists. Many developed countries have sophisticated systems for collecting data on the incidence and causes of FBI. Yet it is known that these data represent only a fraction of the number of cases that occur. Infected individuals may not seek medical advice, and if they do their illness may not be recognized. Therefore, since most FBI are often self limiting, its global burden is often under reported.

The burden of illness pyramid is a model for understanding foodborne disease reporting. This illustrates steps that must occur for an episode of illness in the population to be registered in surveillance (Fig. 5.1).

Starting from the bottom of the pyramid.

1. Many members of the general population are exposed to an organism
2. Some of these exposed persons become ill
3. Some of these ill persons seek medical care
4. A specimen is obtained from some of these persons and submitted to a clinical laboratory
5. A laboratory tests some of these specimens for a given pathogen
6. The laboratory identifies the causative organism in some of these tested specimens and thereby confirms the case
7. The laboratory-confirmed case is reported to a local or state health department. The diagram below displays the disproportio-

nately much lower number of cases reported to the health department and the actual occurrence of foodborne illness. Data from surveillance systems and sentinel sites show only the tip of the iceberg and do not reflect true disease burden of FBI. The reasons are that the sick persons may not seek medical care, may not be tested at laboratory or may not be notified to the relevant health authorities

FBI outbreaks are therefore severely under-reported. It is estimated that only 68% of FBI outbreaks are notified to the Center for Disease Control and Prevention. Even during FBI outbreaks, only a small proportion of the total number of cases is reported. Data from surveillance systems and sentinel sites show only the tip of the iceberg and do not reflect true disease persons may not seek medical care, may not be tested at laboratory or may not be notified to the relevant health authorities (Fig. 5.2).

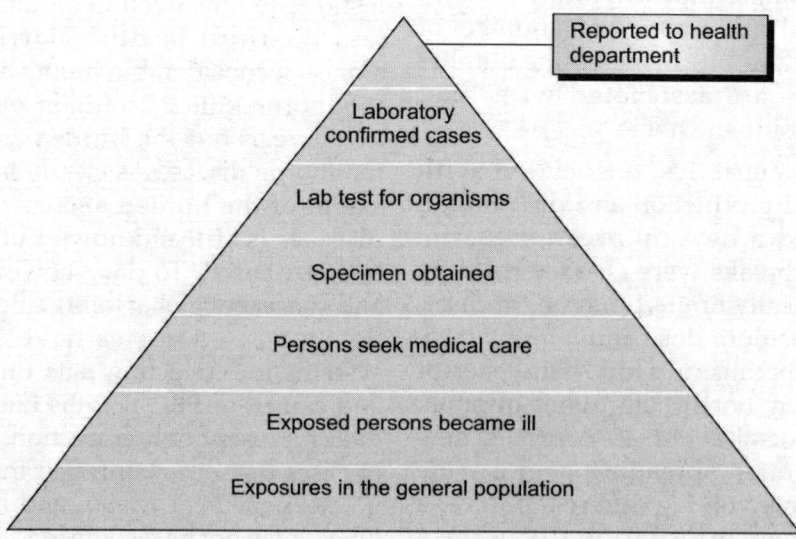

Fig. 5.1: Burden of illness pyramid reflecting the proportion of FBI that make it through each step of the diagnosis and reporting process

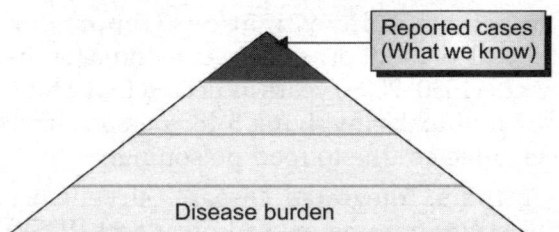

Fig. 5.2: The unknown burden of FBI as only a fraction of the people who experience symptoms from FBI seek medical care

Serious outbreaks of FBI have been documented on every continent in the past decades, illustrating both the public health and social significance of these diseases. Consumers everywhere view these with ever-increasing concern. Outbreaks are likely, to be only the most visible aspect of a much broader, more persistent problem. The heaviest share of the disease burden occurs in poor countries and jeopardizes international development efforts, including the achievement of the Millennium Development Goals.

The real tragedy of these diseases is played out in developing countries, where people are more exposed to hazardous environments, poor food production processes and handling, inadequate food storage and hygiene during food preparation, and poor regulatory standards. The tropical climate of many developing countries also helps pests and naturally occurring toxins to proliferate. People in these regions are therefore, at higher risk of contracting parasitic diseases. When people are malnourished, or living with HIV / AIDS, their immune systems are less able to fight foodborne diseases. In severe famines, there is also an understandable reluctance to discard contaminated or spoilt food. This complicates and compounds the food safety related hazards in such areas.

The global burden of foodborne disease is currently unknown but WHO has responded to this data gap by launching a new initiative to provide better estimates. In 2005 it was reported that 1.8 million people died from diarrheal diseases largely due to contaminated food and water. The estimated 47.8 million cases of foodborne diseases, resulting in 127,839 hospitalizations and 3037 deaths, transmitted through food each year in the USA alone.

The global burden of infectious diarrhoea involves 3–5 billion cases and nearly 1.8 million deaths annually, mainly in young children, caused by contaminated food and water. According to the CDC, an estimated 76 million cases of FBI are reported annually in the United States with approximately 5000 deaths. FBI is estimated to affect more than 76 million people in the United States each year, resulting in 325,000 hospitalizations and 5,200 deaths, but its true incidence is unknown. Because FBI is difficult to diagnose, the vast majority of these illnesses and more than half of such deaths are attributed to "unknown agents".

Every year, over 2 million children die from diarrheal diseases, a considerable proportion of which probably came from food. But the real death toll from across the spectrum of foodborne disease is likely to be much higher. In developing countries, diarrhoeal diseases, particularly infant diarrhoea, area major public health problem. The category "diarrhoea" includes some more severe diseases, such as cholera, typhoid and dysentery—all of which have related faecal: oral transmission pathways. It has been estimated that annually some 1500 million children under five years of age suffer from diarrhoea and over 3 million die as a result. Individual children experience on average 3.3 episodes of diarrhoea each year, though in some areas the number of episodes may exceed 9 and children can be suffering from diarrhoea for more than 15% of their young lives.

It has been reported that in the year 2000 about 2.1 million children died from diarrheal diseases. Many of these cases have been attributed to contamination of food and drinking water. The immediate cause of death from diarrhoeal disease is usually the dehydration that results from the loss of fluid and electrolytes in diarrhoeal stools, but diarrhoea can also have other serious health consequences. It may lead to malnutrition since food intake is reduced either as a result of loss of appetite or the withholding of food, and those nutrients that are ingested are poorly absorbed or simply lost by being swept out with the diarrhoeal stools. Malnutrition in its turn can predispose children to longer episodes of diarrhoea as well as other infections, aggravating the problem still further. This can result in a downward spiral of increasingly poor health which, unless it is broken in some way, will lead ultimately to premature death.

A significant proportion of infant diarrhoeal disease burden in developing countries is associated with contaminated infant weaning foods. Contamination of weaning foods may result from use of contaminated raw foods or contaminated water, but may also be associated with poor standards of hygiene during preparation of the food. This may include handling of foods with unwashed hands, inadequate cooking or re-warming, the use of contaminated feeding utensils, or storage of prepared foods at ambient temperature for prolonged periods.

In India, in the official document of health information of 2004, Government of India recorded 9575112 cases of acute diarrhoeal diseases including gastroenteritis with 2855 deaths and cases of foodborne disease may have been categorized under gastroenteritis.

The scientific investigations/reports on outbreak of foodborne diseases in India for the past 29 (1980–2009) years indicated that a total of 37 outbreaks involving 3,485 persons have been affected due to food poisoning.

In India, Integrated Disease Surveillance Project Programme instead of project (IDSP) reports that food poisoning outbreak reporting increased to more than double in 2009 from 2008 (120 in 2009 and 50 in 2008).

A few instances reported by IDSP are given below.

- Assam (Golaghat), reported 46 cases (vomiting, loose motion and pain abdomen) and 1 death due to acute diarrhoeal disease
- In Sivasagar, 22 cases of vomiting and diarrhoea were reported
- In Punjab (Ludhiana) 1 death and 11 cases of acute diarrhoeal diseases were reported
- 30 cases and 1 death due to food poisoning were reported in Vishakhapatnam. 17 families from the above villages had consumed contaminated food after which cases suffered from diarrhoea.

Statistics from both developed and developing countries show an increasing trend in FBI over recent years. It is estimated that between 6.5 and 33 million cases occur each year in the United States. Although most of these infections cause mild illness, severe infections and serious complications-including death-do occur. It is matter of concern that FBI are a growing public health problem worldwide.

Public Health Impact

The public health challenges of FBI are changing rapidly as a result of newly identified pathogens and vehicles of transmission, changes in food production, and an apparent decline in food safety awareness. They encompass a wide spectrum of illnesses caused by microbial, parasitic or chemical contamination of food. **Illness and death from diseases caused by unsafe food are a constant threat to public health security as well as socio-economic development throughout the world.**

Based on data collected through food net and other sources, the CDC estimates that 48 million Americans are sickened by FBI each year and that children under 15 years of age account for approximately half of all FBI in the U.S. Children under five are particularly vulnerable, experiencing the highest rates of infection for Campylobacter, Salmonella, Shigella, E. coli O157:H7 and other shiga-toxin producing E. coli bacteria (STEC) when compared to all other age categories. It is estimated that five pathogenic bacteria caused 291,162 laboratory confirmed illnesses each year among children less than five years of age, resulting in 102,746 physician visits, 7,830 hospitalizations, and 64 deaths.

Economic Impact of Foodborne Illness

Every illness has an economic cost and same is the case with FBI. Beyond the health impacts, foodborne diseases also affect economic development, and particularly challenge agricultural, food and tourist industries. They not only significantly affect people's health and well-being, but they also have economic consequences for individuals, families, communities, businesses and countries. These diseases impose a substantial burden on healthcare systems and markedly reduce economic productivity. Poor people tend to live from day to day, and loss of income due to foodborne illness perpetuates the cycle of poverty.

FBI have an impact on the public health as well as economy of a country. Financial burden due to FBI arise from a number of different sources and are incurred both by the individual and by society at large. These costs include loss of income by the affected individual, cost of health care, loss of productivity due to absenteeism, costs of investigation of an outbreak, loss of income due to closure of businesses and loss of sales when consumers avoid particular products.

FBI have a negative impact on the trade and industries of the affected countries. Identification of a contaminated food product can result in recalling of that specific food product leading to economic loss to the industry. Foodborne outbreaks may lead to closure of the food outlets or food industry resulting in job losses for workers, affecting the individuals as well as the communities. Moreover, local FBI outbreaks may become a global threat. The health of people in many countries can be affected by consuming contaminated food products, and may negatively impact a country's tourist industry.

Developing countries' access to food export markets depends on their ability to meet the World Trade Organization's regulatory requirements. Unsafe exports can cause severe economic losses. For example, in early 2008, Saudi Arabia refused Indian poultry products valued at nearly US $500,000 following a bird flu outbreak in West Bengal.

In developed countries, efforts to quantify the economic impact of FBI are comparatively recent, but it is clear from these that FBI is a major burden on the economy. The annual economic cost of FBI is calculated by multiplying the cost per case with the expected annual number of foodborne illnesses experienced. It was estimated that in 1999, the US government spent $1 billion on food safety efforts at federal level, an additional $300 million were spent by state governments. Moreover, it is estimated that a total of $152 billion a year is spent on FBI in the US.

Foodborne illnesses are a substantial burden in Australia, with an estimated

5.4 million cases occurring annually, costing an estimated $1.2 billion dollars per year.

In Australia, there are an estimated 5.4 million cases of foodborne illness every year, causing:

- 18,000 hospitalizations
- 120 deaths (0.5 deaths per 100,000 inhabitants)
- 2.1 million lost days off work
- 1.2 million doctor consultations
- 300,000 prescriptions for antibiotics

The annual cost of medical expenses and productivity losses associated with the five most prevalent, diagnosable foodborne illnesses is nearly $7 billion.

The medical costs and value of lives lost from five foodborne infections in England and Wales were estimated at UK £300–700 million annually in 1996.

In India, on the basis of per capita income, the economic burden on people in India affected by an outbreak of *Staphylococcus aureus* food poisoning was found to be higher than in case of a similar outbreak in the US.

Low-income individuals tend to have poorer access to medical care, lower nutritional status, and greater exposure to environmental threats, which can impact their ability to fight foodborne infections.

With increasing burden of FBI, present food safety systems are unable to respond to the existing and emerging challenges to food safety because they do not provide or stimulate a preventive approach. The risk-based approach must be backed by information on the most appropriate and effective means to control foodborne hazards.

BIBLIOGRAPHY

1. A long trail in Spain on Tainted Food. *New York Times*. August 2, 1987
2. Abelson P, Potter Forbes M, Hall G. The annual cost of foodborne illness in Australia. Australian Government Department of Health and Ageing, Canberra: 2006.
3. Angulo F, Voetsch A, Vugia D, Hadler J, Farley M, Hedberg C, Cieslak P, Morse D, Dwyer D, Swerdlow D, FoodNet Working group. Determing the Burden of Human Illness from foodborne diseases: CDC's Emerging Infectious Disease Program Foodborne Disease Active Surveillance Network (FoodNet). Veterinary Clinics of North America: Food Animal Practice 1998; 14: 165–172.
4. CDC, "Estimates of foodborne illness in the United States," 2011, http://www.cdc.gov/foodborneburden/2011-foodborne-estimates.html.
5. Centers for Disease Control and Prevention. FoodNet Surveillance-Burden of Illness Pyramid. 2005; http://www.cdc.gov/foodnet/surveillance_pages/burden_pyramid.htm
6. Council for Agricultural Science and Technology: Risk Characterization: Estimated Numbers of Illnesses and Deaths. In Foodborne Pathogens: Risks and Consequences. Task Force Report Number 122:40–52, 1994.
7. DG Newell, M Koopmans, L Verhoef, et al. "Food-borne diseases—the challenges of 20 years ago still persist while new ones continue to emerge," International Journal of Food Microbiology, vol. 139, supplement 1, pp. S3–S15, 2010.
8. Health Information of India. Central Bureau of Health Intelligence, Directorate General of Health Services, Ministry of Health and Family Welfare, Government of India, New Delhi, 2004
9. Helms, M, Vastrup, P, Gerner-Smidt, P., and Mølbak, K. (2003). Short and long term mortality associated with foodborne bacterial gastro-intestinal infections: registry based study. *BMJ (Clinical Research Ed.)*, 326(7385), 357.
10. Institute of Medicine: Emerging Infections: Microbial Threats to Health in the United States. Washington DC, National Academy Press, 1992.
11. Integrated disease surveillance programme, national centre for disease control, delhi retrieved from www.idsp.nic.in

12. Jones, T. F., Imhoff, B., Samuel, M., Mshar, P., McCombs, K. G., Hawkins, M., Deneen, V., et al. (2004). Limitations to Successful Investigation and Reporting of Foodborne Outbreaks: An Analysis of Foodborne Disease Outbreaks in Food Net Catchment Areas, 1998–1999. *Clinical Infectious Diseases*, *38*(s3), S297–S302. doi:10.1086/381599

13. Mead PS, Slutsker L, Dietz V, McCaig LF, Bresee JS, Shapiro C, Griffin PM, Tauxe RV. Food-related illness and death in the United States. Emerging Infectious Diseases. 1999;5(5): 607–625. [PMC free article] [PubMed]

14. Murray CJL, Lopez AD, (editors). The global burden of disease: a comprehensive assessment of mortality and disability from diseases, injuries, and risk factors in 1990 and projected to 2020. Cambridge (MA): Harvard University Press; 1996.

15. Outbreak of *Escherichia coli* 0157:H7 infections associated with drinking unpasteurized commercial apple juice-British Columbia, California, Colorado, and Washington, October 1996. MMWR 45:975,1996.

16. Outbreaks of *Escherichia coli* O157:H7 infections associated with eating alfalfa sprouts-Michigan and Virginia, June-July 1997. MMWR 46:741–744, 1997.

17. OzFoodNet (2012a) Monitoring the incidence and causes of diseases potentially transmitted by food in Australia: Annual report of the OzFoodNet Network, 2010. Communicable Diseases Intelligence 36(3):E213–E241

18. Parashar, Umesh D.; Hummelman, Erik G.; Bresee, Joseph S.; Miller, Mark A.; Glass, Roger I. (May 2003). "Global Illness and Deaths Caused by Rotavirus Disease in Children". *Emerging Infectious Diseases* 9 (5): 565–72. doi:10.3201/eid0905.020562.

19. Robert, J.A. (1996). Economic evaluation of Surveillance. London, Dept of Public Health and Policy.

20. RV Sudershan, RN Kumar, K1 Polasa, "Foodborne diseases in India—a review," *British Food Journal*, vol. 114, no. 5, pp. 661–680, 2012.

21. Scallan, E.,Mahon, B.E.,Hoekstra,R.M,Griffin, P.M. (2012). Estimates of Illnesses, Hospitalizations, and Deaths Caused by Major Bacterial Enteric Pathogens inYoung Children in the United States. *Pediatric Infectious Disease Journal*

22. Scharff, R. L., McDowell, J., and Medeiros, L. (2009). Economic Cost of Foodborne Illness in Ohio. *Journal of Food Protection*, *72*, 128-136.

23. Sudhakar, P.; Nageswara Rao, R.; Bhat, R. and Gupta, C.P. (1988) The economic impact of foodborne disease outbreak due to Staphylococcus aureus. Journal of Food Protection, 51, (11)

24. Surveillance for foodborne disease outbreaks in the United States, 1988–92. MMWR 45:1–66, 1996.

25. Verhoef, L. P. B., Kroneman, A., van Duynhoven, Y., Boshuizen, H., van Pelt, W., and Koopmans, M. (2009). Selection Tool for Foodborne Norovirus Outbreaks. *Emerging, 2009.*

Infectious Diseases, *15*(1), 31–38. doi:10.3201/eid1501.080673

26. Vogt DU. Food Safety Issues in the 109th Congress. Washington, D.C: Library of Congress, Congressional Research Service; 2005

27. World Health Organization [WHO]. (2011c). Initiative to estimate the Global Burden of Foodborne Diseases: Information and publications. Retrieved from http://www.who.int/foodsafety/foodborne_disease/ferg/en/index7.html

28. World Health Organization. The Global Burden of Disease. 2004 Update. Geneva, 2008. Available from: http://www.who.int/health-info/global_burden_disease/GBD_report _2004 update_full.pdf

29. World Health Organization. WHO consultation to develop a strategy to estimate the global burden of foodborne diseases. Geneva: World Health Organization; 2006. p. vii. from:http://www.who.int/foodsafety/publications/foodborne_disease/fbd_2006.pdf

6

Causes of Foodborne Illness

Ruchi Sharma, Amarjeet Singh, Puja Dudeja

I (Dr Ruchi) was a regular consumer of X brand juice until yesterday, when I had the most disgusting and horrific experience ever. Yesterday, I bought a 20 rupees tetra pack of guava juice from my local supermarket on my way home from work. I immediately started drinking from the tetra pack. But within a few seconds I realized that it tasted slightly off. Then I realized that for some reason the drink was not coming up the straw. By this time I had reached home, and I removed the straw from the pack. Immediately then, to my utter shock, a black wormlike thing came out of the tetra pack! Needless to say, the very sight of this made me sick and I threw up. Thank God for that, otherwise I would have probably had to go the emergency room with food poisoning. There are no excuses for this kind of terrible product quality. I only hope that nobody else will have to go through the traumatic experience that I underwent. Every one of us might have experienced such incident at some point of time in our lives.

This chapter intends to update the readers about the causes of FBI. It will also sound them to remain alert about the infested and rotten products being sold in market so that they can make better choices.

Safe food is essential in helping us to maintain a healthy and productive lifestyle.

With the world's growing population, ensuring the provision of a safe, nutritious and wholesome food supply for all has become a major challenge. Acute FBIes are much more of a concern for government and the food industry today than a few decades ago. Though, significant progress has been made in many countries in making food safer, thousands of millions of people become ill each year from eating contaminated food. The emergence of increased antimicrobial resistance in bacteria causing disease has further aggravated this picture. The introduction of new technologies, including GM food in this climate of concern about food safety is posing a special challenge. This chapter will attempt to increase public awareness regarding the risks posed by pathogenic microorganisms and chemical substances in the food supply chain.

Foodborne diseases encompass a wide spectrum of illnesses and are a growing public health problem worldwide. Foodborne illness result from ingesting contaminated food stuffs. These range from diseases caused by a multitude of micro-organisms to those caused by chemical pollutants. Foods of animal origin, particularly, meat and eggs, are most often implicated in FBI. Desserts, ice cream and confectionery foods are others that need attention. Incidents of such diseases are common both in homes as well as in restaurants.

Some common foods that are more likely to be implicated in FBI:

- Raw foods of animal origin are the most likely to be contaminated; that is, raw meat and poultry, raw eggs, unpasteurized milk, and raw shellfish.
- Filter-feeding shellfish strain microbes from the sea over many month, they are particularly likely to be contaminated if there are any pathogens in the seawater.
- Food supply sources where the vendor mixes together the products of many individual animals, such as bulk raw milk, pooled raw eggs, or ground beef, are particularly hazardous because a pathogen present in any one of the animals may contaminate the whole batch. For example:
  - A single hamburger may contain meat from hundreds of animals:
  - A single restaurant omelette may contain eggs from hundreds of chickens.
  - A glass of raw milk may contain milk from hundreds of cows
  - A broiler chicken carcass can be exposed to the drippings and juices of many thousands of other birds that went through the same cold water tank after slaughter.
- Fruits and vegetables consumed raw are a particular concern. Washing can decrease but not eliminate contamination, so the consumers can do little to protect themselves.
- Unpasteurized fruit juice can also be contaminated if there are pathogens in or on the fruit that is used to make it.

Various Sources and Modes of Transmission of FBI: Foodborne illness FBI affects humans when they eat poorly cooked food derived from infected animals (that is, meat, poultry, eggs, and their by-products). Spread by 'cross-contamination' occurs when germs contaminate ready-to-eat food. For example, when food that will not be cooked further is cut with a contaminated knife or by the hands of an infected food handler. Foodborne illness can spread from person-to-person via the hands of an infected person. It can also be spread from animals to humans.

a. Human-human transmission—an infected food handler/cook can transmit infection like worms, amoebiasis, typhoid to the consumers, e.g. typhoid Mary

b. Animals (direct contact)—pets, livestock and wild animals can also spread foodborne infection to man

c. Consumption of infected food—beef, small ruminant's meat, dairy, pigs meat, poultry meat, eggs, produce (fruits and vegetables), grains and beans, oils and sugar, shellfish, etc.

d. Infected/contaminated water used in food preparation

e. Air (pollutants in air may enter food chain)

f. Soil—when vegetables are grown on grossly infected/contaminated soil (sewage/effluent irrigated)

g. Others (contaminated equipment, etc. used for food processing)

Mechanism of Causation of FBI

Food is best consumed fresh. Time gap between harvesting/food production/processing and consumption is an important factor affecting food safety. Food may become contaminated during production and processing or during food preparation and handling. It is worth mentioning here that the more steps the food goes through before it reaches the plate; the greater are its chances of contamination. Storage of variable duration of raw/cooked food further increases the chances of FBI. Upon close examination of most FBI outbreaks, the evidence strongly suggests that the majority of problems are due to improper food handling practices such as inadequate temperature control in preparing, cooking or storing food and poor personal hygiene of the food handler. Contact between food and pests, especially flies, rodents and cockroaches, is a further important cause of contamination of food.

Various factors are responsible for compromised food safety leading to foodborne illnesses:

1. **Contaminated ingredients:** Micro-organisms like virus and bacteria are everywhere in our life. There are many chances of food being contaminated by them during its production and preparation. This can happen at any point during its production: growing, harvesting, processing, packaging, storing, shipping or preparing. FBI usually arises from improper handling, preparation, or storage of food items. Food may include many ingredients. If any of these is contaminated FBI result. For example, use of contaminated cream to prepare cake.

   a. **During food production and processing—** Many foodborne microbes are present in healthy animals (usually in their intestines) raised for food. They naturally harbour many foodborne bacteria in their intestines that can cause illness in humans. However, these do not cause illness in the animals. Meat and poultry carcasses can become contaminated during slaughter by contact with small amounts of intestinal contents. Similarly, fresh fruits and vegetables can be contaminated if they are washed or irrigated with water that is contaminated with animal manure or human excreta, e.g. sewage water. Alfalfa sprouts and other raw sprouts pose a particular challenge, as the conditions under which they are sprouted are ideal for growing microbes as well as sprouts, and because they are eaten without further cooking. That means that a few bacteria present on the seeds can grow to high numbers of pathogens on the sprouts. Oysters and other filter feeding shellfish can concentrate *Vibrio* bacteria that are naturally present in sea water, or other microbes such as norvo virus that are present in sewage dumped into the sea.

   b. **During food preparation and handling—** The way that food is handled after it is contaminated can also make a difference in whether or not an outbreak occurs. Raw, contaminated ingredients incorporated into foods that receive no further cooking can lead to FBI outbreaks. Between November 1992 and February 1993, a large outbreak of *E. coli* O157 infection in several US states causing more than 700 infections and 4 deaths. Illnesses were traced to thousands of pounds of contaminated hamburger patties that were undercooked at many outlets of one fast-food restaurant chain.

2. **Poor Temperature Control:** Foods not stored properly allows it to remain at temperature suitable for bacterial growth. Food held or stored at warm (10–50°C) temperature allows multiplication of pathogens. This is an important cause of foodborne outbreaks. Many bacterial microbes need to multiply to a larger number before enough are present in food to cause disease. Given warm moist conditions and an ample supply of nutrients, one bacterium that reproduces by dividing itself every half hour can produce 17 million progeny in 12 hours. As a result, lightly

contaminated food left out overnight can be highly infectious by the next day.

Improper reheating of food can also cause microbial growth. Reheated products are those that have been previously cooked, allowed to cool and then reheated before consumption. If the cooling has not been rapid enough, any spores in the food will have had time to germinate. Also, with insufficient cooking due to failure to thoroughly heat or cook food can help bacteria multiply and produce toxins within the food. Many other bacterial toxins are heat stable and may not be destroyed by cooking.

3. **Lack of personal hygiene:** Poor personal hygiene of food handler can result in food becoming contaminated with bacteria. Persons suffering from infections, e.g. vomiting, diarrhoea and FBI can contaminate the food. Most foodborne pathogens are shed in the faeces of infected persons and these pathogens may be transferred to others through food via the faecal-oral route. Bacteria present in infected lesions and normal nasal flora may also be transmitted from an infected employee who practices poor personal hygiene at home and at the workplace to ready-to-eat foods. It is important to remember that anyone who has been in contact with someone suffering from food poisoning can pass on bacteria, even though they show no symptoms themselves. For example, Shigella bacteria, hepatitis A virus and norvo virus can be introduced by the unwashed hands of food handlers who are themselves infected. Any food item that is touched by a person who is ill with vomiting or diarrhoea, or who has recently had such an illness, can become contaminated. When these food items are not subsequently cooked (e.g. salads, cut fruit) they can pass the illness to other people.

4. **Cross-contamination:** Cross-contamination of cooked foods may occur with raw foods, or by employees who mishandle foods, or through improperly cleaned equipment. Pathogens naturally present in one food may also be transferred to other foods during food preparation if same cooking equipment and utensils are used without washing and disinfecting in between, especially in case of ready-to-eat foods. In the kitchen, microbes can be transferred from one food to another food by using the same knife, cutting board, or other utensil to prepare both, without washing the surface or utensil in between. Also, a food that is fully cooked can become re-contaminated if it touches other raw foods or drippings from raw foods that contain pathogens.

For example, if you prepare raw chicken on a chopping board and do not wash the board before preparing food that won't be cooked (such as salad), harmful bacteria can be spread from the chopping board to the food. It can also occur if raw meat is stored above ready-to-eat meals in refrigerator. If juices from the meat drip on to the food below, it can become contaminated.

5. **Poor sanitation:** Sanitation of kitchen, eating place, and neighbourhood or of the ambient atmosphere all affect food-safety adversely. Recently, a number of FBI outbreaks have been traced to fresh fruits and vegetables that were processed under less than sanitary conditions. These outbreaks show that the quality of the water used for washing and chilling the produce after it is harvested is critical.

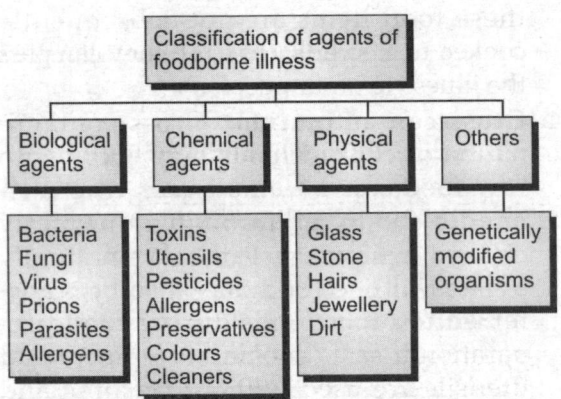

Fig. 6.1: Classification of various agents of foodborne illness

CAUSATIVE AGENTS OF FOODBORNE ILLNESS

A variety of agents can cause FBI. These agents can be classified into four broad groups; Biological, Physical, Chemical agents and other causes (Fig. 6.1). Although most of the FBI are caused by bacteria or viral pathogens, there are certain non-infectious causes such as, chemicals and toxins or physical agents (e.g. glass fragments, bone chips). The potential risk of each agent is assessed based on its likelihood of causing occurrence of illness and its severity. Chemical agents can contaminate the food most likely during pre-harvest phase whereas physical and biological agents are generally implicated during post harvest phase. If these are not handled properly, the food in kitchen is most vulnerable to them.

These agents can cause illness either by infection or intoxication:

a. **Foodborne infections** caused by consuming foods or liquids contaminated with bacteria, viruses, or parasites. These pathogens cause infection. Invading and multiplying in the intestinal tract and releasing a toxin (bacteria only)

b. **Foodborne intoxications** caused by consuming foods or beverages already contaminated with a toxin. Sources of toxins are as follows:

Certain bacteria (pre-formed toxins)

Poisonous chemicals

Natural toxins found in animals, plants and fungi

Table 6.1: Difference between infections and intoxications

	Infections	Intoxications
Cause	Bacteria / Viruses/Parasites	Toxin
Mechanism	Invade and/or multiply within the lining of the intestines	No invasion or multiplication
Incubation period	Hours to days	Minutes to hours
Trans-mission	Can spread from person-to-personvia the faecal-oral route	Not comm unicable
Factors related to food contami-nation	Cross-cont-amination	Inadequate cooking improper holding temperatures

1. BIOLOGICAL AGENTS

FBI can be caused by agents like bacteria, viruses, moulds, or parasites or food contaminated with toxins produced by these organisms. The term foodborne pathogen loosely describes the microbes that are found in animals (in farm/zoo animals and pets) and in the environment (soil, water and air) that make people sick regardless of how they became infected. Usually, infection happens by direct ingestion of a contaminated product.

But it can also happen by contact with other individuals or contact with an animal or pet. Some foodborne microbes make people ill by forming toxins in foods that affect the gut or the neurological system. The dangers of foodborne pathogenic microorganisms have been known for decades. A century ago, cholera and typhoid fever were the main prevalent FBI globally. During last few decades, other foodborne infections have emerged, such as diarrheal illness caused by the parasite cyclospora, and the bacterium vibrio parahemolyticus. The newly identified microbes pose a threat to public health as they can easily spread globally and can mutate to form new pathogens. The increased internationalisation of food production and distribution means pathogens associated with food know no borders.

There are more than 200 known FBI being transmitted through pathogens. However, numerous episodes of FBI and hospitalizations are caused by unspecified agents. Biological agents are the most important cause of FBI for several reasons. They are capable of causing FBI outbreaks that affect large number of people. Microorganisms are frequently, if not always, present as food contaminants. Many are able to multiply in foods. Biological agents present the greatest menace to consumers and should therefore receive utmost priority, while preparing food in kitchen. Despite remarkable advances in food science and technology, FBI is a rising cause of morbidity in all countries. The list of potential foodborne microbial pathogens always keeps increasing.

Following are some examples of FBI outbreak in past:

- Food poisoning affecting 78 personnel was reported in 1998 by the armed forces at high altitude, wherein *Salmonella enteritidis*

was identified as the etiological agent and frozen fowl was the implicated food source for the outbreak.

- A food poisoning outbreak due to *Salmonella paratyphi* A that affected 33 people, due to vegetarian food was reported from Yavatmal (Maharashtra) in 1995.

- Two separate food poisoning outbreaks due to *Salmonella weltevreden* and *Salmonella* wein affecting 34 and 10 people respectively, due to non-vegetarian food (chicken and fish) were reported from Mangalore, 2008–09.

- Environmental studies in the nineties documented the spread of salmonella from the hands of butchers as well as abattoir equipment from Punjab.

- Nearly 8% of eggs and 7% of egg-storing trays from retail markets in Coimbatore were found to be contaminated with Salmonella in 2006.

- Listeria monocytogenes was isolated from 105 (5%) milk samples collected from 52 farms in Maharashtra in 2007. Enterotoxigenic Bacillus cereus was isolated from 29% of fish (finfish, prawns and clams) samples in a study from Cochin in 2009.

- In 1995, outbreaks due to *Clostridium perfringens* (in which mutton and peas were implicated as the food source) and Bacillus cereus (due to a bakery product) were reported.

- In 2002, watermelon was implicated as the food source in another outbreak affecting 6 members of a family.

- A food poisoning outbreak due to *Yersinia enterocolitica* was reported in 1997 from Tamil Nadu affecting 25 people, in which buttermilk was incriminated as the food source.

- An outbreak of foodborne botulism due to *Clostridium butyricum* affecting 34 students from a residential school in Gujarat was reported in 1996, and the food sample found to be contaminated was sevu (crisp made from gram flour)

- An outbreak of *Staphylococcus aureus* food poisoning due to contaminated "bhalla" (a snack made up of urad dal fried in vegetable oil) affected more than 100 children and adults in Madhya Pradesh in 2007.
- A foodborne outbreak affecting 130 nurses from a Delhi hospital, associated with eating salad sandwiches, was diagnosed to be due to Norwalk-like virus in 2002.

Types of biological agents that cause food-borne illness

A. Bacteria

They are single celled organisms with typical dimensions of around 1 micro meter. Bacteria are the most important and well studied foodborne pathogens. A key factor is their ability to rapidly multiply in food, thus increasing the risk of foodborne illness. Most pathogens have optimal growth temperature range from 20 to 45°C (68 to 113°F). However, certain foodborne pathogens (termed psychrotrophs) are capable of growth under refrigerated conditions or temperatures less than 10°C (50°F). Bacteria also vary in their resistance to high salt, sugar, or total solids level as well as the acidity of food products. Bacteria cause foodborne illness by two mechanisms: Infection and intoxication. The latter can also be caused by chemical contaminants and naturally occurring toxins.

i. **Infection:** It occurs when living bacteria are ingested with food in numbers sufficient for some to survive the acidity of the stomach, one of the body's principal protective barriers. The time from ingestion until symptoms occur (Incubation period) is much longer in infection than that of *foodborne intoxications.*

Most common bacterial foodborne pathogens are:

- *Campylobacter jejuni*: It is found in raw or undercooked chicken and unpasteurized milk which can lead to secondary *Guillain-Barré syndrome* and *periodontitis*. The bacteria may also be responsible for some "traveller's diarrhoea".
- *Clostridium perfringens*, the "cafeteria germ": It is found in soil as well as in the intestines of humans and animals. Illness occurs when individuals eat food contaminated with faeces or soil. Foods commonly involved include meats that have been contaminated and then improperly refrigerated, improperly cooked, or inadequately reheated (i.e. stews, meat pies and gravies made from beef, turkey or chicken).
- *Salmonella*: It is a bacterium found in many foods, including raw and undercooked meat, poultry, dairy products, and seafood. Salmonella may also be present on egg shells and inside eggs. Both humans and animals may have the bacteria and still appear healthy. Its *S. typhimurium* infection is caused by consumption of eggs or poultry that are not adequately cooked or by other interactive human-animal pathogens.
- *Escherichia coli*: Common sources of *E. coli* include raw or undercooked hamburger, unpasteurized fruit juices and milk, and fresh produce. Most cases of *E. coli* food poisoning occur after eating undercooked beef (particularly mince, burgers and meatballs) or drinking unpasteurised milk. It includes several different strains, only a few of which cause illness in humans. *E. coli O157:H7* is the strain that causes the most severe illness (hemolytic-uremic syndrome).

Other common bacterial foodborne pathogens are:

- *Bacillus cereus*
- *Listeria monocytogenes*: It has been found in raw and undercooked meats, unpasteurized milk, soft cheeses, and ready-to-eat meats and hot dogs.
- *Shigella*: It is a bacterium that spreads from person to person. These bacteria are present in the stools of people who are infected. If people who are infected do not wash their hands thoroughly after using the bathroom, they can contaminate food that they handle or prepare. Some individuals may be unaware they are transmitting the disease because they do not show signs of having the infection. Water contaminated with infected stools can also contaminate produce in the field.
- *Staphylococcus aureus:* The organism is usually found in humans (i.e. abscesses, infected fingers and eyes, acne and nasal secretions). Foods commonly contaminated include pastries, custards, salad dressings, sandwiches, sliced meat and meat products (i.e. ham, bacon and pressed meats). Milk from cows with infected udders, dried milk, cream and butter are foods that have also been associated with staphylococcal food poisoning.
- *Staphylococcus enteritis*
- Streptococcus
- *Vibrio cholerae*, including O1 and non-O1— It may contaminate fish or shellfish.
- *Vibrio parahaemolyticus*
- *Vibrio vulnificus*
- *Yersinia enterocolitica* and *Yersinia pseudotuberculosis*

Less Common Bacterial agents

- *Brucella* spp.
- *Corynebacterium ulcerans*
- *Coxiella burnetii or Q fever*
- *Plesiomonas shigelloides*

ii. **Intoxication:** With foodborne intoxications, the bacteria grow in the food producing a toxin (*Toxico infections*). In addition to disease caused by direct bacterial infection, some foodborne illnesses are caused by a enterotoxins (an exotoxin targeting the intestines). Enterotoxins can produce illness even when the microbes that produced them have been killed. Here, when the food is eaten, it is the toxin, rather than the microorganisms, that causes symptoms for example—The rare but potentially deadly disease botulism occurs when the anaerobic bacterium *Clostridium botulinum* grows in improperly canned low-acid foods and produces botulin, a powerful paralytic toxin. In general, intoxication is manifested more rapidly after consumption of contaminated food (shorter onset time) than are infections. Included in this group are:

- *Bacillus cereus* (diarrheal-type)
- *Clostridium botulinum*: It is an anaerobic, Gram positive, sporeforming rod. Botulinum toxin is one of the most powerful known toxins: about one microgram is lethal to humans. Botulism is a rare and potentially fatal paralytic illness caused by a toxin produced by the bacteria *Clostridium botulinum*. Foodborne botulism results from contaminated food in which *C. botulinum* spores have been allowed to germinate in low-oxygen conditions. This typically occurs in home-canned food substances and fermented uncooked dishes. Improperly preserved food is the most common cause of foodborne botulism, e.g. fish that has been

pickled without the salinity or acidity of brine that contains acetic acid and high sodium levels, as well as smoked fish stored at too high a temperature, presents a risk, as does the improperly canned food.

Between March 31 and April 6, 1977, 59 individuals developed type B botulism in US. All ill persons had eaten at the same Mexican restaurant and all had consumed a hot sauce made with improperly home-canned jalapeño peppers, either by adding it to their food, or by eating a nacho that had had hot sauce used in its preparation.

• *Clostridium perfringens*: It has been found in meats, stews and gravies. It is commonly spread when serving dishes don't keep food hot enough or food is chilled too slowly.

• *Bacillus cereus*

• Verotoxigenic *E. coli* (*E. coli* O157:H7 and others): It has been found in pork, beef and poultry products. The infection is frequently linked with undercooked meat and occasionally with unpasteurized milk. Person to person spread also occurs. Infection causes hemorraghic colitis, commonly referred to as hamburger disease or barbecue season syndrome. These bacteria produce a poison or toxin which damages the lining of the intestine producing diarrhoea and pain. While most people recover from this disease within two weeks, in a very small number of cases, the toxin results in a serious complication called Hemolytic Uraemic Syndrome (HUS). This illness affects the kidneys and blood and is especially dangerous to young children and the elderly. Death can result from either HUS or the intestinal disease.

• *Vibrio cholerae*

B. Mycotoxins

Mycotoxicoses refers to the effect of poisoning by mycotoxins through food consumption. Mycotoxins are another group of highly toxic or carcinogenic chemical contaminants of biological origin produced by certain species of fungi. Mouldy food can be dangerous to eat if the fungi growing on it are the kind that produces mycotoxins. These are formed as the branching network of fungal filaments (hyphae) spread through the food and break it down. Moulds can grow in drier environments than other microbes and mycotoxins are a problem in products such as nuts and cereals which have been stored in damp conditions. It is easy to mistakenly process and eat contaminated food because it does not look spoilt. Very small amounts of mycotoxins can make people ill.

Five important mycotoxins are aflatoxins, ochratoxins, fumonisins, zearalenone, and trichothecenes. Crops such as peanuts, corn, pistachio, walnuts are susceptible to mycotoxin contamination. Aflatoxins are among the most studied mycotoxins, and the relationship between aflatoxin ingestion and primary liver cancer is well established. Almost all plant products can serve as substrates for fungal growth, and subsequently mycotoxin contamination of human food and animal feed. Animal feed contaminated with mycotoxins can result in the carry-over of toxins through milk and meat to consumers.

Mycotoxins sometimes have important effects on human and animal health. For example, an outbreak which occurred in the UK in 1960 caused the death of 100,000 turkeys which had consumed aflatoxin-contaminated peanut meal. In the USSR in World War II, 5,000 people died due to Alimentary Toxic Aleukia (ALA).

Mycotoxins of importance in India include alfatoxins, fumonisins, trichothecenes, ergot alkaloids and ochratoxins. Inorganic forms of Arsenic predominate in rice and spices, and are a real threat to human health.

The common foodborne mycotoxins include:

- Aflatoxins—originated from *Aspergillus parasiticus* and *Aspergillus flavus*. They are carcinogens frequently found in tree nuts, peanuts, maize, sorghum and other oilseeds, including corn and cottonseeds. The pronounced forms of Aflatoxins are those of B1, B2, G1, and G2, amongst which Aflatoxin B1 predominantly targets the liver, which will result in necrosis, cirrhosis, and carcinoma.

- Altertoxins: Some of the toxins can be present in sorghum, ragi, wheat and tomatoes

- Ochratoxins: They cause kidney disease and are produced in cereals such as maize and barley.

- Patulin: It is associated with mouldy apples and poisoning has arisen from drinking contaminated fresh apple juice.

- Tremorgenic mycotoxins: Five of them have been reported to be associated with molds found in fermented meats. These are Fumitremorgen B, Paxilline, Penitrem A, Verrucosidin, and Verruculogen.

- Trichothecenes: Sourced from Cephalosporium, Fusarium, Myrothecium, Stachybotrys and Trichoderma. The toxins are usually found in molded maize, wheat, corn, peanuts and rice, or animal feed of hay and straw.

Other mycotoxins are—Citrinin, citreoviridin, cyclopiazonic acid, cytochalasins, ergot alkaloids/Ergopeptine alkaloids – ergotamine, fumonisins, fusaric acid, fusarochromanone, kojic acid, lolitrem alkaloids, moniliformin, 3-Nitropropionic acid, nivalenol, oosporeine, phomopsins, sporidesmin A, sterigmatocystin, zearalenone and zearalenols.

C. Virus

Viral infections make up perhaps one-third of cases of food poisoning in developed countries. Approximately 67% of all foodborne illnesses caused by pathogens has viral etiology. Unlike bacteria, viruses cannot multiply in food and do not cause spoilage. Viruses are particulate in nature and multiply only in other living cells. Thus, they are incapable of survival for long periods outside the host. Viruses do not cause any change in the appearance, taste or smell of food and cannot be detected by ordinary laboratory tests.

In US, more than 50% of foodborne illnesses are viral and noro viruses are the most common cause of FBI, causing 57% of outbreaks in 2004. Viruses that infect people must come from other people (transmitted by hands and sneezing,) or by foods contaminated by sewage. For example, the presence of viruses in shellfish, grown in sewage—polluted water, may be significant in causing FBI if the shellfish are eaten raw or undercooked. Foodborne viral infection are usually of intermediate (1–3 days) incubation period, causing illnesses which are self-limited in otherwise healthy individuals; they are similar to the bacterial forms described above.

Viruses involved in FBI are highly contagious, and are usually affecting large numbers of people, such as on cruise ships, in nursing homes, day care facilities, and at banquets.

While greater than 100 types of enteric viruses have been shown to cause, the most commonly implicated viruses in foodborne illnesses are norovirus, hepatitis A virus, hepatitis E virus, rotavirus, and astrovirus, enterovirus. Common foodborne viruses include:

- Rotavirus is highly contagious, requiring only 10 – 100 virus particles to cause an infection. It affects mainly children age 6 to 24 months, and is responsible for 50% of the children hospitalized due to diarrhoea in the United States. It is more prevalent in the winter months and can also infect adults, especially, people at risk. It is commonly caused due to raw, ready-to-eat produce. It can be spread by an infected food handler.
- Noro virus: Noro virus was previously called Norwalk-like virus accounts for most foodborne viral infections. The virus is passed to food mainly from infected food handlers or by cross-contamination, but can be passed in swimming water and drinking water contaminated with human faeces. Outbreaks commonly occur where large numbers of people gather, such as cruise ships, nursing homes, nursery schools, etc. It causes inflammation of the stomach and intestines.
- Hepatitis A: Other people are the only source of the hepatitis virus. The virus is found in faeces in high numbers a week or two before symptoms become obvious. Therefore, it is possible to unknowingly spread the virus to other people through lapses in proper personal hygiene (hand washing). Hepatitis A is a common water contaminant particularly in developing countries. It can be present in food when contaminated water is used to rinse the food, or if the item is prepared by an infected handler.Outbreaks have been associated with contaminated water; food contaminated by infected food handlers, including sandwiches and salads which are not cooked or are handled after cooking; and raw or undercooked shellfish which were grown in sewage—polluted waters.
  Infection causes inflammation of the liver. It is distinguished from other viral causes by its prolonged (2–6 week) incubation period and its ability to spread beyond the stomach and intestines into the liver. It often results in jaundice, or yellowing of the skin, but rarely leads to chronic liver dysfunction. The virus has been found to cause infection due to the consumption of fresh-cut produce which has faecal contamination.
- Hepatitis E: It is a major concern in developing countries. Major outbreaks implicated undercooked pork and dear meat as sources of infection.
- Avian influenza: It is caused due to consumption of undercooked or raw chicken or other fowl.

D. Parasites

Most foodborne parasites are zoonoses. Parasites are of different types and range in size from tiny, single-celled, microscopic organisms (**protozoa**) to larger, multi-cellular worms (**helminths**) that may be seen without a microscope. The size ranges from 1 to 2 µm (micrometers) to 2 meters long. The most common foodborne parasites are protozoa, roundworms, and tapeworms. Any of these organisms can also be transmitted by water, soil, or person-to-person contact. A wide variety of helminthic roundworms, tapeworms, and flukes are transmitted in foods such as:

- Undercooked fish, crabs, and molluscous
- Undercooked meat; raw aquatic plants such as watercress
- Raw vegetables that have been contaminated by human or animal faeces

i. **Parasitic protozoa** are one-celled micro-organisms without a rigid cell wall, but with an organized nucleus. They are larger than bacteria. Like viruses, they do not multiply in foods, only in hosts. The transmissible form of these organisms is termed a *cyst*. Protozoa that have been associated with food and water-borne infections include *Entamoeba histolytica, Toxoplasma gondii,*

Giardia lamblia, Cryptosporidium parvum and Cyclospora cayatenensis. They are spread through water contaminated with the stools of infected people or animals. Foods that come into contact with contaminated water during growth or preparation can become contaminated with these parasites. Food preparers who are infected with these parasites can also contaminate foods if they do not thoroughly wash their hands after using the bathroom and before handling food.

• *Entamoeba histolytica:* Amoebiasis is an intestinal disease caused by the parasite Entamoeba histolytica. The disease is commonly known as "amoebic dysentery" and results when the parasite invades the wall of the large intestine, forming ulcers in the process. Amoebiasis can be transmitted by water and food contaminated with faeces. Amoebiasis is usually spread by direct hand-to-mouth transfer of the parasites, but the parasites can also travel from the faeces to the mouth via food that has been contaminated. Community outbreaks usually involve water supplies contaminated with cysts. A cyst is the form that allows the parasite to survive outside the host for a long time. Amoebiasis can be spread by infected individuals to other people for many years if treatment is not sought. Some individuals may be unaware they are transmitting the disease because they do not show signs of having the infection. These people are called "asymptomatic carriers".

• Giardia lamblia: It is found in humans and animals (wild and domestic). Beavers have been identified as a source of the parasite. The parasite produces cysts which are responsible for the spread of the disease. Faeces containing these cysts can contaminate both water (most commonly) and food. Direct hand-to-mouth transfer of these cysts can occur when individuals do not wash their hands after using the toilet. Giardiasis is an intestinal disease caused by the parasite Giardia lamblia.

ii. **Multi-cellular parasites** are animals that live at the expense of the host. They may occur in foods in the form of eggs, larvae, or other immature forms. Trichinosis has been an important reportable pathogen associated with undercooked pork. Other parasites of concern include flatworms or nematodes (associated with fish) like *Trichinella spiralis*, *Trichuris trichiura*, cestodes or tapeworms (usually associated with beef, pork, or fish) and trematodes or flukes. Here *Trichinella spiralis* is a common type of roundworm parasite. People may be infected with this parasite by consuming raw or undercooked pork or wild game.

E. Prions

The term "prions" refers to abnormal, pathogenic agents that are transmissible and are able to induce abnormal folding of specific normal cellular proteins called prion proteins that are found most abundantly in the brain. Prion diseases may present as genetic, infectious, or sporadic disorders, all of which involve modification of the prion protein (PrP). The abnormal folding of the prion proteins leads to brain damage and the characteristic signs and symptoms of the disease. Prion diseases are usually rapidly progressive and always fatal. **Human Prion Diseases are:**

• Creutzfeldt-Jakob disease (CJD)

• Variant Creutzfeldt-Jakob disease (vCJD)

• Gerstmann-Straussler-Scheinker syndrome

• Fatal familial insomnia

• Kuru

F. Emerging Foodborne Pathogens

Many FBI remain poorly understood, e.g. *Aeromonas hydrophila, Aeromonas caviae,* Aeromonas sobria. Approximately 82 % of the FBI and 25% of the death cases were caused by "unknown pathogens", assumed to be viruses. This percent of outbreaks are caused by unknown sources. The concept of unknown etiology is supported by well-documented foodborne outbreaks of distinctive illness for which the causative agent remains unknown, the large number of outbreaks for which no pathogens is identified and by the large number of new foodborne pathogens identified in recent years. In USA these unknown agents account for approximately 78–81% of FBI (183,000,000 cases annually), for 50% hospitalizations and 64% of deaths.

2. CHEMICAL AGENTS

Chemical agents are also a significant source of foodborne illness, although the effect is often difficult to link to a particular food and may occur long after consumption. Because period of time between exposure to chemicals and effect is usually long, it is difficult to attribute disease caused by long-term exposure to chemicals in food to the actual food in question. There have been long-standing concerns about the chemical safety of food due to misuse of pesticides during food production and storage, resulting in the occurrence of undesirable residues. Similarly, heavy metal contaminants can enter food through soil or water or food contact material, as can other environmental contaminants such as Polychlorinated bi-phenols (PCBs). All can lead to acute or chronic illness.

An excellent example of chemical contamination of food is **1971 Iraq poison grain disaster.** It was a mass methyl mercury poisoning incident where grain treated with a methyl mercury fungicide and never intended for human consumption was imported into Iraq as seed grain from Mexico and the United States. Due to a number of factors, including foreign-language labelling and late distribution within the growing cycle, this toxic grain was consumed as food by Iraqi residents in rural areas. People suffered from parasthesia (numbness of skin), ataxia (lack of coordination of muscle movements) and loss of vision, symptoms similar to those seen when minamata disease affected Japan. 6,530 patients were admitted to hospital with poisoning and the recorded death toll was 650 people, but figures at least ten times greater have been suggested.

India's production of pesticides was 85,000 metric tonnes in 2004, and rampant use of these chemicals has lead to several short-term and long-term effects. The first report of pesticide poisoning in India was from Kerala in 1958, where over 100 people died after consuming food made from wheat flour contaminated with parathion.

In 1997, a foodborne outbreak of organophosphate (malathion) poisoning affected 60 men (and was fatal for one) who ate a communal lunch prepared from food stored in open jute bags which was contaminated with the pesticide sprayed in the kitchen that morning. It is estimated that 51% of food commodities are contaminated with pesticide residues in India.

An outbreak of food poisoning due to epidemic dropsy (mustard oil contaminated with argemone oil) was reported from Delhi in 1998 in which 60 persons lost their lives and more than 3000 cases were hospitalized.

On 16 July 2013, at least 23 students died and dozens more fell ill at a primary school in the village of Dharmashati Gandaman in the

Saran district of the Indian state of Bihar after eating a Midday Meal contaminated with pesticide. Initial indications were that the food was contaminated by an organophosphate, a class of chemicals commonly found in insecticides.

It is estimated that 51% of food commodities are contaminated with pesticide residues in India. Under the Integrated Disease Surveillance Programme (IDSP) in India, food poisoning outbreaks reported from all over India in 2009 increased to more than double as compared to the previous year (120 outbreaks in 2009, as compared to 50 in the year 2008)

The Chemical safety hazards include intentionally added chemicals, unintentionally added chemicals (e.g. cleaners and solvents), allergens and natural toxins (e.g. mycotoxins). Chemicals can also contaminate food through corrosion of metal processing equipment/utensils and residues of cleaning chemicals left on processing equipment. Further, adding too much of an approved ingredient, such as a vitamin A or D in vitamin-fortified products, may compromise the safety of foods.

Factors mentioned below could be one of the possible reasons for chemical contamination of food:

- Indiscriminate spraying of facilities against pests—chemicals can contaminate food if pesticides against insects and rodents are used indiscriminately in a processing facility.
- Mistaken identity of pesticides—Food can become contaminated with pesticides if pesticide container labels are misread or when products are stored in containers that have had another use.
- Spillage of pesticides or other chemicals.
- Raw material contamination with pesticides.

- Corrosion of metal containers/equipment/utensils—metal poisoning can occur when heavy metals leach into food from equipment, containers, or utensils. When highly acidic foods (e.g. citrus fruits, fruit drinks, fruit pie fillings, tomato products, sauerkraut, or carbonated beverages) come into contact with potentially corrosive materials like tin, lead, copper and zinc the metals can leach into the food. These metals can dissolve in acid foods such as fruit juices and produce fast-acting poisons in the body when ingested. Possible sources of contamination include residues migrating into foods from soldered cans, leaching from utensils, contaminated water, glazed pottery, painted glassware and paints.

- Allergens—food allergy is an abnormal response to a food triggered by your body's immune system. Some foods, such as nuts, milk, eggs, or seafood, can cause allergic reactions in people with food allergies.

- Accidentally adding too much of an approved ingredient—some substances, such as preservatives, nutritional additives, colour additives, and flavour enhancers, are intentionally added to food products. For example, some vitamins that are added to fortified foods (such as Vitamin A) are known to be toxic at high doses. One example is the death of 8 children following vitamin A administration in Assam in previous decade. Iron, a necessary dietary component, can also cause severe illness and death if too much is ingested. Reports of stomach upset were reported by girls in many schools of Haryana in 2013 when they were administered iron tablets.

- Residue from cleaning and sanitizing. If equipment and other food handling materials are not rinsed well, then residue from detergents, cleaning compounds, drain cleaners, polishers, and sanitizers can contaminate a food product.

These problems are not confined to food produced on land. They also include algal toxins in fish and the widespread use of chemicals in fish farming. While the importance of chemicals hazards is well recognized, our understanding of their effect in food intolerances and allergies, endocrine system disruption, immune-toxicity, and certain forms of cancer is incomplete. It needs to be stressed that further research is necessary to determine the role of chemicals in foods in the etiology of these diseases. In developing countries, little reliable information is available on the exposure of the population to chemicals in food.

i. **Natural toxins:** Several foods can naturally contain toxins, many of which are not produced by bacteria. These toxins generally occur in raw materials, especially crops and seafood. Plants in particular may be toxic; animals which are naturally poisonous to eat are rare. For example, fish, mushrooms. Many children die or get foodborne illness when they eat poisonous fruit. Such reports are often reported in media. Most animal poisons are not synthesised by the animal, but acquired by eating poisonous plants to which the animal is immune, or by bacterial action. Few examples of natural toxins are:

   • Alkaloids
   • Grayanotoxin (honey intoxication)
   • Mushroom toxins
   • Phytohaemagglutinin (red kidney bean poisoning; destroyed by boiling)
   • Pyrrolizidine alkaloids

   Several kinds of seafood-associated toxins can also cause illness:

   • **Paralytic Shellfish Poisoning (PSP)** is transmitted to humans through mussels,

clams, and scallops that have ingested and concentrated toxic marine protozoa. The toxin is found mainly in coastal waters and is often associated with a red discoloration of seawater due to algal bloom known as "red tide."

• **Diarrhetic shellfish poisoning** is also caused by ingestion of seafood containing toxic marine protozoa. Illnesses have occurred in eastern Canada, Japan and Western Europe.

• **Amnesic shellfish poisoning** can result from eating shellfish that are contaminated with algae that produces domoic acid. It was responsible for over 100 cases and 3 deaths in eastern Canada in a 1987 outbreak.

• **Ciguatera poisoning** is a result of ingestion of ciguatoxin and related toxins, produced in tropical fish, but also implicated in farm-raised salmon. Areas of higher risk are the Pacific and northern Caribbean. However, imported fish have occasionally caused outbreaks in the United States.

• **Scombroid poisoning**, arising from bacterial spoilage of fish and subsequent production of histamine and related compounds, occurs more frequently than other sea food toxin poisonings. Tuna, mackerel, mahimahi and marlin are often implicated.

ii. **Food adulteration:** Consumers, particularly in developing countries, are often exposed to wilful adulteration of their food items by manufacturers/FBOs. This can lead to health hazards and to financial losses for the consumer. Adulteration of milk and milk products, honey, spices, edible oils, and the use of colours to mask product quality to cheat the consumer are quite common. Such episodes invoke public outrage and anger as it violates public trust in the integrity of the food supply.

With 60–70% of the income of middle class families in developing countries being spent on food, food adulteration can impact heavily on both the family budget and the health status of the family members. For example, high fructose corn syrup or cane sugar is used to adulterate honey and water, for diluting milk and alcoholic beverages.

In 2012 a study in India across 33 states and union territories found that milk was adulterated with detergent, fat and even urea, and diluted with water. Just 31.5% of samples conformed to FSSAI standards.

The **2008 Chinese milk scandal** was a food safety incident in China, involving milk and infant formula, and other food materials and components, adulterated with melamine. An estimated 300,000 victims were reported with six infants dying due to kidney damage, and an estimated 54,000 babies being hospitalised.

3. PHYSICAL AGENTS

Foreign objects, or physical safety hazards are the least likely to affect large numbers of people and usually are easily recognized. A physical safety hazard is any extraneous object or foreign matter in food that can cause injury or illness in the person consuming the product. Materials that do not belong to food, like glass or metal cause physical safety hazards. Rocks, metal, wood, and other objects are also sometimes found in raw ingredients. Physical hazards are either foreign materials unintentionally introduced to food products or naturally occurring objects that are a threat to the final consumer. Physical hazards and can cause, cuts to the mouth or throat, damage to the intestine, damage to teeth or gums etc. Poor hygiene practices, whether in housekeeping or amongst the employees, can introduce physical hazards into food. Poor habits and behavior of the food handler in kitchen can introduce foreign bodies into the food. In some instances, food may be handled by person who is ignorant of the basic principles of personal hygiene. Further, contamination can occur during transport, processing, and distribution of foods due to equipment failure, accidents, or negligence as mentioned below.

- Foreign matter in raw materials. Sources of foreign matter in raw materials can include nails from pallets and boxes, ingested metal from animals, harvesting machinery parts, elements from the field, veterinary instruments, caps, lids, closures, dirt, etc.

- Poorly maintained equipment/lines Light fixture breakage-Pieces of equipment can break off and enter food products during processing if equipment is poorly maintained

- Foreign matter introduction during storage—Pests/flies/rodents can enter products during storage, leaving remnants behind

- Human factors—Production line workers can be a major source of contamination. For example, jewellery can fall off or break, fingernails/hairs can break, and pens can fall into food.

4. OTHER CAUSES

Genetically Modified Organisms (GMO) and Novel Foods: Modern biotechnology, also referred to as genetic engineering or genetic manipulation, involves transfer of hereditary material (DNA, RNA) from one organism to another in a way that cannot be achieved naturally, i.e. through mating or cross breeding. Genetic engineering can now transfer the hereditary material across species boundaries. This can broaden the range of genetic changes that can be made to food and can expand the spectrum of possible food sources.

The accelerating pace of developments in modern biotechnology has opened a new era in food production and this may have a tremendous impact on world food supply systems. However, there are considerable differences of opinion among scientists about the safety, nutritional value and environmental effects of such foods. These foods may also give rise to new viruses.

Overall, it is argued that the consequences of some gene transfer methods are less predictable when compared to those of traditional plant breeding methods and considerable scientific evidence will be needed to clear these foods from points of view of nutrition, food safety and impact on the environment.

Conclusion

In 2020, the world population is projected to reach 7.6 billion, an increase of 31% over the mid-1996 population of 5.8 billion. Approximately 98% of the population growth occurring during this period will take place in developing countries. While urbanization is a global phenomenon, it has been estimated that between the years 1995 and 2020 the developing world's urban population will double, reaching 3.4 billion. Such population growth poses great challenges to food safety. Further extension of improved agriculture and animal husbandry practices; use of measures to prevent and control pre- and post-harvest losses; more efficient food processing and distribution systems; introduction of new technologies including the application of biotechnology, and others will have to be exploited to increase food availability to meet the needs of growing populations.

Growing urbanization and associated changes in the way food is produced and marketed have led to a lengthening of the food chain and potential for introducing or exacerbating FBI.

Increasing industrialization and urban living has meant that the food chain has become longer and more complex, increasing opportunities for contamination. It also means that more people are likely to be affected by a single breakdown in food hygiene mechanism.

In poorer countries increased urbanization and rapid population growth have not been matched by development of the health-related infrastructure, basic sanitation, and this has led to increased risk of contamination of the food and water supply.

Increasing affluence in other areas has also led to greater consumption of foods of animal origin such as meat, milk, poultry and eggs. These foods are recognized as more common vehicles of foodborne pathogens and this situation can be exacerbated by the methods of intensive production required to supply a larger market. There is greater international movement of both foods and people. For example, exotic Salmonella serotypes have been introduced into Europe and the United States as a result of the importation of animal feeds. A number of outbreaks of illness associated with imported foods have also been recorded.

Tourism is one of the world's major growth industries and every year more and more people travel abroad where they are exposed to increased risk of contracting foodborne illness.

Changing lifestyles also means that food preparation may be in the hands of the relatively inexperienced as more mothers go out to work and more people eat pre-prepared foods, meals from catering establishments or food from street vendors.

It is important to monitor and investigate the FBI in order to control and prevent them. The contamination of food is influenced by multiple factors and may occur anywhere along the food chain. Good agriculture practice and good manufacturing practice should be adopted to prevent introduction of pathogens into food products. Most FBI can be tracked to infected food handlers. Therefore, it is important that strict personal hygiene measures should be adopted during food preparation. The consumers should also take precautions for prevention of foodborne illness. These include cooking food at appropriate temperatures and following standard hygiene practices, proper storage and prevention of cross-contamination of food.

Thus, integrated intervention strategies are required to prevent FBI at community level. Successful implementation of these interventions requires intersectoral collaboration including agriculture industry, food industry and health care sector.

BIBLIOGRAPHY

1. Atreya, CD. Major foodborne illness causing viruses and current status of vaccines against the diseases. Foodborne Pathogens and Disease, 1(2), 89–96; 2004 doi:10.1089/153531404323143602

2. Bakir F, Damluji SF, Amin-Zaki L, et al. (July 1973)."Methylmercury poisoning in Iraq" (PDF). Science 181 (4096): 230–41.doi:10.1126/science.181.4096.230. PMID 4719063

3. Botulism Type B: Epidemiologic Aspects of An Extensive Outbreak". Aje.oxfordjournals.org

4. Das, M.; Khanna, SK. "Clinicoepide-miological, Toxicological, and Safety Evaluation Studies on Argemone Oil". *Critical Reviews in Toxicology* **27**(3): 273–297 1997. doi:10.3109/10408449709-089896. PMID 9189656.

5. Dubois E, Hennechart C, Deboosère N, et al. (April 2006). "Intra-laboratory validation of a concentration method adapted for the enumeration of infectious F-specific RNA coliphage, enterovirus, and hepatitis A virus from inoculated leaves of salad vegetables". Int. J. Food Microbiol. **108** (2):16471. doi:10.1016/j.ijfoodmicro.2005.11.007.PMID 1638 377

6. Eisenberg WV. Sources of food contaminants. Principals of Food Analysis for Filth, Decomposition and Foreign Matter, FDA Technical Bulletin No.1, Washington, DC (1981): 11–25.

7. Mead PS, Slutsker L, Dietz V, et al. Food-related illness and death in the United States, Emerg Infect Dis 5:607–25; 1999

8. Mead PS, Slutsker L, Dietz V, McCaig LF, Bresee JS, Shapiro C, Griffin PM, Tauxe RV. Food-related illness and death in the United States. Emerging Infectious Diseases 1999; 5:607–25.

9. Sen, Sunrita; Fiedler, Doreen (19 July 2013). "Corruption, poor quality taint India school meal scheme". Business Recorder (Karachi).

10. Sinha, Kounteya (10 January 2012). "70% of milk in Delhi, country is adulterated". *Times of India*.

11. Tuttle J, Gomez T, Doyle MP et al. Lessons from a large outbreak of *Escherichia coli* O157:H7 infections: insights into the infectious dose and method of widespread contamination of hamburger patties. *Epidemiol Infect* 1999;122: 185–92.

7

Food Adulteration: Prevention Through Detection

Shalini Dwivedi, Neha Chanana

"Don't give into temptation for sweets. These could be adulterated"

Such headlines have become a part of our daily newspapers informing us about the sale of adulterated sweets during festival seasons like *diwali.* Unscrupulous *halwais* often use *khoya* adulterated with urea, starch, etc. to prepare sweets which makes them unfit for consumption. This is not all. There are many similar instances to talk about, e.g. 40% of the ghee available in the market is adulterated with vegetable fat and high concentration of fatty acids. Still higher percentages (65%) of mustard seeds samples were adulterated with *Argemone mexicana* in a study conducted by Department of food technology and quality control. Moreover, the incident of adulterated milk being sold in different outlets of country was much talked about recently. It is estimated that 70% of the milk supplied all over the country is adulterated with detergents, urea, water, etc. Even there have been TV discussions focusing on FSSAI report on the availability and sale of milk adulterated with water, detergents and starch, etc. by profit making companies.

Such incidences surely make us wonder, *what exactly is food adulteration and how does it occur?*

Food adulteration is the addition or removal of any substance to or from food such that the natural composition and quality of food substance is affected. A related term 'adulterant' means any material which is or could be employed for making the food unsafe or substandard or misbranded or containing extraneous matter. Food is considered to be adulterated, if there is evidence of:

i. Substandard quality

ii. Substitution by cheaper substance

iii. Abstraction of any constituent of article

iv. Preparation or storage in unsanitary conditions

v. Presence of poisonous ingredients

vi. Use of colouring agents and/or preservatives in excess of prescribed limits

vii. Quantity or purity below the prescribed standards

For example, milk adulteration involves adding water and removing fats from milk. Often adulterants like detergents, starch, wheat flour are added to milk. The addition of starch and extraction of fat lowers the nutritional content of milk and makes it even unfit for consumption. Similarly other such

56

examples of food adulteration are given below:

i. Adulteration of fats and oils: *Ghee* is often mixed with hydrogenated oils and animal fats. Synthetic colours and flavours are added to other fats to make them appear like *ghee*.

ii. Food grain adulteration: This involves mixing of sand or crushed stones to increase the bulk of food grains. Pulses are adulterated with plastic beads that resemble the pulse grains in colour and size.

iii. Other adulterants: Brick powder is added to chili powder, used tea leaves are added to the fresh ones to increase the bulk.

These are clear examples of food being adulterated intentionally (intentional adulterants). This is mostly done by Food Business Operators (FBO), grocery merchants for petty monetary gains.

However, there is other side of the coin which is often ignored with respect to food adulteration. Food can also get adulterated unintentionally due to inappropriate food handling. This includes use of pesticides and fertilizers during farming or harvesting or improper storage, processing, packaging and transportation methods. Such adulterants which incidentally get mixed during the food handling process are known as incidental adulterants (Fig. 7.1).

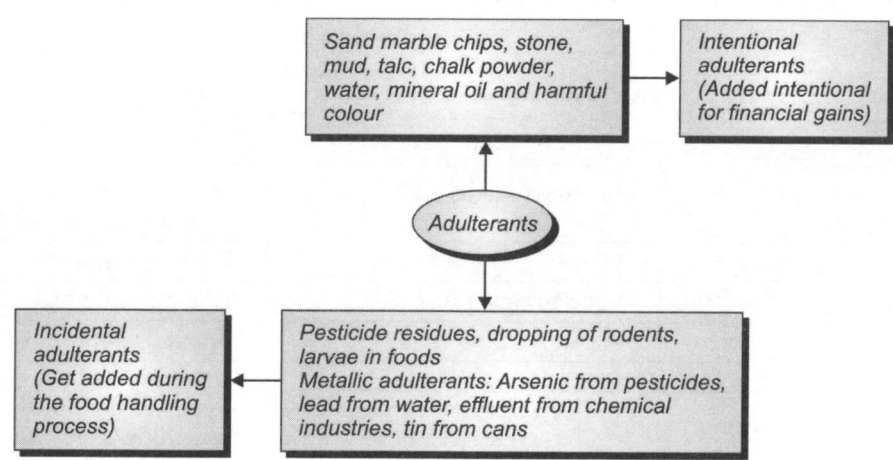

Fig. 7.1: Types of adulterants

Why and how can the adulterants be detected?

These adulterants whether intentional or incidental can cause detrimental effects which may prove fatal in serious cases. The adulteration of mustard oil with argemone oil in 1994 posed serious cases of epidemic dropsy and even death among people in National Capital of Delhi. This is an example of how serious the situation can become if common mustard oil which is used daily in our kitchen for cooking becomes adulterated.

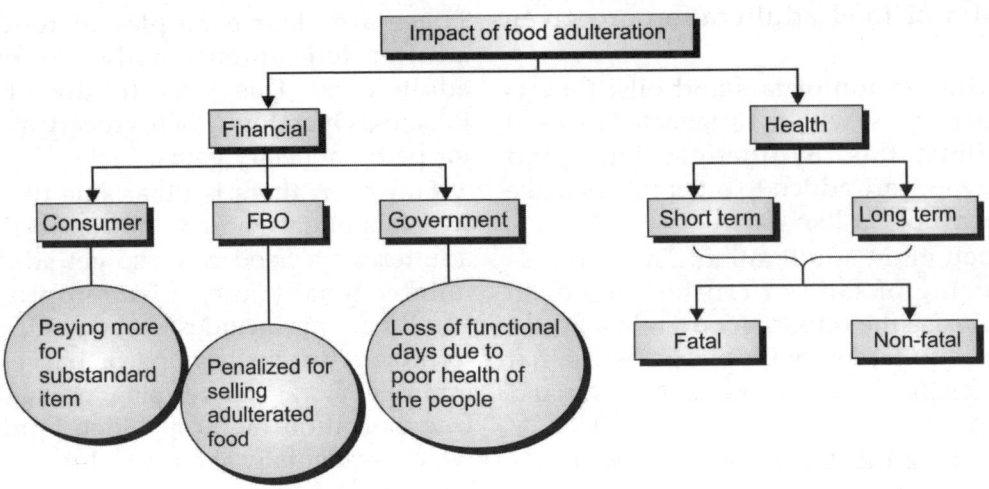

Fig. 7.2: Impact of food adulteration

The impact of food adulteration is grievous as it affects the finances as well as health of the individual and the country at large. Impact on health can be short term and long term (Fig. 7.2).

The harmful effects of major adulterants are shown in Table 7.1.

Table 7.1: Harmful effects of major adulterants		
Adulterant	*Food commonly involved*	*Health hazards*
Argemone seeds Argemone oil	Mustard seeds, edible oil, fats	Epidemic dropsy, Glaucoma, Cardiac arrest
Rancid oil	Oils	Destroys vitamin A and E
Sand, marble chips, stones, filth	Food grains, pulses, etc.	Damages digestive tract
Methanol	Alcoholic liquors	Blurred vision, blindness, death
Barium	Foods contaminated by rat poisons (Barium carbonate)	Violent peristalisis, arterial hypertension, muscular twitching, convulsions, cardiac disturbances
Bacillus cereus	Cereal products, custards, puddings, sauces	Food infection (nausea, vomiting, abdominal pain, diarrhoea)
Salmonella spp.	Meat and meat products, raw vegetables, salads, shell-fish, eggs and egg products, warmed-up leftovers	Salmonellosis (food infection usually with fever and chills
Non-permitted colour or permitted food colour beyond safe limit	Coloured food	Mental retardation, cancer and other toxic effect

Contd.

Table 7.1: Harmful effects of major adulterants (Contd.)

Adulterant	Food commonly involved	Health hazards
Monosodium glutamate (flour)(beyond safe limit)	Chinese food, meat and meat products	Brain damage, mental retardation in infants
Food flavours beyond safelimit Artificial sweetners beyond safe limit	Flavoured food Sweet foods	Chances of liver cancer Chances of cancer
Pesticide residues (beyond safe limit)	All types of food	Acute or chronic poisoning with damage to nerves and vital organs like liver, kidney, etc.

Thus, keeping in view the harmful effects of consuming adulterated food, it is essential that these adulterants are timely detected in food items so as to protect the health of the consumer. This can be done using rapid and simple methods so that their consumption can be avoided and help in reducing foodborne illness. Many of these tests can be carried out easily at home by the consumers in their daily lives using simple and common laboratory chemicals and apparatus. Listed below are some such methods/tests for common adulterants.

i. *Simple visual tests:* These tests are done at the level of procurement or receiving of the food items. Such tests are carried out through naked eye examination and do not require any specialized laboratory equipment. As the only instrument used here is our eyes, experience is required for this test so that the adulterated food items can be sorted out. This is the most cost effective test as only after simple training anyone can conduct this test. This test can be conducted by even by a person who can just read and write, can be carried out easily at home by the lady of the house or by even children. For example, chocolates should not contain any vegetable oil according to the food standards. Therefore, just by looking at the ingredients on the wrapper of the chocolates we can make out if the chocolate is adulterated with vegetable oil or not.

ii. *Simple physical tests:* These tests are based on physical properties of the food items like specific gravity, pH, magnetism, etc. These tests are easy to perform. As no specific apparatus is required, these tests can be carried out by middle income families in their houses, by children in school labs, e.g. iron filings mixed in tea can be detected by a magnet.

iii. *Simple chemical tests:* These tests are of 2 types viz. home based which can be carried out easily at home and there are other sophisticated tests. Trained personnel are required for conducting sophisticated tests. Since these tests require specific apparatus, it can be carried out only in government accredited labs.

These tests will help in preventing FBI. The details of the tests are given in Annexure A7.

Moreover, there are other preventive measures which have been taken by the government so as to curb food adulteration.

Preventing Food Adulteration

The government has taken steps to prevent food adulteration with the Food Safety Act, 2006 through which the producer and the supplier are liable to punishment in case of involvement in act of food adulteration. Moreover, the Department of Health through health education, has taken steps to spread awareness among the consumers regarding the ill-effects of food adulteration. Role of different stakeholders in prevention of food adulteration is given in Fig. 7.3.

A. Legislative Control

Legislative procedures for food adulteration can be traced back to the times of *Kautilya*. Adulteration was considered the gravest of socio-economic crime. Stringent laws, regulation and procedures were elaborated by *Kautilya* to ensure protection of king from any poisoning attempts through Royal Kitchen.

In modern India, first preventive measure for adulteration was taken in 1860. But it was under Indian Penal Code and was enforced by the Municipality.

1899: First food adulteration bill was introduced in Bombay. In 1937 Central Advisory Board of Health and Food Adulteration committee were inaugurated.

1944: Central committee for food standards started working.

In 1947 Indian Standards Institute (ISI) was established. ISI certification Marks Act came into being from 1952. Apart from ISI Agricultural Marketing Association or Agmark, Army Services Corps or ASC, Fair Average Quality (FAQ) standards, Food Adulteration Committee (FAC) recommended different standards.

Until 1954, several states formulated their own food laws. Then Government of India appointed the Central Advisory Board and the

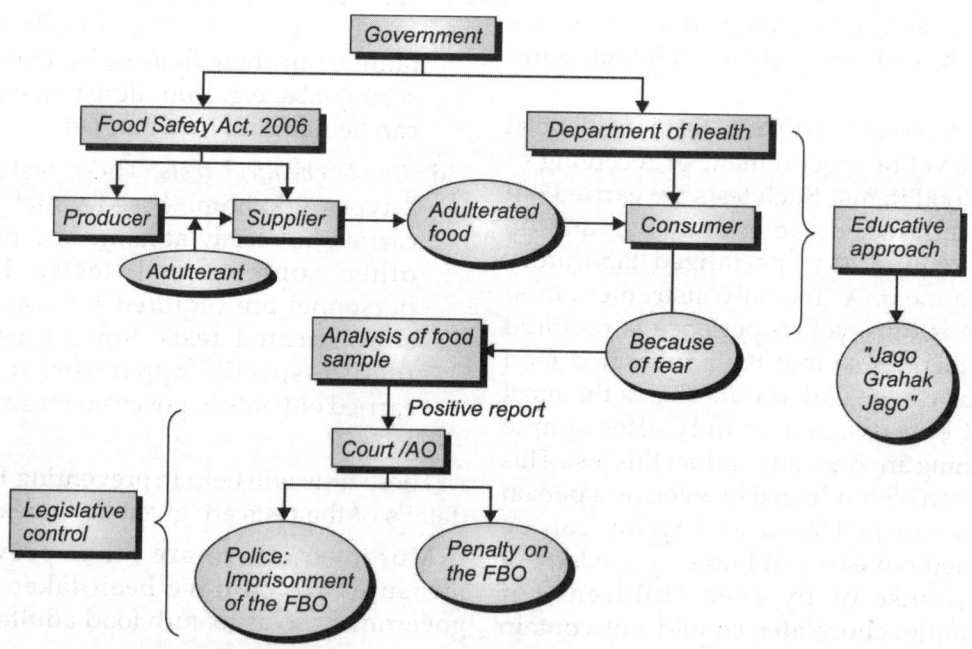

Fig. 7.3: Role of stakeholders in prevention of food adulteration in India

Food Adulteration Committee in the years 1937 and 1943 respectively.

1954: Prevention of Food Adulteration Act came into being. The main objectives of PFA were to have a central legislation to bring uniformity in food laws, protection of public from poisonous and harmful foods, prevention of the sale of substandard foods, protection of the interests of the consumers by eliminating fraudulent practices and to prevent, curb and check the rampant adulteration of food stuffs. This act prohibited the manufacture, sale and distribution of not only the adulterated food but also foods contaminated and misbranded food. The central food laboratory at Calcutta and the Central Food Technological Research Institute at Mysore tested the foods for adulteration. The state government had a food testing laboratory with public analyst and staff to test the suspected foods.

Subsequently the government passed other orders and acts such as the Fruit Products Order, 1955; Vegetable Oil Products (Control) Order, 1947; Meat Food Products Order, 1973; Milk and Milk Products Order, 1992 and Edible Oil Packaging Regulations, 1998. Thus, there were multiple food laws, standard setting and enforcement agencies for covering different aspects of food product businesses. These varied quality/safety standards led to confusion in ensuring and enforcing food safety. Hence, there was a need for a central legislation which consolidated all the laws related to food. Eventually, the Food Safety and Standards Act was promulgated in 2006.

Under this act the reporting system regarding food adulteration has been ensued through diverse and varied mechanisms. The Food Safety Officer (FSO) now has the power to inspect any place where any adulterant is manufactured. He can collect food samples he suspects to be adulterated and send them for analysis.

Moreover, the Food Business Operator (FBO) can also inform the concerned authorities in case he believes the food under his possession to be contaminated or adulterated, in any form. The consumer can also get the food article analyzed from the food analyst if he suspects the food item sold to him is adulterated in any form.

Penalties on the FBO for Food Adulteration

The Food Safety and Standards Authority of India (FSSAI) under the food safety and standards act poses heavy penalties on the person involved in the import, manufacture, storage, distribution or sale of any adulterant. The person is liable to:

i. Penalty not exceeding 2 lakh rupees if the adulterant is not injurious to health.

ii. Penalty not exceeding 10 lakh rupees if the adulterant is injurious to health.

The court can issue the following prohibition orders on the convicted FBO:

i. A prohibition on the use of the process of treatment for the purposes of the food business;

ii. A prohibition on the use of premises or equipment for the purposes of food business or any other food business of the same class or description.

The court if satisfied of the offence committed by the FBO, may impose prohibition, by an order, on the FBO participating in the management of any food business, or any food business of a class or description specified in the order.

The court or the Adjudicating Officer (AO) can also order for cancellation of license, recall of food from market and also publish the name and place of residence of the person held guilty, the offence and the penalty imposed in newspapers or any other widely circulated means of communication.

Liabilities on the FSO

The FSO is also guilty of an offence under the act and is liable with a penalty extending to one lakh rupees if he vexatiously and without reasonable ground seizes any article of food or adulterant.

B. Health Education Approach

The government has also taken measures for educating the masses on the ill effects from consuming adulterated food items through electronic and print media regularly. Under the famous brand of "JAGO GRAHAK JAGO" the public has also been made aware of their rights and duties with respect to food adulteration, e.g. whom to report in case they suspect the food item to be adulterated, etc. Regular messages on television and radio have to a larger extent helped in awaking the masses about the social evil of food adulteration.

Challenges associated within the current system for preventing food adulteration and measures to overcome them

The government has taken measures to curtail food adulteration; however, there are certain challenges which still need to be met in order to control food adulteration. Some of them are as follows:

i. There is lack of continuous monitoring of the food handling units by the FSOs. This is majorly because of lack of adequate number of FSOs in the country. Therefore, the filling up of the vacant posts of FSOs should be taken up on priority basis.

ii. Lack of implementation of strict punishments for the offenders. In the span of last 10 years hardly 10% of the offenders have received punishment in cases of adulteration.

iii. Lack of good testing facilities: The present condition of food testing laboratories of the country is not satisfactory as majority of the posts in these laboratories are lying vacant. Moreover, the central and the state labs set up under FSSAI are short of sophisticated instruments to monitor pesticide residues in food samples or for analyzing samples of carbonated beverages. Thus, the government needs to set up better testing facilities in terms of increased number of accredited labs with good infrastructure as well as trained human resource for testing of food adulterants.

iv. Awareness rallies or antifood adulteration drives organized by schools can help in educating children as well as adults regarding the effects of food adulteration and its preventive measures.

v. Training programmes for the public regarding basic tests for detecting food adulteration need to be organized.

vi. Some FSOs for their financial benefit support the offenders. Therefore, a continuous check in this regard on the FSO is required. Appropriate punishment for the FSO found guilty should also be implemented.

vii. Lack of involvement of consumer forum and NGOs. In the present scenario there is no NGO in place working to prevent food adulteration. The involvement of the NGOs sector will further help in curtailing food adulteration.

Thus, preventing food adulteration is the duty of every individual of the country. Provided the community is aware about the simple methods of detection of adulterants in food, the harmful effects caused by consuming adulterated food items can be reduced. Moreover, the strict implementation of legislative procedures can help in curbing the practice of food adulteration.

Annexure A7

Table A7.1: Simple visual tests for detecting adulterants

Food	Adulterant	Method of detection
Pulses (whole and split)	Khesari dal	It has edged type look showing a slant surface on one side and square in appearance in contrast to other dals
Black pepper	Papaya seeds	Papaya seeds can be distinct from pepper as they are of oval and shrunken shape, and greenish brown or brownish black colour.
Wheat, Rice and other grains	Dust, crystals, stone, weed seeds, spoiled grain, straw, insects and rodent and their hair and excreta	These can be examined visually to for any foreign matter, damaged or discoloured grains, insect and rodent contamination, etc.
Mustard seeds	Argemone seed	It resembles like mustard seed but shows a difference on close examination. Surface of argemone seed is rough while mustard seeds are smooth. When mustard seed is crushed, it is yellow inside whereas argemone seed is white.
Cloves	Extracted oil	Extracted cloves are small in size and of shrunken look. The characteristic pungent taste of original clove is less found in extracted cloves.

Table A7.2: Simple physical tests for detecting adulterants

Food	Adulterant	Method of detection
	Water	Measures the specific gravity with a lactometer by immersing it in milk kept in a deep vessel. The normal value lies between 1.028 and 1.032. Lower value indicates added water. But this is not a foolproof method as in addition to water, sugar, urea may have been added to the milk to increase its specific gravity.
Milk	Developed acidity	Place a test tube containing 5 ml of the milk sample in a boiling water bath and hold for about 5 minutes. Remove the tube and rotate in an almost horizontal position. The film of milk on the side of the test tube is examined for any precipitated particles. Formation of clots is indicative of developed acidity in the milk due to microbial spoilage. Such milk is unsuitable for consumption.
Tea leaves, suji	Iron fillings artificial color	Easily separated by passing a magnet over surface of food. Put the tea leaves on a moistened blotting paper. Artificially dyed tea will impart color to the moistened blotting paper immediately.

Contd.

Table A7.2: Simple physical tests for detecting adulterants (Contd.)

Food	Adulterant	Method of detection
Powdered spices	Salt	Take a pinch of sample and taste it. Presence of salt can easily detected by taste.
Coffee	Chicory	Sprinkle coffee powder on the surface of water in a glass. Coffee floats while chicory starts sinking leaving a trail of color, due to a large amount of caramel.
Asafoetida	Soap stone and earthy matter	Take a small portion of sample with water in a test tube and shake well. Allow to settle. Soap stone and other matter will settle at the bottom of the tube.
Saffron	Dried maize tendrils	Original saffron will not easily break like artificial one. Artificial saffron is prepared by soaking maize tendrils in sugar and colouring by coal tar colours. The artificial colour will dissolves in water. Pure saffron when dissolved in water will give its colour as long as it lasts.
Common salt	White stone powder	Mix a teaspoon of salt sample in a glass of water. If chalk is present solution will be white and insoluble impurities will be down at bottom.
Honey	Sugar solution	Take a cotton wick and dip in honey sample and burn. Presence of water will not allow the sample to burn, if it goes it will produce a cracking sound.
Sugar	Saccharin	Taste a small quantity. It leaves a lingering sweetness on tongue for a fixed time and a bitter taste at the end.
	Chalk powder	Dissolve 10 gm of sugar sample in glass. Leave it for some time. If the sample is added with chalk, it will settle down.

Table A7.3: Simple chemical tests for detecting adulterants

Food	Adulterant	Method of detection
Milk	Starch	Take small amount of milk in a test tube. Add few drops of iodine solution.Formation of blue colour indicates the presence of starch in the sample. Starch is added to increase the SNF(solid not fat) of milk
	Urea	Take 5 ml milk in a test tube. Mix with 5 ml paradimethyl amino benzaldehyde solution. If solution turns yellow, the sample is adulterated with urea.
	Vanaspati	Take 3 ml of sample to be tested in a test tube. Add 10 drops of hydrochloric acid. Mix one teaspoonful of sugar into it. Wait for 5 minutes. Then examine the colour of mixture.
	Formalin	Put 10 ml of milk in a tests tube and add 5 ml of conc. HSO from the wall of tube without mixing. If violet blue

Contd.

Table A7.3: Simple chemical tests for detecting adulterants (Contd.)

Food	Adulterant	Method of detection
		ring appears at the middle of these two layers then it indicates the presence of formalin in sample. Formalin is added to increase the shelf life of the product.
Rabri	Blotting paper	Take 5 ml rabri in a test tube. Add 3 ml of HCl and 3 ml of distilled water into it. Mix the content with a glass rod. Take out the rod and examine. Presence of fine fibres on glass rod indicates the presence of blotting paper in rabri.
Khoa	Starch	Boil a teaspoon of khoa with 2 teaspoon water. Cool the content and add few drops of iodine solution to it. If blue colour appears, it shows the presence of starch.
Ghee	Vanaspati	Mix 1 teaspoon melted sample with equal quantity of conc. HCl in a test tube and shake well for 2 minutes. Now add a pinch of sugar into it. Shake well to completely mix the content and wait for 5 minutes. If crimson colour appears in lower layer, it shows the presence of vanaspati.
	Mashed potato, sweet potato	Take a small amount of melted sample in test tube and few drops of iodine solution to it. If violet blue colour appears, it indicates the presence of starch.
Mustard oil	Argemone oil	Take 1 teaspoon of oil in a test tube and add same amount of concentrated nitric acid and well. Red to brown colour in lower layer shows the presence of argemone oil.
Wheat, rice and other grains	Ergot (fungus which contains poisons)	Ergot seeds are purple black long size grains and mostly found in bajra. Put some grains in a glass container. Add water containing 20% salt solution. If ergot is present it will float over on the surface while real grains settle down.
Black pepper	Papaya seeds	Leave the sample in alcohol or rectified spirit. Black pepper seeds settled down while papaya seeds float.
Powdered spices	Starch	Take a small portion of sample on a dish and add few drops of iodine solution to it. If blue colour comes it indicates the presence of starch in the sample.
Turmeric powder	Yellow soap Powder or Chalk powder	Take a small amount of sample in a test tube and add some water and mix well. Now add few drops of concentrated HCl. If it produces bubbles, then it confirms the presence of yellow soap powder or chalk
	Coloured saw dust	Put a teaspoon turmeric powder in a test tube and add few drops of concentrated HCl. Instantly appeared pink colour

Contd.

Table A7.3: Simple chemical tests for detecting adulterants (Contd.)

Food	Adulterant	Method of detection
		which suddenly goes on mixing with water, shows the presence of turmeric. If the colour remains, it confirms the presence of an artificial colour metanil yellow which is a non-permitted coal tar colour.
Silver foil	Aluminum foil	Take a small portion of foil in tube and add concentrated HCl to it. Stir the content properly.If the foil dissolves, it shows the sample is aluminium foil.
Sweetmeats, ice cream and beverages, pulses, spices	Metanil yellow	Extract colour with lukewarm water from food samples and add a few drops of concentrated hydrochloric acid. A magenta colour indicates the presence of metanil yellow.
Pulses, whole and split, besan	Khesari dal	Put a sample in dilute hydrochloric acid. Pink colour develops indicating the presence of khesari dal.
Silver foil	Aluminum foil	To metal foil add 2 drops of concentrated nitric acid in a test tube. The silver foil will completely dissolve whereas the aluminium foil remains undissolved.

BIBLIOGRAPHY

1. Kalyan Bagchi, Prevention of food adulteration: some thoughts.Health and Population - Perspectives and Issues 1984; 7(3):167–175.
2. Radomir Lasztity, Marta Petro-Turza, Tamas Foldesi. History of food quality standards.Iin food Quality and Standards., 2004. Eolss Publishers, Oxford, UK.
3. V. Lakshmi et al. Food Adulteration. International Journal of science inventions today. 2012; 1(2): 106–113.

8

Clinical Features, Diagnosis and Management of Foodborne Illness

Ruchi Sharma, Rohit Tewari, Amarjeet Singh

Often there are cases in medical emergencies which present with features of partial paralysis or stroke. Patient may complain of double vision and may have trouble in swallowing. For attending doctor it is difficult to imagine that the culprit may be consumption of home-canned food. In such case, possibility of botulism needs to be ruled out first. Therefore, it is essential to have a high index of suspicion for FBI for early diagnosis, better evaluation and management of such cases. Hence, in the initial assessment of patients with suspected foodborne illness (FBI), the history of food consumed, its type and source is very important.

The global supply of food has led to an increasingly connected planet, not only in terms of food products but also in terms of risks for FBI. These illnesses comprise a broad spectrum of diseases and are responsible for substantial morbidity and mortality worldwide. Food isn't sterile; it comes from animals or grows in soil. If we are healthy adults, our immune systems can deal with small numbers of bacteria and viruses. But at higher levels of concentration they can make us quite sick. The outcome of exposure to foodborne diarrhoeal pathogens depends on a number of host factors including pre-existing immunity, the

ability to elicit an immune response, nutrition, age and non-specific host factors.

For most adults in the industrialized world, incidents of FBI are unpleasant but are generally mild and self limiting. Indispositions that are restricted to gastroenteritis are not usually life threatening. Vulnerable people such as pregnant women, the elderly or people with poor immune systems who are already very sick or weak for some other reason can get very ill or even die from FBI. These vulnerable groups constitute quite a large proportion of the population. For many of them theses diseases can be fatal.

FBI outbreak generally refers to an incident in which two or more persons experience a similar illness after ingestion of a common food and epidemiological analysis implicates the food as the source of the illness. The time of onset (incubation period), duration of illness, clinical symptoms, history of recent travel, or antibiotic use, as well as presence of blood or mucus in the stool, recent meals (including type of food, especially raw or uncooked food, unpasteurised milk or food products), cooking and refrigeration as well as details of others affected by similar symptoms can provide valuable clues to cause of FBI. This chapter provides detailed clinical features of FBI.

FBI are infections or irritations of the gastrointestinal (GI) tract caused by food or beverages that contain harmful bacteria, parasites, viruses, or chemicals. The GI tract is a series of hollow organs joined in a long, twisting tube from the mouth to the anus. These different diseases have many different symptoms, so there is no one "syndrome" that is FBI. However, as the microbe or toxin enters the body through the GI tract, and often causes the first symptoms there, so nausea, vomiting, abdominal cramps and diarrhoea are common symptoms in many FBI. Signs and symptoms may start within hours after eating the contaminated food, or they may begin days or possibly even weeks later.

Foodborne illness may be:

- Acute: It lasts from a few hours to days.
- Acute or chronic
- Chronic: This manifests as carrier state of the disease in the individual. Here the individual does not manifest disease symptoms, but is a source of infection to others. The pathogenic bacteria are present in various secretions of the carrier e.g. in stool of salmonella carrier. The carrier state can be of short or long duration (temporary or chronic carrier). The carrier state usually ceases spontaneously after several weeks or a few months, but some individuals may become chronic carriers (e.g., for periods exceeding a year, for agents such as *Salmonella typhi*). A number of chronic sequelae may result from foodborne infections, including ankylosing spondylitis, arthropathies, renal disease, cardiac and neurologic disorders, and nutritional and other malabsorptive disorders (incapacitating diarrhea).

Different degrees of severity are observed, from a mild disease which does not require medical treatment to the more serious illness requiring hospitalization, long-term disability and/or death. Most FBI are acute, meaning they happen suddenly and last a short time. Usually most people recover on their own without treatment. Rarely, FBI may lead to more serious complications. Each year, an estimated 48 million people in the United States experience a FBI (3,000 deaths in the United States annually).

Foodborne illness may be:

- Self-limiting
- With complications
- Fatal

Less commonly, neurologic symptoms may develop, such as blurry vision, dizziness or tingling in the arms. In some instances, the most life-threatening problems occur several days after the start of intestinal symptoms and these can include kidney failure, meningitis, arthritis, and paralysis, depending on the type of foodborne microbe involved. Often symptoms can range from mild gastroenteritis to life threatening neurologic, hepatic and renal syndromes that could lead to death. FBI can also lead to residual illness ranging from arthritis, gastrointestinal symptoms, kidney pain and other debilitating problems that can plague victims for the rest of their lives.

Acute gastroenteritis, presenting with any combinations of the key symptoms of abdominal pain, vomiting, diarrhoea and fever can be readily related to the suspicion of a FBI episode. Many bacterial agents as well as norovirus can cause acute gastroenteritis. Bloody diarrhoea may be the presentation of bacterial agents known to attack lower intestinal tracts through breaching of mucosal barriers. A number of bacterial and viral agents, such as *Salmonella typhi* and *Listeria monocytogenes* are known to cause systemic illness and even fatal complications. Further cases can be prevented should the source of infection be identified and removed.

Biochemical or chemical agents in food, such as ciguatoxin and pesticides could cause neurological symptoms. Most patients may experience transient symptoms but fatal complications can occur. Chemical food poisoning usually results in burning sensations in the chest, neck and abdomen. Some chemicals are extremely poisonous and if ingested may result in severe vomiting within a few minutes. Exposure to chemicals in food can result in acute and chronic toxic effects ranging from mild and reversible to serious and life threatening. These effects may include cancer, birth defects and damage to the nervous system, the reproductive system and the immune system. Toxic metals if ingested in sufficient quantities can cause metallic taste in mouth, vomiting and abdominal pain, usually within a few hours. Diarrhoea may also occur.

Every year new information on food safety emerges. We must update ourselves with most current information on FBI, their clinical features, diagnosis and management.

Pathogenesis

Organisms causing FBI include bacteria, viruses and parasites. However, the mechanism may vary from ingestion of preformed toxin to enteroinvasion to infection by toxigenic organism. Onset and clinical features of some contaminants are given in Table 8.1.

Bacterial properties that are responsible for the pathogenesis of enterocolitis include bacterial adhesion and replication, bacterial enterotoxins and bacterial invasion. The fluid stream in the gut is powerful and will wash away any organism, unless the organism has the ability to adhere to the mucosa. Adherence of enterotoxigenic organisms like E. coli and Vibrio is mediated by plasmid encoded adhesins. Wiry, rigid projections on the organism's surface, called pili, express these adhesion proteins. Adherence causes changes in the enterocyte membrane with destruction of the brush border and changes in the underlying cell cytoplasm. Bacterial enterotoxins are polypeptides that cause diarrhoea. These toxins may be either secretagogue toxins as in Vibrio and Enterotoxigenic E. coli, where the toxin causes intestinal secretion of fluids and electrolytes without causing tissue damage or they may be cytotoxins, exemplified by Shiga toxin produced by Shigella and Shiga like toxins produced by Enterohemorrhagic E. coli which cause direct tissue damage by epithelial cell necrosis. Both enteroinvasive E. coli and Shigella have the ability for epithelial cell invasion, apparently by microbe stimulated endocytosis, which is followed by intracellular proliferation, cell lysis and spread. Clinical presentation of FBI is given in Table 8.2.

Shigella bacillary dysentery: The transmission of this disease is remarkable for the fact even up to 10 ingested organisms can cause clinical illness. The bacteria invade the intestinal epithelial cells and remain restricted to the lamina propria. Subsequently, the bacteria evade the phagolysosome in the epithelial cell, multiply in its cytoplasm and then destroy the cell. This causes dysentery.

Salmonellosis and typhoid fever: After crossing the epithelial layer of the intestine, the bacteria are phagocytosed by macrophages in the lamina propria. Now they spread throughout the body via lymphatics and multiply in the lymph nodes. After this multiplication they enter the bloodstream via the thoracic duct and cause a transient bacteremia by which the bacteria colonise the liver, spleen, lymph nodes and bone marrow. Subsequently, after further multiplication the bacteria cause a secondary and a heavier bacteremia which corresponds with the onset of clinical illness. From the blood stream bacteria localise in the lymphoid tissue and other organs. The principal lesions are hyperplasia and necrosis of the lymphoid tissue like the peyers patches in the intestine which may ulcerate generating classic typhoid ulcers.

Campylobacter enteritis: The organism has flagella which not only confers the property of motility, but also the ability to penetrate mucus covering the enterocyte. This invasion into the epithelium can lead to diarrhoea, dysentery as well as febrile illness akin to enteric fever.

Vibrio: The bacteria are noninvasive. They remain in the lumen and secrete a virulence phage encoded enterotoxin. Flagellar proteins involved in motility and attachment are essential for colonisation. The toxin causes massive secretory diarrhoea which overwhelms the reabsorptive capacity of the colon, and large volumes of "rice water" stools are passed resulting in dehydration and electrolyte imbalance.

Clostridium botulinum: The organism is non-invasive. Disease is caused due to ingestion of preformed neurotoxin, which after absorption from the intestine reaches the peripheral cholinergic synapses through the circulation. It then acts by blocking release of the neurotransmitter acetylcholine at synapses and neuromuscular junctions resulting in a flaccid paralysis.

Table 8.1: Onset and predominant upper GI, lower GI, neurological and allergic symptoms with selected foodborne contaminants

Time of onset to symptoms and predominant symptoms	Associated organism or toxin
Upper gastrointestinal tract symptoms occur first or predominate (nausea, vomiting)	
Less than 1 h Nausea, vomiting, unusual taste, burning of mouth.	Metallic salts
1–2 h Nausea, vomiting, cyanosis, headache, dizziness, dyspnea, trembling, weakness, loss of consciousness.	Nitrites
1–7 h, mean 2–4 h Nausea, vomiting, retching, diarrhea, abdominal pain, prostration	*Staphylococcus aureus* and its enterotoxins
5 to 6 h Vomiting or diarrhea, depending on whether diarrheic or emetic toxin present; abdominal cramps; nausea	*Bacillus cereus* (emetic toxin)
6–24 h Nausea, vomiting, diarrhea, thirst, dilation of pupils, collapse, coma	Amanita species mushrooms
Lower gastrointestinal tract symptoms occur first or predominate (abdominal cramps, diarrhoea)	
2–36 h, mean 6–12 h Abdominal cramps, diarrhoea, putrefactive diarrhea	*Clostridium perfringens, Bacillus cereus* (diarrheic form), *Strep-*

Contd.

Table 8.1: Onset and predominant upper GI, lower GI, neurological and allergic symptoms with selected foodborne contaminants (Contd.)

Time of onset to symptoms and predominant symptoms	Associated organism or toxin
associated with *Clostridium perfringens*; sometimes nausea and vomiting.	*tococcus faecalis, S. faecium*
12–74 h, mean 18–36 h Abdominal cramps, diarrhoea, vomiting, fever, chills, malaise, nausea, headache, possible. Sometimes bloody or mucoid diarrhea, cutaneous lesions associated with *V. vulnificus*. *Yersinia enterocolitica* mimics flu and acute appendicitis.	*Salmonella* species (including *S. arizonae*), *Shigella*, enteropathogenic *Escherichia* coli, other Enterobacteriaceae, *Vibrio* parahaemolyticus, Yersinia enterocolitica, Aeromonas hydrophila, Plesiomonas shigelloides, Campylobacter jejuni, *Vibrio cholerae* (O1 and non-O1) *V. vulnificus, V. fluvialis*
3–5 days Diarrhea, fever, vomiting abdominal pain, respiratory symptoms.	Enteric viruses
1–6 weeks Diarrhea, often exceptionally foul-smelling; fatty stools; abdominal pain; weight loss.	*Giardia lamblia*
1 to several weeks Abdominal pain, diarrhoea, constipation, headache, drowsiness, ulcers, variable; often asymptomatic.	*Entamoeba histolytica*
3–6 months Nervousness, insomnia, hunger pangs, anorexia, weight loss, abdominal pain, sometimes gastroenteritis	*Taenia saginata, T. solium*
Neurological symptoms occur (visual disturbances, vertigo, tingling, paralysis)	
Less than 1 h Gastroenteritis, nervousness, blurred vision, chest pain, cyanosis, twitching, convulsions	Organic phosphate
Less than 1 h Excessive salivation, perspiration, gastroenteritis, irregular pulse, pupils constricted, asthmatic breathing	Muscaria-type mushrooms
Less than 1 h Tingling and numbness, dizziness, pallor, gastric haemorrhage, desquamation of skin, fixed eyes, loss of reflexes, twitching, paralysis	Tetradon (tetrodotoxin) toxins
1–6 h Tingling and numbness, gastroenteritis, dizziness, dry mouth, muscular aches, dilated pupils, blurred vision, paralysis	Ciguatera toxin Chlorinated hydrocarbons

Contd.

Table 8.1: Onset and predominant upper GI, lower GI, neurological and allergic symptoms with selected foodborne contaminants (Contd.)

Time of onset to symptoms and Predominant symptoms	Associated organism or toxin
Nausea, vomiting, tingling, dizziness, weakness, anorexia, weight loss, confusion.	
2 h to 6 days, usually 12–36 h Vertigo, double or blurred vision, loss of reflex to light, difficulty in swallowing, speaking, and breathing, dry mouth, weakness, respiratory paralysis	*Clostridium botulinum* and its neurotoxins
More than 72 h Numbness, weakness of legs, spastic paralysis, impairment of vision, blindness, coma	Organic mercury
More than 72 h Gastroerteritis leg pain, ungainly, high-stepping gait; foot, wrist drop.	Triorthocresyl phosphate
Allergic symptoms occur (facial flushing, itching)	
Less than 1 h Headache, dizziness, nausea, vomiting, peppery taste, burning of throat, facial swelling and flushing, stomach pain, itching of skin	Histamine (scombroid)
Less than 1 h Numbness around mouth, tingling sensation, flushing, dizziness, headache, nausea	Monosodium glutamate
Less than 1 h Flushing, sensation of warmth, itching, abdominal pain, puffing of face and knees	Nicotinic acid
4–28 days, mean 9 days Gastroenteritis, fever, edema about eyes, perspiration, muscular pain, chills, prostration, labored breathing	Trichinella spiralis
7–28 days, mean 14 days Malaise, headache, fever, cough, nausea, vomiting, constipation, abdominal pain, chills, rose spots, bloody stools	*Salmonella typhi*
10–13 days Fever, headache, myalgia, rash	*Toxoplasma gondii*
Varying periods (depends on specific illness) Fever, chills, head-or joint ache, prostration, malaise, swollen lymph nodes, and other specific symptoms of disease in question	*Bacillus anthracis, Brucella melitensis, B. abortus, B. suis, Coxiella burnetii, Francisella tularensis, Listeria monocytogenes, Mycobacterium tuberculosis, Mycobacterium species, Pasteurella multocida, Streptobacillus moniliformis, Campylobacter jejuni, Leptospira species*

Table 8.2: Clinical presentation of foodborne illness due to various causative agents

Clinical presentation	Causative agents
Acute gastroenteritis	Vibrio parahaemolyticus, Non-typhoidal Salmonella spp. Staphylococcus aureus Bacillus cereus Vibrio cholerae O1/ O139 Clostridium perfringens Noro virus
Bloody diarrhea	Shigella spp. Campylobacter E. coli O157: H7 Entamoeba histolytica
Systemic illness	Salmonella typhi Salmonella partyphi Listeria monocytogenes Hepatitis A virus Hepatitis E virus
Neurological involvement	Clostridium botulinum Ciguatoxin Shellfish poisoning Scombroid fish poisoning Puffer fish poisoning Mushroom poisoning Clenbuterol pesticides

Symptoms of FBI can range from mild to serious and can last from a few hours to several days. Summary of FBI is given in Table 8.3

FBI may also lead to dehydration, Haemolytic Uremic Syndrome (HUS), and other complications.

Dehydration

When loss of fluids from body as vomiting, diarrhoea is more than the fluid replaced through consumption. When dehydrated, the body lacks enough fluid and electrolytes—minerals in salts, including sodium, potassium, and chloride—to function properly. Infants, children, older adults, and people with weak immune systems have the greatest risk of becoming dehydrated.

Signs of dehydration are:

- Excessive thirst
- Infrequent urination
- Dark-coloured urine
- Lethargy, dizziness, or faintness

Signs of dehydration in infants and young children are:

- Dry mouth and tongue
- Lack of tears when crying
- No wet diapers for 3 hours or more
- High fever
- Unusually cranky or drowsy behaviour
- Sunken eyes, cheeks, or soft spot in the skull

Also, when people are dehydrated, their skin does not flatten back to normal right away after being gently pinched and released.

Severe dehydration may require intravenous fluids and hospitalization. Untreated severe dehydration can cause serious health problems such as organ damage, shock, or coma—a sleep like state in which a person is not conscious.

Haemolytic Uremic Syndrome (HUS)

HUS is a rare disease that mostly affects children younger than 10 years of age. HUS develops when toxins from *E. coli* bacteria lodged in the digestive tract enter the bloodstream. The toxin destroys red blood cells, leading to formation of blood clot in the lining of the blood vessels.

In the United States, *E. coli* O157:H7 infection is the most common cause of HUS, but infection with other strains of *E. coli*, other bacteria, and viruses may also cause HUS. A recent study found that about 6 percent of people with *E. coli* O157:H7 infections developed HUS. Children younger than age 5 have the highest risk, but females and people age 60 and older also have increased risk.

Symptoms of *E. coli* O157:H7 infection include diarrhoea, which may be bloody, and abdominal pain, often accompanied by nausea, vomiting, and fever. Up to a week after *E. coli* symptoms appear, symptoms of HUS may develop, including irritability, paleness, and decreased urination. HUS may lead to acute renal failure, which is a sudden and temporary loss of kidney function. HUS may also affect other organs and the central nervous system. Most people who develop HUS recover with treatment. Research shows that in the United States between 2000 and 2006, fewer than 5 percent of people who developed HUS died of the disorder. Older adults had the highest mortality rate—about one-third of people age 60 and older who developed HUS died.

Acute FBI may also lead to chronic or long lasting health problems.

C. botulinum and some chemicals affect the nervous system, causing symptoms such as:

- Headache
- Tingling or numbness of the skin
- Blurred vision
- Weakness
- Dizziness
- Paralysis

Studies have shown that some children who recover from HUS develop chronic complications, including kidney problems, high blood pressure, and diabetes.

Other Complications

Some FBI lead to other serious complications. For example, *C. botulinum* and certain chemicals in fish and seafood can paralyze the muscles that control breathing. *L. monocytogenes* can cause spontaneous abortion or stillbirth in pregnant women.

Chronic disorders due to FBI includes:

- **Reactive arthritis,** a type of joint inflammation that usually affects the knees, ankles, or feet. Some people develop this disorder following FBI caused by certain bacteria, including *C. jejuni* and *Salmonella*. Reactive arthritis usually lasts fewer than 6 months, but this condition may recur or become chronic arthritis.

- **Irritable bowel syndrome (IBS),** a disorder of unknown cause that is associated with abdominal pain, bloating, and diarrhoea or constipation or both. Foodborne illnesses FBI caused by bacteria increase the risk of developing IBS.

- **Guillain-Barré syndrome,** a disorder characterized by muscle weakness or paralysis that begins in the lower body and progresses to the upper body. This syndrome may occur after FBI caused by bacteria, most commonly *C. jejuni*. Most people recover in 6 to 12 months.

- Paralysis of the muscles that control breathing from the toxin produced by *Clostridium botulinum* and certain chemicals in fish and seafood.

- Spontaneous abortion or stillbirth in pregnant women caused by *Listeria monocytogenes* infections.

Table 8.3: Co-relates of foodborne illness

Causative agent	Clinical presentation	Incubation period/latency of onset	Laboratory test	Associated foods
Bacteria				
Bacillus cereus	Abdominal cramps, watery diarrhea	10–16 hours	Stool culture	Meats, stews, gravies, vanilla sauce.
Bacillus cereus (performed toxin)	Sudden onset of severe nausea and vomiting. Diarrhoea may be present.	1–6 hours	Stool culture	Improperly refrigerated cooked and fired rice, meat
Campylobacter jejuni	Diarrhoea which may be blood stained, abdominal pain, fever and vomiting.	2–5 days	Stool culture; Campylobacter requires special media and atmosphere to grow.	Raw and undercooked poultry, unpasteurized milk, contaminated water
Clostridium perfringens	Sudden onset of colic followed by dirrhea; nausea is common, vomiting and fever are usually absent	Usually 10–12 hours, range from 6 to 24 hours	Stool culture and detection of enterotoxin	Inadequately cooked or reheated meat and meat products (e.g. stew, meat pie, gravies made of beef or chicken)
Clostridium botulinum	Diplopia, blurred vision, and bulbar weakness. Symmetric paralysis may progress rapidly	Usually 12–36 hours	Stool culture and detection of enterotoxin.	Inadequately cooked or reheated meat and meat products (e.g. stew, meat pie, gravies made of beef or chicken)
E. coli O157	Acute bloody diarrhea and abdominal cramps with little or no fever. May be complicated with hemolytic uremic syndrome	Range from 2–10 days, with a median of 3–4 days	Stool culture, requires special media to grow. Detection of Shiga toxin produced from culture isolates	Undercooked beef, and unpasteurized milk. Others include raw fruits and vegetables, contaminated water
Listeria monocytogenes	Fever, muscle aches, and sometimes nause or diarrhoea. An invasive disease manifests most commonly as meningitis or septicemia. Infection	9–48 hours for gastrointestinal symptoms; 2–6 weeks for invasive disease	Blood or cerebrospinal fluid culture. Stool culture may not be helpful in sporadic case	Fresh soft cheeses unpasteurized milk, inadequately pasteurized milk

Contd.

Table 8.3: Co-relates of foodborne illness (Contd.)

Causative agent	Clinical presentation	Incubation period/latency of onset	Laboratory test	Associated foods
	during pregnancy may result in abortion, premature delivery, stillbirth, or neonatal meningitis.		because asymptomatic fecal carriage occurs.	
Non-typhoidal *Salmonella* spp.	Diarrhoea, abdominal pain, vomiting and fever	Usually about 12–36 hours, can range from 6–72 hours		Inadequately cooked meat and poultry. Conta minated raw egg and egg products, milk and milk products, foods contaminated by food handlers
Salmonella typhi (typhoid fever)	Fever, headache, malaise, anorexia, relative bradycardia and splenomegaly, non-productive cough the rose spots on the trunk. Diarrhoea is uncommon and vomiting is not severe.	Usually 8–14 days, range from 3 to over 60 days	Blood, stool, or urine culture. Widal test may be helpful	Contaminated water and food, in particular contaminated shellfish, raw fruits and vegetables, and raw milk.
Salmonella paratyphi (paratyphoid fever)	Similar to typhoid fever but tends to be milder	1–10 days	Blood, stool, or urine culture. Widal test may be helpful	Contaminated water and food, in particular contaminated shellfish, raw fruits and vegetables, and raw milk.
Shigella spp.	Abdominal cramps, fever and diarrhea and stool typically contain blood and mucus	Usually 1–3 days, range from 12 to 96 hours, up to 1 week for *S. dysenteriae*	Stool culture	Usually person-to-person spread through food or water contaminated with human fecal material
Staphylococcus aureus	Sudden onset of severe nausea and vomiting, abdominal pain. Diarrhoea and fever may be present	1–6 hours	Stool culture Detection of staphylococcal enterotoxins in food remnants	Food contami nated by food handlers with skin infection or nasal carriers, especially those food involving manual handling

Contd.

Table 8.3: Co-relates of foodborne illness (Contd.)

Causative agent	Clinical presentation	Incubation period/latency of onset	Laboratory test	Associated foods
				and no reheating afterwards (e.g. sandwiches, cakes)
Vibrio cholerae O1or O139	Acute painless watery diarrhoea with or without vomiting	Usually 2–3 days, range from a few hours to 5 days	Stool culture; *Vibrio cholerae* requires special media to grow and confirmation by Public Health Laboratory	Contaminated water and food such as raw or undercooked seafood
Vibrio parahaemolyticus	Watery diarrhoea and abdominal pain. May also have nausea, vomiting and fever	Usually 12–24 hours, can range from 4–30 hours	Stool culture	Inadequately cooked seafood (e.g. crab, shrimp, clams) or other food cross-contaminated by seafood
Hepatitis A	Fever, malaise, anorexia, nausea, abdominal discomfort, tea-coloured urine and jaundice	Usually 28–30days but may range from 15–508 days	Serum for IgM anti-HAV	Food and water contaminated by food handler, such as shellfish
Hepatitis E	Fever, malaise, anorexia, nausea, abdominal discomfort, tea-coloured urine and jaundice	Usually 26–42 days, range from 15–64 days	Serum for IgM anti-HEV or HEV by polymerase chain reaction (PCR)	Contaminated water and food
Norovirus	Usually symptoms of nausea, vomiting, diarrhoea, abdominal pain, low-grade fever and malaise	Usually 24–48 hours, range from 10–50 hours	Stool for norovirus by RT-PCR	Shellfish such as oyster, faecal contaminated foods
Parasite				
Entamoeba histolytica	Diarrhoea (may be bloody), frequent bowel movement, lower abdominal pain	Usually 2–4 weeks, can be as short as 2–3 days	Examination of stool for cysts and trophozoites. Demonstration of trophozoites in tissue biopsy or ulcer scrapings	Contaminated food and drinking water

Contd.

Table 8.3: Co-relates of foodborne illness (Contd.)

Causative agent	Clinical presentation	Incubation period /latency of onset	Laboratory test	Associated foods
			by culture histopathology	
Biochemical				
Ciguatera toxin	Numbness in limbs, face, tongue or perioral area, heat/cold reversal. May have gastrointestinal symptoms include diarrhoea, abdominal pain, nausea and vomiting	Usually 1–6 hours, may be up to 36 hours	Detection of toxin in fish	Large coral reef fish such as a variety of grouper, hump head wrasse and black fin red snapper
Mushroom poisoning	Vomiting, diarrhoea, confusion, visual disturbance, salvation, diaphoresis, hallucinations, disulfiram-like reaction, confusion, visual disturbance	<2 hours	Mushroom identification and/or demonstration of the toxin	Wild mushrooms
Puffer fish (Tetrodotoxin) poisioning	Paraesthesias, vomiting, diarrhoea, abdominal pain, ascending paralysis, respiratory failure	<30 minutes	Detection of tetrodotoxin in fish	Puffer fish
Scombroid poisoning (amnesic shellfish poisoning)	Flushing, rash, abdominal pain and neurological problems such as confusion, memory loss, disorientation, seizure and coma	Shellfish	Vomiting, diarrhoea	24–48 hours
Chemical				
Clenbuterol poisoning	Tremor, palpitation, headache, nausea, weakness, dizziness and nervousness	Usually about 1 hour, range from 10 minutes to 8 hours	Detection of clenbuterol in urine or food remnant	Clenbuterol contained pig or pork
Pesticides (organophosphates) poisoning	Numbness, weakness, dizziness, abdominal pain, nausea and vomiting	4–12 hours, can happen within few minutes	Serum pseudocholinesterase level may be lowered. Detection of organophosphates in food remnant	Inadequately soaked or rinsed contaminated leafy vegetables

Diagnosis of Foodborne Illness

It is important to maintain high index of suspicious if patient presents above compatible symptoms with history of relevant food intake. Diagnosis of foodborne diseases depends not only on the clinical presentations, but also a positive history of intake of relevant food, and the capability of the laboratory to detect the pathogen. For example, patients suffering from ciguatera fish poisoning usually have neurological involvement like limb numbness and may have characteristic sensation of heat/cold reversal.

Listerosis is associated with cheese and unpasteurized milk while *E. coli* O157 infection is related to undercooked beef. History of consumption of tuna fish may raise the suspicion of scombroid poisoning; raw egg and under-cooked chicken for non-typhoidal salmonellosis; and raw oyster for norovirus; fried rice for *Bacillus cereus*.

Knowing the symptoms of FBI can help save your life; it can also keep you from spreading diseases to others. To diagnose the FBI, ask about symptoms, foods and beverages recently consumed, and medical history. Perform a physical examination to look for signs of illness.

1. **Relevant history of food consumption in terms of physical signs and symptoms**
   i. History of consumption of high risk foods like raw eggs, uncooked poultry, unpasteurized milk, raw eggs, mayonnaise, cream, pastry, cut salads, fish and canned foods
   ii. Specific food-handling practices, cooking preferences–home cooked and eating outside or intake of leftover food
   iii. Type of water supply
   iv. H/o travel (domestic and international)
   v. H/o recent group gatherings, visitors, social events
   vi. H/o contact with animals
   vii. Chronic illness, immune suppression, pregnancy
   viii. Recent changes in medical history, regular medications
   ix. Occupational history
   x. H/o similar illness in family, friends and colleagues
   xi. H/o reported similar illness from place of consumption
   xii. H/o bottle feeding in small child
   xiii. H/o antibiotic intake
   xiv. H/o allergies, recent immunizations
   xv. The respondent's thoughts on what caused their illness

Consideration also should be given to exogenous factors such as the association of the illness with emotional stress, sexual habits, exposure to other ill persons, recent hospitalization, child care center attendance, and nursing home residence.

1. **General examination:** Measure the patient's blood pressure, pulse, weight, and temperature and perform a physical examination to look for:
   - Signs of dehydration—extreme thirst, dry mouth, dizziness or light headedness and little or no urination
   - Diarrhoea for more than 2 days in adults or for more than 24 hours in children
   - Bloody diarrhoea
   - Diarrhoea and vomiting leading to dehydration
   - A fever higher than 101.5°F (38.6° C)
   - Signs of rash, myalgia, arthralgias
   - Inability to speak or see
   - Trouble swallowing
   - Double vision
   - Seizures (fits)
   - Changes in mental state, such as confusion

2. **Systemic examination**:
   - GI symptoms—weight loss, severe pain/tenderness in the abdomen or rectum
   - Nervous system symptoms—paresthesias, weakness, paralysis and malaise
   - Signs of HUS
   - Stools containing blood or pus
   - Stools that are black and tarry
   - Respiratory symptoms—difficulty in breathing/coughing, etc.
   - Profuse sweating
   - Cardiovascular symptoms—high blood pressure

In addition to foodborne causes, a differential diagnosis of GI tract disease should include underlying medical conditions such as irritable bowel syndrome; inflammatory bowel diseases such as Crohn's disease or ulcerative colitis; malignancy; medication use (including antibiotic-related *Clostridium difficile* toxin colitis); gastrointestinal tract surgery or radiation; malabsorption syndromes; immune deficiencies; and numerous other structural, functional, and metabolic etiologies.

The differential diagnosis of patients presenting with neurologic symptoms due to a foodborne illness FBI is also complex. Possible food-related causes to consider include recent ingestion of contaminated seafood, mushroom poisoning, and chemical poisoning. Because the ingestion of certain toxins (e.g. botulinum toxin, tetrodotoxin) and chemicals (e.g. organophosphates) can be life-threatening, a differential diagnosis must be made quickly with concern for aggressive therapy and life support measures (e.g. respiratory support, administration of antitoxin or atropine), and possible hospital admission.

3. **Laboratory testing** (*see* Chapter 9 on Lab diagnosis of foodborne illness)

Management of Foodborne Illness

Most episodes of gastroenteritis are self-limiting. Symptoms are usually mild to moderate and last for a few days only. Fluid and electrolyte replacement and other symptomatic treatment are sufficient in most cases. Initial treatment of patients with foodborne illness FBI should focus on assessment and reversal of dehydration, either through Oral Rehydration Therapy (ORT) especially in children, or through IV fluids in seriously dehydrated cases. Routine use of antidiarrheal agents is not recommended because many of these agents have potentially serious adverse effects in infants and young children.

- **Specific treatment**
  a. The choice of antibiotics, if indicated, should be based on clinical presentation, organism detected in clinical specimens and susceptibility test results.
  b. Specific treatment in case of pesticide poisoning with chelating agents like burnt charcoal may be done based on epidemiological and clinical features, under medical supervision.
  c. Suspected cases of botulism are treated with botulinum antitoxin.

- **General treatment**

The following steps may help relieve the symptoms of FBI and prevent dehydration in adults:

a. ORS is drink of choice to prevent dehydration

b. Homemade butter milk and lemon water can be good option to prevent dehydration and vomiting

c. In case of diarrhoea 2 tsf isabgol mixed with curd can be of help

d. Sip small amounts of clear liquids or sucking on ice chips if vomiting is still a problem

e. Gradually reintroduce food, starting with bland, easy-to-digest foods such as curd, rice, potatoes, toast or bread, cereal, lean meat and bananas

f. Avoid fatty foods, sugary foods, dairy products, caffeine, and alcohol until recovery is complete

g. Hospitalization may be required to treat life threatening symptoms and complications, such as paralysis, severe dehydration, and HUS

h. Avoid self medication and use of antibiotics

Infants and children present special concerns. Infants and children are likely to become dehydrated more quickly from diarrhoea and vomiting because of their smaller body size. The following steps may help relieve symptoms and prevent dehydration in infants and children:

a. Give oral rehydration solutions to prevent dehydration

b. Give food as soon as the child is hungry

c. Give infants breast milk or full strength formula, as usual, along with oral rehydration solutions

Older adults and adults with weak immune systems should also drink oral rehydration solutions to prevent dehydration.

- **Control of spread**

    a. Bacteria, virus and parasite may be excreted from infected patients who should be reminded to observe good personal hygiene. Adequate infection control measure should be implemented on all gastroenteritis cases. Patient should be nursed with standard precautions: hand washing; wearing gloves for contact with blood, excretions, secretions and contaminated items; eye protection and gown for splashes of blood, secretions and excretions.

    b. For patients with diarrhea, isolation is required insome cases. For example, additional preventive measure against secondary person-to-person spread is necessary for patients suspected to be affected by noro-virus. The virus is highly contagious and infection can be caused by as few as 100 particles. Apart from faecal-oral route, norovirus can also be transmitted through droplet from patients's vomitus. Delay or inappropriate management of patients may result in secondary spread in hospital setting or in the community.

    c. Prompt disinfection of the environment (including toilet used by patients or the environment soiled with patients' vomitus) using 1,000 ppm hypochlorite solution (one part of 5.25% hypochlorite solution added to 49 parts of water) is important to prevent nosocomial outbreak of norovirus. Patients suspected with norovirus should be isolated or cohorted until symptoms subsided for 48 hours.

It is advisable to ask the patients contacts (other family members, friends, schoolmates, colleagues or residents of elderly homes) for similar symptoms. Early notification of such incident, without waiting for the laboratory result, will facilitate public health actions to prevent further spread of the outbreak.

- **Preventive measures:** Food safety education is a critical pre-requisite to prevent Foodborne outbreaks by education of food-handlers and the community about proper practices in cooking and storage of food, and personal hygiene. Hand washing is one of the key interventions, not just by food handlers, but also by the community at large. Environmental measures include discouraging sewage farming for growing vegetables and fruits.

BIBLIOGRAPHY

1. Andary MT. Guillain-Barré syndrome. Emedicine. http://emedicine.medscape.com/article/315632-overview. Updated August 26, 2011.
2. Burns B. Reactive arthritis in emergency medicine.Emedicine. http://emedicine.medscape.com/article/808833-overview. Updated February 1, 2010.
3. Gould HL, Demma L, Jones TF, et al. Hemolytic uremic syndrome and death in persons with *Escherichia coli O157:H7* infection, Foodborne Diseases Active Surveillance Network sites, 2000–2006. Clinical Infectious Diseases 2009:49 (10):1480–1485.
4. Scallan E, Griffin PM, Angulo FJ, Tauxe RV, Hoekstra RM. Foodborne illness acquired in the United States—unspecified agents. Emerging Infectious Diseases 2011;17(1):16–22.
5. Spiller R, Aziz Q, Creed F. Guidelines on the irritable bowel syndrome: mechanisms and practical management. *Gut*. 2007;56 (12):1770–1798.
6. US Food and Drug Administration, Center for Food Safety and Applied Nutrition. Foodborne Pathogenic Microorganisms and Natural Toxins Handbook: The Bad Bug Book. 2nd edition: 245–48.

9

Lab Diagnosis of Foodborne Illness

Rohit Tewari, Ruchi Sharma

Foodborne diseases are responsible for high levels of morbidity and mortality not only in the general population, but also for high risk groups like infants and children, the elderly and the immunocompromised. The organisms responsible may be bacterial, viral or parasitic (Table 9.1). In addition, some noninfectious agents like metals, chemicals and toxins are also involved (Table 9.2). Diagnosing a FBI depends to a large extent on the patient's clinical presentation; the differential diagnosis considered and sound clinical judgment. This is supported by laboratory testing. Signs and symptoms which should prompt laboratory evaluation are bloody diarrhea, diarrhea leading to dehydration, presence of fever, sudden onset of symptoms, severe abdominal pain and neurologic involvement in the form of paraesthesias and motor weakness.

In addition to food borne causes, due consideration must be given to all other diseases and conditions which can have similar symptoms like irritable bowel syndrome, inflammatory bowel disease, medication use, etc. Some other factors like travel, occupation, emotional stress, exposure to other ill people, etc. also need to be given due consideration and excluded. A small group of patients would present with neurologic symptoms and possible food related causes would include recent consumption of contaminated sea food, mushroom poisoning and chemical poisoning.

Microbiology Testing

The objective of laboratory evaluation in such a case would be to try and isolate an etiological agent in addition to the other supportive investigations in the form of routine hematology and biochemistry to look for infection, and complications like dehydration with electrolyte disturbances. Hence clinicians and public health care professionals should understand routine specimen collection and testing procedures as well situations where special tests may be required. Some tests like toxigenicity testing, serotyping and molecular testing may be available only in a large public health laboratory.

Stool cultures are indicated if the patient is immunocompromised, febrile, has bloody diarrhea, has severe abdominal pain, or if the illness is clinically severe or persistent. Stool cultures are also indicated if many fecal leukocytes are present, which indicates diffuse colonic inflammation and is suggestive of invasive bacterial pathogens such as *Shigella*, *Salmonella*, and *Campylobacter* species, and invasive *E. coli*. In most laboratories, routine stool cultures are limited to screening for

Table 9.1: Laboratory testing of foodborne pathogens

Etiological agent	Laboratory testing
Bacteria	
Bacillus anthracis	Blood culture
Bacillus cereus (diarrhoeal toxin)	Testing not necessary since the disease is self limiting. Food material and stool may be tested for toxin in outbreaks
Bacillus cereus (preformed enterotoxin)	In routine, this is a clinical diagnosis. Clinical laboratories do not usually test for this organism. If required, food and stool samples can be sent to reference lab for culture and toxin production
Brucella abortus, B. mellitensis and B. suis Campylobacter jejuni	Blood culture and serology. Stool culture. The orga nisms can be grown in campy-lobacter specific media
Clostridium botulinum, children and adults (preformed toxin)	Stool, serum and food can be tested for toxin. Reference labs can also perform stool and food culture
Clostridium botulinum (infants)	Stool, serum and food can be tested for toxin. Reference labs can also perform stool and food culture
Clostridium perfringens toxin	Stool can be tested for enterotoxin and also cultured for the organism
Enterohemorrhagic E. coli, including O157; H7 and other shiga toxin producing E. coli	Stool culture. 0157; H7 requires specific media. Toxin testing for Shiga toxin
Enterotoxigenic E. coli	Stool culture. ETEC needs special lab techniques for identification
Listeria monocytogenes	Blood or CSF cultures. There may be asymptomatic carrier status, hence stool culture is usually not helpful. Antibody testing for listerolysin O.
Salmonella spp	Stool culture and blood culture
Shigella spp	Stool culture
Staphylococcus aureus	Usually the diagnosis is clinical. If indicated, stool, vomitus and food can be tested for toxin and cultured.
Vibrio cholerae	Stool culture
Vibrio vulnificus	Stool, wound or blood cultures
Yersinia enterocolitica and Y. pseudotuberculosis	Stool, vomitus or blood culture
Viruses	
Hepatitis A	Increase in enzymes (transaminases) and bilirubin. Positive IgM antihepatitis A antibodies

Contd.

Table 9.1: Laboratory testing of foodborne pathogens (Contd.)

Etiological agent	Laboratory testing
Norwalk like viruses	Usually a clinical diagnosis. Negative bacterial cultures and a 4 fold increase in antibody titres to Norwalk virus
Rota virus Other viruses (astrovirus, adenovirus, calcivirus, parvovirus)	Viral detection in stool by immunoassay. Viral identification in early acute stool samples, serology
Parasites	
Cryptosporidium parvum	Examination of stool for *Cryptosporidium parvum*
Cyclospora cayetanensis	Stool examination
Entamoeba histolytica	Stool examination for cysts
Giardia lamblia	Stool examination for ova
Toxoplasma gondii	Presence of toxoplasma specific IgM antibodies. In addition demonstration of the organism can be attempted in specimens like bronchoalveolar lavage or lymph node aspirate
Trichinella spiralis	Serology, demonstration of larvae in muscle biopsy. Eosinophilia

Table 9.2: Laboratory testing of foodborne agents

Etiological agent	Laboratory testing
Antimony	Identification of metal in beverage or food
Arsenic	Identification of metal in urine. Eosinophilia may be present
Cadmium	Identification of metal in food.
Ciguatera fish poisoning	Consistent history with radioimmunoassay for toxin in fish
Copper	Identification of metal in beverage or food
Mercury	Analysis of blood and hair
Mushroom toxins	Typical history with identification of the mushroom or demonstration of toxin
Pesticides	Chemical analysis of food and blood.
Shellfish toxins	Detection of toxin in shellfish. Techniques like high performance liquid chromatography (HPLC) may be used
Tin	Chemical analysis of the food

Salmonella and *Shigella* species, and *Campylobacter jejuni/coli*. Cultures for *Vibrio* and *Yersinia* species, *E. coli* O157:H7, and *Campylobacter* species other than *jejuni/coli* require additional media or incubation conditions and therefore require advance notification or communication with laboratory and infectious disease personnel.

Stool examination for parasites generally is indicated for patients with suggestive travel histories, who are immunocompromised, who suffer chronic or persistent diarrhea, or when the diarrheal illness is unresponsive to appropriate antimicrobial therapy. Stool examination for parasites is also indicated for gastrointestinal tract illnesses that appear to have a long incubation period. Requests for ova and parasite examination of a stool specimen will often enable identification of *Giardia lamblia* and *Entamoeba histolytica*, but a special request may be needed for detection of *Cryptosporidium parvum* and *Cyclospora cayetanensis*. Each laboratory may vary in its routine procedures for detecting parasites.

Blood cultures should be obtained when bacteremia or systemic infection are suspected.

Direct antigen detection tests and molecular biology techniques are available for rapid identification of certain bacterial, viral, and parasitic agents in clinical specimens. In some circumstances, microbiologic and chemical laboratory testing of vomitus or implicated food items also is warranted.

The commonly collected important specimens for clinical microbiology testing are stool, blood and vomitus.

Clinical signs and symptoms help in adopting a particular line of testing. Naked eye examination of the stool helps to characterise the faeces, e.g. presence of blood and mucus would make one suspect Shigella. Gram staining may help indicate possible primary pathogens like gram-positive cocci.

Motility testing is helpful in cholera and it can be made more specific by neutralization with specific antiserum. Protozoa and helminths can be excluded with a saline wet mount and iodine mount preparation. Selective media are used for isolation of specific organisms being looked for. However, it should also be kept in mind that all the common selective media are also mildly inhibitory to other pathogenic organisms and hence it is useful to inoculate an enrichment medium and a non-inhibitory medium like MacConkey's agar simultaneously. In acute stages of the disease, pathogens are present in large numbers and hence can usually be isolated. However, in later stages the normal flora may predominate and hence enrichment media may be required.

Shigella group of organisms: Portion of the stool, especially containing the blood and mucus is selected and plated on MacConkeys agar, Salmonella–Shigella agar and lysine desoxycholate agar. Direct microscopic examination of the specimen is useful to pick up large numbers of fresh macrophages and few bacteria. A small portion of the sample can also be inoculated in a tube of selenite F broth which can be subcultured after 6 hrs. Nonlactose fermenting colonies appearing can be screened for motility and biochemical testing. Further identification can be carried out serologically.

Salmonella group of organisms: The stool usually does not contain the pus and mucus and hence there is no special portion of the stool to be selected for sampling. In such a situation, a well mixed sample is good enough and is inoculated on the media already listed in the above paragraph. Use of enrichment media is helpful. The nonlactose fermenting colonies are picked up and processed with biochemical reactions. Further typing is done by serology.

Vibrio cholerae: The stool is often watery and direct microscopy may reveal organisms with darting type of motility. Inoculation is performed on alkaline peptone water, alkaline bile salt agar and thiosulfate-citrate-bile salt (TCBS) sucrose-agar. Further identification is done by serotyping.

B. cereus: Primary medium for inoculation is blood agar incubated in air/CO_2. Polymyxin, egg yolk, mannitol, bromothymol blue agar (PEMBA) may also be used.

Clostridial organisms: Anerobic culture media are inoculated.

Staphylococcus: Blood agar, chocolate agar and high salt milk agar are inoculated.

Campylobacter: These are motile gram-negative curved rods and can be seen in faeces by grams or wrights stain. In dark field microscopy, darting motility can be demonstrated. The organisms can be grown in campylobacter specific media.

E.coli: If above mentioned pathogens are not identified, then *E.coli* present predominantly may be subjected to toxigenicity testing for identification of pathogenic strains.

Vomitus: This specimen needs to be transported to the lab as soon as possible and neutralised by adding a drop of an indicator like phenol red and N/10 NaOH. Subsequently it should be centrifuged and the deposit can be innoculated on blood agar, desoxycholate citrate agar and tetrathionate broth. One plate of blood agar should be subjected to anaerobic culture. Organisms which could be found are Staphylococcus, *Clostridium perfringens, Salmonella* spp. and Shigella spp.

Blood: Blood culture can be performed in suspected salmonellosis in appropriate liquid media.

Serum: This is useful for performing serological tests for viral infections, like hepatitis A and E (for viral antigen and for IgM and IgG antibody) and Norwalk virus. Biochemical testing on serum for bilirubin, liver enzymes like ALT and AST is helpful in viral hepatitis.

10

Investigation of an Outbreak of Food Poisoning

Sonika Raj, Amarjeet Singh, Puja Dudeja

The terms foodborne diseases, including foodborne intoxications and foodborne infections, cover illnesses acquired through consumption of contaminated food. These are also frequently referred to as food poisoning. It is defined under the Food Safety and Standards Act, 2006 as "any disease of an infectious or toxic nature caused by or thought to be caused by the consumption of food or water". An outbreak or epidemic is defined as the occurrence in a community of cases of an illness clearly in excess of expected numbers. The difference between an epidemic and an outbreak is the scale. While an outbreak is usually limited to a small focal area, an epidemic covers large geographic areas and has more than one focal point. For field epidemiological purposes another definition of an outbreak is: occurrence of two or more epidemiologically linked cases of a disease of outbreak potential (e.g. food poisoning, measles, cholera, dengue, Japanese encephalitis (JE), acute flaccid paralysis (AFP) or plague). So, food poisoning outbreaks are defined as the occurrence of two or more cases of a similar illness resulting from ingestion of a common food or when observed number of cases of a particular disease exceeds the expected number.

Food poisoning is an important cause of morbidity and mortality worldwide. There is not a single week when we do not come across a large outbreak of food poisoning in India. Some happen after wedding parties or religious festivals and most of them in schools. The scientific investigations/reports on outbreak of FBI in India for the past years indicated that a total 3,485 persons have been affected due to food poisoning. Under the Integrated Disease Surveillance Programme (IDSP) in India, food poisoning outbreaks reported from all over India in 2009 increased to more than double as compared to the previous year (120 outbreaks in 2009, as compared to 50 in 2008). Sudershan et al. investigated the cases of food poisoning in Hyderabad from 2003–2005 and found that there were 10 outbreaks of food poisoning involving 996 persons. The patients aged between 5 and 9 years were 7%, 10 and 14 years were 19%, 15 and 19 years were 17%, 20 and 29 years were 24%, 30 and 39 years 11%, 40 and 49 years were 9%, and more than 50 years were 9%. The type of food responsible for these outbreaks were *lauki/ghiya ki kheer* (sweet prepared from desiccated milk and bottle guard), milkshake, chicken *biryani* (dish prepared from rice and chicken), fruit salad, butter milk, mango juice, and jaggery rice. Table 10.1 shows some of the food poisoning outbreaks in India. It becomes quite

Table 10.1: Some food poisoning outbreaks in India

State/UT	Year	No. of cases	Population affected	Food item involved
Gujarat 1996	1996	34	Children	Sevu (crisp made from gram flour)
Tamil Nadu	1997	25	All	Butter milk
Kashmir	1998	78	Army personnel	Frozen Fowl
Karnataka	2008	34	All	Non-vegetarian food (chicken and fish)
Maharashtra	2010	13	School Children	Mid day meal
Karnataka	2012	46	Nurses	Kerala matta rice
Bihar	2013	23	Children	Rice and soybean-potato curry
Andhra Pradesh	2013	17	Boys	Egg curry
Gujarat	2014	40	Glass factory workers	Tomato curry
Maharashtra	2014	19	Adults	Juice
Tamil Nadu	2014	40	Girls	

clear that it affects people from all age groups, gender and can be caused from almost all the types of food articles.

Estimates of the burden of food poisoning are complicated by a number of factors:

- Different definitions of acute diarrheal illness used in various studies
- Most diarrheal illness is not reported to public health authorities
- Only few illnesses can be definitively linked to food.

So, rapid detection and detailed knowledge of burden of FBI will reduce the risk of spread of disease and demonstrate the real impact of unsafe food on economic growth and development.

Objectives of Investigation of an Outbreak of Food Poisoning

FBD outbreaks are investigated to prevent both ongoing transmissions of disease and similar outbreaks in the future. Specific objectives include:

- Control of ongoing outbreaks;
- Detection and removal of implicated foods;
- Identification of specific risk factors related to the host, the agent and the environment;
- Identification of factors that contributed to the contamination, growth, survival and
- Dissemination of the suspected agent;
- Prevention of future outbreaks and strengthening of food safety policies;
- Acquisition of epidemiological data for risk assessment of foodborne pathogens;
- Stimulation of research that will help in the prevention of similar outbreaks.
- To check for any malafide intentions/ sabotage

The investigation and control of FBI outbreaks require multi-disciplinary approach with skills in the areas of clinical medicine, epidemiology, laboratory medicine, food microbiology, food chemistry, food safety, food control risk communication and risk management. Investigating an epidemic involves a series of steps. These steps are not necessary to be undertaken in the same sequence. In fact, different situations will require changes in this order.

A full investigation of a FBI outbreak will normally include (Fig. 10.1)
– Epidemiological investigations
– Environmental and food investigations
– Laboratory investigations

Steps in investigation of an outbreak of food poisoning

1. **Confirm the existence of an outbreak:** An outbreak of food poisoning has three characteristics—Firstly, cases usually have gastrointestinal tract symptoms. Secondly, a large number of cases should occur in a small period of time. Thirdly, the cases should have some history of sharing at least one common meal. If these three criteria are fulfilled, then it is an outbreak of food poisoning.

2. **Verification of diagnosis:** Secure complete list of people involved and are interviewed as soon as possible. This also helps in developing the epidemiological case sheet.

The interviews should include questions about:
• Demographic details, including occupation
• Clinical details including signs and symptoms, date of onset, duration and severity of symptoms
• Visits to health care providers or hospitals
• Laboratory test results
• Contact with other ill persons
• Food consumption history
• Respondent's thoughts on what caused their illness
• Date of exposure to suspected foods

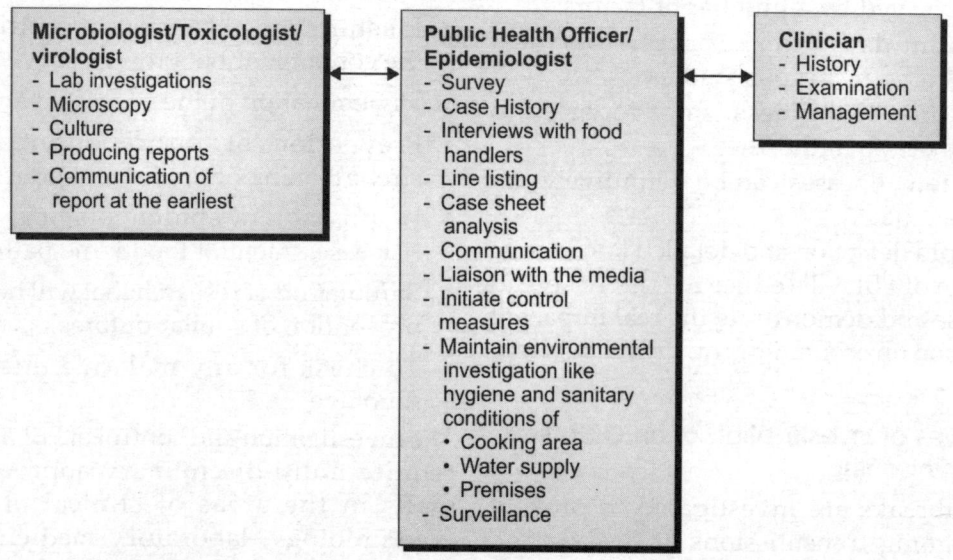

Fig. 10.1: Experts involved in investigation of outbreak of food poisoning

It also include visit to various sites like:

- Hospital/clinic where people are admitted
- Kitchen place where food was cooked

Clinical specimens (e.g. faecal samples, vomitus) from cases should be collected at the time of first contact. In case of food poisoning, microbiological diagnosis is always desirable to confirm the diagnosis but it is usually not possible because the procedures are highly specialized, takes time and not available at all places. Therefore, it becomes important to go for the "clinicoepidemiological" profile of the outbreak.

Let us consider a following **hypothetical example** to understand the investigation of an outbreak of food poisoning.

The girls hostel no. 4 in university had a farewell party on Sunday evening, May, 17. Out of total of 180 girls in the hostel, 110 girls attended the party. There were two messes in hostel and a canteen run by a private contractor. From the hostel, 30 girls complained of vomiting and diarrhea on Monday morning to hostel warden. They were referred to university's health centre. The in-charge Medical Officer (MO) was directed to investigate the matter.

Preliminary findings were that out of 30 cases, all the girls had 6–8 watery stools, 25 girls (83.3%) had abdominal cramps. Ten girls (33%) also complained of vomiting. None of them had fever, blood/mucus in stools and any dehydration. The first 3 cases had onset of symptoms between 9 and 10 am on May 18, 5 between 10 and 11 am, 8 cases between 11am to 1 pm and similarly all had onset in similar hourly periods. The last case had onset at 3 pm. By 10 am, 15 cases had onset of GI symptoms. No food sample of any meal had been kept as all meal was finished.

After verification of diagnosis, MO calculated the percentage of cases with a particular symptoms or signs and arranged those in a table in decreasing order as shown in Table 10.2. Organizing in this way will help in determining whether the outbreak was caused by intoxication or an enteric infection. For example:

- If the predominant symptom is vomiting without fever and the incubation period is short (less than 8 hours), intoxication by, for example, *Staphylococcus aureus*, *Clostridium perfringens* or *Bacillus cereus* are likely.
- Fever in the absence of vomiting and an incubation period of more than 18 hours points to an enteric infection such as *Salmonella*, *Shigella*, *Campylobacter* or *Yersinia*.

Table 10.2: Frequency of signs and symptoms among cases

Signs and symptoms	n = 30
Watery stools	30
Abdominal cramps	25
Nausea/vomiting	10
Fever	0
Blood/mucous in stools	0

The frequency of gastrointestinal symptoms in the present outbreak are compared with expected ones to confirm the diagnosis of etiological agent. Table 10.3 shows the major etiological agents of food poisoning with their median incubation period and major symptoms. According to the percentage of various signs and symptoms, investigating officer can make a tentative diagnosis of the etiological agent and can initiate control measures accordingly.

Table 10.3: Etiological agents of food poisoning with their median incubation period and major symptoms

Etiological agent	Median incubation (range)	Nausea /Vomiting	Loose-motions	Abdominal Cramps	Fever	Blood/ Mucus in stools	Neuro-muscular symptoms
Staph. aureus and B. cereus-1	2 hr (1 – 6)	+++	+	+	–	–	–
C. perfringens and B. cereus-2	13 (10 –18)	+	+++	+++	–	–	–
Salmonella spp	18 (16 – 24)	++	+++	++	+++	–	–
Shigella spp	24 (12 – 96)	+++	+++	+++	+++	+++	–
Insecticides /mushroom	Few mts.	++	+	–	–	–	+++
C. botulinum	24 (18 – 36)	+	–	–	–	–	+++

(+++ : occurs in 80 to 100%; ++ : 40 to 60%; +: 20 to 30%; – : 0 to 5% patients)

Then the MO should draw an epidemic curve, plotting the number of cases along the y-axis and the time according to hourly period on x-axis. The unit of time is usually based on the apparent incubation period of the disease and the length of time over which cases are distributed. As a rule of thumb, the x-axis unit should be no more than one-quarter of the incubation period of the disease under investigation. The peak of epidemic curve must be noted. Also note the first case and the last case. From the peak of the curve and by the known median incubation period of the suspected etiologic agent and also the minimum and maximum incubation periods from the first and last case respectively are noted. The meal which was consumed where all these three time periods converge gives the meal which was most probably contaminated and needs to be further investigated.

As evident from clinical profile, the entire epidemic was clearly due to either C perfringens or B. cereus type-2. If only the time of onset of illness is known then the shape of the epidemic curve suggests a point-source outbreak, inferences about the average incubation period and thus the suspected time of exposure may be drawn from the epidemic curve (Fig. 10.2). Here the further course of action for MO is to:

• Identify the median time of onset of illness.
• Calculate the time between occurrence of the first and last case (width of the epidemic curve).
• Count back this amount of time from the median to obtain the probable time of exposure.

If the organism and the time of onset of illness are known and the shape of the epidemic curve suggests a point-source outbreak, the probable time of exposure may be determined from the epidemic curve. The epidemic curve plotted with the above data showed a classical "common vehicle point exposure" curve typical of food poisoning (Figs 10.2 and 10.3). Going by the median and range of incubation

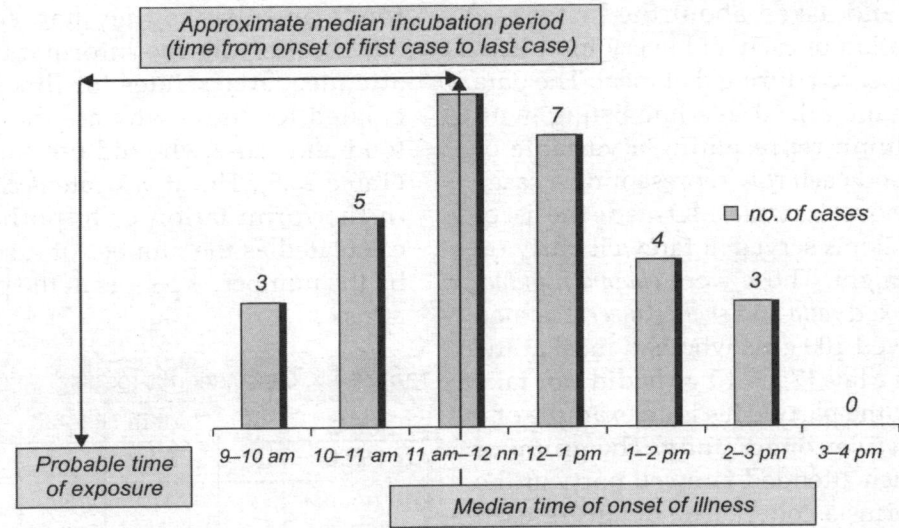

Fig. 10.2: Determining the median incubation period and probable time of exposure in a point-source outbreak

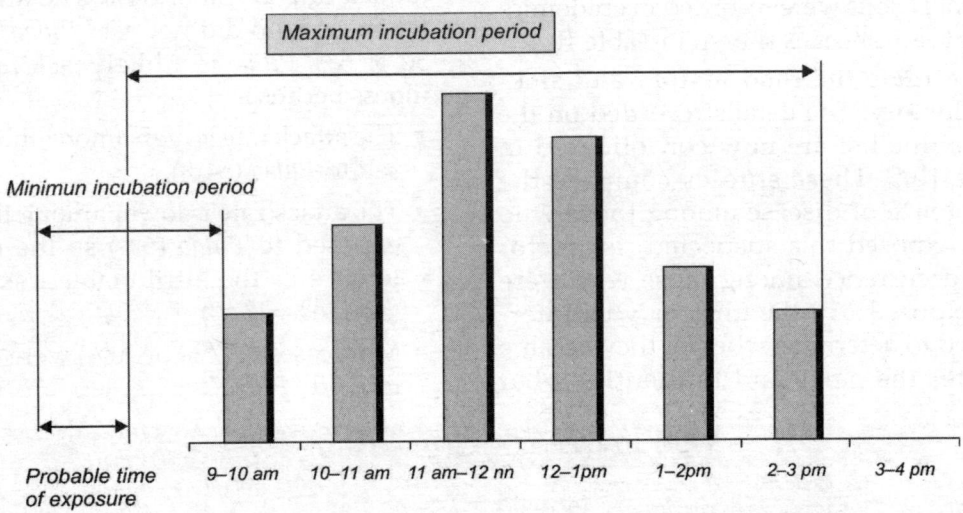

Fig. 10.3: Determining the probable period of exposure in a point-source outbreak with known pathogen

periods of these etiological agents, the suspicion converged to meal which was eaten at about 9 pm on May 17.

3. **Take the Food History about the Suspected Meal:** Now after the meal which

was most probably associated with the epidemic has been identified, a list of each and every food item which was served during that meal is made. After that every person who attended that meal is interviewed and asked about the history of consumption of each and every food item that was served during that meal. The data can be summarized in a line listing, with each column representing a variable of interest and each row representing a case. Like in above example, MO made the list of the food items served at farewell party on Sunday night. Those were *chappati, pulao, rajma*, mixed *raita* and *shahi paneer*. He then interviewed 100 girls who had meal at the party on May 17, as 10 girls did not take dinner at the party. This is an example of a **retrospective cohort study**. The group of girls which attended farewell party at the hostel forms a cohort. The details of each and every food items whether eaten or not, irrespective of the sickness status was taken from the cohort. If they had developed symptoms, details of symptoms were enquired. Details were recorded in epidemiological case sheet as shown in Table 10.4.

4. **Consolidate the food history and sickness history:** The details recorded on the above line list are now consolidated in Table 10.5. These studies compare the occurrence of disease among those who were exposed to a suspected risk factor with occurrence among those who were not exposed. All the girls may be interviewed to determine whether they became ill after the party and to identify what foods and drinks they had consumed. After collecting the information of each attendee, attack rates for illness are calculated for those who ate the particular food and those who did not eat that food (Table 10.5). The attack rate is a key factor in the formulation of hypotheses. It is calculated as the number of cases divided by the number of people in the population at risk.

Table 10.5: Calculation of food specific attack rate

Exposure	ill	Not ill	total	Attack rate
Ate *Pulao*	27	23	50	54%
Did not eat *Pulao*	3	47	50	6%
Total	30	70	100	30%

A total of 27 girls who ate "*Pulao*" fell ill (attack rate 27/50 or 54%. The attack rate for those who did not eat "*Pulao*" was 3/50 or 6%. *Pulao* is a likely risk factor for illness because:

• The attack rate is high among those exposed to *Pulao* (54%);

• The attack rate is lower among those not exposed to *Pulao* (6%) so the risk difference or the attributable risk is high (54%–6%=48 %)

• Most cases (27/30 or 90%) were exposed to food "*Pulao*".

Table 10.4: Line Listing (N = 100, n = 30)											
Serial No.	Name and Personal particulars	Developed illness (Yes /No)	If Yes, Presence of symptoms (Yes/No)					History of eating food items				
			Watery stools	Nausea/ vomiting cramps	Abdo-minal	fever	Mixed raita	Rajma paneer	Shahi paneer	Chap-pati	Rice
1	AK	Yes	Yes	Yes	Yes	No	Yes	No	Yes	No	Yes
2	SS	No	–	–	–	–	Yes	Yes	No	Yes	Yes
100	MK	Yes	Yes	No	Yes	No	Yes	Yes	Yes	Yes	Yes

Similarly, we have to calculate the attack rates of different food items. In addition, a ratio of the two attack rates, known as the relative risk (RR), can be canculated.

Relative risk (RR) =

Attack rate for those who ate food "A" / Attack rate for those who did not eat food "A" = 54%/6% = 9

Relative risk has no units and is a measure of the strength of association between the exposure and the disease. In the above example, the relative risk associated with eating *Pulao* is 9. This means that persons who ate rice dish were 9 times were more likely to develop illness than those who did not. This is quite a high risk. Then we can make the summary Table 10.6 showing the attack rates, relative risk and attributable risk. Now, the food item which shows the maximum Attributable risk is the food item which most probably was involved in the transmission of the epidemic.

However, in many circumstances, no clearly defined cohort of all exposed and unexposed can be identified or interviewed. In such situations—when cases have already been identified during a descriptive study and information has been gathered from them then, case control study can be an efficient study design. Like in above example, if out of 100 girls who attended the party, 50 girls were not available for interrogation as they had left the hostel after the party. Then in such instances, those who ate the meal and become ill (cases) and a "random sample" of those who ate the meal but did not become ill (controls) are interrogated about their food intake history.

In a case-control study, the distribution of exposures among cases and a group of healthy persons ("controls") are compared with each other. Controls must not have the disease in question but should represent the population from which the cases come. In this way, controls provide the level of background exposure that might be expected among cases. The questionnaire used for the controls is identical to that administered to the cases, except that questions about the details of clinical illness may not pertain to the controls (Table 10.7).

In this example, 90% of all cases had consumed "*Pulao*" compared with only 32.8%

Table 10.7: Case-control study

Exposure	Cases	Controls	Total
Ate *Pulao*	27	23	50
Did not eat *Pulao*	3	47	50
Total	30	70	100
Percentage exposed	90%	32.8%	50%

Table 10.6: Attack rates, relative risk and attributable risk w.r.t each food item

Food item	Those who ate food item				Those who did not eat food item				Relative risk	Attributable risk
	Fell sick	Did not fall sick	Total	Attack rates	Fell sick	Did not fall sick	Total sick	Attack rates		
Pulao	27	23	50	54%	3	47	50	6%	9	48%
Raita	10	50	60	20%	5	35	40	12.5%	1.6	7.5%

of the controls. This suggests that consumption of "*Pulao*" is associated with illness in one way or another. In contrast to a cohort study, attack rates (and therefore relative risk) cannot be calculated since the total number of persons at risk is unknown. Here we would calculate the Odds ratio and not the Attributable risk. The food item showing highest Odds Ratio would be taken as the suspect item.

$OR = 27*47/3*23 = 1269/69 = 18.4$

Thus, in this example, an exposure odds ratio of 18.4 for "*Pulao*" can be interpreted as: the odds of having been exposed to the contaminated food in those who developed the disease was 18.4 times that of people who did not eat "*Pulao*". This odds ratio means that there is a very strong association between being developing an illness and consumption of "*Pulao*".

5. **Undertake food and sanitary investigation:** An extensive assessment of the sanitary history and food hygiene of each constituent of the suspected food item is done to find out how and why a particular food item got contaminated and most importantly, to institute corrective action to avoid similar recurrences in future. Such investigations should endeavour to clarify the actual conditions at the time foods were prepared. Each suspect food item that has been (or could be) implicated in the outbreak should be thoroughly investigated and detailed history of each and every constituent should be taken as given below.

- Interviewing employees who may have had a role in the processing or preparation of suspected foods;

- A review of the overall operations and hygiene at the point of procurement and storage

- Sanitary conditions and temperature at the time of cooking
- Specific assessment of procedures undergone by a suspect food
- The suspect food should be fully described in terms of:
  - All raw materials and ingredients used
  - Sources of the ingredients
  - Physical and chemical characteristics, including pH
  - Use of leftover foods in processing
- Assessment of the water system and supply
- Personnel records (including who was working when and absenteeism)

The amount of physical evidence may diminish rapidly with time after an outbreak has been identified, therefore, associated food investigations and appropriate food and environmental samples should be taken as soon as possible.

The following questions may help to focus an efficient food investigation:

- What are the known reservoirs or common sources of the suspected pathogen?
- What type of environment does it survive in?
- Where and how could the food have been contaminated?
- What environmental conditions support the growth and spread of suspected pathogen?
- Where are the opportunities for cross-contamination of a pathogen in this environment or establishment?

Like in the given case "*Pulao*" was identified as the possibly contaminated item. MO inquired about the list of all items used in making the dish. He undertook a detailed sanitary history of each and every item used in making "*Pulao*" from the point of procurement till the final cooking, storage and serving. He also observed the hygienic conditions present in the kitchen and surrounding areas. He found that all the edibles like rice, pulses, fruits and vegetables were bought by the hostel mess in-charge from the grain market Chandigarh. However, dairy products like milk, *paneer*, butter, cheese and eggs were supplied by a contractor. On the day of party in the hostel, cook washed rice and vegetables properly in running water. He then cut the vegetables to be used for making *pulao* at 4 pm. He then prepared the pulao at 5pm. As there was no space in refrigerator, the cook kept the *pulao* out at room temperature for 4 hours. Then at around 8 pm he garnished that with dry fruits. And at 9 pm it was only warmed and not reheated thoroughly and served for dinner. This seemed to be a case of *Bacillus cereus* food poisoning as it normally occurs in rice dish and when food is improperly cooked or reheated. Cooking temperatures less than or equal to 100°C allows some *B. cereus* spores to survive. This problem is compounded when food is then improperly refrigerated, allowing the endospores to germinate. Germination and growth generally occurs between 10 and 50°C.

6. **Initiate control measures**: Steps for immediate control measures and long term prevention should not wait for the final proof of the cause of epidemic but should start immediately and continue concurrently as the investigations proceed. These measures are generally directed towards the source of infection, towards channels of transmission, protection of susceptible population and for developing a long term early warning system. Make specific recommendations based on the findings. Continued monitoring of the control mea-

sures is essential to ensure that the measures are effective.

Here in above case, The MO made the following recommendations based on his findings

- All meals should be served freshly cooked and hot.

- If cooked item is to be stored, the storage should be in a deep freeze and should not be stored at room temperature for more than 4 hours.

- Cooked food item should be thoroughly reheated before serving.

- No leftover food should be used.

- Keep hot foods above 60°C and cold foods below 4°C to prevent formation of spores.

- Proper cleaning and disinfection of food contact surfaces by sodium hypochlorite or other approved sanitizers.

- Wash hand, utensils and food contact surfaces with hot soapy water before food preparation and after using toilets.

- Specific precautions should be taken for milk/meat based products.

- Hostel—Mess in-charge should check implementation of these recommendations

One of the goal of this investigation is to identify "contributing factors" related to contamination, proliferation or amplification of a pathogen that played a role in the occurrence of an outbreak. One of the easiest ways to prevent food poisoning associated with Bacillus spp. is by ensuring that foods are cooked thoroughly and cooled rapidly. Reheating of the food should be done thoroughly and the interval between cooking and eating should be as short as possible.

Factors Contributing to Contamination of Foods

- Raw foods can be contaminated at their source with pathogens with *Salmonella, Campylobacter, Clostridium perfringens, Yersinia enterocolitica, Listeria monocytogenes, Staphylococcus aureus*, etc.
- Use of non potable water in food preparation.
- Obtaining food from unsafe sources.
- Infected food handlers (e.g. nasal carriers of *Staphylococcus aureus*, persons in the incubatory phase of hepatitis A)
- Improperly cleaned equipment like slicers, grinders, cutting boards, knives, storage containers
- Contaminated food or ingredients insufficiently heat-processed
- Cross contamination by worker's hands, cleaning cloths or equipment, from raw foods to cooked foods

Factors Affecting Survival of Pathogens in Food

- Food cooked or heat-processed for an insufficient time or at an inadequate temperature.
- Previously cooked food reheated for an insufficient time or at an inadequate temperature.

Factors Affecting Microbial Growth in Food

- Cooked food left at room temperature (5–60°C) that permitted multiplication of bacteria
- Improperly cooled food
- Inadequate or slow fermentation

BIBLIOGRPAHY

1. Bhalwar R. Investigations of an epidemic. Textbook of Public Health and Community Medicine. Bhalwar R, Vaidya R, Tilak R, Gupta R, Kuntr R. Department of community Medicine, AFMC, Pune and WHO, New Delhi.

2. Disease alerts/outbreaks reported and responded to by states/UTs through Integrated Disease Surveillance Programme (IDSP). Accessed from www.idsp.nic.in/.

3. Foodborne disease outbreaks: guidelines for investigation and control. World Health Organization 2008. ISBN 978 92 4 154722 2

4. Food-Borne Diseases. CD Alert Directorate General of Health Services, Government of India. December 2009 Vol.13: No. 4.

5. Management of Outbreaks of Foodborne Illness in England And Wales. Food Standards Agency 2008 London.

6. Standard Operating Procedure for the Inves-tigation of Food Poisoning Outbreaks. Perth Metropolitan Area Edition 2003 Accessed from http://www.public.health.wa.gov.au/

7. Sudershan RV, Kumar RN, Polasa K, "Foodborne diseases in India—a review," British Food Journal.2012; 114(5):661–680.

8. Sudershan RV, Naveen Kumar R, Kashinath L, Bhaskar V, Polasa K. Foodborne Infections and Intoxications in Hyderabad India. Epidemiology Research International. 2014. Accessed from http://dx.doi.org/10.1155/2014/942961

9. WHO Initiative to Estimate the Global Burden of Foodborne Diseases. Department of Food Safety, Zoonoses, and Foodborne Diseases Health Security and Environment.2007. Accessed from http://www.who.int/food-safety/publications/foodborne_disease/burden_nov07/en

11

Farm to Fork Surveillance: The Missing Link in Food Safety

Neha Chanana, Jaideep, Amarjeet Singh

*Food is a necessity, without which
there is no intensity…*
*It provides us energy, it provides us nutrition;
that's the mandate of all its provisions….*
*Watching over its production to consumption,
Is the basis for our healthy function!!!*

Food contamination can occur at any stage of the *"farm to fork"* chain, i.e. from production to consumption of food. Therefore, continuous watchfulness to ensure food safety throughout this chain from the production of food till it reaches the consumer holds great public health importance.

The familiar case of food poisoning from consuming *kuttu ka atta* during *navratra* festival in 2011 which affected parts of Delhi, Punjab and Himachal Pradesh brings to light the importance of scrutinizing the complete process of preparing food from production till consumption, in case of outbreaks, to search for the cause of contamination. *Kuttu ka atta* is prepared from buck wheat and is considered a staple diet for those observing fast during the nine days of *navratra* festival celebrated in northern India. The information collected from the cases as well as their relatives indicated the consumption of unpacked and unsealed flour. To locate for the product in the food chain, shops selling the *atta* in the locality were raided. Samples were also collected. Upon further tracing a mill in east Delhi, declared as the manufacturer of spurious buck wheat flour, was sealed. The reports of samples suggested that flour was adulterated with old husk of cereal and pulses unfit for consumption.

Following this revelation an awareness campaign was launched regarding the adulteration. This incidence brings to light the importance of traceability in searching for the cause of breach in food safety and thereafter withdrawing the product from the supply chain to control the situation.

Present System of Surveillance

Surveillance is the ongoing systematic collection, collation, analysis and interpretation of health data and the dissemination of information to those who need to know in order that action be taken. Here, the basis for surveillance is a desire to control and prevent FBI.

In India the Integrated Disease Surveillance Programme (IDSP) under NRHM tackles the information about various diseases. In addition the Food Safety and Standard Authority of India (FSSAI) has been established and made overall responsible for food safety activities in the country.

Under the FSSAI, there is an organizational set up from the central to the field level (Fig. 11.1).

At the national level: The FSSAI has the powers and functions assigned to it under the Food Safety and Standard Act (FSSA), 2006. The food authority works with the Chief Executive Officer (CEO) as its chairman. The CEO has the administrative control over the officers and other employees of the food authority. The CEO is assisted by a central advisory committee which ensures close cooperation between the food authority and the enforcement agencies and organizations operating in the field of food safety. The committee also advises the food authority in identifying potential risk related to food safety. The CEO also has various officers under him out of which Chief Surveillance Officer and the Chief Enforcement Officer are mainly concerned with matters of surveillance. The Chief Surveillance Officer is mainly concerned with epidemiology and clinical surveillance. Matters related to compliance to standards, inspection and prosecution are under the jurisdiction of the Chief Enforcement Officer.

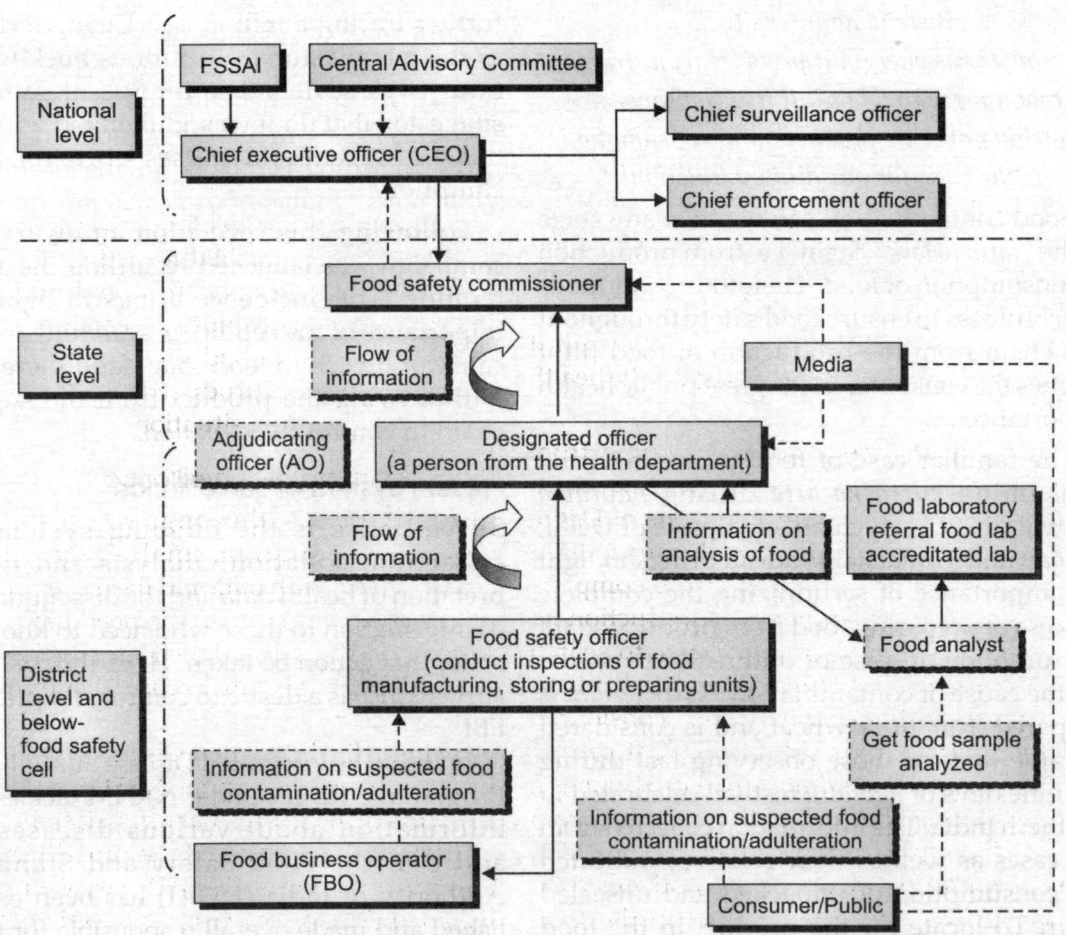

Fig. 11.1: Major people involved in food safety surveillance under FSAAI

At the state level: There is a food safety office in every state/UT with its organizational hierarchy. The office is headed by the food safety commissioner. He has the following powers:

i. Prohibit the manufacture, storage, distribution or sale of any article of food in interest of public health

ii. Carry out surveys of food processing and manufacturing units so as to find out their compliance to standards with regards to various food articles as notified by the food authority.

iii. Conducting training programmes for personnel engaged in food safety.

iv. Ensure the uniform and efficient implementation of the laid down standards and other requirements under the FSSA (2006)

v. Sanction prosecution of offences punishable with imprisonment as mentioned under the act.

At the district level and below: The commissioner is assisted by a designated officer in each district who in turn is supported by the food safety officers positioned in the local areas.

The designated officer is personnel from the Department of Health and is the in-charge of the local food safety administration. He performs the following functions:

i. Issue or cancel a license of FBO.

ii. Prohibit the sale of any food article which is not in adherence with the provisions made under the act.

iii. Receive reports and samples of articles of food from the food safety officer and get them analyzed from the referral or accredited lab under the supervision of the food analyst[1].

iv. Recommend the commissioner to launch prosecution in case of contraventions punishable with imprisonment.

v. Sanction or launch prosecutions in cases of contraventions punishable with fine.

vi. Maintain record of all inspections and actions taken thereof by food safety officers while performing their duties.

vii. Get investigated any written complaint in respect of any contravention of the provision of this act.

The Food Safety Officers (FSOs) are assigned local areas for the purpose of performing functions under the act. The FSO has the following powers and duties:

i. To take samples of any food or substance intended for sale for human consumption. He can enter and inspect any place where food is manufactured, stored or prepared as well as where any adulterant is manufactured or kept.

ii. To seize any food article that appears to be in contravention of this Act. If the article seized is of perishable nature, the FSO after issuing a written notice to the FBO, destroys the article.

iii. To seize and submit for analysis any adulterant found in possession of a manufacture or distributor of any food article or in any of the premises occupied by him as such and for the possession of which he is unable to account to the satisfaction of the food safety officer.

The FSOs conduct routine checks for food samples of eating establishments. They also undertake sample surveys in case of complaints from the consumer as well as during festival seasons. It is not uncommon to hear about the adulterated *mithai* being sold

---

[1] Food safety analyst is appointed by the commissioner of food safety

especially during festival season. Moreover, the Food Business Operator (FBO) can also inform the concerned authorities in case he believes the food under his possession to be contaminated or adulterated, in any form.

All this information and records collected from the field by the FSO is sent to the designated officer at the district level office who maintains the records and reports the issues to the food safety commissioner.

The Adjudicating Officer (AO) is also a part of the surveillance system. If during surveillance a manufacturer/distributor/seller is found guilty of manufacturing/distributing/selling contaminated or adulterated food article, which on consumption causes grievous injury or death of the consumer, has to face penalties by law. All such legal matters are handled at the district level by the AO. The AO or the court makes sure that the victim is compensated from the offender with a sum of:

i. Not less than 5 lakh rupees in case of death to the kin next to the victim.

ii. Not exceeding 3 lakh rupees in case of grievous injury.

iii. Not exceeding 1 lakh rupees in other cases of injury.

The court or the AO can also order for cancellation of license, re-call of food from market and also publish the name and place of residence of the person held guilty, the offence and the penalty imposed in newspapers or any other widely circulated means of communication.

Under IDSP: This monitors data related to major food and water borne diseases like food poisoning, acute diarrheal infections, etc. along with other communicable as well as non-communicable diseases. The administrative structure under IDSP at the national, state and district level has 2 bodies viz. the surveillance committee and surveillance unit.

The surveillance committee takes the policy decisions. It monitors and coordinates with the stakeholders. Surveillance unit is responsible for implementing various activities under the project.

At the national level: The Central Surveillance Unit (CSU) is headed by the project director. National Project Officer (NPO) is the technical in-charge of the project. The unit has consultants and supporting staff to assist in implementation of project activities at the central level. The unit also has technical and administrative officers for better coordination and timely action. The CSU acts as the control room at the central level for coordinating responses to requests received from states during epidemics and disasters. It also forms and supervises the movement of the rapid response team at the central level to supplement the efforts of states during food and water borne epidemics.

At the state level: The State Surveillance Unit (SSU), located in the state Department of Health and Family Welfare, is headed by the State Surveillance Officer (SSO). He is assisted by the rapid response team representatives, consultant (training and technical), consultant (procurement and finance), data manager and data entry operators. The SSU collates and analyses data on foodborne illnesses received from the districts and transmits the same to the CSU. It also coordinates the activities of the rapid response team and sends them to the district whenever need arises. The SSU monitors and reviews the activities of the district surveillance unit. It coordinates the activities of the state public health laboratories and the medical college laboratories and sending regular feedback to the district units on the trend analysis of data received from them.

At the district level: The District Surveillance Unit (DSU) is headed by the District Surveillance Officer (DSO). He is assisted by the data entry operators and supported by a rapid response team. The DSU collates and analyzes all data being received from various service providers within the districts and transmitting the same to the State/Central surveillance units. Data is received every week from the village and block level to the district level IDSP division. The private practitioners, hospitals, nursing homes, labs are also kept in the loop of reporting information through prescribed formats.

In case of foodborne outbreaks, the district rapid response team consisting of the epidemiologist, microbiologist and clinicians comes into action.

As a response to emergency, preliminary information is collected on the number of people exposed, number of people affected, clinical symptoms, the setting in which the outbreak occurred, etc. Then, after visiting the site information is collected from cases and noncases that were exposed. Immediate measures are taken to manage the cases and masses are educated to prevent the further spread of the disease. The environmental inspection like the structural and operational hygiene of the food premise along with collection of sample of food article is done. After analysis of the collected data and reports of samples, a hypothesis about the likely source of outbreak is developed. Steps are taken for eliminating the source like removing flour from the market in case of food poisoning caused by *kuttu ka atta*. The report so generated is shared with the DSO who further takes action on the same.

Thus, these two departments are conducting surveillance for controlling and preventing foodborne illnesses within their own set protocols. However, there are certain challenges, as described below, which still need to be met for the efficient surveillance for food safety.

CHALLENGES

- Lack of proper guidelines for investigating foodborne disease outbreaks. A quick reference step by step guide needs to be developed focusing on the concept of traceability from table to the farm to identify the source of disease and plan for public health action that needs to be taken to control the illness.

- The role of FSO finds no mention in case of FBI outbreaks. The FSO should be a part of the rapid response team of the district.

- The FSO is only concerned with the surveillance of eating establishment. The concept of farm to fork with respect to surveillance in food safety is still missing. To generate this concept in the present surveillance system, intersectoral coordination with the department of agriculture, department of animal husbandry, dairying and fisheries, department of food processing, department of environment is required. The designated officer should work in close collaboration with a surveillance officer from each of the above mentioned departments.

Thus, to mend these loopholes, a surveillance system focusing on the concept of farm to fork is needed. As yet no such specific surveillance plan is in place.

The FSSAI should take steps in this regard and strengthen the surveillance system for the control and prevention of FBI.

A suggested surveillance plan for food safety is as follows:

Step 1: Establishment of a surveillance system: A separate system for surveillance is not required. The existing mechanism should be strengthened. The designated officer at the district level should be the overall in-charge for farm-to-fork surveillance. There will be

| Primary soure | Data collected by the food inspector | Lab reports of food samples |

Fig. 11.2: Source of primary data collection

close coordination with IDSP and other departments at the district level for sharing of information.

Step 2: Defining the objective of the surveillance system: To reduce the burden of foodborne illness. Having an efficient surveillance system will help in identifying the location of the cause and source of contamination. This would help in establishing control and preventive measures for foodborne illness.

Step 3: The organization and structure of food safety surveillance: Surveillance is required at each step from the production of the raw material till it reaches the consumer. Therefore, each department involved in this farm to table chain viz. Department of agriculture; department of animal husbandry, dairying and fisheries; department of environment; department of health and department of food processing along with private sector hospitals, clinics, labs and vetenary practitioner shall designate a surveillance officer who will work in close alliance with the designated officer of food safety. There will be sharing of information between different departments. All these records will be shared with the designated officer. After analysis, of these records necessary action will be taken.

Step 4: Defining FBI: Foodborne illness or food poisoning often present with the following symptoms: nausea, vomiting, diarrhea, fever and pain in abdomen

Step 5: Collection of information:

i. Source of data and method of data collection: Data required for surveillance for food safety can be collected from various sources.

   a. Primary data should be collected through active surveillance by the FSO (Fig. 11.2).

   b. Secondary data should be collected from records of the concerned departments (Fig. 11.3).

ii. Frequency of reporting: A weekly reporting mechanism should be developed. In case of outbreaks daily reporting should be done.

iii. Method of data transmission: The transmission of report should be made online through emails. Each department involved should also upload the data on their respective websites.

Step 6: Data analysis, interpretation: Data should be analyzed and interpreted and at the district level regularly.

Step 7: Plan for action: The information interpreted from the data should be used for taking action as and when required. The masses should also be educated on a regular basis regarding the cause and prevention of foodborne illnesses through media. In case of FBI outbreak, FSO should work in collaboration with the district rapid response team of IDSP, i.e. epidemiologist, microbiologist, clinician, etc. for tracing the cause and source

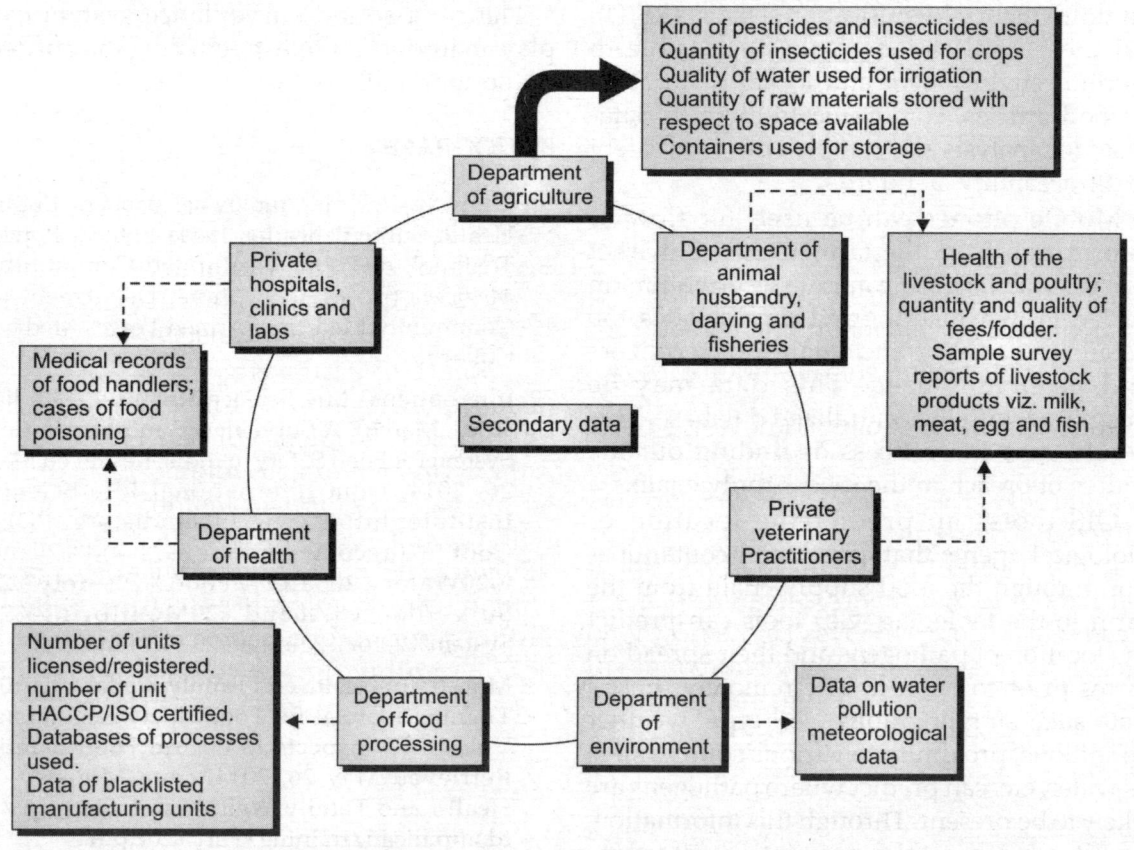

Fig. 11.3: Sources of secondary data collection

of illness and taking appropriate public health action for example *in food poisoning outbreak due to consumption of kuttu ka atta, the mill was sealed and the adulterated flour removed from the market. The masses were educated on the effects of consumption of the spurious flour.*

The designated officer should disseminate the prepared reports documenting the source and cause of illness and the action taken to the food safety commissioner at the state level and the CEO at the national level as well as with stakeholders of other departments. Furthermore, the role of information technology in maintaining these reports and records should be harnessed. All this information should be maintained in a database electronically. There should be sharing of information between all the concerned departments involved in the food supply chain through e-mails as well uploaded online on the respective departmental sites. Also, the laboratory reporting of food samples that are analyzed can be made online. This would enhance transparency as well as easy access of data.

Other modern solutions which can further enhance surveillance in the farm to table chain is the use of mobile, GIS, remote sensing and

Radiofrequency Identification sensors (RFID). All these applications make it possible to monitor environment and location variables of food articles, communicate them to databases for analysis and comply with food safety and traceability standards.

Mobile phones can be used for flow of information from the farmers to the market wherein the farmers can give information on variety grown, planting and harvest dates, the details of quantity and quality of fertilizers and pesticides used. This data may be integrated with the centralized database. This would ease the process of finding out the source of breach in the food supply chain.

GIS tools can predict the location of biological agents that cause food contamination through the food supply chain from the farm to the table, e.g. GIS tools can predict the location of pathogens and their spread on farms prior to harvest. The remotely sensed data such as typography, soil type, weather conditions, proximity to various sources such as water, etc. can predict where pathogens are likely to be present. Through this information, the farmers can take preventive measures such as draining standing water, etc. and thus prevent the crop from infestation. Similarly, remote sensing techniques can be used to identify pests and pathogens at the level of storage, processing, packaging, and transportation till it reaches the consumer.

RFID tag technique can be used for detection of food freshness and bacterial growth at all steps of the *'farm to fork'* chain.

Step 8: Review and evaluate the surveillance system developed: Yearly evaluation of the surveillance system is recommended.

Hence, a sound surveillance system can play an important role in reducing the burden of foodborne illnesses.

BIBLIOGRAPHY

1. Bhalwar, R. Epidemiological Basis of Public Health Surveillance for Disease. In B. Rajvir, Textbook of Public Health and Community. *Medicine* (pp.) (2009). Pune: Deparment of Community Medicine, Armed Forces Medical College.

2. International Life Sciences Institute – India. (2007, March). A Surveillance and Monitoring System for Food Safety in India. Retrieved May 26, 2014, from International Life Science Institute: http://www.ilsiindia.org/PDF/Conf.%20recommendations/Food%20and%20Water%20Safety/Food%20safety%20Surveillance%20and%20Monitoring%20System%20for%20India%20-%20Final.pdf

3. Ministry of Health and Family Welfare. (2010). Training Manual for Food safety Regulators-Vol V: Key aspects to ensure Food Safety. Retrieved May 26, 2014, from Ministry of Health and Family Welfare website http://fda.up.nic.in/training/Part%202.pdf

4. Ministry of Law and Justice (2006, August 23). *Food Safety and Standards Act*. Retrieved June 2, 2014, from Food Safety and Standards Authority of India: http://www.fssai.gov.in/Portals/0/Pdf/FOOD-ACT.pdf

5. Park, K. Surveillance. In K. Park, Park's Textbook of Preventive and Social Medicine (pp. 36–37) 207. Jabalpur: Banarsidas Bhanot.

6. World Health Organization. (2002). *WHO Global Strategy for Food Safety*. Geneva: World Health Organization. http://www.who.int/foodsafety/publications/general/en/strategy_en.pdf

12

Pathogens' Toxic Assault on My Gut: A Harrowing Ordeal with Gastroenteritis

Ishwarpreet Kaur

My heart was pounding fast as I stepped on the weighing scale. As expected the reading on the scale was good enough to make me rethink over my lifestyle. Motivated to lose weight I planned a new diet and exercise plan for myself. With great enthusiasm, the same evening, I did abdominal crunches. Next morning, as per the plan I first took a glass of lime water with honey and did some yoga. All this while I felt uneasy, I related it to the previous day's exercise. And I did not pay much heed to it. After around half an hour I took a glass of warm milk. As I was sipping over the milk I felt some discomfort and nausea. But again I did not take those signs seriously and continued with my routine. As I was getting ready, suddenly I felt a griping pain in abdomen. I had a sudden urge to rush to the washroom. I was terrified when I passed watery stools which did not stop for next 15 min. I felt totally drained off and exhausted. Once I came back to my room I again experienced cramps and gushing pain. Then again after some time I was sitting on the commode. I was left with no energy and felt dizzy. Seeing my plight my mother gave me a concoction of herbal tea. It did work for some time but again I was back on the toilet seat, passing loose watery stools. Even taking small sips of water resulted in repetition of the cycle.

This continued till afternoon and I felt that my intestines were all clear now.

As the day progressed, I felt some relief. I finally had a small bowl of boiled rice with dal. Diarrhoea completely stopped, but I felt cramps during the day. I thought it was a self liming condition and that the worst was over. I again linked the cramps to previous day exercises. I was absolutely fine during the day and slept early. Next morning at around 5 o' clock unbearable abdominal pain started again. To get some relief I took herbal tea and applied hot water pouch for fomentation locally. With relief in pain I fell asleep again. But as I woke up, again the loose motions started. Then I decided to visit a doctor. I somehow gathered energy and went to the doctor. As I told him about my condition he questioned me, *"bahar ka kya khaya tha"*.

This question made me wonder what had probably caused this infection. In retrospect, I realised that two days back I had family dinner at a local restaurant. There I had both vegetarian and non vegetarian (chicken) in my meal. I told the doctor about the restaurant meal. He examined me and prescribed medicines and gave some dietary tips also.

I just forgot about the cause/reason of infection, and focused on starting my treatment and get better. I immediately went to the chemist and took my medicines. But, still there was no relief in symptoms and in next half an hour vomiting started. My whole body was aching and I felt week and drowsy. I tried taking ORS, but I could not have more than a quarter of cup. Diarrhea continued till evening. As I took the second dose of medicine, diarrhea and vomiting stopped. I felt better, and for change I tried going for a walk. But soon while walking I realised my body was not ready for it. I came back and took rest. Slowly the symptoms disappeared. To overcome my weakness I took medicines, rest and gradually increased my food and fluid intake.

Now as I am recuperating, I try to figure out the exact source of infection. All I can logically think of is the chicken serving; I had in my meal at the restaurant. The reason being simple, out of all the recipes chicken and fish were the ones served as separate pieces. Out of these two non-vegetarian recipes, I just had chicken. The pathogens must have been sparsely distributed within the chicken dish. The portion/pieces which I got must have

been the tainted one. Improper/under cooking must have led to the survival of germs. And as these entered my body within two days (incubation time) they multiplied and finally made me sick.

Had it been vegetable recipes it would have probably affected others too. But as no one else fell sick the reason was most probably "chicken". One thing is clear that high temperatures during summer season provides ideal growth environment for germs.

As I felt very weak, I decided to weigh myself again. This time I was amused to see that in these two days I had lost two kilos of weight. I thought this could be because of water loss from my body. After this experience, all I can suggest the readers is to avoid non-vegetarian and raw foods from outside during and rains. And one must remember that they should not follow my 'regime' to loose 2 kg body weight within 2 days.

Remember the adage/cliché, "prevention is certainly better than cure!"

Section III

Food Safety: Farm to Fork

13

Understanding Food Safety: 'Farm to Fork' Concept

Puja Dudeja, Amarjeet Singh

Food Safety is an issue that matters to all of us. It is not a new idea or a new suggestion. Man has always been concerned with food safety. In Indian culture, highest grade of hygiene has been associated with kitchen. Elaborate codes of conduct for food handlers/cooks have been prescribed for ensuring food hygiene in Indian civilization. In modern context, the concept of food safety is described by the 'Farm to Fork' model. This concept captures the essence of food safety. This forms the backbone of food safety.

The perception of food safety varies from person to person. Different people interpret it differently. In our homes kitchen food safety is generally limited from kitchen to table. The lady-handling kitchen processes/cleans/prepares food, which is brought from the market. She is not much concerned about the journey of food before it reaches kitchen.

The Food Business Operators (FBOs) have a wider responsibility as they have to struggle hard for satisfaction of consumers at low cost and also provide safe food. They need to ensure food safety all along the farm to fork chain. Food safety can be compromised anywhere in the entire food supply chain. This is the food supply chain approach or the farm to fork model to ensure safe food. This implies that the responsibility of providing safe food

to the consumer is shared among all those involved with production, processing, trade, cooking, serving, etc.

The concern of government is to make sure that food we eat is safe. The increasingly complex nature of hazards have triggered the need for an integrated approach to food safety in the entire production chain that is from farm to fork. The concept of food safety is not restricted to kitchen only. This approach places the primary responsibility of food safety on all contributors in the product supply chain. Simultaneously, individual chain participants look for assurance of safety of safety of products supplied by the preceding chain participant. These trends place greater attention on compliance with food safety measures at the farm level, it being one of the important stages affecting the level of food safety of products consumed at the end of the chain. For example, the wholesalers believe that the product from the farm is safe. The retailers rely on wholesalers and so on.

Food safety can be compromised anywhere in the entire food supply chain. There have been discussions about the paradigm of farm to fork model in the food safety context. This concept implies that the responsibility of provision of safe food to the consumer is shared among all those involved with food material/product at various levels for example production, processing, trade, cooking, serving, etc. It also portrays the journey/history of seeds from the time they sown in the fields to their growth, harvest, storage, transportation, cooking and the way they are served on the table (Fig. 13.1). For food from animal sources, the same concept can be explained as 'Stable to Table'. The benefit of this concept is that in case of a problem developing, such as a food 'poisoning' incident, the food, its processing and its source can more easily be traced, investigated and effectively corrected. The reverse 'Fork to Farm' shall help in traceability and recall in case of an outbreak. In the long run it benefits the FBOs and raises the quality and standard of food served by them. It translates into better marketing opportunities for all viz. farmer, wholesaler, retailer, FBO, etc. The umbrella concept takes care of all the hazards biological, chemical and physical. It forms a platform where one of the basic

principles of public health practice, i.e. intersectoral coordination to prevent FBI can be seen. Following the concept of food safety, help the regulators in easy stepwise monitoring of the entire process. Last but not the least it solves many issues of global trade of various products.

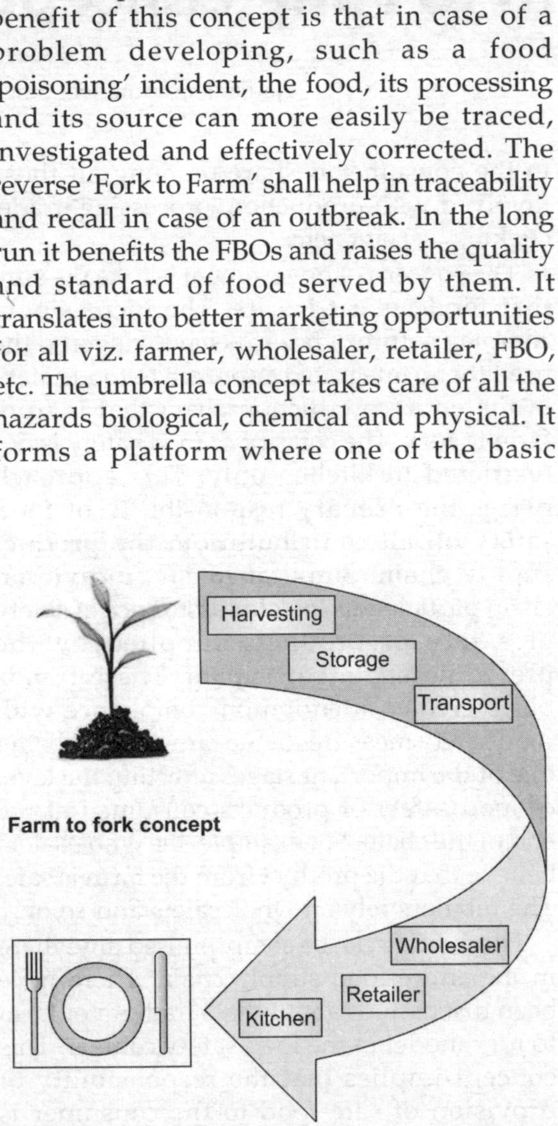

To understand the farm to fork concept we can broadly categorize food sources as in Fig. 13.2.

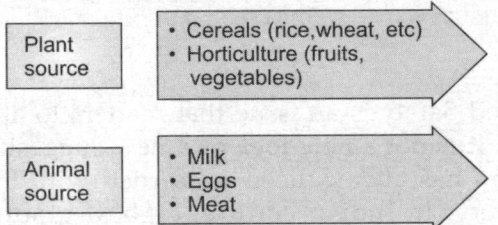

Fig. 13.2: Categorization of food sources

Let us begin our journey from the farm and understand how and what activities can compromise food safety at the farm level.

Let us take a few examples. Cereals, grains, spices, herbs, etc. can be contaminated by use of pesticides, heavy metals, mycotoxins and radionuclides at the farm. Agrochemicals like pesticides when used at the farm level are supposed to be degraded after their application on the crops but that does not happen in all the cases. Small amount of residues remain mixed with food and make it unsafe. Meat and poultry products can also have pesticides, heavy metal, dioxins and environmental chemical residues in them. They can also become unsafe with antibiotics, veterinary drugs and microbes. Various processes during their travel from poultry farm like quality of feed, veterinary treatments, hygiene sanitary and phytosanitary conditions, processing of meat, preservatives used for storage, handling and packaging affect food safety.

Similarly milk and milk products can be contaminated depending upon the animal

Farm to fork concept

Fig. 13.1: Farm to fork concept of food safety

husbandry practices at the farm, quality of feeds, veterinary treatments, environmental conditions, storage, packing and transportation. Veterinary drugs and antibiotics used for the animals continue their presence as residues. Fruits and vegetables can be contaminated with pesticides at the farm level and with microbes and other chemicals during storage, transportation and handling. Safety of processed and packed food safety is compromised with use of additives, banned dyes, toxic chemicals like residual solvents, acrylamide, benzene, PAHs, melamine, etc. Persistent agricultural pollutants, leechables from packaging material, microbes, and degraded products harm the food product during processing and packaging. Even water used for irrigation, washing and cooking of raw materials can be contaminated with pesticides, heavy metals like lead arsenic, radionucleides, surface active agents, environmental chemicals and pathogens. Use of such water in turn can make food unsafe. Let us now understand how contamination can occur during the journey from farm to fork and what are risks involved.

Food Safety at Farm

Roti, kapda aur makan (Food, shelter and clothing) have been the three basic necessities of mankind since ages. Everyone needs to eat. Therefore, amongst these requirements the need for food dominates. That is why agriculture has been one of the greatest revolutions of humankind. Indian society is still predominantly agrarian. Indian agriculture is demographically the broadest economic sector and plays a significant role in the overall socio-economic growth of the country. According to Food and Agriculture (FAO) statistics, India is the world's largest producer of many fresh fruits and vegetables, second largest producer of wheat and rice, third largest producer of several dry fruits, roots and tuber crops, pulses, farmed fish, eggs, coconut and sugarcane. India is also one of the world's five largest producers of livestock and poultry meat.

Famous Bollywood movies like Mother India, *Upkar*, *Karan Arjun*, Lagaan have showcased farming as the main profession. Almost every superstar has once in his career has been associated with the fields be it the old ones like Manoj Kumar or the recent ones like Amir Khan, through his famous episode of poison on plate in serial *satyamev jayate*. News related to compromised food safety at farm level also catch media attention time-to-time (Fig. 13.3).

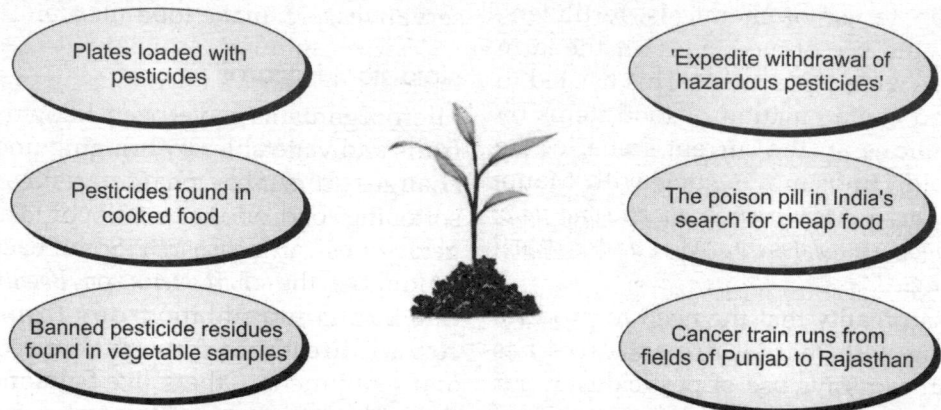

Fig. 13.3: News highlights related to compromised food safety at farm level

Various factors that can affect food safety at the farm level are quality of seeds, soil composition and preparation methods, crop management, water quality, irrigation techniques, management of pests, harvesting techniques and equipment, storage and transport.

Good Agricultural Practices (GAP) at the farm aim at ensuring food safety during pre-production, production, harvest and post harvest stages. They also help protect the environment and the safety of workers. Farmers must apply GAP to produce safe food. Often, it is more difficult for small farmers to comply with GAP than it is for large farmers. The hazards at the farm level can be classified as under:

a. Chemical
b. Biological
c. Physical

Chemical Hazards

After Independence our country witnessed green revolution in 1960s. In due course, India has become self sufficient in agrarian production. However, our less educated farmers do follow a mix of both traditional and modern practices. To meet the ever increasing demand of grains and livestock products for the nation and to attain food security, practices like indiscriminate use of chemicals, fertilizers, pesticides and use of medicines for the farm animals are widely prevalent. This has led to widespread contamination of food items by these chemicals. In the current scenario, the popular old Hindi movie song with Manoj Kumar as hero 'Mere desh ki dharti sona ugle, ugle heere moti' translates into 'Mere desh ki dharti pesticides ugle'!

It is a sad reality that the need to produce more along with market driven forces has resulted in excessive use of pesticides in our farms. This has led to the deterioration of nutritive value of food products. Compromised safety of food products from such farms affects the health of people, the environment and other living beings.

Other non-pesticide chemicals that can make food unsafe at farm level are lubricants, cleaners sanitizers, paints, fertilizers, etc. These can enter due to any oil leaks in nearby area. Field equipment when meets the produce can contaminate produce with grease or paint. Similarly contaminated picking containers, transportation vehicles that previously were used for chemicals can harm the produce. Heavy metal residues (arsenic, lead, cadmium) can be present in the food due to continued use of fertilizers (including compost). In case the farm is near a busy road lead contamination can occur from car exhaust fumes. The water used for irrigation can also have high levels of heavy metals like arsenic in it. The soil at farm may be naturally or from previous use or leakage from industrial sites be contaminated with heavy metals.

Natural toxins—allergens, mycotoxins, alkaloids, enzyme inhibitors occur in field due to unsuitable storage conditions leading to mould on produce, use of injectable chemicals in the produce to keep them fresh, intentionally added substances like added colour, mineral oil, carbides (for ripening), preservatives, etc. make food unsafe.

Biological Hazards

Microorganisms have been known to spoil fruits and vegetables by bringing undesirable changes in quality characteristics such as softening, bad odour and flavor. The pathogenic ones can cause FBI. Some bacteria can be found in the soil (Listeria spp, Bacillus cereus) and can contaminate crops through soil contact directly or through dirty containers and equipment. Others like Salmonella pass through the intestinal tract of animals and humans and can contaminate fruits and vegetables through manure, contaminated water, and humans handling produce.

Fruit and vegetables can act as a vehicle to pass some parasites from one host to another, animal to human or human to human. Cysts of *Giardia* can survive and remain infectious for up to seven years in the soil. Water contaminated with faecal material, infected food handlers and animals in the field or packing shed can be vehicles for contamination of produce with parasites.

Fungus/moulds can affect food crops for example aflatoxins are naturally occurring toxins produced by many species of the fungus *Aspergillus*. Aflatoxins are toxic and among the most carcinogenic substances known. Crops susceptible to Aspergillus infection include cereals, oilseeds and spices. The toxin can also be found in the milk of animals that are fed contaminated feed. Biological hazards can harm the farm depending on various factors given below:

How is the Produce Grown?

Produce that are grown in or close to the ground like carrot have a higher risk of contamination than produce grown well above the ground (grapes).

Contact with Water

Produce grown in frequent contact with water can have a higher risk, for example, rice.

Type of Produce Surface

Produce with a large uneven surface (lettuce) have a higher risk than produce with a smooth surface (apple).

Way of Human Consumption

Produce that is eaten raw (leafy vegetables) has a higher risk than produce that is cooked (potato). Those with an edible skin (grape) have a higher risk than those with an inedible skin (banana).

Physical Hazards

Physical hazards are foreign objects that can cause illness or injury to consumers. These can occur during production and post-harvest handling. Types of physical hazards include glass, wood, metal, plastic, soil and stones, personal items like jewelry, hair clips, other like paint flakes, insulation, sticks, staples, weed seeds, toxic weeds. Foreign objects from the environment can enter while harvesting of ground crops during wet weather, dirty harvesting and packing equipment, picking containers, packaging materials, stacking of dirty containers on top of produce, broken lights above packing equipment and areas, inadequate cleaning after repairs and maintenance.

Other Hazards

Potential food safety hazard may also occur through application of technologies of which the hazards may not yet be fully understood (e.g. genetically modified plants or nano-technology). Research is still ongoing to better understand the hazards associated with this.

Not all hazards are applicable to all the fields. Hazard identification specific to each field is an important activity that can prevent contamination later. Specific actions need to direct to reduce or eliminate these hazards.

Methods to Reduce Hazards at Farm Level
For Food of Plant Origin

Consumer concerns for food safety are increasing day-by-day. Farm to Fork is a holistic concept to make available safe food. Most of the countries world over are adopting it. This model encompasses the flow of food from the time seed is planted until the finished product reaches consumer plate. Understanding the flow of product allows for ensuring safety at all the levels. Farm-to-table (or farm-to-fork) refers to, in the food safety field, the stages of the production of food:

harvesting, transportation, storage, processing, packaging, sales, and consumption. Farm-to-table also refers to a movement concerned with producing food locally and delivering that food to local consumers. It starts with the quality of seed, quality of soil, farming practices, use of pesticides/ fertilizers in farm, quality of water used, transportation, food processing units, storage, and handling of the food and food products.

For Food of Animal Origin

For the animal food, the major components of food safety at the farm level include animal welfare, use of medicines, animal feed along with the control and treatment of the animal including zoonotic diseases, slaughter house and transport system. Good animal welfare equates to good food safety. Food animal markets, worldwide, are well known for spreading disease from farm back to farm or further down the 'farm to fork' food chain. Therefore, this is an area where strict monitoring and additional measures such as "rest days" (during which markets are kept empty while they are thoroughly cleansed) are also required. Public health, animal health and animal welfare are indeed interrelated and require a holistic approach. As an example of this, stressed animals are more likely to develop diseases, which will require veterinary treatment. However, this may increase the presence of drug residues in the animal produce, which in turn may affect public health.

While treating the animal, the practice of using the right medicine for the right species and at the correct dose should be followed strictly. The maintenance of proper records of medicines given to individual animals is important to ensure that any animal or the milk sent to the market has drug residues below the permitted limit. The period required to achieve it following treatment to the animal

is known as the milk and meat "withdrawal period". The animal feed containing toxic materials, will affect the health of the animals. In some cases, the toxin in the animal feed can be absorbed by the animal and then passed on to the consumer in the milk, meat or eggs produced.

In this regard, Bovine Spongiform Encephalopathy (Mad Cow Disease) is a well known example of contamination in animal feed. This hogged the limelight in 1990s due to the episode in UK. The disease is transmitted to human beings by eating food contaminated with the brain, spinal cord or digestive tract of infected carcasses also even after it had gained the attention of health authorities a decade back. In 2009, it killed 166 people in UK and 44 elsewhere. Similarly other diseases like avian influenza (H5N1 infection), tuberculosis (TB) and brucellosis can also affect both animals and man. Efforts have to be made so that such diseases do not enter the food chain. For food of animal origin this concept envisages the process of following the food from the farm where the animal is born through to the dinner plate and on to the fork.

An important step in animal food production is ante and postmortem inspection. Ante-mortem inspection identifies animals not fit for human consumption. Here animals that are down, disabled, diseased, or dead are removed from the food chain and labeled "condemned." Other animals showing signs of being sick are labeled "suspect" and are segregated from healthy animals for more thorough inspection during processing procedures. In order to safeguard the public, additional checks are to look for substances such as growth promoters, hormones, antibiotics or chemicals used legally or illegally in the production of the meat. A high general level of hygiene in a slaughterhouse is vital. While processing the food products, the screening has to be repeated for serious pathogens (like Salmonella) and Mad Cow Disease that might have escaped recognition earlier. During storage and transport, food has to be correctly handled at the correct temperature and maintained hygienically. The importance of food handling is highlighted by the fact that the United States Department of Agriculture (USDA) estimates that 85% of food poisoning cases could be avoided if people just handle food properly. Figure 13.1 describes the generic farm to fork concept model.

Today, we expect the food that they eat to be safe, regardless of its source. To achieve this public expectation and worthy objective, an enhanced integration and comprehensiveness in the food safety system is necessary. Firstly, this requires a greater clarity and acceptance of the role and responsibilities of each of the partners in this chain of farm to, including governments, industry, health professionals, educators, the mass media and, finally, the consumers. Secondly, it will require the complementary resolve and action of all stakeholders in making their necessary and timely contributions to the effort.

In the preindustrial era, man used to grow crops on fields. They stored the grains at home and consumed the same so the distance travelled by the food from farm to fork was minimum. However, with industrialization, increased trade and travel the food after production was transported miles of distances. Now, it frequently crosses international borders before consumption. Since there are numerous possible routes for introduction and transmission of pathogens in food, ensuring food safety needs a coordinated multidisciplinary and inter sectoral approach. It involves collaboration between department of agriculture, medicine, veterinary science, food processing, transport industry, packaging industry, retail sector, government and research fraternity and consumers. Government has a key role in setting and providing legislation that lays down minimum food safety standards at all levels. Governments must ensure that these are implemented through training, inspections sand enforcement. The promulgation of the new FSSA 2006 is an attempt by the government to achieve food safety. Although food safety legislation affects everyone in the country, it is particularly relevant to anyone working in the production, processing, storage, distribution and sale of food, no matter how large or small the business. This includes non-profit making organizations also. It has laid down science-based standards for articles of food. The Act also seeks to regulate their manufacture, storage distribution, sale and import, to ensure availability of safe and wholesome food for human consumption and for matters connected therewith or incidental thereto. To understand and control all of the major factors that affect food safety 'from farm to plate', it is necessary to integrate more effectively the various disciplines (veterinarians, botanists, molecular biologists, microbiologists, inspectors, physicians, etc.) that are involved in the study, investigation and control of FBI.

14

Good Agricultural Practices

Puja Dudeja, Amarjeet Singh

Indian agriculture has a rich historical past. Hymns in Rigveda describe plowing, sowing, irrigation, fruit and vegetable cultivation. An ancient Indian Sanskrit text, Bhumivargaha, classified agricultural land into twelve categories: Urvara (fertile), ushara (barren), pankikala (muddy), maru (desert), aprahata (fallow), jalaprayah (watery), kachchaha (land contiguous to water), sharkara (full of pebbles and pieces of limestone), shadvala (grassy), nadimatruka (land watered from a river), sharkaravati (sandy), and devamatruka (rainfed). Archaeological evidence suggests that rice was grown along the banks of the Indian river Ganges in the sixth millennium BC. Thousands of years ago, Indian farmers used to domesticate cattle, buffaloes, sheep, goats, pigs and horses The farmers used traditional methods of cultivation.

However, over past fifty years Indian population has tripled. To meet the food requirements of the increasing population and save them from starvation increase in farm production was the need of the hour. Norman Borlaug, titled as the "Father of the Green Revolution" introduced the concepts of high-yielding varieties of cereal grains, increase of irrigation infrastructure, advancement of management techniques, distribution of hybridized seeds, use of synthetic fertilizers and pesticides to farmers in developing countries. India too successfully implemented it, which led to rapid growths in farm productivity and enabled us to become self-sufficient by the 1970s. However, this historical revolution created some problems also. For example, high yield was associated with land de-gradation. Also, there was increase in number of weeds. There was evidence of chemicals in water and crops making them unsafe. Today, India is among the top three global producers of many crops, including wheat, rice, cotton, pulses, peanuts, fruits and vegetables. Worldwide, India has the largest herds of buffalo and cattle. It is also the largest producer of milk. It has one of the largest and fastest growing poultry industries. India's basic strength lies in its farms. With this huge farm productivity it becomes imperative that the safety and quality of farm produce is ensured at all stages of production. We need to balance the requirements of food security and safety both. The solution to this complex problem is by adopting Good Agricultural Practices (GAPs). To increase the quantity and quality of food in response to growing demand it is required to increase the agricultural productivity. Good agricultural practices, often in combination with effective input use, are one of the best ways to increase productivity and improve quality.

GAPs enhance the safe production and good quality of food. These practices are usally environmentally safe and ensure that the final product is appropriate handled, stored and transported. When GAPs are put in practice in true spirit it can be assured that the food will meet quality and safety standards at the time of harvest. GAPs protect food at the primary stage of production from contamination by the following:

- Physical hazards like rocks, dirt, sand filth, putrid and decomposed materials
- Toxic chemical hazards and contaminants from the environment like heavy metals, environmental pollutants and industrial chemicals
- Excessive or unsafe levels of agricultural chemical residues as pesticides, fertilizers, veterinary drugs and other chemicals
- Contamination or damage by pests, vermin and other insects
- Biological contamination by mould, pathogenic bacteria or viruses which can cause spoilage, crop damage and FBI or chronic health hazards in humans

According to FAO and GAP are "practices that address environmental, economic and social sustainability for on-farm processes, and result in safe and quality food and non-food agricultural products" (Fig. 14.1).

The international market is becoming competitive. The developed countries have become more demanding, critical and stringent when it comes to accepting export of food from developing countries. To have a good standing of our farm produce in the international market Indian GAP have been formulated. Adopting theses practices will ensure a safe and sustainable farm produce.

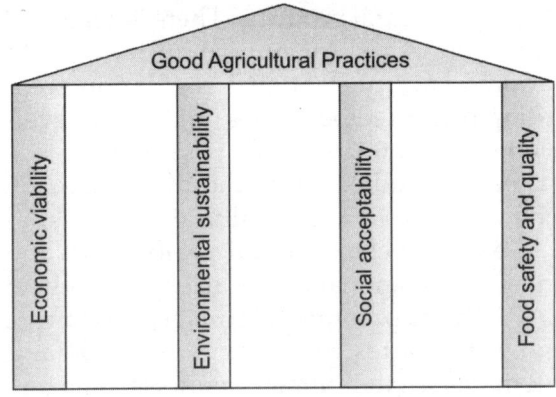

Fig. 14.1: Four main pillars of GAP

INDGAP defines certain minimum standards with a well defined system of accreditation mechanism and implementation of GAP. These standards are voluntary and non-discriminatory to the growers. INDGAP has different modules for all farm, crops, fruits and vegetables, combinable crops, green tea and coffee. Broad outline of various aspects which need to be managed are:

1. Site history and management
2. Soil management
3. Soil mapping
4. Plant nutrition management and fertilizers
5. Irrigation and fertigation
6. Integrated pest management
7. Plant protection products
8. Traceability
9. Complaints management
10. Visitors safety
11. Record keeping
12. Health welfare and safety of workers
13. Environmental conservation
14. Waste and pollution management

The potential benefits of GAP are significant improvement in quality and safety of food and other agricultural products. There is a marked reduction in risk of non-compliance with national and international regulations regarding permitted pesticides, maximum levels of contaminants (including pesticides, veterinary drugs, radionuclide and mycotoxins) in food and non-food agricultural products, as well as other chemical, microbiological and physical contamination hazards. Adoption of GAP helps to promote sustainable agriculture and contributes to meeting national and international environment and social development objectives.

However, there are various challenges related to GAP. The most prominent is a definite increase in cost of production. There is lack of harmonization between existing GAP-related schemes and availability of affordable certification systems which often leads to increased confusion and certification costs for farmers and exporters. There is a high risk that small-scale farmers will not be able to seize export market opportunities unless they are adequately informed, technically

prepared and organised to meet this new challenge. It is required that governments and public agencies play a facilitating role in this aspect. However, at times it has been experienced that compliance with GAP standards does not promote all the environmental and social benefits which are claimed.

Some key points for adopting GAP are:

- Selecting the right type of land to be cultivated for food crop production.
- Planting the best-quality seeds and of the most appropriate varieties.
- Use of authorized and acceptable chemical inputs (fertilizers, pesticides) as per approved directions (e.g. concentration, frequency, timing of use).
- Controlling the quality of irrigation water (in case of use).
- Use of appropriate harvesting and on-farm storing and handling techniques.
- Use of suitable methods for shipping of produce to markets or food processors.

15

Good Animal Husbandry Practices

Puja Dudeja, Amarjeet Singh

Survival of the fittest; as per this theory of Charles Darwin man has overcome/mastered control over other animals. Man's capacity to develop/design instruments/implements/weapons helped man control animals. This contributed to a large extent in evolution of livestock over the past 12,000 years. As per historical evidence, goat and sheep were the first species of animals domesticated for human use. This was followed by domestication of pig. About 8000 years ago cow was the last major food animal that humans domesticated. This led to introduction of milk as a useful foodstuff. Goat, sheep, reindeer and camel milk were also used. Other species used for food were avian, amphibian, fish, and various arthropods.

Kamdhenu an Indian mythological figure symbolizes livestock as source of wealth. Since time immemorial livestock have formed an integral component of India's agricultural and rural economy. They have been supplying energy for crop production in terms of draught power and organic manure, and in turn deriving their own energy requirements from crop byproducts and residues. The livestock sector has been growing faster than crop sector in our country. India's livestock sector is one of the largest in the world. It has 56.7% of world's buffaloes, 12.5% cattle, 20.4% small ruminants, 2.4% camel, 1.4% equine, 1.5% pigs and 3.1% poultry. The Department of Animal Husbandry and Dairying (AH and D) now renamed as Department of Animal Husbandry Dairying and Fisheries (DADF) is one of the Departments in the Ministry of Agriculture. It has emerged by converting two divisions of the Department of Agriculture and Co-operation namely Animal Husbandry and Dairy Development into a separate Department. The Fisheries Division of the Department of Agriculture and Cooperation which was a part of the Ministry of Food Processing Industries were later transferred to this Department w.e.f. 10th October, 1997. The Department is responsible for matters relating to livestock production, preservation, protection from disease and improvement of stocks and dairy development. It also looks after all matters pertaining to fishing and fisheries both inland and marine. Various animals and their foods designed from them are given in Table 15.1.

Table 15.1: List of animals and food obtained from them

Animal	Food
Dairy	Fluid and dried milk, butter, cheese and curd, casein, evaporated milk, cream, yoghurt and other

Contd.

Table 15.1: List of animals and food obtained from them (Contd.)

Animal	Food
	fermented milk, ice cream, whey
Cattle, buffalo, sheep	Meat (beef, mutton), edible tallow
Poultry	Meat, eggs, duck eggs (in India)
Pig	Meat
Fish (aquaculture)	Meat
Horse, other equines	Meat, blood, milk
Micro-livestock (rabbit, guinea pig), dog, cat, bulls	Meat
Insects and other invertebrates (e.g. vermiculture, apiculture)	Honey, 500 species (grubs, gras-shoppers, ants, crickets, termites, locusts, beetle larvae, wasps and bees, moth caterpillars) are a regular diet among many societies

With modernization, there have been advances in bio-chemical and mechanical technologies. The link between livestock and crops have weakened and livestock are now more viewed as source of food. Demand for animal food products has been responsive to globalization. Human's lifestyle has changed dramatically over the time. The consumption of meat and meat production has increased radically through out the world and is expected to increase in future also. There is changing trend that is occurring globally in how people eat. With globalization the quality of life of people has improved substantially. The standard of living has become better. As the economic status of people changes, the food consumption pattern changes as well. Coupled with this increase in information technology, international trade, advertise-ments and modern lifestyle have made the people to shift to nonvegetarian diet. Meat and fish consumption has increased remarkably since they are desirable, although expensive, food sources . The modern life style with high Per capita Purchasing Power (PPP) has increased the meat production and con-sumption.

Apart from this there has been increase in awareness about the quality and content of animal foods. Sometimes back newspaper headlines read 'Indian badminton women team for the first time in semifinals for the world cup'. Below the news there was an advertisement with the same player regarding eggs *'sunday ho yaa Monday roj khayen anday'* meaning whether it is Sunday or Monday one must take eggs'. Similarly, another advertise-ment highlighted the importance of butter and milk in the same paper. Not only in news-papers such health advertisements bombard us with messages on education and awareness about the importance of good quality com-plete protein foods of animal origin. Hence, there has been an increasing demand of protein rich foods of animal origin in the society.

Animal proteins are the rich source of essential amino acids in the diets of human being. They supply about 20–25 percent of total daily protein requirement. Milk is one of the most important sources of animals protein in the diets of predominately-vegetarian population of Indians. The other protein rich foods of animal origin are the meat of different animals as chicken and fish and eggs.

The food processing industry has also lengthened the list of items in the menu of non-vegetarian platter. There are newer non-vegeterain products to pamper the taste buds. Better international trade has ensured the availability of these products all over world through the year and made them even more popular. International food chains like KFC, McDonalds, Subway, Al Kabir, Pizza Hut have modified the tastes of these nonvegetarian items as per the requirement/culture/ traditional taste of the each country.

However, besides being a source of food, animals are also source of Zoonoses, i.e those diseases and infections the agents of which are naturally transmitted between [other] verte-brate animals and man. Zoonoses make up more than 60 percent of all human infectious diseases and more than 70 percent of all emerging infectious diseases. These diseases occur most frequently in Asia and Africa, where people tend to live in close proximity to their livestock. Limited resources in these regions hinder both surveillance and public health response for such a disease outbreak response.

The zoonoses can be classified as:

1. Anthropo-zoonoses: infections which are transmitted to man from animals like brucellosis from cattle, bird flu, swine flu

2. Zooanthroponoses: infections which are transmitted from man to lower vertebrate animals

3. Amphixenoses: infections may be trans-mitted in either direction.

Frequent outbreaks of various pandemics which were unknown previously like SARS, avian flu, swine flu, Bovine Spongiform Encephalopathy (BSE) has raised a question about the safety of non vegetarian food or food of animal origin. This matter concerns us too. The H5N1 pandemic originated from China. The H5N1 virus was the first documented infection of pigs by any H5 subtype of avian influenza virus. Historically, densely populated southern China has been

a breeding ground for new influenza viruses, because of the large numbers of animals and people living in close proximity. The backyard animal husbandry of pigs, chickens, and ducks; and the presence of live-animal markets are other strong factors for such occurrences. All these factors were implicated in the SARS outbreak of 2003 also.

For the safety of the animal food, it is important that only healthy animals enter the food chain. Pandemic H1N1 is an example how important human and animal health are linked. There are many other zoonotic diseases of concern as Bovine tuberculosis, rabies, listeriosis, salmonellosis, babesiosis, brucellosis, anthrax, etc. There have been many examples on record where lapses in animal food safety have taken place leading to human suffering. In 2008 in Canada 28 people were affected by outbreak of Listeriosis in meat products. This happened as most of these animals were reared in filthy and unhygienic conditions. These animals were being fed continuously with low doses of antibiotic and other growth promoters.

Another issue of concern is the indis-criminate use of veterinary drugs which are available everywhere without a prescription and their use is rampant in poultry and milk for extra profits. The chicken produced under such practices is not fit for human con-sumption. Good hygiene, proper feed, appropriate husbandry and good manage-ment practices are by far the best strategies to be adopted for safety of food of animal origin.

In this context at the international level animal feeding and live stock products standards have been developed. Agencies like codex, ISO, OIE, FAO, WHO have all made contribution for providing directions on safety of food of animal origin. With initiative of FAO, a web site titled 'International Portal on

Food Safety, Animal and Plant Health ' has been developed (www.ipfsaph.org) to provide guidance on food safety. They have also developed a manual for good practices for meat industry.

Good Animal Husbandry (GAH) or Good Veterinary Practices (GVP) have been established to assure consumers that foods derived from animals meet acceptable levels of quality and safety. These practices are the guiding principles in professional veterinary practice for the care and treatment of animals, including animals used for human food production. The important components of GAH practices for safe food of animal origin can be understood better by using farm to fork model (Fig. 15.1).

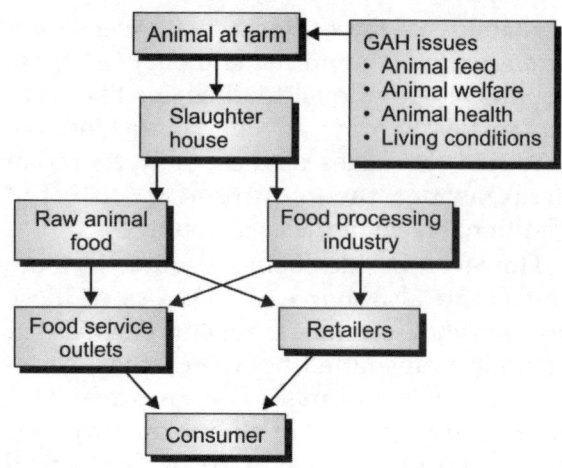

Fig. 15.1: Farm to fork model for food of animal origin

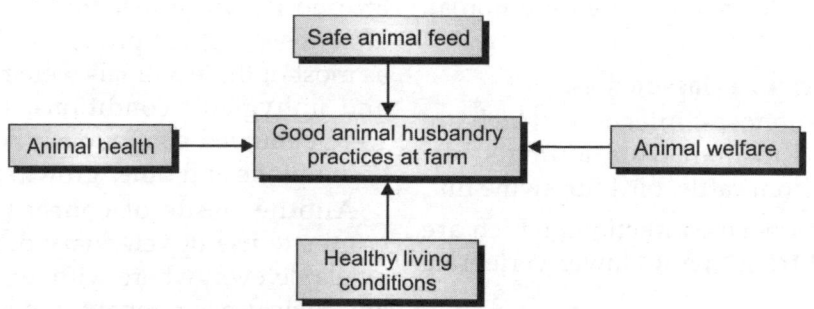

Fig. 15.2a: Good animal husbandry practices at farm

Good Animal Husbandry Practices at Farm
Safe Animal Feed

The first and foremost good practice at farm is provision of safe feed to the animals. Various threats in animal feed are given in Figs 15.2 a and b. Safe animal feed will provide us with safe animal food. Microbial contamination, antibiotic residues and adulteration animal feed is rampant. The subtherapeutic feeding of antibiotics as antibiotic treatment of diseased animals are current practices being followed. This low dose antibiotic feeding to animals has resulted in development of antibiotic resistance of zoonotic pathogens. Many antibiotics added to animal feed are also used in human medicine, and antibiotic-resistant bacteria could develop and cause infections in animals and humans.

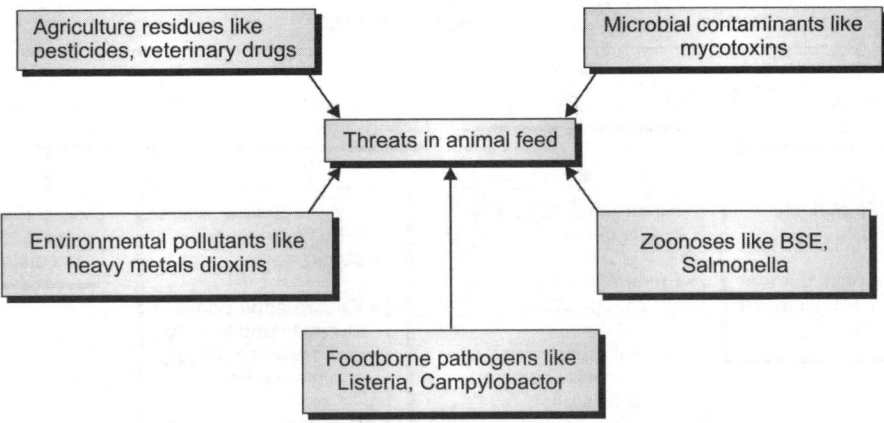

Fig. 15.2b: Threats in animal feed

Improved nutritional content of livestock feed may lead to direct health benefit for animals, improving their welfare, and also indirectly to benefit for consumers. Residues of antibiotics used in livestock or added to feed have been found in food-producing animals including dairy cows. Among these drugs are chloramphenicol and sulphamethazine. Alternatives to the prophylactic feeding use of antibiotics to maintain animal health include the modification of production systems. These modifications include reduced animal confinement, improved ventilation and improved waste treatment. Various hormones have been given to animals like oxytocin to improve the output of milk from the animal. Residues of such chemicals have also been detected in the milk.

Certain disease outbreaks like BSE in animals raise serious public concern for human health. A rare Creutzfeldt-Jakob disease (CJD) emerged among beef-importing nations in 1996. Eating beef infected with BSE, popularly known as mad cow disease was the cause behind CJD infection. Although unproven, public perceptions include the proposition that the disease has entered cattle from feed containing bone meal and offal from sheep afflicted with the similar disease, scrapie. All three diseases, in humans, cattle and sheep, exhibit common symptoms of sponge-like brain lesions. The diseases are fatal, their causes are unknown, and there are no tests to detect them. Good feeding practices provide the herd with adequate feed and water, keep the animals healthy, preserve water supplies and animal feed materials from chemical contamination, prevents micro-biological or toxin contamination or unintended use of prohibited feed ingredients or feeds contaminated with chemical pre-parations. Contamination of feed ingredients at times occurs due to improper storage in compounded feed/feed ingredients especially during warm and humid climate and needs to be controlled to maintain quality of feed. Whole grains can be adulterated with substandard/cut grains stones, sand, iron fillings and shriveled/insect-eaten grains. Urea and or ammonium salts are sometimes mixed in oilseed cakes to increase the protein contents. Various oil seed cakes and brans

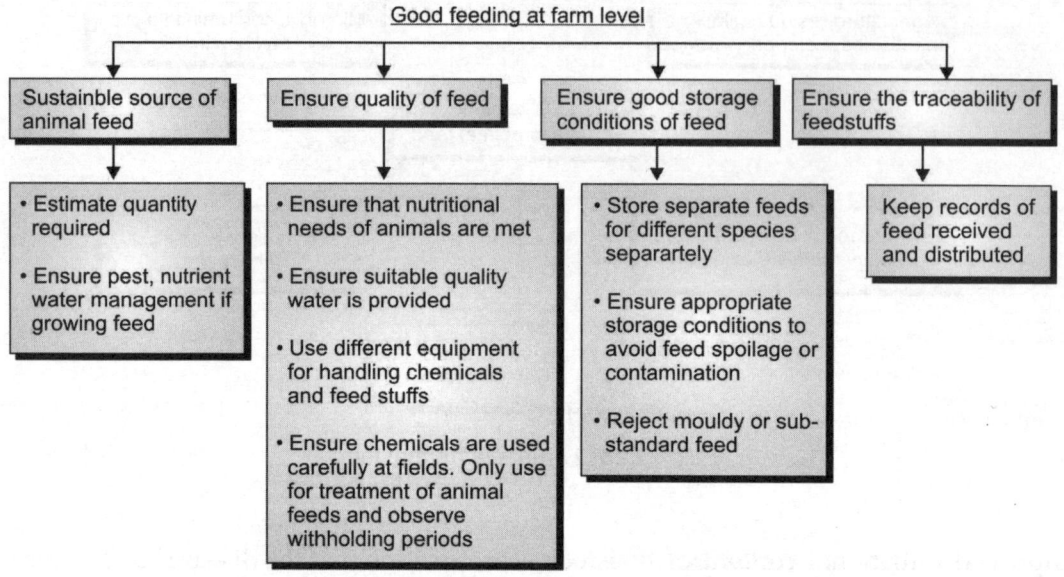

Fig. 15.3: Good feed practices for livestock at farm level

may contain fibrous material such as saw dust, husk and hulls. Generally adulterants used in feed ingredients include excess moisture to increase weight which can lead to fungus infestations and low shelf life of feed.

The occurrence and detection of antibiotic residues in milk continue to be a concern for dairy industry, consumer and the Government with emerging issues of development of resistance to multiple antibiotics. Presence of antibiotic residues is also linked to cause allergic reactions in human beings even if they are present at very low level (1 ppb), carcinogenicity and spread of bacterial resistance to antibiotics.

In India, many of these antibiotics are being used in an "extra label" fashion and furthermore, due to the malpractices new generation antibiotics recommended for human use are also being used in animal for disease control, for which no safe levels have been recommended in milk and are leading to development of antibiotic resistance. Good Feed Practices at farm can be classified in Fig. 15.3.

As per FSSA 2006 no article of food shall contain pesticides, veterinary drugs and antibiotic residues and microbiological counts in excess of such tolerance limits as may be specified by regulations.

Animal Welfare

Looking after the welfare of animals is both an ethical and medical issue. Welfare activities will ensure good health of the animals at the farm. Animals should be reared in humane conditions. Animal welfare at the farm includes freedom from the following Table 15.2.

Table 15.2: Animal welfare activities

Freedom from	Action to be taken
Thirst, hunger and malnutrition	Making sure animals have enough food and water of the right quantity and quality to meet their daily needs
Discomfort	Providing an environment that gives shelter and shade; sheds for newborn calves
Pain, injury or disease	Vaccination programme, drenching for internal parasites; pain relief for lame cows; safe facilities for handling animals
Freedom to express normal behaviour	Providing natural conditions
Fear and distress	Not keeping animals away from the herd, handling animals calmly and gently during milking

Animal Health

Outbreaks of diseases like foot and mouth disease, Influenza, etc. have been documented in farm-reared animals. Many human foodborne illnesses result from pathogenic bacteria of animal origin. Examples include Listeria and Salmonellae which found in dairy products and Salmonellae and Campylobacter found in meat and poultry. Dairy cattle in intensive production systems, commonly suffer from metabolic diseases with underlying nutritional aetiology, conditions which predispose animals to infections, disease and wider range of ailments, including infertility, lameness and welfare problems. The majority of human infectious diseases have their origins in animals, and many important zoonotic diseases in livestock not only adversely affect productivity and welfare, but may also be transmitted to man either directly or via animal products and the environment. Sustainable future farming systems must mitigate against risks such as avian influenza, Q-fever, *E.coli* O157, bovine tuberculosis, and various helminthiasis. It is of utmost importance that animal health is looked after. This will enhance herd immunity, reduce stress, help in early disease detection, ensure food safety, traceability and prevent occurrence of chemical residues in milk. Varoius practices to ensure good animal health at farm are given in Flowchart 15.1.

Presently there are a total of 8,732 veterinary hospitals/polyclinics and 18,830 veterinary dispensaries in the country providing services for the large livestock population. These numbers are grossly inadequate. Whatever exists also has poor infrastructure in terms of wrecked buildings, lack of equipment, etc. The polyclinics, wherever established, lack the adequate infrastructure for surgical interventions and diagnostic imaging. Presently the country is facing acute shortage of manpower to manage these institutions and provide required services. Diagnostic facilities too in terms of good clinical laboratories, equipment, quick and quality diagnostics and the human resource having expertise in these areas are practically non-existent. Several diagnostic kits required for national surveillance and monitoring are imported at a huge cost.

Surveillance and monitoring of livestock diseases is a major component of good animal health. The present disease reporting is neither timely nor complete. Due to this delay, many times animal disease out-breaks assume serious proportions before control and containment steps can be initiated. An authentic epidemiological data for realistic assessment of the prevalence and emergence of these diseases in different agro-climatic zones is essential not only for identification and prioritization of the most important diseases but also their prevention and control

Flowchart 15.1: Good animal health practices

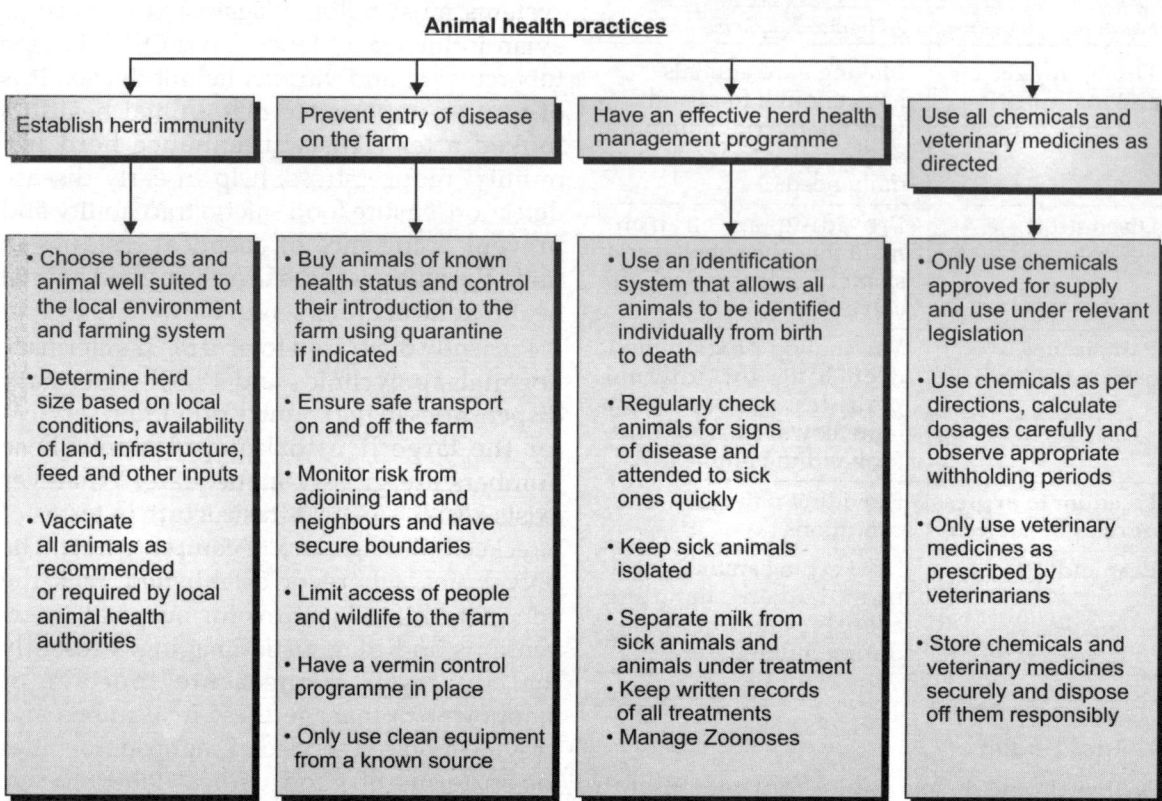

by making judicious use of available resources. Recording the incidence of diseases is essential for estimating the economic loss, conducting risk analysis and also for obtaining disease free status of the country.

National Disease Control Program involves the vaccination of all susceptible livestock against major infectious diseases. Except for (foot and mouth diseare) FMD vaccine, production of most other vaccines is with the state biological units. The biological production centers available in the Govt. sector (both state and central) are old, obsolete, do not comply with the GMP requirements, and are not having contemporary technologies and infrastructure to meet the requirements. The availability of qualified man power to run these institutions are also not available. Assistance is provided to State Governments for control of economically important diseases of livestock and poultry by way of immunization. Foot and Mouth Disease Control Program (FMD-CP) has been implemented in 221 districts in Phase I with 100% central funding towards cost of vaccine, maintenance of cold chain and other logistic support to undertake vaccination. The program has lead to reduced incidence of disease. The ultimate objective of government is to eradicate this disease from the country in a time-bound manner on the lines of Rinderpest eradication. Similarly, National Control Program on Brucellosis initiated in 2010 which envisages mass screening of cattle and buffaloes to ascertain exact incidence of the disease and vaccination of all female calves using S-19 vaccine is in progress.

Animal Living Conditions

Healthy living conditions at the farm will prevent the occurrence of various diseases in the farms at the farm. It will also reduce stress levels in the animals. A rough check list for ensuring health living conditions are given below:

✓ Clean water
✓ Clean feed
✓ Clean housing
✓ Good ventilation
✓ Appropriate temperature
✓ Consistent feeding
✓ Vaccination
✓ Quarantine sick animals
✓ Balanced diet
✓ Pest control
✓ Sound control
✓ Avoid overcrowding

The location of stable/coop/enclosure/pen/shed should be suitably selected with an appropriate drainage system, good lighting and ventilation. There should be adequate space for each animal to avoid overcrowding. Implement practices to reduce, reuse or recycle farm waste. Good pest control is required to provide safe environment to animals.

Some important good animal husbandry related measures are:

- Only healthy animals are slaughtered for the purpose of human food:
- Any drug used in the control of animal disease is safe for its intended use and used according to approved directions (i.e. appropriate amounts, frequency and timing), and residues of such drugs do not remain in the edible tissues at unsafe levels when the food is made available for consumption.
- Chemicals utilized in animal husbandry (e.g. dips for insect pest control) are safe for their intended uses and used according to instructions (i.e. appropriate levels, frequency and timing), and residues of such chemicals do not remain in the edible tissues at unsafe levels when the food is made available for humans.
- Live animal inspection and handling are properly conducted before slaughter, and carcass inspection and handling after slaughter.
- Appropriate temperature controls, storage conditions, handling and butchering techniques and sanitary conditions are maintained during processing and butchering to prevent post-slaughter contamination.
- Safe shipping and handling practices to prevent any unnecessary exposure of the product to contamination.

16

Good Manufacturing Practices

Puja Dudeja, Amarjeet Singh

Food processing in not a recent phenomenon. Processing of food dates back to the prehistoric ages when raw food items were subjected to fermenting, sun drying, preserving with salt, and various types of cooking such as roasting. There is archeological evidence of salt-preservation for foods that constituted warrior and sailors' diets until the introduction of canning. Processing methods like making of pickles (using high salt solution), *morraba* (using high sugar solution), chutneys, drying grapes as *kishmish*, dried dates, etc. have been common practices in Indian households. These tried and tested processing techniques continued until the advent of the industrial revolution. Examples of ready-meals also date back to the preindustrial revolution period. Both during ancient times and in today's modern society these are considered as processed foods.

Modern food processing technology was developed in the 19th and 20th centuries as a solution to diet and nutrition related problems of military personnel. In 1809, Nicolas Appert invented a hermetic bottling technique that would preserve food for French troops which ultimately led to the development of tinning, and later canning in 1810 by Peter Durand. Initially these foods were expensive. Also the lead used in cans made the canned goods

potentially harmful to the consumer. But in due course of time these became popular around the world. In 1864, Louis Pasteur gave the method of pasteurization which contributed to improved quality of preserved foods. This was the beginning for wine, beer, and milk preservation. After the second world war there were advances in food processing area like spray drying, juice concentrates, freeze drying and the introduction of artificial sweeteners, colouring agents, and preservatives as sodium benzoate. In the late 20th century, products such as dried instant soups, reconstituted fruits and juices, and self cooking meals were developed.

With the modernization forces in full swing in Western countries there was a pursuit of convenience foods. The marketing strategies of food processing companies targeted the middle-class working wives and mother with frozen foods. These had the benefits of being free from toxins, had ease of cooking, transportation, and less susceptibility to early spoilage than fresh foods and increased food consistency. As these food had a long shelf life they have contributed to the success of long voyages.

The processing industry also contributed immensely to a variety of foods. This brought a paradigm shift in the diet of people from traditional to modern diet. Food consumption patterns changed from fresh, unprocessed, unbranded food products to processed, packaged and branded products. The concept of convenience of foods was introduced which gradually became an important feature of modern diet. Convenience foods are processed foods, which are so prepared and designed that they provide ease of preparation and consumption. Though meals served in a restaurant do meet this definition, nevertheless, the term is seldom applied to them. Convenience foods include prepared foods such as ready-to-eat foods, frozen foods prepared mixes such as idli, upma, vada mix, etc. The types of convenience foods can vary by country and geographic region. In simple words they typically cost more money and less time compared to home cooking. Some examples of these foods which have flooded the supermarkets are beverages such as soft drinks, juices and milk; fast food; nuts, fruits and vegetables in fresh or preserved states; processed meats and cheeses; and canned products such as soups and pasta dishes, frozen pizza, potato chips cookies.

With improved long distance transportation, this modern diet was available all around the world. Transportation of more exotic foods gives the modern eater easy access to a wide variety of food unimaginable to their ancestors.

Indian lifestyle has also undergone many changes. Today Indian households too welcome food with convenience in cooking and purchase. This is the result of both working women culture in present society coupled with increased and easy availability of processed and convenience foods. Use of processed foods benefitted the working couples. It gave free time for the working lady which she could spent with her family which she would otherwise spent in kitchen. She could manage her time better by fulfilling her duty of providing food to all the members of the family in less time. They want to spend less time in kitchen. The increasing prevalence of nuclear families, rising disposable income, more bachelors staying away from home for work have also contributed to rise in consumption of processed foods in our country.

More convenience has been added to such foods by offering home delivery/doorstep delivery by supermarkets for such products reducing the gap between the shelves of supermarkets and our refrigerators.

Revolution in packaging industry like retort packages have made it possible for Indians who go abroad for short term assignments or those who are settled there to take these packages along. These provide them with Indian food at lesser price as the restaurants serving Indian foods are quite expensive abroad.

Indian Processed food industry has been divided into two main segments

- **Regional Indian Processed food** like MTR, Kohinoor foods, ITC, Haldiram, Tasty Bites, Priya. These make rice dishes like (*Bisibele Bhath, Rajma Chawal, Sambar Rice Jeera Rice*), soups, ready to eat south Indian dishes like (*Avial, Kesari Bhath, Khara Bhath*), north Indian dishes like (*Alu Muttar, Chana Masala, Dal Fry, Dal Makhani , Navatan Kurma, Paneer Buttar Masala*), Frozen Foods (*Masala Dosa, Alu Curry Rava Idli, Punjabi Chole*, Paratha Palak, Paneer), Instant Sweet Mixes (Gulab Jamun), Vermicelli and Instant Snacks.

- **Multinational Companies (MNCs)** like Quaker (oats), Kellogs (corn flakes)

Over the past few years processed food market has been able to construct a huge consumer base in urban India. In India urban people nowadays are cooking less traditional food less frequently. Members of the family are often eating processed food, e.g. the breakfast menu has changed *from prantha, subzi, puri,* milk, etc. to oats, cornflakes, museli, processed cheese, sandwich spreads, juices, etc. Even during lunch hours the *chhole kulche sold* by the street vendor has become a regular feature.

The processed food industry is the fastest growing industry in India and is expected to be the world's largest food factory in the times to come. However, there are many food safety concerns of processed food. For example, use of food additives. The health risks of any given additive vary greatly from person to person, use of sweeteners, preservatives, stabilizers are permitted only at specified levels for use in food products which if not strictly adhered to can cause harm to the consumer.

Food processing is typically a mechanical process that utilizes large mixing, grinding, chopping and emulsifying equipment in the production process. These processes inherently introduce a number of contamination risks. Ministry of Food Processing, Government of India is responsible for looking after Food Processing Sector. This sector includes the food processors, suppliers of equipment, raw materials, ingredients, packing materials, processing aids, pestcides, fertilisers and cleaning chemicals. However, it does not include primary production, transportation, storage and retail. It has divided food processing units into various subsectors (Fig. 16.1).

Food Processing Industry

To ensure safety of during processing Good Manufacturing Practices (GMPs) have been defined. These are referred to as practices and procedures performed by a food processor which can affect the safety food product. GMPs refer to the people, equipment, process and the environment in the production process. It is a term that is recognized worldwide for the control and management of manufacturing, testing and overall quality

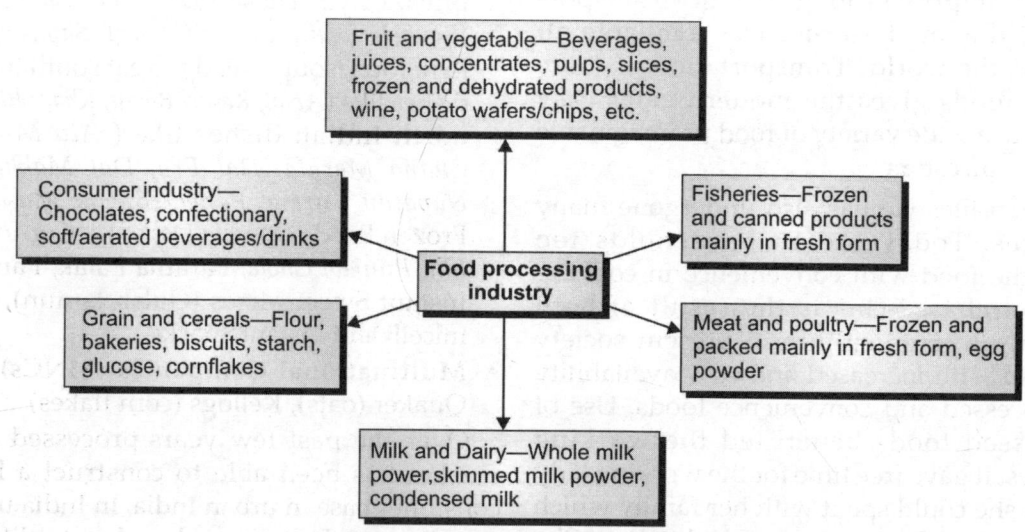

Fig. 16.1: Subsectors of food processing industry

control of food and pharmaceutical products. The focus of GMP is primarily at diminishing the risks inherent in any food or pharmaceutical production. These were first given by US Food and Drug Administration. The various components of GMP are given in Fig. 16.2.

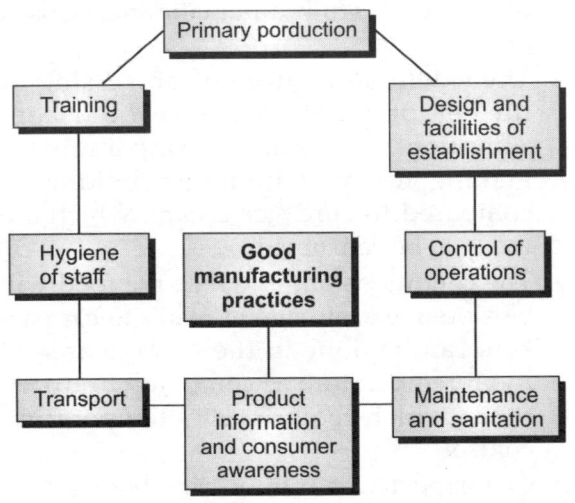

Fig. 16.2: Components of GMP

Product information and consumer awareness GMPs have been developed to reduce the risk of contamination in the food processing sector. The design, documentation and implementation of an organisation's GMPs are influenced by the specific needs of the products handled and the processes employed. GMPs is a set of procedures which if adopted and implemented in true spirit will lead to production of safe and wholesome foods. The various components of GMP are described in succeeding paragraphs.

Hygiene of Staff

Hygiene of all the personnel carrying out activities in the food processing/manufacturing plant has a impact on the food safety.

Organization needs to ensure that those who come directly or indirectly into contact with food do not contaminate the food product being manufactured in the plant. Their personal hygiene behavior and method of operating should comply with the instructions issued to them.

Food handlers need to maintain a high degree of personal cleanliness and, where appropriate, wear suitable protective clothing, head covering, and footwear. Cuts and wounds, where personnel are permitted to continue working, shall be covered by suitable waterproof dressings. Personnel shall always wash and disinfect their hands when personal cleanliness may affect food safety.

It is mandatory that annual medical check up of workers is carried out and records are maintained. People known, or suspected, to be suffering from, or to be a carrier of a disease or illness likely to be transmitted through food, shall not be allowed to enter any food handling area if there is a likelihood of their contaminating food. Any person so affected shall immediately report illness or symptoms of illness to the management. In case a staff member in food operations is suffering from jaundice/diaorrhea/vomiting/fever/sore throat with fever/visibly infected skin lesions (boils, cuts, etc.)/discharges from the ear or eye it shall be reported to management. The concerned officials may then examine any need for medical examination and/or possible exclusion from food handling can be considered. Vaccination of food handlers against typhoid is recommended as a good practice for the food handlers. People engaged in food handling activities shall refrain from behavior, which could result in contamination of food. For example, smoking, spitting, chewing or eating, and sneezing or coughing over unprotected food. Personal items as jewellery, watches, pins, flowers or other items shall not be worn or brought into food handling areas if they pose a threat to the safety and suitability of food. Visitors to food manufacturing, processing or handling areas shall, where appropriate, wear protective clothing and adhere to the other personal hygiene provisions.

Training of Staff

All the staff working in the unit shall have adequate and appropriate education, training and skill about the plant operations with a special emphasis on critical points where food safety can be compromised. It is of utmost importance that extra attention is paid for training of those personnel who come directly or indirectly into contact with food.

To ensure that this aspect of GMPs is in place, the management must identify the training needs of personnel whose activities have an impact on food safety and quality. Apart from conducting training periodic assessments of the effectiveness of training and instruction, is essential. Routine supervision and checks during working hours can ensure that methods learnt during training are being implemented effectively. It is necessary that the personnel are aware of the relevance and importance of their individual activities in contributing to GMP system. All personnel shall be aware of their role and responsibility in protecting food from contamination or deterioration. Apart from food handlers those individuals who handle strong cleaning chemical or other potentially hazardous chemicals ought to be trained in safe handling techniques.

It is the job of the management to ensure that training programmes are routinely reviewed and updated where necessary. Records of all training and related actions should be maintained. The kind and level of training depends upon following factors:

- The nature of the food and in particular its ability to sustain growth of pathogenic or spoilage, micro-organisms. For example, curd is more susceptible than tinned fruits
- The manner in which the food is handled and packed, including the probability of contamination. For example, whether direct contact with hands is required in the process or it is fully automatic and a closed process
- The extent and nature of processing or further preparation before final consumption. For example, using a mix for making idli will undergo cooking as compared to curd/ice-cream, which are ready to be consumed.
- The conditions under which the food will be stored, e.g. storage of ready to eat packets can be done in the storage area at room temperature as compared to frozen foods which require strict temperature control.
- The expected length of time before consumption, e.g. storage of tetra packs can be done for months at room temperature as compared to curd/yogurt which need to be consumed within a specified duration.

Food Safety Implications of Lack of Hygiene and Training of Staff

Laxity in hygienic practices of food handler in a food processing plant is one of the leading causes of food contamination. The main challenge here is to motivate employees to comply with hygienic practices. Training is an important step but is often not enough to ensure employee compliance. There are wide gaps in training and implementation. Various methodologies can be adopted to ensure employee compliance. For example, using a CCTV to check hand washing practices by employees, use of thumb sensors at hand washing area to cross check number of times hand washed by a particular employee. Other indirect methods are counting number of soaps/amount of liquid soap/number of towels dispensed by the end of the day. In case sophisticated gadgets are not applicable, a strict monitoring of all workers by a senior employee can serve the purpose. To ensure compliance a strict action against defaulters to set an example can make the employees understand that there is no scope of complacency in hygienic practices.

Effective training is crucial to ensure that sanitation standards are met. The problem here is that outside agencies conducting training for employees are generic in approach. The content of training needs to be customized and tailor made to suit that particular manufacturing unit. Other impediments to effective training might include training the wrong people, not training enough people, or not providing enough training.

Documentation Requirements for GMP

The GMP policy of the manufacturing unit along with the related objectives is an essential document in case the manufacturing unit has adopted GMP. The policy document is expected to have the details of various processes, production system and operation of the unit. It is required that any change proposed in the food manufacturing system is reviewed by the GMP experts of the unit prior to implementation to determine its effect on food safety system. The relevant versions of GMP documents should be available at various points of use in the unit.

Records shall be established and maintained to provide evidence of conformity to requirements and evidence of the effective operation of the GMP system. Records shall remain legible, readily identifiable and retrievable. A documented procedure shall be established to define the controls needed for the identification, storage, protection, retrieval, retention time and disposition of records.

For example, for cleaning of the establishment and equipment following records shall be maintained.

i. Areas, items of equipment and utensils to be cleaned

ii. Responsibility for particular tasks

iii. Method and frequency of cleaning and

iv. Monitoring arrangements.

Food Safety Implications of Lack of Correct Documentation

Documentation in every aspect of the process, activities, and operations involved with manufacture of the food product is a handy mechanism in ensuring safety of food. It not only works as a preventive tool but also helps in pinpointing where things went wrong and fixing accountability.

In case a manufacturing unit undergoes plant renovations, it is vital here that various Standard Operating Procedures (SOPs) are revised and documented alongside. For example, if a new equipment is installed then the number of mops/time allotted for cleaning the plant should increase simultaneously.

Infrastructure Requirements for GMP

Food safety risks in a manufacturing unit depend upon the nature of the food item and the process which are undertaken. The management needs to make certain that the premises, equipment and facilities shall be located, designed and constructed to ensure that contamination is minimized. The design and layout of equipment should permit appropriate maintenance and prevent cross contamination. There shall be a unidirectional flow of process and materials. The surfaces and materials, in particular those in contact with food, should be non-toxic and durable.

Appropriate facilities for light, temperature, humidity, pest management ensure smooth implementation of GMP.

Location of Manufacturing Unit

Food manufacturing unit should ideally be located away from environmentally polluted areas and industrial activities which pose a serious threat of contamination of food; harborage of pests and accumulation of wastes. Wherever the unit is located effective measures shall be taken to prevent contamination from neighbouring surroundings.

Food Safety Implications of an Unclean Location

In case the surrounding areas of a food-manufacturing unit are filthy there ought to be breeding of flies, presence bad odour and occurrence of pests. All these in vicinity can affect the safety of food in the plant.

Premises and Rooms: Design and Layout

The internal design and layout of food establishments shall permit good food hygiene practices between and during operations by foodstuffs. There shall be smooth flow of material during production to prevent cross contamination between products.

Structures within food establishments shall be built of durable materials and be easy to maintain, clean and where appropriate, able to be disinfected. In particular the following specific conditions should be satisfied, where necessary, to protect the safety and suitability of food:

- The surfaces of walls, partitions and floors should be made of impervious materials with nontoxic effect in intended use
- Walls and partitions should have a smooth surface up to a height appropriate to the operation

- Floors should be constructed to allow adequate drainage and cleaning
- Ceilings and overhead fixtures should be constructed and finished to minimize the build up of dirt and condensation, and the shedding of particles
- Windows should be easy to clean, be constructed to minimize the build up of dirt and where necessary, be fitted with removable and cleanable insect-proof screens. Where necessary, windows should be fixed, doors should have smooth, non-absorbent surfaces, and be easy to clean and, where necessary, disinfect
- Doors between external environment and the production areas should be provided with air curtains or suitable means to prevent entry of air from the external environment into the production area
- Working surfaces that come into direct contact with food should be in sound condition, durable and easy to clean, maintain and disinfect. They should be made of smooth, non-absorbent materials, and inert to the food, to detergents and disinfectants under normal operating conditions.
- Potable water (IS 10500) and effective waste disposal system shall be provided.

Equipment

The main objective of good manufacturing practices related to equipment are to ensure that:

✓ Harmful or undesirable micro-organisms or their toxins are eliminated or reduced to safe levels or their survival and growth are effectively controlled and

✓ Temperature and other conditions necessary to food safety and suitability can be rapidly achieved and maintained.

Equipment shall be so located that it permits adequate maintenance and cleaning to prevent cross contamination. It should function in accordance with its intended use and facilitate good hygiene practices, including monitoring. Equipment and containers (other than once-only use containers and packaging) coming into contact with food, shall be designed and constructed to ensure that, where necessary, they can be adequately cleaned, disinfected and maintained to avoid the contamination of food. Equipment and containers shall be made of materials with no toxic effect in intended use. Where necessary, equipment shall be durable and movable or capable of being disassembled to allow for maintenance, cleaning, disinfection, monitoring, and facilitate inspection for pests.

Equipment used to cook, heat treat, cool, store or freeze food shall be designed to achieve the required food temperatures as rapidly as necessary in the interests of food safety and suitability, and maintain them effectively. Such equipment shall also be designed to allow temperatures to be monitored and controlled. Where necessary, such equipment shall have effective means of controlling and monitoring humidity, air-flow and any other characteristic likely to have a detrimental effect on the safety, quality or suitability of food.

Food Safety Implications

✓ In case the design of building or of equipment is not suitable then niche environments can develop. These are sites within the manufacturing environment where bacteria can get established, multiply, and contaminate the food processed. These sites may be impossible to reach and clean with normal cleaning and sanitizing procedures. For example, hollow rollers on chapatti conveyors, the spaces between close-fitting metal-to-metal or metal-to-plastic parts, worn or cracked rubber seals around doors, and on-off valves and switches. Identification of niches in a manufacturing unit needs to be done by an experienced and dedicated employee with an in-depth understanding of the equipment and food contact surfaces. Microbiological sampling of the environment and equipment can detect a niche. Further, sanitary equipment design can also help prevent niches. Proper maintenance to keep equipment parts from providing potential niches is also essential. If the pieces of equipment are located too close to each other or too close to the wall then also proper cleanliness cannot performed.

Good hygienic design of equipment prevents or minimizes microbiological contamination of food. The materials used for food processing equipment should be easily cleanable. For effective cleaning and sanitation, all parts of the equipment should be readily accessible. Another way to improve equipment hygiene is by applying antimicrobial coatings on equipment parts.

✓ *Reactive rather than routine/predictive maintenance:* Most of the times the maintenance programmes are a knee jerk reaction to a breakdown. They are reactive in nature, i.e. "run it 'til it breaks." Reactive maintenance can result in food contamination before a failure is identified. If regular maintenance of equipment is not done then niches can develop or controls can become defective. For example, tea/coffee vending machines. It is recommended that they are thoroughly cleaned at the end of the day.

Containers for Waste and Inedible Substances

For making GMP work it is mandatory that containers for waste, by-products and inedible or hazardous substances, shall be specifically identifiable, suitably constructed and, where appropriate, made of impervious material.

Containers used to hold dangerous substances shall be identified and, where appropriate, be lockable to prevent malicious or accidental contamination of food.

Food Safety Implications

If containers with hazards materials/waste are not identifiable easily then unintentional mixing/cross contamination with food can take place compromising food safety.

Facilities (Water Supply)

An adequate supply of potable water with appropriate facilities for its storage, distribution and temperature control, shall be available whenever necessary to ensure the safety and suitability of food. Potable water shall be as specified in the latest edition of Guidelines for Drinking Water Quality (IS 10500) or water of a higher standard or as specified by applicable statutory or regulatory requirement. Non-potable water (for use in, for example, fire control, steam production, refrigeration and other similar purposes where it would not contaminate food), shall have a separate system. Non-potable water systems shall be identified and shall not connect with, or allow reflux into, potable water systems.

Food Safety Implications

Mixing of potable and non-potable water can lead to cross contamination and hence compromised food safety.

Other Facilities

Adequate drainage and waste disposal systems and facilities shall be provided. They shall be designed and constructed so that the risk of pest ingress contaminating food processing area or the potable water supply is avoided.

Food Safety Implications

Inadequate drainage can lead to breeding of pests and hence affect food safety.

Air Quality and Ventilation

Adequate means of natural or mechanical ventilation shall be provided in the manufacturing unit. Ventilation systems shall be designed and constructed so that air does not flow from contaminated areas to clean areas. Ventilation should be provided in particular to:

✓ Minimize air-borne contamination of food, for example, from aerosols and condensation droplets

✓ Control ambient temperatures,

✓ Control odours which might affect the suitability of food.

✓ Control humidity, where necessary, to ensure the safety and suitability of food.

Food Safety Implications

Adequate ventilation will help to control high humidity conditions. This will make the working environment comfortable for the workers inside the manufacturing unit. In such a well ventilated comfortable area workers are less likely to sweat or touch body parts and hence less chances of contamination of food.

Lighting

Adequate natural or artificial lighting shall be provided to enable the undertaking to operate in a hygienic manner. Where necessary, lighting shall not be such that the resulting colour is misleading. The intensity shall be adequate to the nature of the operation. Lighting fixtures shall, where appropriate, be protected to ensure that food is not contaminated by breakages. It is recommended that there shall be lighting of following intensity in various areas:

a. 500 lux (minimum) in working area

b. 110 lux (minimum) in storage rooms

c. 600 lux (minimum) in inspection areas

Food Safety Implications

It is difficult for the workers to identify any foreign material like metal fragments, flaking plaster, paint, debris, etc. in an environment without adequate lighting.

Maintenance, Cleaning and Sanitation

The food manufacturing unit shall establish effective systems to:

a. Ensure adequate and appropriate maintenance and cleaning

b. Control pests

c. Manage waste

d. Monitor effectiveness of maintenance and sanitation procedures

Establishments and equipment shall be kept in an appropriate state of repair and condition to facilitate all sanitation procedures. Cleaning shall remove food residues and dirt, which may be a source of contamination. The necessary cleaning methods and materials will depend on the nature of the food business. At certain places disinfection may be necessary after cleaning. Cleaning chemicals shall be handled and used carefully and in accordance with manufacturers' instructions and stored, where necessary, separated from food, in clearly identified containers to avoid the risk of contaminating food.

Cleaning shall be carried out by the separate or the combined use of physical methods, such as heat, scrubbing, turbulent flow, vacuum cleaning or other methods that avoid the use of water, and chemical methods using detergents, alkalis or acids. Cleaning procedures should involve removing gross debris from surfaces, applying a detergent solution to loosen soil and bacterial film and hold them in solution or suspension. This shall be followed by rinsing with water, to remove loosened soil and residues of detergent. Some processes are dry wherein entry of water can cause spoilage/infections and therefore need dry methods of cleaning such as flushing with sugar or salt, after brushing down or vacuum cleaning to remove carry over flavours. Disinfection can be by use of alcohol based non-toxic microstat/cidals.

Cleaning and disinfection programmes shall ensure that all parts of the establishment are appropriately clean, and shall include the cleaning of cleaning equipment. These procedures shall be continually and effectively monitored for their suitability and effectiveness and documented.

Food Safety Implications

Biofilms are a serious food safety hazard in any manufacturing unit. These occur when bacteria form a slime layer upon a surface and provide an environment for pathogens to proliferate. This adhesion of pathogenic bacteria to a biofilm is a microbial safety hazard because the biofilm can detach and become a significant source of food contamination. This can be handled by cleaning to remove biofilms. However, attached bacteria may survive conventional cleaning methods. Most of the times there is a mind set to take the cleaning step lightly in routine thinking that the next step of sanitizing will take care of all pathogens. Hence, adequate cleaning preferably scrubbing prior to sanitizing is of paramount importance to control this problem. Another method to tackle this issue is coating drains and equipment parts with antimicrobial material that can counteract biofilms. However, this does not eliminate the need for proper cleaning and sanitizing.

Older Equipment

Effective cleaning is paramount to controlling allergen contamination. As we learnt under microbial hazards all parts of the equipment should be readily accessible and visible for cleaning and sanitation. Further, equipment surfaces should not harbor allergens. Thus, sanitary equipment design is necessary to ensure proper removal of allergens from equipment.

Poorly Maintained Equipment and Lines

Pieces of equipment can break off and enter food products during processing if equipment is poorly maintained. Routine or preventive maintenance and other periodic checks of equipment can minimize the risk from this safety issue. Risk is further minimized with the use of metal detectors and X-ray machinery. Proper calibration of equipment and minimizing contact between pieces of machinery is also helpful.

Lighting Fixture/Other Glass Breakage

Glass can be controlled by having a glass breakage policy, such as throwing away all food and containers within 10 feet of the incident. Light fixtures can be protected so that if they break, the glass does not spill out. Capping equipment should be properly calibrated and lines should be monitored for evidence of glass breakage. X-ray technology can also be helpful in identifying glass pieces in food.

Pest Control Systems

Pests pose a major threat to the safety and suitability of food. Pest infestations can occur where there are breeding sites and a supply of food. Good hygiene practices shall be employed to avoid creating an environment conducive to pests. Good sanitation, inspection of incoming materials and good monitoring shall be implemented to minimize the likelihood of infestation and thereby limit the need for pesticides.

Buildings shall be kept in good repair and condition to prevent pest access and to eliminate potential breeding sites. Holes, drains and other places where pests are likely to gain access shall be kept sealed. Animals shall, wherever possible, be excluded from the grounds of factories and food processing plants. It is recommended to guard openings like windows, exhaust fans with fly-proof mesh and provide double doors or fix strip curtains or air curtains on entrance.

The availability of food and water encourages pest harbourage and infestation. Potential food sources shall be stored in pest-proof containers and/or stacked above the ground and away from walls. Areas both inside and outside food premises shall be kept clean. Where appropriate, refuse shall be stored in covered, pest-proof containers.

Establishments and surrounding areas shall be regularly examined for evidence of infestation. Pest infestations shall be dealt with immediately and without adversely affecting food safety or suitability. Treatment with chemical, physical or biological agents suitable for use in food and beverage industry shall be carried out without posing a threat to the safety or suitability of food. All Chemicals used for Pest Control or Cleaning/Sanitation shall be supported by "Material Safety Data Sheets (MSDS)".

Waste Management

Suitable provision shall be made for the removal and storage of waste. Waste shall not be allowed to accumulate in food handling, food storage, and other working areas and the adjoining environment. Waste stores shall be kept appropriately isolated and clean and regularly inspected. Waste bins used should be self closing with foot-operated lids.

Sanitation systems shall be monitored for effectiveness, periodically verified by means such as audit preoperational inspections or, where appropriate, microbiological sampling of environment and food contact surfaces and regularly reviewed and adapted to reflect changed circumstances.

Personal Hygiene Facilities and Toilets

Personal hygiene facilities shall be available to ensure that an appropriate degree of personal hygiene can be maintained and to avoid contaminating food. Such facilities shall be suitably located and should include:
✓ Adequate means of hygienically washing and drying hands, including wash basins and a supply of hot and cold (or suitably temperature controlled) water.
✓ Lavatories of appropriate hygienic design; and adequate changing facilities for personnel.

Product Information and Consumer Awareness

The organization shall determine requirements for delivery and post-delivery activities, statutory and regulatory requirements related to the product. All raw material, ingredients and product-contact materials shall be suitable so as to meet food safety requirements.

As per specifications of incoming material, those known to contain parasites, undesirable micro-organisms, pesticides or other toxic, decomposed, extraneous matter, which would not be reduced to acceptable levels through sorting and or processing, shall not be accepted. All incoming material shall be inspected and sorted before processing. Where necessary, laboratory tests shall be made to establish fitness for use. Stocks shall be subject to effective stock rotation. All records in respect of raw materials shall be maintained. Rejected material if required, be stored away from accepted material.

Raw material shall be used in first-in–first-out (FIFO) and FEFO (First expiry–first out) basis according to plant specified product rotation/inventory control schedule. Raw material shall be stored at temperatures that maintain product condition. Frozen material to be kept frozen if required. The package pallet integrity must be maintained throughout storage period to maintain condition of material. Product identity in storage should allow for in plant tracking system.

All food products shall be accompanied by or bear adequate information to enable the next person in the food chain to handle, display, store, prepare and use the product safely and correctly. Flow diagrams shall be prepared for the products or process categories covered by the GMP system. Flow diagrams shall provide a basis for evaluating the possible occurrence, increase or introdduction of food safety hazards.

The characteristics of end food products shall be described in documents to the extent needed to ensure safe and quality food including information on the following as appropriate product name or similar identification, composition, biological, chemical and physical hazard specification and allergens relevant for food safety, intended shelf life and storage conditions.

The packaging design and materials shall provide adequate protection for products to minimize contamination, prevent damage and accommodate appropriate labelling. The material shall be non-toxic so as not to pose a threat to safety and suitability of food under the specified conditions of storage and use. Where appropriate, reusable packaging shall be suitably durable and easy to clean and be able to disinfect where necessary.

Control of Operations

FBOS shall produce food as per its specification and reduce the risk of unsafe food by ensuring control of operations. Apart from building design and equipment process controls needs to be established. For example, temperature control. Based on the nature of food operations undertaken adequate facilities shall be available for heating, cooling, refrigeration and freezing food, for storing refrigerated or frozen foods, monitoring food temperatures and when necessary, controlling ambient temperatures to ensure safety and suitability of food. Temperature recording devices shall be calibrated at stipulated intervals. Specific process steps which may contribute to food hygiene are chilling, thermal processing, irradiation, drying, chemical preservation, and vacuum or modified atmosphere packaging.

Prevention of Microbiological Cross Contamination

Raw unprocessed food shall be effectively separated either physically or by time, from ready to eat foods, with effective intermediate cleaning and where appropriate disinfection. Access to processing areas may need to be restricted or controlled. Where risks are particularly high, access to such processing area shall be made through a changing facility. Personnel may need to put on protective clothing including footwear and wash their hands before entering. Surfaces, utensils, equipment, fixtures a fitting as applicable, shall be thoroughly cleaned and where necessary disinfected after raw food, particularly meat and poultry has been handled or processed.

Food safety implications: Fruits can contaminated by direct or indirect contact with animal feces. Various studies have shown that pathogens can infiltrate fruit through damaged or decayed areas or through the flower end of the fruit. To overcome these problems best control practices—such as not using dropped fruit, removing damaged fruit, and washing/brushing fruit prior to processing or by pasteurization. These are important in industries like pickle and juice industry.

Many pathogens, like *E. coli* and *Salmonella*, enter the food processing environment through raw materials contaminated with those pathogens. Manufacturing units which use animal-derived products or products at risk of cross-contamination by animal feces are more at risk. Such hazards can be minimized by minimizing the risks of raw material contamination (i.e., ensuring that raw material suppliers comply with good agricultural practices) and others (i.e. irradiation, pasteurization). In case of meat or poultry chilling of raw material, spray-washing/warm water wash, steam vacuuming/pasteurization, feed ingredient control spraying of chicks, use of starter culture can be done. For dairy products pasteurization is a fool proof method to eliminate any kind of contamination. Vegetables and fruits can be washed with chlorine water/ozone treatment/scrubbing to remove contaminants. In case of eggs shell pasteurization/washing/spraying of chicks/feed ingredient control/use of salmonella-free chicks are few options.

Sources of foreign matter in raw materials can include nails from pallets and boxes, ingested metal from animals, harvesting machinery parts, elements from the field, veterinary instruments, caps, lids, closures, etc. Foreign matter in raw materials can be controlled with raw material inspections. X-ray technology is also available to examine incoming material.

Physical and Chemical Contamination

Systems shall be in place to prevent contamination of foods by foreign bodies such as glass or metal from machinery, dust, harmful fumes and unwanted chemicals. In manufacturing and processing, suitable detection and screening devices shall be used as necessary.

The food products at any stage during its manufacture should be stored under appropriate conditions to maintain its safety and suitability. Adequate facilities for the storage of food, ingredients, and non-food chemicals (e.g. cleaning materials, lubricants, fuels) shall be provided. Where necessary, separate, secure storage facility for cleaning materials and hazardous substances shall also be available.

Transport

Products shall be adequately protected during transport. The type of conveyances or containers required depends on the nature of food and conditions under which it is to be transported. Where necessary, conveyances and bulk containers shall be designed and constructed so that they do not contaminate foods or packaging, can be effectively cleaned and where necessary disinfected. It should permit effective separation of different food or food from non-food items during transport, provide effective protection from contamination including dust and dirt. There should be maintenance of effective temperature, humidity and other conditions necessary to protect food from harmful or undesirable micro-organisms and deterioration likely to render it unsuitable for consumption.

Conveyances and containers for transportation of food shall be maintained at an appropriate state of cleanliness repair and condition. Where the same conveyance is used for different food or non-food items, effective cleaning and where necessary, disinfection shall be in place between loadings. Where appropriate, these should be designated and marked for food use only and used for that purpose only.

Post-processing contamination: Products can also be contaminated if the post-processing environment, utensils, or equipment have been contaminated with a pathogen. This issue is especially relevant to the pathogen *Listeria monocytogenes*, due to its hardiness and pervasiveness in the environment. Effective controls against post-process contamination include eliminating the pathogen from the post-processing environment by using environmental sampling to eliminate niches, effective sanitation, and various in-package pasteurization methods. Contamination can be prevented by application of HACCP, regular environmental sampling and testing, in-package steam hot water treatment, pasteurization, irradiation or use of preservatives.

Traceability

The FBO shall ensure that effective traceability procedures are in place from raw material to finished products and to the consumer as appropriate, so as to deal with any food safety hazard and enable the complete, rapid recall of any implicated lot of product from market. The organization shall identify product status with respect to inspection and testing.

Traceability records shall be maintained for a defined period for system assessment to enable the handling of potentially unsafe products and in the event of product withdrawal. Records shall be in accordance with statutory and regulatory requirements and customer requirements and may, for example, be based on the end product lot identification.

Mistaken identity of pesticides: Food can become contaminated with pesticides if pesticide container labels are misread or when products are stored in containers that have had another use. Bihar mid day meal tragedy which took away lives of 23 children happened by using pesticide instead of cooking oil. The best way to control the risk of mistaken identity is to store pesticides away from food ingredients, keep an inventory of pesticides, and store the products in their original containers.

Corrosion of metal containers/equipment/utensils: Metal poisoning can occur when heavy metals leach into food from equipment, containers, or utensils. When highly acidic foods (e.g. citrus fruits, fruit drinks, fruit fillings, tomato products, or carbonated beverages) come into contact with potentially corrosive materials, the metals can leach into the food. One solution to the problem is to use appropriate, non-corrosive materials in food processing.

Residue from cleaning and sanitizing: If equipment and other food handling materials are not rinsed well, then residue from detergents, cleaning compounds, drain cleaners, polishers, and sanitizers can contaminate a food product. This problem can best be controlled by properly training personnel about cleaning and sanitizing.

Accidentally adding too much of an approved ingredient. Some substances, such as preservatives, nutritional additives, color additives, and flavor enhancers, are intentionally added to food products. But adding an approved ingredient in inordinate amounts by accident—such as adding too much nitrite to cured meat—can result in a toxic product. To tide over such problems it is recommended that nitrite be stored in a locked cabinet and weighed and bagged separately before being added to any product. Nutritional safety issues can also arise when product labels' nutrition information is incorrect. Thus, it can be dangerous to public health when too little or too much of a specified nutrient is added. For example, malnutrition can occur if infant formula does not deliver the expected nutrient content during its shelf life. There are also many examples of nutritional food safety issues arising when too much of a nutrient gets added to a product unintentionally. For example, some vitamins that are added to fortified foods (such as vitamin A) are known to be toxic at high doses. Another example of iron, a necessary dietary component, can cause severe illness and death if too much is ingested. Controlling chemicals by keeping an inventory of additives minimizes the occurrence of this type of contamination.

Natural toxins: Food can be contaminated with naturally occurring chemicals that cause disease. Toxins such as mycotoxins and marine toxins are naturally produced under certain conditions. Given that these toxins generally occur in raw materials, especially crops and seafood, manufacturers should require suppliers to certify that the products they purchase are free from natural toxins.

There is a wide range of other issues related to the safety and wholesomeness of food in addition to GMPs. These should be considered in addition to the problems identified at the food processing level when evaluating the effectiveness of food GMPs. For effective implementation of GMPs there are 10 basic principles which are record keeping, follow of procedures, traceability, designing facilities and equipment, maintaining facilities and equipment, validating work, job competence, cleanliness, component control and auditing for compliance.

GMP System Verification

The organization shall conduct internal audits at planned intervals to determine whether the GMP system conforms to the planned arrangements and to the requirements of this Indian Standard, and is effectively implemented and updated.

An audit programme shall be planned, taking into consideration the importance of the processes and areas to be audited, as well as any updating actions resulting from previous audits. The audit criteria, scope, frequency and methods, shall be defined.

Top management shall ensure that the GMP system is continually updated. In order to achieve this, the GMP team shall evaluate the GMP system at planned intervals.

Withdrawals

A stable system for withdrawal helps to facilitate the complete and timely withdrawal of lots of end products which have been identified as unsafe. The management shall appoint personnel having the authority to initiate a withdrawal and personnel responsible for executing the withdrawal. There should be a documentation procedure in place for withdrawals and the sequence of actions to be taken.

Withdrawn products shall be secured or held under supervision until they are destroyed, used for purposes other than human consumption, determined to be safe for human consumption, or reprocessed in a manner to ensure their safety.

GMPs aim at having a quality approach to manufacturing and reducing contamination in food processing. These are comprehensive and address issues like record keeping, personnel qualifications, sanitation, cleanliness, equipment verification, etc. The requirements are open ended and allow manufacturers to decide how best to implement necessary controls. They are not only flexible but also require that the manufacturer interpret the requirements suited to his business. The top management of the food business shall ensure that requirements which are applicable to the nature of food operations, are complied with in addition to those applicable statutory and regulatory requirements.

17

Safe Transportation of Food: The Spokes of Food Safety System

Neha Chanana, Mamta Bansal, Puja Dudeja, Amarjeet Singh

It is a common sight to see trucks loaded with fruits, vegetables, poultry, grains or other food products on the city roads or train wagons carrying grain stocks or fruits to long distances across the country.

To make food accessible for the population residing away from the farms transportation of food is necessary. Moreover, the type of crops and animal food production best suited for a region depend on its climate, topography, soil and other factors, e.g. the famous apples from Jammu & Kashmir can be enjoyed by people staying in some other part of the country, say Delhi, or even other countries like UK/USA.

But what if apples from J & K decay on the way and rotten apples reach Delhi.

No matter what the quality of food may be at the time of harvest, temperature and handling during transportation affect safety of food.

Thus, transportation of food is one of the most delicate links in the farm to table chain. Improper transportation can make safe food also unsafe for consumption. Therefore, safe transportation of food items is crucial.

Several factors need to be monitored while the food is on the move from one place to another. The temperature of the food items, distance travelled duration of transportation, storage of food items as well as the packaging of food need to be checked for ensuring safe transportation of food products. Moreover, to keep food safe during transit, the vehicles by which food is transported also needs attention.

Different Modes/Vehicles of Transportation

Generally, the raw materials are transported in trains, planes, ships or big vehicles like trucks, trolleys, etc. from the farm to the cities from where these are further shifted to the *mandis*. After procurement from the market the raw food items are transported in small vehicles like auto, rickshaw, tempo, cars, etc. to the retailers/supermarkets in AC malls/ kitchen of restaurants, hospitals, etc. The further transit of food along the supply chain from the retailer to our homes is done through hawkers in open carts and *thellas*. The food from supermarkets in AC malls is generally brought either manually/2 wheelers/4 wheelers like auto or personal cars (Figs 17.1 to 17.3). On the other hand, for non vegetarian food, poultry is usually transported in crates from the farm where they are reared to the place of slaughter.

Moreover, with urbanization, the home delivery two wheelers of various restaurants

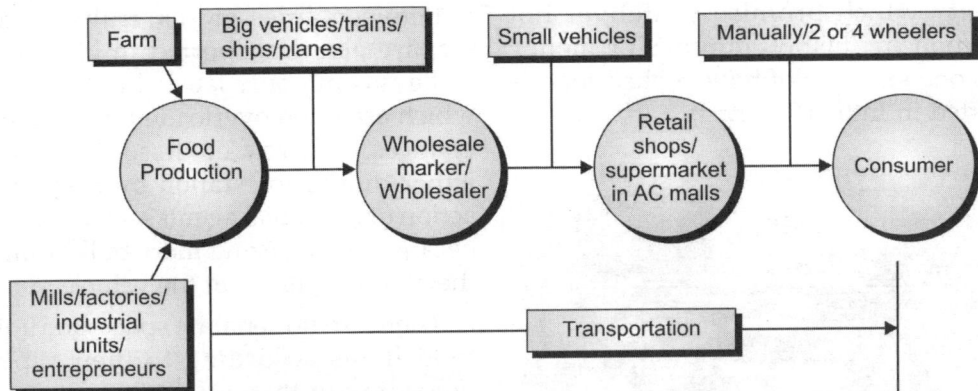

Fig. 17.1: Transportation of raw material

Fig. 17.2: Vehicles/ manual transportation of raw food

Fig. 17.3: Transport of poultry

in the city which promise to deliver hot cooked food in "just 30 minutes or money back" is one example of how cooked food is transported in India (Fig. 17.4).

Fig. 17.4: Home delivery vehicle

Present Food Transportation Scenario in India

In India, food items are transported through all modes of transportation utilizing almost every kind of vehicle.

A lot of food items are imported and exported through airways and maritime. The famous dates of dubai available in the market are imported using planes and ships as a means of transportation.

For longer distances within the country, railways is used as the mode of transport wherein the goods train popularly known as *maal gaadi* carries raw materials like grains and raw vegetables like potatoes, onions, etc. However, these food items are sometimes carried in the same bogie in which animals are transported. All the more sometimes, grains/ onions/potatoes are loaded in a separate boogie in passenger trains as well. These trains are not cleaned and disinfected properly after each journey because of which food items/ grains get contaminated with pests and insects. Moreover, there is no temperature control mechanism for food items transported over long distances in trains.

Between cities grains/fruits/vegetables, etc. are also transported in big trucks and trolleys commonly labeled as 'goods carrier' which are often overloaded due to which the chances of decay of these items increases either due to infestation by pests or by the action of microbial agents and sometimes also because of squeezing more bulk compared to the space available in the vehicle.

There are no separate spaces for unloading food items according to their type. After unloading in the *mandis,* the dairy products like *paneer* are kept with fruits and vegetables which again provide optimal conditions for decaying of these food items.

Within the city, when fruits and vegetables are transported in open carts, rickshaw and *thellas,* there are no means of temperature control so as to maintain freshness of these fruits and vegetables. When the temperature is optimal for the growth of microbial agents especially in summers and rainy season, these food items decay quickly while on the *thella* itself even before it reaches the consumer. For example, the *paneer* available with the vegetable vendor on the *thella* is kept usually in a metal box. In case of home delivery of cooked food from restaurants, the metal box which is expected to keep food hot during transport is seldom cleaned. Moreover, the cleanliness of hot and insulated bags in which pizzas are delivered is also likely to be of doubtful quality.

Thus, during transportation, the quality of food items may deteriorate as they are exposed to various underlying risk factors. Few of these are listed below:

In-transit Risks

i. Improper refrigeration or temperature control of food products especially perishable food items like meat, chicken, dairy products.

ii. Improper management of transportation units or storage facilities to preclude cross contamination

iii. Improper packing or storage facilities, including incorrect use of packing materials during transportation.

iv. Poor pest control in transportation units or storage facilities

v. Improper loading practices, conditions or equipment. For example, loading of different kinds of foods together like loading fruits and vegetables with dairy products and chicken without covering, or carrying food items in the same vehicle in which an animal is being transported.

vi. In case food items are imported/exported across cities or countries, they decay if left unattended for days together at the loading or unloading dockyards/rail yards. These conditions make food items susceptible to decay.

vii. Poor transportation unit design, construction and maintenance.

Organizational Risks

i. Lack of driver/employee training and/or supervisor/manager/owner knowledge of food safety.

ii. Poor employee hygiene.

These risk factors increase the chances of chemical contamination as well as growth of microbial agents and other pests which under ideal conditions of their growth deteriorate the quality of food items. These risk factors have been neglected over time by the authorities and make food unsafe for consumption at the receiver end. In this context, certain preventive measures need to be taken so as to avoid food from becoming spoiled during transportation. Some of the suggested measures are described below.

PREVENTIVE MEASURES

In view of both in-transit or organizational risk factors, adopting certain measures would help in making transportation safe for all kind of food items.

A. In-transit Measures

General Measures

Following measures are to be taken during loading, unloading and movement of the food items. However, the shipment or the cargo should be carefully examined to be in good condition by the transporter before uploading.

i. The loading and unloading areas should be cleaned, disinfected (where appropriate), configured and properly maintained to prevent product contamination. These areas should be designed in such a way that cleaning becomes accessible to all areas. Moreover, an effective, systematic program for preventing environmental contamination and infestation by insects needs to be established for these loading and unloading areas.

ii. The vehicle should not be used to transport anything other than food, especially carrying animals or chemicals. In case of bulk transport, vehicles should be designated and marked *"for food use only"*.

iii. Keep loading time as short as possible to prevent temperature changes (increases or decreases) that could threaten the safety or quality of food products.

iv. If feasible, vehicle should be restricted to a single commodity. This reduces the risk of cross contamination from previous or other carried cargoes.

v. The vehicles used for transportation of foods must be maintained in good repair and kept clean. They should be leak proof and rust resistant.

vi. Time required for transportation should be minimum

vii. Stops on the way should be avoided.

viii. All foods during transportation must be kept covered and in such a way as to limit pathogen growth or toxin formation by controlling time of transportation, exposure, temperature control and using safe water for cleaning, etc.

ix. Proper labeling and/or signage and/or transporter instructions should be ensured.

x. Instructions and contact information should be clearly mentioned with the shipment. For example, "Store in a cool dry place", etc.

xi. Direct handling of food should be minimal. It should be ensured that utensils or containers used for transporting food are clean and sanitized properly.

Specific measures: Proper transportation of food items is dependent on the kind or type of food being transported. Raw material transportation will be different from transportation of cooked food.

Raw Food Items

i. Perishable food/frozen items:

a. **Loading/unloading areas** should have adequate temperature control capacity.

b. **Vehicles:** Refrigerated trucks (Fig. 17.5) carrying perishable food items need be pre-cooled for at least 1 hour before loading to remove residual heat caused by the insulation and inner lining of the trailer. For pre-cooling, the doors should be closed and the temperature setting of the unit should be no higher than 26°F. The doors of refrigerated trucks need to be closed so that there are no air leaks.

c. **Temperature maintenance during transportation:** Frozen items, e.g. for meat, chicken and dairy products, temperature maintenance needs special care. The temperature should be <4°C.

Warm meat meant for immediate sale need to be transported in hygienic and sanitary condition in clean insulated containers with covers (lids) to meat shops or selling units.

Fig. 17.5: Refrigerated truck carrying food

ii. The tanks used for carrying raw milk must be cleaned and disinfected after every use.

iii. Sealed/packaged/tinned like jam, ketchups, etc. and bottled beverages/water should be transported such that their seal remains intact and undamaged.

iv. Appropriately packaged products should have sufficient air circulation to maintain the temperature.

If different kinds of raw food items are to be transported in the same vehicle such as meat or dairy products with fruits and vegetables, then all these items should be covered in plastic bags so as to avoid cross-contamination.

Cooked Food

Transporting cooked food includes deliveries from restaurants, cafes, take away outlets, superfast short distance trains, food served in air flights, weddings, religious occasions, etc. In such situations, food should be completely covered. If distance is more, insulated packaging or containers can be used. Here again, the temperature control of the food transported should be taken into account.

i. Cooked food to be served hot should be kept at a temperature of at least 60°C to prevent microbial growth.

ii. Cooked food to be served cold should be kept below 5°C to prevent growth of pathogens.

iii. The home delivery box on 2 wheelers, that transports ordered cooked food from the restaurants/cafes to homes, need to be cleaned after every delivery so as to remove any food item or oil in the box.

Transportation of Poultry

Poultry has to be transported from the farm to slaughter houses, the poultry crates should be in good repair. There should be no crate/cage damage that would cause any injury to poultry or allow crates to open accidentally. Transport crates should not be over-filled and enough space should be provided to allow all poultry to lie-down.

B. Organizational Measures

i. There needs to be appropriate documentation accompanying each load (i.e. time of dispatch of the cargo, time in transit and time of receiving, tanker wash record, seal numbers, temperature readings, etc.)

ii. A documentation of the food received especially food spoiled during transportation needs to be maintained.

iii. The Standard Operating Procedures (SOPs) or Instructions for each designation involved in transportation like the driver, loader, unloader, etc. needs to be available handy. It should be ensured that appropriate implementation of these SOPs/instructions is done.

iv. Vendor or food transporter certification programs can be organized frequently by Food Standards and Safety Authority of India (FSSAI) wherein the Food Business Operators (FBOs) involved in transportation can be updated on the methods of safe transportation of food. Such certificate programmes can be made mandatory for renewal of transporters' licenses.

v. There should also be awareness and training programmes for drivers on food handling during transportation that will further help in transporting food safely.

vi. Good communication between shipper, transporter and receiver would cut down on the errors that can be made while transportation and thus make transit of food safer. For example, the guidelines on refrigeration of frozen food items issued by the shipper if propagated properly through the supply chain would cut down on the errors that could spoil frozen food during transportation.

vii. Frequent monitoring of the transporting system by Food Safety Officer (FSO) needs to be incorporated as a measure for improving the transport system of food items.

viii. The transporter should be penalized if food is found to be spoiled during transportation because of his ignorance.

The food we consume travels miles before it reaches us. The milk, cheese, bread, meat, rice, flour, pulses, fruits and vegetables that make our meals for the day and provide us energy travel long distances sometimes in trucks, ships, trains, and planes also before it is served on our tables. Thus, chances of consumption of unsafe food and also wastage due to decaying can be reduced through good food transportation practices.

18

Safety of Food in Retail Shops

Ishwarpreet Kaur, Mamta Bansal, Puja Dudeja

Ensuring safety of raw food and processed food products is important as it is liable to bacterial, chemical and physical contamination/spoilage. Hygiene and safety standards must be maintained at every step of food journey from farm to fork (Fig. 18.1). Once the food item is harvested or processed its journey towards consumption starts. All the parties involved during this transit have to consider the possible hazards and risks of food safety. Ideally everyone in between the chain must take balanced responsibility in maintaining food safety. However, retailers have a dual responsibility. They have to ensure safety of food both while receiving from wholesaler as well as at the time of sale to the consumers. Any shortfall can lead to contamination of the food. Generally people get their food supply from retail stores. Only less than 1% ever get it directly from farms or wholesalers. So, practically the consumer has a direct chance of inspecting the safety standards maintained by the retailers only before buying food stuffs.

Retail stores: The word 'retail' is of French origin that refers to 'sale in small quantities'. Retail stores are business establishments that sell goods and merchandise to consumers or other businesses. Retail marketing is comprised of the activities related to selling products directly to consumers through

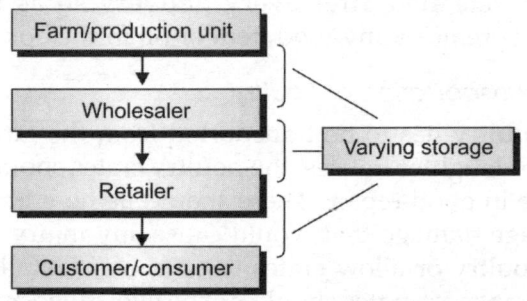

Fig. 18.1: Journey of foods from farm/factories to customers' plate

channels such as stores, malls, kiosks, vending machines or other fixed locations.

Grocery store: It is a retail store that primarily sells food. Traditional grocery retail stores or 'kirana' shops offered non-perishable foods only. A grocer used to deal in comestible dry goods such as spices, grains, sugar, tea, coffee, etc. (Fig. 18.2).

But new and modern grocery stores offer wide range of food products including perishable foods like fresh produce, meat, poultry, sea food, dairy products, delis, bakeries, and frozen foods. These stores vary in size and style from convenience stores to mega marts. The scale of stores also varies according to the location like metros, cities, and villages. These are:

Fig. 18.2: Traditional style grocery shops

a. **Green grocer:** It is a small grocery store that mainly sells fruits and vegetables.

b. **Convenience store:** It is a small store that stocks a range of everyday items such as groceries, snack foods, candy, toiletries, soft drinks, and newspapers. For example, Reliance fresh (Fig. 18.3).

c. **Health food store:** It is a type of grocery store that primarily sells health foods, organic foods, local produce, and often nutritional supplements. Nowadays private companies like Fortis hospital has their own retail health stores and even herbal/ayurvedic health food are sold at outlets under the brand name Patanjali (Figs 18.4a and b). Apart from this government run 'Khadi' shops also sell health food products.

d. **Supermarket:** It is a large form of the traditional grocery store. It is a self-service

Fig. 18.3: Modern grocery shops

Fig. 18.4a: Showing Patanjali shop sign board

Fig. 18.4b: Patanjali food products

shop offering a wide variety of food and household products, organized into aisles. The supermarket typically comprises meat, fresh produce, dairy, baked, canned and packaged goods and other various non-food items such as kitchenware, household cleaners, pharmacy products and pet supplies. Examples of well known store are Subhiksha, Walmart, etc.

e. **Hypermarket:** It is a superstore combining a supermarket and a department store. The result is an expensive retail facility carrying a wide range of products under one roof, including full groceries lines and general merchandise.

Importance of food safety in retail shops: Shift in grocery retail trends from non-perishable to perishable food items have necessitated the need for good food safety standards and practices. Irrespective of the type or location of grocery store, safety of food is of utmost importance. The aim is to avert any possible food hazards and to safeguard food quality. Trained store operators and managers can help maintain food safety and prevent any lapse in food quality and hygiene standards that can compromise the health of customers. Contaminated and spoilt food can result in FBI. Any such incidents of will bring bad name and affect the reputation of the retail store. Due to new stringent laws the retail operator at fault can also face penalty and punishment for selling adulterated, expired date items and substandard foods.

Critical control points specific to operations of retail establishments: The three major risk points identified in retail establishments that can produce greatest food hazards are:

1. Related to sourcing unsafe food source
2. Related to storage and processing—improper storage or holding time and temperature, unhygienic and unsanitary premises
3. Related to cross contamination—contaminated containers and equipment, poor personal hygiene of personnel

To fight back the risk of FBI it is vital to keep a strict vigilance over the source from where food is procured. Poor quality, adulterated, or contaminated food will further deteriorate during the lag time at grocery store and cause major health threats. Unhygienic handling of food and surroundings can result in cross contamination. Poor storage condition will result in fast multiplication of bacteria and concurrent spoilage of food.

Measures taken for ensuring control over critical control points and thereby maintaining food safety: To achieve safe standards in retail sectors some basic good retail practices can be followed. According to the Food Safety and Standard regulations (FSSR) 2011, retailers need to ensure that food sold is safe to eat and they are following standard system of food and environment hygiene. Some of the good retail practices are as follows:

1. **Procure supplies from approved source or authorized dealers:** It is recommended that the supplies should be bought from reliable and reputed sellers who possess food business license and follow good food

safety system like HACCP. Even small grocery store/kirana shops and convenience stores must follow minimum level of food safety and hygiene code.

2. **Proper checks at receiving area**

   a. Trailer inspection—trailers in which food items are delivered should be examined and must be suitable for carrying food. It must be clean and safe to use.

   b. Product inspection—it must be done before accepting the delivered lot. The samples of food items must be checked for contamination, ripped or damaged packages, off odours, excess purges. Any doubtful food item should be both returned to the manufacturer or destroyed and documented in a log book.

   c. Special attention must be paid while unloading perishable food items. For example, meat and poultry products should be promptly moved into the refrigerator or frozen storage.

   d. Hygiene and adequate sanitation must be ensured at loading area. The area must be cleaned periodically and waste disposed off efficiently.

   e. Records of receipts and invoices from whom items have been purchased must be maintained and retained. In case of any problem the manufactures can be traced.

3. **Storage conditions (Time and temperature control) and display**

   a. Storage conditions in retail stores are to be hygienic and safe. The area should be clean and away from washrooms, garbage, locker rooms or chemical storage rooms. The wooden boxes for storage must be routinely inspected for loose nails or broken boards. The amount of items stored has to be strictly in accordance with space available.

b. Box placement—no food item should be placed on floor. The minimum height at which food items are to be kept is 6 inches above the ground (Fig. 18.5).

Grocery store has display coolers and freezers (Fig. 18.6). Even in chest freezers storage boxes are not to be placed against the wall or directly on floor. For sufficient air flow make use of pallets and maintain approximately a distance of 4 inches (10 cm) between the products and cooler walls. For air space between boxes use dividers between layers of pallet. Always remember that food items placed higher above or close to the door will encounter warmer temperatures.

In India generally the business operators in small retail stores do not maintain

Fig. 18.5: Food items to be stored at least 6 inches from ground

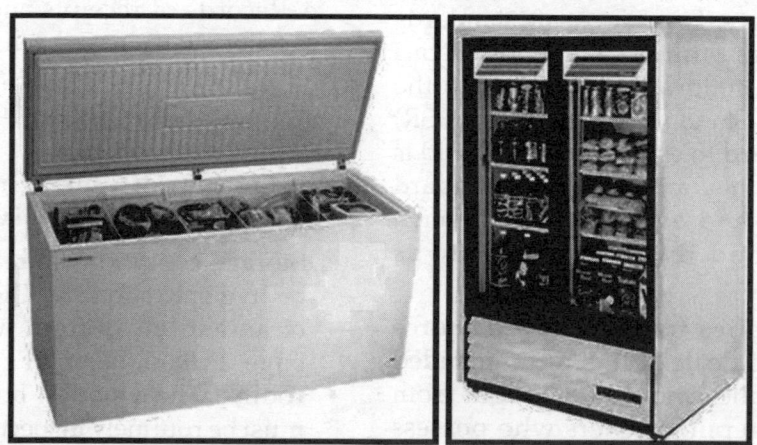

Fig. 18.6: Chest freezer and display coolers

hygienic conditions. They keep food boxes on floor and even outside the shops. It is common to see chest freezers and display coolers kept outside store in direct sunlight (Figs 18.7a and b). This is purposely done to publicise and attract attention as these coolers are provided by the manufacturing companies for storing their foods. But this is an unsafe habit as the functioning of coolers and freezers is greatly affected during harsh weather conditions.

Separate storage is required for vegetarian and non-vegetarian food items.

The cold storage cases must be as close to the billing counter so that they are the last items to be purchased to prevent temperature abuse.

c. Storage temperature and time—to keep food safe it is important to keep it out of temperature danger zone. The food temperature danger zone refers to temperature between 5° and 63°C. During this temperature zone the harmful bacteria can quickly grow on food and can make a person sick. High risk foods that hit this danger zone mean you should discard it. High risk or hazardous food products include meat, poultry, fish, sea food, dairy, chutneys, gravies and egg products. The temperature at which food items will be stored depends on the perishability of

Figs 18.7a and b: Chest freezers kept outside shops and in unhygienic surroundings

the products. Correct storage temperatures will ensure increased shelf life. For
packaged food items follow the instructions on the cover. Nonperishable foods
can stay good at room temperature while
perishable foods items must be kept at
below 4°C, preferably between –1° and
–2° C. Frozen meat temperature should
be –18°C while frozen fish must be kept
at –21°C. Storage time for foods in
refrigerator and freezer shown in
Table 18.1.

To prevent freezer burns the meat should
be properly wrapped. The temperature of

Category	Refrigerator (40°F/4.5°C or below)	Freezer (0°F /-18°C or below)
Raw eggs in shell	3 to 5 weeks	Do not freeze.
Fresh poultry	1 to 2 days	9 months
Fish	1–2 days	2–4 months
Fresh pork, beef	3–5 days	6–12 months
Pies Custard and chiffon	3 to 4 days	Do not freeze.
Hot dogs	1 week	1 to 2 months
Butter	1–3 months	6–9 months
Butter milk	7–14 days	3 months
Cheese (hard)	6 months	6 months
Cheese (soft)	1 week	6 months
Cottage cheese	1 week	Do not freeze
Milk	7 days	3 months
Yogurt	7–14 days	1–2 months
Canned vegetables	1 year	–
Frozen vegetables	–	8 months
Fruit juices in cartons	3 weeks	8–12 months
	Raw	Blanched/cooked
Salads	3 – 5 days	Do not freeze
Beans	3 – 4 days	8 months
Carrot	2 weeks	10–12 months
Celery	1 – 2 weeks	10–12 months
Lettuce	3 – 7 days	Do not freeze
Spinach	1 – 2 days	10–12 months
Squash	1 week	10–12 months
Tomatoes	2 – 3 days	2 months

Table 18.1: Storage time for foods in refrigerator and freezer

cases must be monitored using food grade thermometer thrice a day to ensure safe and shelf life. Display cases and freezers must be cleaned at least once a week and inspected routinely.

Frozen storage—for maintaining food safety in case of frozen foods it is important to emphasise on proper cold chain management. 'cold chain management' refers to maintaining appropriate temperature of foods from receiving through processing, transport, storage and retailing. Cold storage units/cabinets should be designed and operated so as to maintain a product temperature of –18°C or colder with a minimum of fluctuation. A critical temperature abuse situation may jeopardize food safety. To prevent such a situation it is important to entrust that cold storage units have the following:

1. Adequate refrigeration/freezer capacity. For better cooling do not load the freezer beyond its capacity. If possible alarm system should be installed that can warn the business operators about drop in temperature of freezers and refrigerators so that prompt action can be taken.

2. Stock placed in a manner so that the circulation of cold air is not affected

3. Rotation of products helps them to leave the cold store on a "first in-first out" basis.

4. Provision to control and record temperatures on a regular basis (using thermometers, temperature probes and/or recorders, temperature indicators and time-temperature indicators, etc.)

5. Prevent loss of cold air and introduction of warm and humid air; and leaks of any refrigerant are prevented.

6. The open display area is not subject to abnormal radiant heat (e.g. direct sunlight, strong artificial light or in direct line with heat sources); and never be stocked beyond the load line.

7. At the time of defrosting frozen foods should be moved during defrost cycles to a suitable cold store. It is also important to clean the freezers regularly.

8. In case of electricity failure prior provision of electricity back up should be available. If the product temperature exceeds 4°C for a significant period of time then the product must be disposed of. During light cut, which is very common in India, keep refrigerator and freezer doors closed as much as possible. Refrigerator can keep food cold for about four hours if its unopened. A full freezer can keep an adequate temperature for about 48 hours if the door remains closed. Once power is back check the temperature. If the freezer thermometer reads 40°F or below, the food is safe and may be refrozen. Records of these measurements should be kept. Hazardous foods should not be allowed to remain in the danger zone for more than a total of 2 hours.

9. For sampling of frozen food products for temperature measurement in retail display cabinets, one sample should be selected from each of three locations representative of the warmest points in the cabinets.

Since only expiry or best before product dates are not a sure shot guide for safe use of a product, consult to ensure food safety and quality (Tables 18.1 and 18.2).

Frozen food and power outages: Thawed or partially thawed food in the freezer may be safely refrozen if it still contains ice crystals or is at 40 °F or below. Partial thawing and refre-ezing may affect the **quality** of some food, but the food will be **safe to eat.** If you keep an appliance thermometer in your freezer, its easy to tell whether food is safe. When the power comes back on, check the thermometer. If it eads 40°F or below, the food is safe and can be refrozen.

Note: Always discard any items in the freezer that have come into contact with raw meat juices.

To evaluate each item separately use this chart (Table 18.2) as a guide.

Table 18.2: A guide to when to save and when to dispose the frozen food			
Food categories	*Specific foods*	*Still contains ice crystals and feels as cold as if refrigerated*	*Thawed and held above 40°F for over 2 hours*
Meat, Poultry, Seafood	Beef, veal, lamb, pork, and ground meats	Refreeze	Discard
	Poultry and ground poultry	Refreeze	Discard
	Variety meats (liver, kidney, heart, chitterlings)	Refreeze	Discard
	Casseroles, stews, soups	Refreeze	Discard
	Fish, shellfish, breaded seafood products	Refreeze. However, there will be some texture and flavor loss	Discard
Dairy	Milk	Refreeze. May lose some texture	Discard
	Eggs (out of shell) and egg products	Refreeze	Discard
	Ice cream, frozen yogurt	Discard	Discard
	Cheese (soft and semi-soft)	Refreeze. May lose some texture.	Discard
	Hard cheeses	Refreeze	Refreeze
	Shredded cheeses	Refreeze	Discard
	Casseroles containing milk, cream, eggs, soft cheeses	Refreeze	Discard
	Cheesecake	Refreeze	Discard
Fruits smell, or	Juices	Refreeze	Refreeze. Discard if mold, yeasty sliminess develops.
	Home or commercially packaged	Refreeze. Will change texture and flavor.	Refreeze. Discard if mold, yeasty smell, or sliminess develops
Vegetables	Juices	Refreeze	Discard after held above 40°F for 6 hours

Contd.

Table 18.2: A guide to when to save and when to dispose the frozen food (Contd.)

Food categories	Specific foods	Still contains ice crystals and feels as cold as if refrigerated	Thawed and held above 40°F for over 2 hours
	Home or commercially packaged or blanched	Refreeze. May suffer texture and flavor loss.	Discard after held above 40°F for 6 hours.
Breads, Pastries	Breads, rolls, muffins, cakes (without custard fillings)	Refreeze	Refreeze
	Cakes, pies, pastries with custard or cheese filling	Refreeze	Discard
	Pie crusts, commercial and homemade bread dough	Refreeze. Some quality loss may occur	Refreeze. Quality loss is considerable
Other	Casseroles—pasta, rice based	Refreeze	Discard
	Flour, cornmeal, nuts	Refreeze	Refreeze
	Breakfast items—waffles, pancakes, bagels	Refreeze	Refreeze
	Frozen meal, entree, specialty items (pizza, sausage and biscuit, meat pie, convenience foods)	Refreeze	Discard

Refrigerated food and power outages: Its food in the refrigerator safe during a power outage? It should be safe as long as power is out **no more than 4 hours**. Keep the door closed as much as possible. Discard any perishable food (such as meat, poultry, fish, eggs, and leftovers) that has been above 40 °F for over 2 hours. **Note:** Always discard any items in the refrigerator that have come into contact with raw meat juices. To evaluate each item separately use this chart (Table 18.3) as a guide.

Table 18.3: A guide to when to save and when to dispose the refrigerated food

Food categories	Specific foods	Held above 40°F for over 2 hours
Meat, Poultry, Seafood	Raw or leftover cooked meat, poultry, fish, or seafood; soy meat substitutes	Discard
	Thawing meat or poultry	Discard
	Salads: Meat, tuna, shrimp, chicken, or egg salad	Discard
	Gravy, stuffing, broth	Discard
	Hot dogs, sausage	Discard
	Pizza—with any topping	Discard
Cheese	Soft Cheeses	Discard

Contd.

Table 18.3: A guide to when to save and when to dispose the refrigerated food (Contd.)

Food categories	Specific foods	Held above 40°F for over 2 hours
	Hard Cheeses: Cheddar, Colby, Swiss, Parmesan, provolone, Romano	Safe
	Processed Cheeses	Safe
	Shredded Cheeses	Discard
	Low-fat Cheeses	Discard
	Grated Parmesan, Romano, or combination (in can or jar)	Safe
Dairy	Milk, cream, sour cream, buttermilk, evaporated milk, yogurt, eggnog, soy milk	Discard
	Butter, margarine	Safe
	Baby formula, opened	Discard
Eggs	Fresh eggs, hard-cooked in shell, egg dishes, egg products	Discard
	Custards and puddings, quiche	Discard
Fruits	Fresh fruits, cut	Discard
	Fruit juices, opened	Safe
	Canned fruits, opened	Safe
	Fresh fruits, coconut, raisins, dried fruits, candied	Safe
	Fruits, Dates	
Sauces, Spreads, Jams	Opened mayonnaise, tartar sauce, horseradish	Discard if above 50°F for over 8 hrs.
	Peanut butter	Safe
	Jelly, mustard, catsup, olives, pickles, soy	Safe
	Fish sauces, oyster sauce	Discard
	Opened vinegar-based dressings	Safe
	Opened creamy-based dressings	Discard
	Spaghetti sauce, opened jar	Discard
Bread, Cakes, Cookies, Pasta, Grains	Bread, rolls, cakes, muffins, quick breads, tortillas	Safe
	Refrigerator biscuits, rolls, cookie dough	Discard
	Cooked pasta, rice, potatoes	Discard
	Pasta salads with mayonnaise or vinaigrette	Discard
	Fresh pasta	Discard
	Cheesecake	Discard
	Breakfast foods—waffles, pancakes, bagels	Safe
Pies, Pastry	Pastries, cream filled	Discard

Contd.

Table 18.3: A guide to when to save and when to dispose the refrigerated food (Contd.)

Food categories	Specific foods	Held above 40°F for over 2 hours
	Pies—custard, cheese filled, or chiffon; quiche	Discard
	Pies, fruit	Safe
Vegetables	Fresh mushrooms, herbs, spices	Safe
	Greens, pre-cut, pre-washed, packaged	Discard
	Vegetables, raw	Safe
	Vegetables, cooked; tofu	Discard
	Vegetable juice, opened	Discard
	Baked potatoes	Discard
	Commercial garlic in oil	Discard
	Potato salad	Discard
	Casseroles, soups, stews	Discard

d. FIFO (first in – first out) or FEFO (first expiry – first out) Storage should be: done in organized manner that facilitates FIFO and FEFO rotation system. The products that will expire first are to be sold first. This will help customers to receive fresh and safe products. Products must be moved from storage to display in accordance to FIFO program. "Best before" dates of the products should be monitored daily to ensure any outdated product is not on display.

e. Routine package condition monitoring should be performed. The products must be withdrawn in case of any leakage, tear, purges, spoilage or if labels are missing or not readable.

4. **Personal hygiene, premises cleaning and sanitation practices:**

a. Sanitation: Cleaning of premises (floor, walls, windows, containers, equipments) must be done using appropriate chemicals. After cleaning, the surfaces should be sanitized to kill all the bacteria. While cleaning special care must be taken to prevent contact of food and packaging material with cleaning agents to prevent chemical hazards.

The garbage bins should have lids and windows and doors must have screens to prevent entry of pests and flies.

b. Basic personal hygiene with special emphasis to hand hygiene. Employees should not be allowed to work when ill. Any sores or cuts must be covered with dry, tight fitted bandages and gloves worn when hands are affected. Hair and beard nets are also recommended.

c. Pest control—Pest like cockroaches, weevils, rats, mice, fruit flies, etc. cannot only damage the food contents but can also spread diseases. It is important that all crevices and cracks in walls should be sealed. The food items should be kept covered and properly stored.

5. **Packaging**: A packaging material should be safe and free of any possible contamination as it comes in contact with the food products. Put raw food in individual plastic bags and pack them separately. In case store sells meat and poultry products, label the bag with the type of food it carries.

6. **Training and certification of managers:** The managers and the employees recruited should be professionally trained or should be given on the site training in food safety and hygiene. Staff with required skills and knowledge appropriate to their work must be employed to ensure that safety and quality of foods is not adversely affected during handling. They should also be aware of the importance of maintaining temperature control for frozen foods.

Customers' responsibility: An aspect that often receives insufficient attention is the consumer responsibility. Consumers should be knowledgeable about correct behaviour in-store, as well as their responsibility to extend the cold chain. Consumers often abuse food in retail stores, such as damaging packaging and products, and leaving cold chain products in other areas.

Consumers must ensure that they keep the products back in place from where they have picked instead of leaving them anywhere in case of no purchase. Buy cold or frozen food at the end of your shopping to keep raw meat, poultry and seafood cold. Refrigerate or freeze them as soon as you reach home from the grocery store. Keep your raw meat, poultry, fish and seafood away from other food in your grocery cart in order to prevent cross contamination from raw food to ready-to-eat food.

A brief Reminder to Safeguard Food during Storage

- Refrigerate or freeze perishables right away.
- Keep your appliances at the proper temperatures.
- Check storage directions on labels.
- Use ready-to-eat foods as soon as possible.
- Be alert for spoiled food.
- Clean the refrigerator regularly and wipe spills immediately.
- Keep foods covered.
- Check expiry dates.
- Do not store food under the sink.
- Check canned goods for damage.

19

Safety Aspects of Food Packaging

Ruchi Sharma, Sonia Puri, Alka Ahuja

Ever since ancient man felt the need for storing food for the next day's consumption, technology of food packaging in its primitive form emerged. Food packaging dates back to ancient times. Food was packed even with mummies. In prehistoric era, people hunted for their food and consumed the same soon after. Later they realized that they could keep their food longer if they protected it in some sort of packets or pouches. As he gathered experience with vulnerability of such protection, the concept was innovated with the formation of a container made from plant leaves using tiny needles of a tree stem. With the progress of civilisation, the concept of protecting food products was developed more and more in response to increasing needs. Hence, men started making packets from plant leaves, animal skins as well as other materials available to them. Later, he started using containers made from coconut shells and dried vegetable skins for storing water. Ceramic jar or earthen pots are still used in villages for packing or storing food items.

The first packages used were the natural materials that were available at the time which includes baskets of reeds, wooden boxes, pottery vases, wooden barrels, woven bags, etc. Use of paper for packaging dates back to 1035 AD, when a Persian traveler visiting markets in Cairo noted that vegetables, spices and hardware were wrapped in paper for the customers. The use of tinplate for packaging dates back to the 18th century. The first corrugated box was produced commercially in 1817 in England. Gair discovered that by cutting and creasing he could make prefabricated paperboard boxes. Packaging advancements in the early 20th century included bakelite closures on bottles, transparent cellophane over wraps, increased processing efficiency as well as improved food safety. Some other materials such as aluminum and several types of plastics were also incorporated into packages.

Packaging problems that were encountered in World War II led to Military Standard or "mil spec"(takes in to consideration packaging of barrier materials, field rations, antistatic bags, and various shipping crates) regulations being applied to packaging, designating it "military specification packaging". This officially came into being around 1941, due to operations in Iceland experiencing critical losses, eventually attributed to bad packaging. Plastic wraps were invented in 1953. The first plastic wrap on the market was made of Poly Vinyl Chloride (PVC). It usually forms a seal without clinging to itself or to the container.

It has a low permeability to oxygen, water vapors and flavor however there has been concerns in recent years about its toxicity.

Packaging is the technology of enclosing or protecting products for distribution, storage, sale, and use. It also refers to the process of design, evaluation, and production of packages. In today's context, packaging is both a symbol of society's consumption habits and reflection of its progress. The user expects the packaging to have better strength, be easier handled, be lighter, more aesthetic and safer from the point of view of hygiene. In addition to its standard attributes, today's packaging is also expected to contribute to protecting the environment, besides being friendly to human health.

Also, in 21st century food packaging has evolved as a specialized industry. Currently the packaging sector accounts for about 2% of the Gross National Product (GNP) developed countries. About half of this market is related to food packaging.

Need for Food Packaging)

Food packaging developed mainly due to the need for storing food and carrying eatables while commuting to places of work, school, market, etc.

Today, food packaging is done at all different levels starting from manufacturing unit, transport units, departmental stores, and small grocery shop or even at household level. In the earlier times people used to carry the food items purchased from grocery shops in homemade cloth bags. In some areas, these items were used to be wrapped inside the leaves, e.g. banana leaves that maintained their freshness. Mothers used to wrap the *chapattis* in the muslin cloth to keep these fresh. Later on they started using paper bags or newspapers and other papers too to wrap the food items. But it was realized that print ink got leached on the food items and

eventually the use of newspaper declined (but this practice still persists). Now in the upper and middle income group families we see the use of aluminum foil in the lunchboxes that keeps the food fresh. Wrapping food in aluminum foil protects it from both light and oxygen. But it is reactive so, it can not be used with acidic foods like tomatoes or berries. This gives an unpalatable metallic taste to such acidic food items. It is also used for cooking, freezing, wrapping, storing, etc. If its fairly clean after one use, foil can also be re-used after washing (unlike plastic wrap). However, it should not be used in microwave/oven.

Changing trends in packaging technology: In response to the changing trends, the traditional packaging technologies have either been improved or a few new technologies have progressed to match new requirements.

Plastic wrap has the advantage of being transparent, so we do not have to open it to find out what's inside. Plastic wrap is good for making an air-tight seal on bowls or containers that do not have a lid. **Re-sealable plastic bags** are very handy for keeping things like cold cuts or cheese fresh in the refrigerator. These airtight bags can also be used to prevent food from getting soggy in humid areas. Heavier-weight plastic bags are sold for storing food in the freezer. These are particularly suitable for freezing batches of soup or saucy dishes. Development of plastics received an impetus in last 30 years as natural materials such as wood was not available in adequate quantities; man-made materials such as steel, glass were found to be very expensive or could not provide economically viable solutions for packaging of products of mass consumption. But recently, concern about diffusion of substances like Bisphenol A and DEHA (diethylhexyl adipate) from plastic containers has been expressed, as many studies have found them to be carcinogenic.

Earlier only the steel tiffin's were used to carry food. But now various types of lunch-boxes are available that keep food hot and fresh. Also people are opting for packed food materials or ready to eat food. In industries also, we see that various types of food packaging are employed starting from the wax coated paper that is used to pack biscuits to the large cartons to store small bottles of pickles/jams. Many more examples can be cited from our daily lives like getting a pizza placed on a square cardboard box (*gutta*), wrapped trays or *thaali* containing lunch or dinner, wheat or rice in *bori* or the *thelas* the jute bags, etc. Figures 19.1 to 19.4 show some of the food articles which are packed and Fig. 19.5 gives commonly used packed food items.

Fig. 19.1: Bamboo baskets used for packaging

Fig. 19.2: Wooden baskets used to pack fruits and vegetables

Figs 19.3a and b: Jute sacks used to store and transport fruits and vegetables

Fig. 19.4: Modern plastic packaging for confectionery

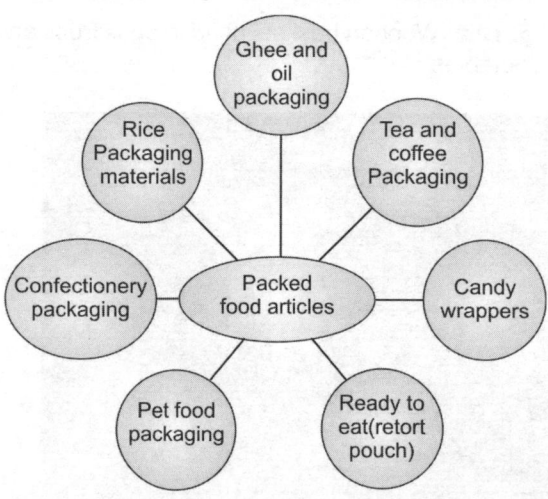

Fig. 19.5: Examples of packed food articles

But packaging of food articles should be done in an appropriate way so that chances of contamination, reaction with packed material, theft, etc. is avoided. It should be done in layers as depicted in Fig. 19.6.

Packaging in appropriate way or as per the norms so recommended by Food Safety and Standards Authority of India offers many advantages is given in Fig. 19.7.

Safe Food Packaging

Safe food packaging helps to store food and beverages in a hygienic yet convenient manner. Packaging helps in containing, protecting, and preserving, transporting food items. It also helps in informing and selling the items. Whenever we purchase something from market whether it is a food article or some other material, we need some packaging material for their transportation to other place. But while carrying a food article in a packet we must ensure that we are carrying in a safe and hygienic packet. When evaluating the safety of food packaging, the *Food and Drug Administration* **(FDA)** uses the criterion of **'GRAS', or 'Generally Regarded as Safe'**. Some materials such as certain types of plastic

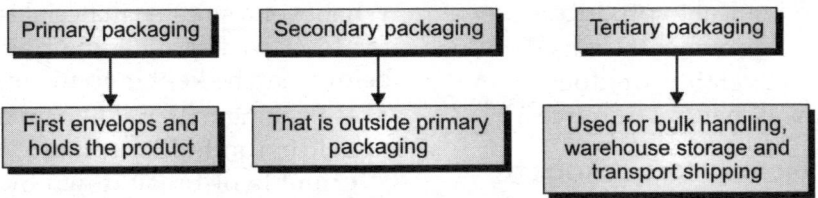

Fig. 19.6: Layers of packaging

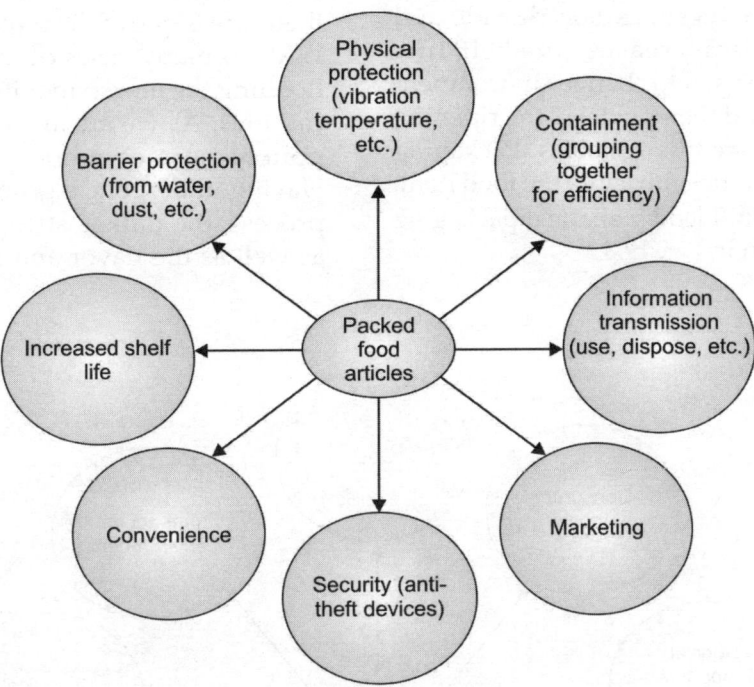

Fig. 19.7: Advantages of food packaging

and Styrofoam can release toxins when they are heated and care must be taken to ensure safety of food in them. Packaging materials which is irradiated inside its package can transfer unsafe nonfood substances into the food.

Food packaging makes use of a variety of substances, including dyes for printing colorful labels, and glues and adhesives for keeping packaging closed. In order to protect consumers effectively, some agency should individually certify each of these food packaging materials subjecting them to rigorous testing protocols.

Key points to be considered while designing packaging material are:

- Is it easy to handle and open?
- Is it a convenient shape, easy to stack?
- Which colors will be used on the packaging?
- What size of print should be used (so that the consumers can read it easily)?

- Will it be economical to produce?
- What about environmental considerations (will it be recyclable or does it make minimum use of natural resources)?

Choosing the material for packaging

Packaging acts as a physical barrier to protect food from contamination. It also preserves the nutrition value of food by preventing interaction with oxygen, carbon dioxide and moisture. It also increases its shelf life. Packaging can be used to change environment inside the pack and thereby delay the ripening of fruits or spoilage of vegetables. Packaging should be able to efficiently close the food item. Properties to be fulfilled by an ideal packaging material are given in Fig. 19.8.

Whenever a suitable material is to be selected for food packaging some points mentioned below may be kept in to the mind.

- Packaging material like tin may react with acidic foods like tomato. Others might crumble or break down over a period of time and mix with the food content within.
- Milk once exposed to air is most susceptible to bacterial attack. Loose milk therefore is one of the most unsafe options. With so many cases of milk adulteration flooding the newspaper its time to be extra cautious. As we are aware that the plastic pouches are commonly used in India. Having undergone a grueling sterilization process, the milk is stripped off microbes as well as the flavor and nutrient content.

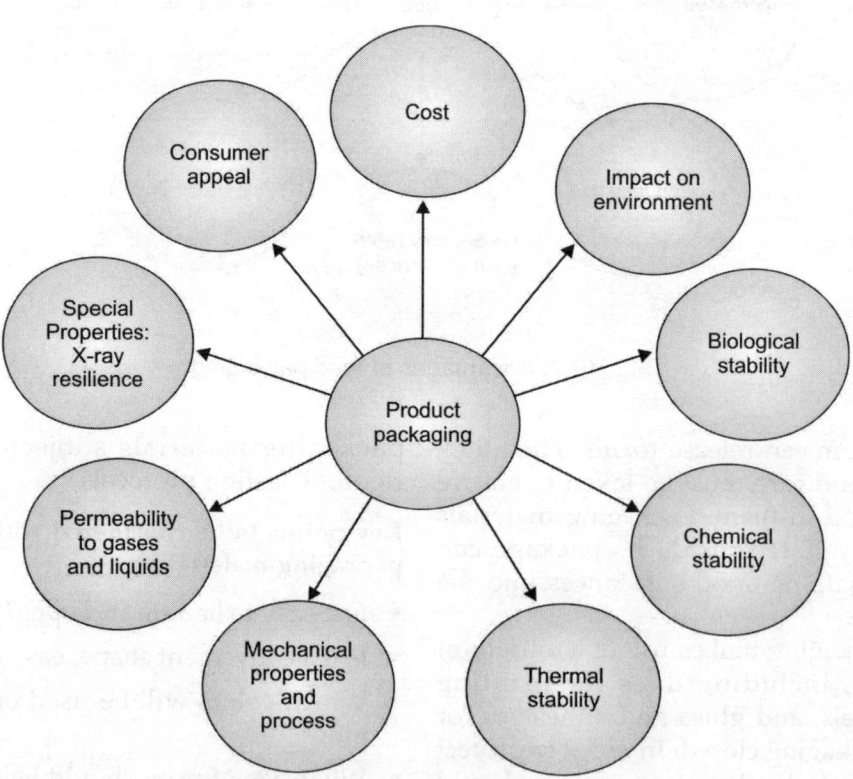

Fig. 19.8: Properties of packaging material

Carton milk, on the other hand, undergoes UHT treatment where the flavors and nutrient content of milk are preserved while doing away with any sort of microbes.

- Sometimes the packaging material may not be convenient to use for the particular food item. Opening cans and tins might become a complete hassle sometimes. With cartons too, the ones with the cap are easier to use than the ones that need to be cut open. Even them safe food packaging needs to be a higher priority than convenience of use.

Types of Packaging

Food items play an important and an integral part of our daily life. Packaging of the food used affects its safety, freshness and shelf life. The type of packing depends on various factors, e.g. type of food item, the process of production, and quality of the food.

Materials used for Packaging

a. **Glass:** Even in earlier times when the concept of safe food packaging was not prevalent, people were conscious about the safety of the food items they consumed, e.g. Cow used to be moved from house to house in nearby area to distribute milk by milkman or people used to bring milk from nearby dairy in their own utensils. Today milk is transported in metal cans with lids for home delivery. Glass milk bottles were also used earlier to increase the shelf life. Similarly, small sachets of coffee, spices, jams, ketchup, etc. are also available in glass bottles.

Glass is heavy, brittle, and non-degradable; it is mainly used for production of bottles and jars and is used to store liquids and sauces (Fig. 19.9). The lid of the jar or the bottle's cap keeps the food product fresh; these containers are to be handled with proper care as the glass can break easily if mishandled, e.g. cold drinks, pickles, jams, ketchups, squashes, etc.

b. **Aluminum:** Aluminum containers are inexpensive and are made out of recycled materials (Fig. 19.10). We are aware about the use of aluminum foil for keeping food warm and fresh. It is also used for packaging of sweets, cheese, coffee, tea, etc. aluminum cans are used to protect beverages like juices, beer and soft drinks for longer periods. Metal cans have an inner coating to prevent the food from reacting with the metal and getting contaminated while the heating process is on.

If steel is also the component in aluminum cans, it cannot be easily recycled (reprocessing needs high amount of energy), is

Fig. 19.9: Glass jars and bottles

Fig. 19.10: Aluminum containers

relatively expensive, although it has a very good barrier properties. It is used mainly for can production, metallic trays and forms for ready-to-cook food that are resistant for high and low temperatures and thus can be applied for frozen and heated meal. Use of plastics containing a thin layer of aluminum metal has improved barrier properties to moisture, oils, air, odors. These are usually used to pack snacks.

c. **Plastic:** Plastic comes in a wide variety of containers, wraps, buckets, bags, etc. Buckets are used to store items in bulk like wheat, cereals, sugar, rice, etc. Some common forms of plastic packaging include plastic wrap, plastic food pouches and plastic bags, etc. based on the concept of **Modified Atmosphere Conditions (MAP)** technology. It allows control over atmospheric conditions inside the bag that slows downs the deterioration process.

Plastic wraps are thin films with a sticky tape which are used to seal containers after putting the product inside it.

Plastic food pouches are lightweight, small to fit in the pocket and reusable (Fig. 19.11). Plastic bags increase the shelf life and maintain the freshness of the product. Items that are extremely moisture free can be stored in plastic bags for long without adopting MAP technology.

Plastic containers are used to store food items when the amount is in bulk. For example, these days Tupperware plastic containers are much in fashion. However there is a problem in storing food products in plastic containers. Plastic containers or buckets cannot stop the entry of oxygen into the container. There is a slow transmission of oxygen through the polyethylene walls into the container. For this oxygen barrier bags made out of plastic or

Fig. 19.11: Modern plastic based packing for food products

metal are used to increase the shelf life of the product.

d. **Paper and cardboard:** Paper is prepared from cellulose based materials (from wood), it is permeable for air, water vapor and oxygen has low tear strength; a wide range of bags and boxes for different applications are prepared from this, which are mainly used for carrying dry food such as sugar, salt, flour, bread, cakes, etc. Paper can also be used to make lightweight cartons that are used as a colorful outer cover for products packed in plastic or metal containers. Paper waste can be burned (with energy recovery), recycled or biodegraded during composting in environment. Various types of cans are made out of cardboard to store snacks, spices, nuts, etc. (Fig. 19.12).

Fig. 19.12: Cardboard and paper packages

Product Labeling and Traceability

Pre-packaged foodstuffs must comply with compulsory harmonized standards on labeling and advertising. The details that must appear on packaging include the name under which the product is sold, a list of ingredients and quantities, potential allergens (products which may cause allergies), the minimum durability date and conditions for storage. In many countries, nutritional information is also required. To ensure the safety of the domestic and global food supply, government regulations and brand protection demands from customers are on the rise. To address these growing requirements, packaged food supply chain has introduced traceability systems.

Modern Packaging Concept

Packaging plays a key role in ensuring food safety and providing convenience to consumers. Consumers increasingly demand food, which retains the natural flavour, colour and texture and contains fewer additives such as preservatives. In response to these needs, one of the most important recent developments in the food industry has been the development of minimal processing technologies designed to limit the impact of processing on nutritional and sensory quality and to preserve food without the use of synthetic additives. When production site is far away from consumption site, the need for safe food packaging increases. While tracking and shelf-life extension technologies are employed in the food service industry to reduce the risk of foodborne illness, proper heating and heat retention continue to be the main challenges. Carbon dioxide absorbers and emitters may be added to suppress microbial growth in certain products such as fresh meat, poultry, cheese, and baked goods (L'opez-Rubio, et al. 2004)

As we know, excess moisture in packed foods can have detrimental effect on food, e.g. caking in powdered products and moistening of hygroscopic products such as sweets and candy. In this case moisture control agents can help to control water activity, thus reducing microbial growth.

Humidity controllers can help to maintain relative humidity in-package (about 85% for cut fruits and vegetables), reduce moisture loss, and retard excess moisture.

Antimicrobials in food packaging are also used to enhance quality and safety by reducing surface contamination of processed food; however they are not a substitute for good sanitation practices.

Ethylene accelerates respiration, resulting in maturity and senescence of body tissues. Removing ethylene from a package environment helps extend the shelf life of fresh produce.

Self-heating packaging employs calcium or magnesium oxide and water to generate an exothermic reaction. It has been used for plastic coffee cans, military rations, and on-the-go meal platters.

Packaging provides protection chemical, biological, and physical. Chemical protection minimizes compositional changes those are triggered by environmental influences such as exposure to gases (typically oxygen), moisture (gain or loss), or light (visible, infrared, or ultraviolet). Many different packaging materials can provide a chemical barrier. Glass and metals provide an early absolute barrier to chemical and other environmental agents, but few packages are purely glass or metal since closure devices are added to facilitate both filling and emptying. Closure devices may contain materials that allow minimal levels of permeability. For example, plastic caps have some permeability to gases and vapors, as do the gasket materials used in caps to facilitate closure and in metal can lids to allow sealing after filling. Plastic packaging offers a large range of barrier proper ties but is generally more permeable than glass or metal. Biological protection provides a barrier to microorganisms (pathogens and spoiling agents), insects, rodents, and other animals, thereby preventing disease and spoilage. In addition, biological barriers maintain conditions to control senescence (ripening and aging). Such barriers function via a multiplicity of mechanisms, including preventing access to the product, preventing odor transmission, and maintaining the internal environment of the package.

While some innovations in packaging have stemmed from unexpected sources, most have been driven by changing consumer preferences. The new advances have mostly focused on delaying oxidation and controlling moisture migration, microbial growth and volatile flavors and aromas. This focus parallels that of food packaging distribution, which has driven change in the key areas of sustainable packaging, use of the packaging value chain relationships for competitive advantage, and the evolving role of food service packaging. Nanotechnology has potential to influence the packaging sector greatly. Nanoscale innovations in the forms of pathogen detection, active packaging, and barrier formation are poised to elevate food packaging market.

a. **High barrier packaging materials:** High barrier is the term, which has been associated with those plastics whose permeability is low enough to significantly prolong the shelf-life of food. Barrier packaging is one of the industry's most exciting developments, which has witnessed internal and external coatings, oxygen absorption and other barrier systems competing for a wide range of food and beverage applications.

Barrier packaging is designed to keep oxygen, moisture and carbon dioxide out of the packaging to preserve the flavour, colour, odour and freshness of its contents. With gas-fill techniques, barrier packaging retains CO_2 or N_2 to protect the product and extend its useful shelf-life.

b. **Tetra packs:** These offer processing and packaging solutions for dairy, beverages,

Fig. 19.13: Aseptic tetra packs for dairy products

cheese, ice creams, and prepared food including distribution tools like accumulators, cap applicators, conveyors, crate packers, film wrappers, etc. (Fig. 19.13).

c. **Pouches:** Various types of pouches are available such as spout pouches, zipper pouches and printed stand up pouches, reusable pouches, etc. these are prepared from high quality plastic/paper material that are durable and environmental friendly. Their stylish design enhances the appeal of the items. Moreover, they also contain the information like manufacturing date, expiry date, nutrient content, logos, and messages, etc. examples, Juices, milk, wines. Pouches can also be used to keep food and vegetables in refrigerators. They are commonly used at homes to carry *cream, tea, dahi,* etc.

d. **Retort packaging:** The retort pouch was invented by the United States Army Natick R and D Command, Reynolds Metals Company, and Continental Flexible Packa-

ging. The retort pouches used in packaging need to be kept in boiling water prior to serving the food. These "heat and eat" products are preferred for the ultimate convenience they offer to the consumer.

They are shelf-stable, which means that they can be stored at room temperature without requiring refrigeration and, in their packed form, remain fresh for over one year, and all this without any added preservatives. The production of shelf-stable products is attained by application of heat that kills organisms, which if not destroyed, will multiply and produce enzymes, which may decompose the food and in some cases produce food poisoning toxins.

The main requirement of plastic material is not only to withstand rigors of heating and cooling process, but also to maintain the overpressure correctly. In this way the internal pressure developed during processing is balanced by the pressure of heating system. A **retort pouch** is made up of plastic and metal foil laminate pouch, with 3 or 4 wide seals usually created by aseptic processing, allowing for the sterile packaging of a wide variety of drinks, that can range from water to fully cooked, thermostabilized meals such as that can be eaten cold, warmed by submersing in hot water, or through the use of a heater, lighter in weight and less expensive to ship (Fig. 19.14).

In this technique, food which is first prepared (raw or cooked) is then sealed into the retort pouch. The pouch is then heated to 240–250°F (116–121°C) for several minutes under high pressure, inside retort or autoclave machines. This process reliably kills all commonly occurring microorganisms (particularly *Clostridium botulinum*), preventing it from spoiling.

e. **Aseptic packaging:** This is a technique in which the contents of a package and the packaging itself are sterilized separately.

Aseptic packaging system basically comprises of the following:

• Sterilisation of the products before filling

• Sterilisation of packaging materials or containers and closures before filling

• Sterilisation of aseptic installations before operation (UHT unit, lines for products, sterile air and gases, filler and relevant machine zones).

• Maintaining sterility in this total system during operation, sterilisation of all media entering the system, like air, gases, sterile water.

• Production of hermetic packages

Recent efforts of introducing aseptically packaged milk and milk derivatives in flexible pouches has tremendous scope to improve the markets for aseptically packed products for extended shelf-life at affordable prices. Similar development on aseptic flexible pack for fruit juices and

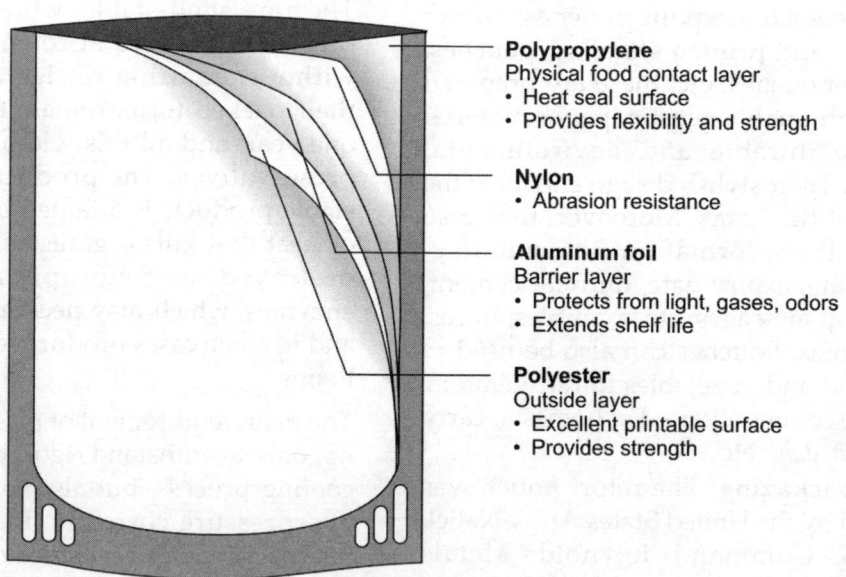

Polypropylene
Physical food contact layer
• Heat seal surface
• Provides flexibility and strength

Nylon
• Abrasion resistance

Aluminum foil
Barrier layer
• Protects from light, gases, odors
• Extends shelf life

Polyester
Outside layer
• Excellent printable surface
• Provides strength

Fig. 19.14: Retort pouch and its different layers

fruit drinks can also result in rapid expansion of markets for such products, as there is growing awareness for consuming them for reasons such as better health and fitness.

f. MAP/CAP (Modified Atmosphere Packaging/Controlled Atmosphere Packaging) system.

The change of the normal composition of air surrounding the product (within the pack) is one of the most effective methods of extending the shelf-life. MAP is a system where there is no control of the final atmosphere inside the package. The air inside the package is replaced partially or completely either with a single gas or mixture of gases or with vacuum, while, in CAP system, the atmosphere within the package is adjusted or controlled to specific requirement. Packaging materials with suitable barrier properties are used.

Generally, the most commonly used gases are:

• Carbon dioxide to retard the growth of micro-organisms, taking enough care to prevent CO_2 absorption and collapsing of the pack.

• Nitrogen is largely used to displace oxygen and delay oxidation. It is also used as a filler gas to prevent vacuum to occur.

• Oxygen is reduced to as low as possible, except in some specific applications, e.g. in meat where colour retention is desired. Right gas mixture has to be established for each product taking into account the product nature and its requirements, potential gas absorption and proper use of barrier packaging material. With this technology, fresh products can have a shelf-life extended by a few days to a few weeks.

For MAP of perishables, such as fresh fruits and vegetables (Fig. 19.15), which respire and meat products which need to retain the red colour, polymeric films are most suitable for use due to their permeability. Conventional materials like glass and metal are non-permeable to moisture vapour and gases and, therefore, would not be suitable for this application.

Fig. 19.15: Polymeric films for packaging to increase shelf life of food products

g. **Active and intelligent packaging:** Active packaging performs an additional role, other than just exhibiting itself as an inert barrier to external influences. Active packaging is based on the principle to address the issues of oxygen and moisture control. Moisture-control agents suppress water activity, serving to remove fluids from meat products, prevent condensation from fresh produce, and curb the rate of lipid oxidation. It can also involve the use of antimicrobials and ethylene absorbers. These designs that involve intelligent packaging facilitate the monitoring of food quality. Time temperature indicators (TTIs), ripeness indicators, biosensors, and Radio Frequency Identification (RFID) are the examples of intelligent packaging components.

h. **Micro oven able packaging:** Foods packaged for microwave heating and cooking have become among the fastest growing categories of convenience foods. A micro-wavable package is one, which can be used for cooking or heating the product in a microwave oven. The package alters the heating pattern of the contained food, by releasing or trapping the water vapour inside the package, thereby cooking the product under controlled temperature and pressure.

Future Trends in Food Packaging

The changing food consumer preferences, needs and demands in India will be one of the most powerful forces for change in the food industry. Increasing incomes, increasing literacy rates, smaller family sizes, women entering the work force, urbanisation and increasing concern about health and hygiene will be motivating consumer changes in attitudes. Therefore, many innovative trends are seen to be occurring in foodservice. As consumers continue to spend heavily on food service products, the role of packaging in ensuring food safety and providing convenience will only increase. Hence, innovations in the area of heat and heat retention can reduce the safety risk associated with improper cooking. The high demand for on-the-go meals has led to a significant increase in the variety of foods packaged for in-transit dining.

Use of Nanotechnology

Nanotechnology involves characterization, fabrication and/or manipulation of structures, devices or materials that have at least one dimension in 1–100 nm length range. Nanocomposites are materials that are made up of nanoparticle components. Polymer nanocomposites are created by dispersing an inert, nano scaled filler through a polymeric matrix.

It helps in the following:

1. Enhance polymer barrier properties makes it stronger
2. Stronger
3. More flame resistant
4. Possesses better thermal properties (melting points, degradation and glass transition temperatures)
5. Alterations in surface wet ability and hydrophobicity

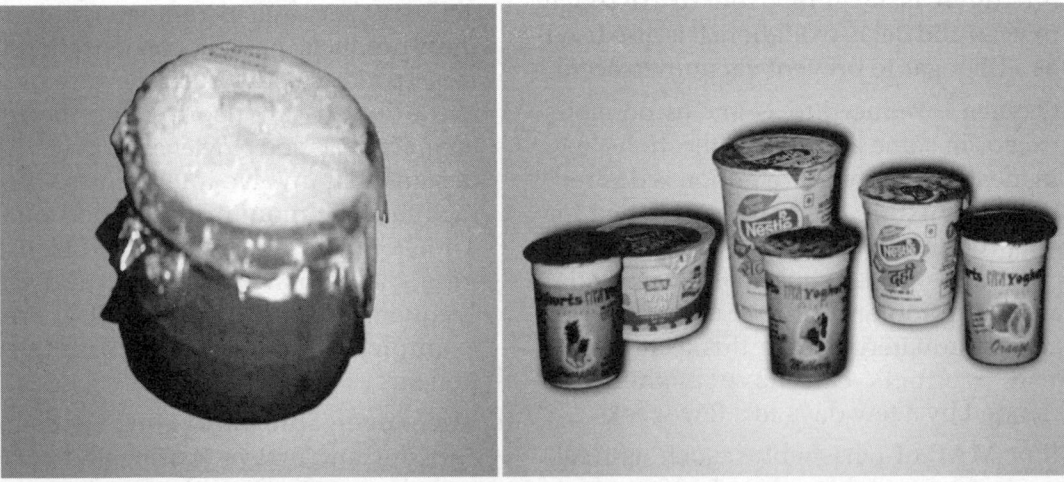

Fig. 19.16: Yoghurt packed in traditional earthen pot and injection moulded plastic containers

Nano-sized innovation could produce remarkable new packaging concepts for barrier and mechanical properties, pathogen detection, and active and intelligent packaging. Researchers have discovered that carbon nano-tubes exert powerful anti-microbial effects direct contact with aggregates of carbon nano-tubes proves to be fatal for *Escherichia coli*. The theory is that the long, thin nano-tubes puncture *E. coli* cells, causing cellular damage.

Some Toxic Effects of Packaging Materials

- Polyethylene Tetraphthalate (PET) is used in soft drinks, water and the juice as well as in peanut butter jars. This plastic may leach the chemical known as antimony trioxide. Liquids left in the container for long time have the potential to leave antimony in the liquid. The toxins Di(2-ethylhexyl) adipate (*DEHA*) released cause liver problems, other possible reproductive difficulties, and is suspected to cause cancer in humans.

- Polyvinyl chloride used in clear food packaging, cooking oil containers has been described as one of the most hazardous product. It is an endocrine disruptor.

- Polystyrene used in egg cartons, disposable coffee containers and packaging of cheese and meat in supermarkets, leaches chemical styrene which can cause developmental and reproductive problems.

- Polycarbonates used in baby bottles, various food and drink containers contain a chemical known as Bisphenol-A (BPA) BPA related chemicals can cause permanent alterations of breast and prostate cancer, insulin resistance and chromosomal damage linked to recurrent miscarriage

- The mass transfer of components between and within the food and packaging leads to the loss of volatile flavors and aromas from food. Migration of packaging components to food must be considered with toxicological risk analysis. Most incidences of migration occur in plastic packaging systems; thus, the most commonly studied migrants are plastic monomers, dimers, oligomers, antioxidants, plasticizers, and dye/adhesive solvent residues.

- Unacceptable odor pick-up can be avoided by proper package wrap in high-barrier materials. The use of high-barrier packaging materials can also prevent the absorption of other nonfood odors.

Packaging material should be safe as well as the area where that material is stored must also be clean and should ensure safety from dirt, filth on packets or cans etc. Rats, cockroaches in the store rooms where these packaging materials are stored, e.g. boxes placed in a sweet shop to pack sweets may be dangerous. They may contaminate the boxes with their urine; stool, etc. they may be harmful to health of the persons who consume materials stored in that boxes.

A series of inspection carried out by railway board on various trains in February, 2013 highlighted pathetic conditions of catering services in Indian railways.

Overall Sustainability

Safe packaging relates to sustainability, addressing security, safety, and health, for the whole of the consumer market. If the packaging meets legislative and consumer demands, we can expect such packaging to be:

- Beneficial throughout its life cycle
- Designed to meet market criteria for performance and cost

- Manufactured using clean production technologies and best practices
- Made from materials suitable in all probable life scenarios
- Physically designed to optimize materials and energy usage
- Effectively recovered mechanically, biologically or as energy

Green Packaging

The user expects the packaging to have better strength, be easier handled, lighter, more aesthetic and safer from the point of view of hygiene. In addition to its standard attributes, today's packaging is also expected to contribute to protecting the environment, besides being friendly to human health (Fig. 19.17). Therefore, one of the key trends within food packaging is sustainable packaging. A food packaging is sustainable if it is safe and healthy for individuals and communities throughout its life cycle; it is sourced, manufactured, transported, and recycled using renewable energy. It should be biodegradable and not harm environment *as well*.

Conclusion

Packaging is more than a just a container with a label stuck on it. It is a vital opportunity to

Fig. 19.17: Green packaging made from plants

build new brands or reinforce and add value to a positive experience of an existing product or brand. Packaging maintains the benefits of food processing. It enables food to travel to long distances. Along with it this provides necessary information labeled on the packaging material. However this technology must balance with other issues like material costs, as well as regulation on safe disposal of the discarded packaging material like glass, paper, plastic, etc. Goal of food packaging should be to store food in a cost effective way that satisfies industry requirement, consumer desires, maintains food safety and minimizes environmental impact.

20

Good Cooking Practices in Kitchen

Puja Dudeja, Neeti Rustagi, Ranabir Pal

All of us have seen our mothers/sisters/wives cook in the kitchen. In earlier days, the women used to spend most of their time in cooking food for all members of the family. It was a time consuming task requiring a lot of effort. Rather it was more or less a whole day preparation of raw material sifting, washing, drying, grinding grains, cooking perse, serving along with cleaning of kitchen and washing of dishes. All members of the family consumed hot freshly cooked food. The practice of storing and reusing leftover food for days together was not there. Hygienic preparation and food safety of home cooked meals reduced chances of Foodborne Illnesses (FBI). Eating out culture did not exist.

With changing role of women in society and more number of women going out to work the concept of cooking has drastically changed. Since these working females have less time to work in kitchen they usually hire domestic help for cutting/chopping vegetables, serving food and cleaning of kitchen and for washing of utensils. Some of the double income working couples with a comfortable salary even give the task of cooking to the housemaid. Simultaneously universal availability of refrigerator to store food (raw and left over) and microwave to thaw and reheat food has also contributed to comfort of the lady who has to handle both her office and kitchen. Many other appliances like food processors are also being used in the kitchen. However, these times saving methods and gadgets have an impact on the safety of food. As judicious and correct use of these equipment (refer to Chapter 21 on equipment maintenance), temperature control and other safe cooking practices need to be followed to ensure that food in all kitchens is safe for consumption by the family members.

With changing times eating out culture has increased exponentially in our society. Initially the Eating Establishments (EE) used to be busy on weekends. But gradually as more people started adopting the culture of going out to eat the EE started remaining occupied all days of the week. It is a common sight to see people waiting outside an EE after 9.30 pm for their turn to eat food!

In a typical kitchen of an EE, it is the chef or senior most food handler who cooks and there are other food handlers to assist him. They also handle various kitchen equipment required in cooking. To prevent contamination of food it is vital that these people working in kitchen are not suffering from any infectious disease, have good personal hygiene, practice good hand washing, and are well trained regarding food safety issues.

Similarly the equipment used in kitchen like utensils, mixer, grinder, *chapatti maker*, toaster, grill, *tandoor*, water purifier, etc. are in a good state of repair and are hygienically maintained. The present chapter deals with various cooking practices/mechanisms which will ensure food safety. In the journey of food from farm to fork let us understand food safety processes in kitchen.

There are recipes for different food items and even different recipes for one dish. However, preparation of any dish involves some common generic steps. Though for different dishes the sequence of steps varies.

1. Preparation of raw material
   a. Cleaning
   b. Peeling, cutting/chopping
   c. Thawing in case of frozen raw material
2. Cooking
   a. Conventional cooking: temperature and time control method
3. Hot and cold holding
4. Cooling and storing
5. Transport
6. Serving
7. Reheating of leftover food

Preparation of Raw Material

Food Safety precautions have to be observed at each of the above steps to avoid FBI. Details are given below.

Cleaning

The practice of preparation of raw material is different in a home kitchen and in a commercial or public EE. In an urban middle income home the raw fruits and vegetables are procured either from vegetable market/ green grocer/supermarket. Sorting of rotten ones is done mainly at the time of procurement. These are then stored in the refrigerator. When required they are washed, peeled, chopped and cooked. Practices vary as per the prevailing custom in each home. Most of the people store fruits and vegetables in refrigerator without washing. While some discerning people may religiously wash, dry and then store these items.

However, in EE the procedure is slightly different. In case of a star rating EE, which has adequate space, equipment and utensils, procedures are slightly different from a small EE. This also requires training of food handlers. Raw material reaches the kitchen following safe transportation practices (refer Chapter 17 on good transport practices). At the time of receipt, fruits and vegetables are sorted out at in the reception area and cleaned of dust and dirt. This is done in three stages. The dirt is removed first, then it is rinsed with clean water and at third stage it is sanitized/ washed with 50 ppm chlorine solution for five minutes or potassium permanganate solution. Special precautions are taken for the cleaning of green leafy vegetables. In case these are not washed properly these can be a source of infectious disease like neurocysticercosis in vegetarians (a disease which occurs by consuming improperly cooked pork in nonvegetarians).

It is a common practice to remove physical impurities like mud, stones, grit, insects, etc. from cereals and pulses by handpicking before cooking. Some inedible parts of vegetables and fruits are also removed. They are then stored at the appropriate temperature (less than 5°C) in a refrigerator (*see* storage practices in refrigerator in Chapter 21).

This practice is not followed in smaller EEs, which do not have adequate capacity refrigerator. FBOs of smaller EEs procure raw material form vegetables market in the morning. After sorting, these are washed and chopped early in the morning for the days requirement. Food safety issue in these EEs is that cut vegetables and fruits if kept for a long time they can easily be contaminated as bacteria from outside can go to the inside part. Hence, if such a practice is followed it is mandatory that these cut vegetables are cooked, cooled and then stored in refrigerator.

Once I visited an EE which was following this practice of chopping veggies in the morning so that the waste generated like peels, discarded vegetable pieces could be cleared by the mid morning waste lifting team. Lack of storage space in the kitchen refrigerator made them use the sponsored refrigerator of aerated drinks instead. They lacked containers too. So, they were using empty cartons instead of food grade storage containers (Fig. 20.1). Such practices by FBOs to cut the cost should strongly be discouraged as this compromise safety of the food.

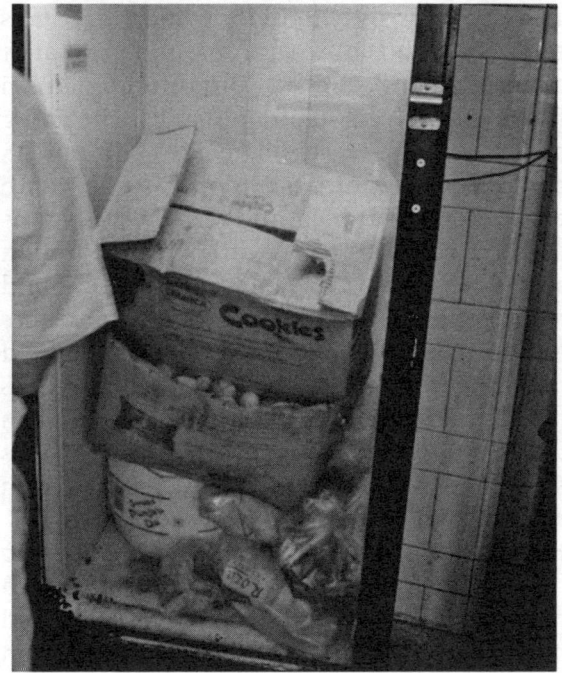

Fig. 20.1: Cut vegetables in cartons for storage in refrigerator (an unhygienic practice)

Cutting and Chopping

This process generally needs a chopping board and sharp knives. In some EEs food processors may be used for this process. After having ensured the cleanliness of raw fruits and vegetables, the food safety issue in next step of cutting and chopping is related to the hygiene of cutting board and knives. Contaminated equipment can be a source of cross contamination and make raw material unsafe. The cutting boards are available in different colours (red, green, blue, yellow, white) so that they can be distinguished for different foods. Generally, green colour is preferred for vegetables and fruits, red for mutton, blue for fish, yellow for poultry and white for ready to eat precooked food. Each EE can develop its own system of identification. Similarly the knives can be identified for vegetarian and nonvegetarian foods by the colour of handle or shape of the blade. This practice of keeping separate chopping boards and knives for vegetarian and non-vegetarian foods and or cooked and raw foods strongly prevents cross contamination and helps in maintaining the safety of food.

- *Safety concerns for chopping boards*
  - ✓ Separate for vegetarian and nonvegetarian items
  - ✓ Separate for cooked and raw foods
  - ✓ Always use a clean cutting board for food preparation.
  - ✓ After each use and before next step while preparing food, clean the cutting boards thoroughly in hot, soapy water, then rinse with water and air dry or pat dry with clean paper towels.
  - ✓ After cutting raw meat, poultry or seafood on your cutting board, clean thoroughly with hot soapy water, then disinfect with chlorine bleach or other sanitizing solution and rinse with clean

water. To disinfect cutting board, use chlorine solution or readymade sanitizers. Flood the surface with the solution and allow it to stand for several minutes. Rinse with water and air dry or pat dry with clean paper towels.

✓ All cutting boards eventually wear out. Discard cutting boards that have become excessively worn or have hard-to-clean grooves. These grooves can hold harmful bacteria that even careful washing will not eliminate.

Thawing

Freezing does not destroy microorganisms. If a contaminated food is frozen all the organisms are not destroyed during freezing. In addition, organisms do not grow in frozen food. But, when such food is thawed they multiply rapidly.

It is imperative that all frozen foods are thawed properly before cooking them. Thawing frozen food correctly is important for keeping food safe to eat. The temperature of food should not exceed 5°C during the thawing process. Hence appropriate planning is required before cooking frozen foods. Freezing food keeps a check on most bacteria from multiplying, but it does not kill them. If food is allowed to enter the temperature danger zone of 5°–60°C, bacteria will grow rapidly.

There are four ways for thawing food: in a refrigerator, under cold running water, in a microwave, or as part of the cooking process. For refrigerator thawing put the packages of frozen food in a pan. It is imperative to use a pan as the juices can drip on other foods. After some time change the drip pan when liquid is visible in the pan. Another way is to put frozen food completely submerged under clean, drinkable running water. The water tempe-

rature should be 20°C or below. The water should be at sufficient velocity as to agitate and float off loose particles in an overflow. In case frozen food is to be cooked immediately (within 2 hours of thawing) use a microwave for thawing.

Sometimes thawing can be done as as a part of cooking process for certain foods like frozen patties, nuggets, pizza, soup, and vegetables. The process of thawing frozen food needs to be monitored by checking the temperature of food using a thermometer. This is done at the end of refrigerator thawing. For thawing in running water, temperature of the food every 30 minutes is required to be checked. The cleanliness of the thermometer in use cannot be ignored in these steps.

Cooking

Conventional Cooking: Temperature and Time Control Method

The most likely time when food becomes contaminated is during preparation and after cooking. Food can be contaminated from surfaces, utensils, clothing, sinks, chopping boards, hands, waste or unclean equipment, by using contaminated foods as eggs with dirty or cracked shells and from pests such as cockroaches, flies or rats in food preparation areas. Before we understand the time and temperature control for safe cooking practices it should be noted that practice of tasting food with fingers should be strictly discouraged as it can compromise food safety. Instead a clean tasting spoon can be used. In small EEs sometimes a ladle is used to put the food on the surface of hand which is then tasted. Freshly cooked food should never be used to 'top up' containers. A clean container for each new batch of food that is prepared or cooked must be used. Always separate raw food from food that is ready to eat. All foods must reach a temperature of 75°C during cooking.

At this stage we first need to understand time and temperature control principle of food safety.

Temperature control is the use of temperature to protect the safety of food and minimise the growth of bacteria. This means keeping chilled food at 5°C or below, and hot food at 60°C or above. Most types of micro-organisms that cause food poisoning grow in potentially hazardous foods at temperatures between 5°C and 60°C. This temperature range is called the '**temperature danger zone**'. To measure temperature of the food, a food grade thermometer is required. The practice of using a thermometer is not common in our country. Even the star rating hotels do not possess or use such a gadget.

So, it is advisable to cook any food with liquid consistency till it boils. Boiling ensures sufficient rise in temperature to kill harmful bacteria. In case a thermometer is present it must be a food grade thermometer, which is accurate to +/–1°C. The thermometer must have a probe so that the internal temperature of food can be measured. The thermometer must be kept safely in the kitchen and must be cleaned and sanitized before every use. This is important to prevent contamination from one food to another.

Foods that are not potentially hazardous may become potentially hazardous if they are altered in some way. For example, custard powder, is not potentially hazardous because it is too dry for bacteria to grow. But, the custard becomes potentially hazardous when milk is added. Most raw whole fruit and vegetables are not potentially hazardous because they do not allow any food-poisoning bacteria to grow. But, when they have been cut, bacteria may be able to grow on the cut surface, and so cut fruit and vegetables should be stored chilled.

There are occasions when it is impractical to keep the food at 5°C or below, or 60°C or above. Some examples are buffets at parties/meetings, etc. In such situations the food will be safe for a short time, unless it has been contaminated during handling. The maximum time that potentially hazardous food can be in this temperature danger zone of 5°C–60°C is 4 hours. After 4 hours, any remaining food must be thrown away. The 4 hours must include any time that the food was between 5°C and 60°C during handling, during preparation and processing, after processing, during transport and, in the case of buffets, the time setting up. If it happens so that the food is required to be held between 5°C and 60°C and then refrigerated then it must not be in 5°C and 60°C for longer than 2 hours.

Another common practice in our households and EEs is that food handlers insert the finger into the dish to check the right temperature. It is unhygienic. This can transmit harmful bacteria from fingers to the food. In case the food is to be tasted or temperature is to be observed, a clean spoon needs to be used.

In case a microwave is being used for cooking following precautions must be ensured for safety of food:-

- Stir food between cooking
- Cover and cook to prevent loss of moisture
- Let the food stand covered for two minutes after cooking so that temperature equilibrium is maintained.
- Heat the foods to an additional 14°C above recommended temperature in conventional cooking.

Hot and Cold Holding

Food once cooked can be consumed immediately or later. There are certain situations when food needs to be displayed for hours together before consumption, for example, in buffets, hostel messes, parties, religious occasions, trade fairs, festivals, etc. In such situations, there is a time gap between the consumption of food by the first person and last one. During this time duration, the food needs to be held at a specified temperature for a certain duration to maintain the safety of food. Here also the principle of temperature danger zone (5°– 60°C) applies. This will prevent the growth or development of bacteria. The golden principle for holding cooked foods is to keep hot foods hot and cold foods cold. Hot-holding equipment must be able to keep foods at a temperature of 65°C or higher, and cold-holding equipment must be capable of keeping foods at a temperature of 5°C or colder.

When holding hot foods for service, following guidelines help to ensure safety of food:

- Stir at regular intervals, as it will help distribute heat evenly throughout the food.
- Cover as covering will help retain heat and eliminate potential contaminates from falling into the food.
- Use food thermometer to measure the food's internal temperature every two hours.
- Discard any hot food after four hours if it has not been maintained at a temperature of 60°C or higher.

Another important precaution that must be observed is, not to use hot-holding equipment to reheat. In case food needs to be reheated it should first be heated to an internal temperature of 75°C and then transferred to the hot-holding equipment. Also, one should never mix/top up freshly prepared food with foods being held for service as this practice can result in contaminated foods.

When holding cold foods for service, following guidelines help to ensure safety of food:

- Protect all foods from possible contamination by covering them.

- Use food thermometer to measure the food's internal temperature every two hours.
- Never store food items directly on ice. All food items, with certain exceptions, should be placed in pans or on plates when displayed. Ice used on a display should have a proper drainage system. All pans and plates should be sanitized after each use.

In case there is a situation when one has to deal with questionable hot and cold-holding practice, the issue must be resolved in favor of food safety. It is better to discard foods than risk health or safety. The best one way to avoid discarding too much food is to prepare and cook only as much food as required.

Cooling and Storing

In case the food is not to be consumed, immediately after cooking it must be cooled and then stored in the refrigerator. Food when hot should never be placed in the refrigerator. It must be cooled quickly and then stored for use later. Cooked food certainly should not be left longer than one to two hours at room temperature before being placed in the refrigerator. This is so because bacteria start to develop within two hours and then spread rapidly. It is a myth that keeping hot foods in refrigerator will warm other items or reduce the working efficiency of refrigerator. Foods taking longer than two hours to cool can be cooled using methods give in succeeding paragraph. However if these cannot be followed still food must be placed in the refrigerator to avoid spoilage.

There are different ways of cooling food so that it can be put inside the refrigerator. The simplest is to divide food into smaller portions in shallow so that heat loss is facilitated. Another way is to place the cooked food in a sealed container and then run under cold water. However, in case the container is not closed properly food can be contaminated. If feasible cooked food can be kept over an ice bucket also.

Transport

Refer to chapter on safe transport practices and safety of home delivered foods.

Serving

Food can be contaminated during serving through the person doing service, environment (dust, flies, etc.) or through the utensils being used. The famous dictum of 'There are many slips between the cup and the lip' applies here too. Before serving food the food handler or the service staff should practice personal hygiene as washing of hands with soap, ensuring cleanliness and paring of nails, etc. The serving dishes need to be checked for any dirt or remains of detergent or left over food. It is a common practice in our country to wipe the cleaned and freshly washed utensils with a mop (cloth piece) to dry it. This practice can recontaminate the washed utensil. It needs to be discouraged. It is not uncommon to see this practice being done in our homes too!

Even if the serving dishes are dry enough to be served, they are mopped in front of the customer to assure and impress him of the cleanliness practices being followed. The serving dishes must be handled in a way to avoid touching of eating surface or mouth contact surface. This will prevent contamination from hands. For example, the plates must be handled from the bottom or edge, cups by handles or bottom, spoons by handles. Do not serve food in chipped or cracked dishes as these areas can attract dust, food particle accumulation and latter growth of bacteria. Never pick up glasses or cups by inserting fingers inside to hold them.

Handling of food with bare hands should be avoided. *Chappatis, paranthas, papads*, sugar cubes, cheese, garnishes, salads should not be touched. Always use a service spoon, pair of tongs, or a fresh disposable glove to pick and serve them. It is a common practice to distribute food with bare hands during religious occasions. This can also contaminate food in case the hands of person serving are not clean or not washed properly.

Reheating of Leftover Food

Leftover or surplus foods must be stored safely in case they are required to be used in future. Those foods which are highly perishable and have been in danger zone of temperature for more than two hours should be discarded. Most of other foods can be kept for a day if they have not been handled much, stored at correct temperature and reheated properly. Some foods contain spores which are not destroyed during normal cooking process. When these foods are kept in danger zone the spores germinate and bacteria began to multiply till food is refrigerated where they remain dormant. When such food is reheated and passes through temperature danger zone these bacteria multiply. Hence, it is important to reheat the food to 75°C and kept at a temperature of more than 65°C until it is served. In our homes we commonly use our microwaves to warm the leftover food taken out of refrigerator. Simply warming the food before serving will be doing harm as bacterial growth is favored in these temperatures. Another important food safety feature is that left over food must never be mixed with fresh food. High risk foods (discussed later in this chapter) should not be reheated more than once.

Key Points for Good Cooking Practices

- Do not prepare food long before serving time
- Do not store perishable foods at room temperature for more than 4 hours

- Ensure appropriate cooling of food before storage
- Provide sufficient temperature for reheating
- Thaw frozen meat and poultry well before cooking
- Prevent cross contamination between raw and cooked food during handling of various food items

In addition to above any lady of the house/FBO/food handler should be ever vigilant while handling high risk foods. They are called so as microorganisms grow easily in them. These foods include:

✓ Raw and cooked meat

✓ Dairy products and dairy-based desserts such as fruit cream, custard

✓ Cooked rice

✓ Foods containing raw eggs like mayonnaise sandwiches, salad dressings

✓ Salads and salad ingredients

✓ Bread, toast, rolls, sandwiches, pizzas

✓ Baked goods

✓ Garnishes such as lemon wedges, or pickles on plates

✓ Fruit or vegetables for mixed drinks

✓ Ice

✓ Paneer as an ingredient

✓ Seafood

For handling potentially hazardous food it is important not to use bare hands. Such foods at risk can be handled by using forks and spoons, napkins, spatulas, tongs, etc. If such foods at risk are stored then it is imperative to monitor the temperature of refrigerators and deep freezers or of the display units to ensure they keep food between 0°C and 5°C. A record of such checks should be maintained. In case if high-risk foods have been left in the temperature danger zone (5–60°C) for up to

two hours the food should be reheated, refrigerated or consumed. If they have been left in the temperature danger zone **for longer than two hours**, but **less than four hours**, they should be consumed immediately. In case the time in danger zone exceeds four hours the food must be thrown and discarded.

Over and above special care must be observed while making following preparations of food.

Stuffed Preparations

While making any stuffed preparations like *paranthas*, rolls, pattis, dosas, etc. the material for stuffing should be cooked first and then filled. This is so because the stuffing slows the process of heat penetration of these food items and the temperature in the centre may not reach at desired levels of safety. Another food safety issue is that food items are exposed to bare hands while preparing them. The external temperature of the food gives a false sense of security, as it is hot from outside.

Often chopped slices of chicken, meat, etc. (cooked/raw) are kept in refrigerator to be used as stuffing when required. These are minced and mixed with other ingredients to make the stuffing. Mincing such potentially hazardous food items can transfer bacteria from surface to inside so that they get distributed throughout the entire mass of mince. Minced meat is also handled more and spoils much faster. Hence, minced meat must be thoroughly cooked at sufficiently high temperatures.

Coated Preparations

Many food items are coated in batter. This batter can be made of *besan*, corn floor, bread crumbs, etc. At times food items are dipped in egg also before cooking. Examples are *pakoras*, cutlets, etc. These coatings act as insulators and reduce the transfer of heat to the food being cooked. The batter for coating is generally handled with bare hands for mixing. Often in commercial EEs the batter is left as such for hours. This way batter can get contaminated with microorganisms and spoil the entire food preparation.

Salads and Sandwiches

When salad and sandwiches are prepared highly perishable food are used like raw fruits and vegetables, sandwich or salad dressings with raw eggs and poultry. Such food items must be stored in the refrigerator immediately after preparation. If bread is used it acts as a insulator and prevents the chilling of items inside or between bread.

Meat

Meat items are high-risk food and are a common cause of FBI. Meat for cooking must be bought from a reputable butcher or retail shop. In case it is in frozen form the use by date must be confronted to. While storing them in refrigerator keep them away from ready to eat foods so that juices from meat do not cross contaminate these foods. These must be purchased in the end of shopping spree and kept first in the refrigerator. It is preferred to keep meat on the lowermost shelf of the refrigerator. Thaw as discussed earlier. Keep a separate chopping board and knife. During cooking of meat ensure that high enough temperature is reached that it is cooked not only from outside but also from inside. In case there are small pieces they must be moved so that each piece is cooked evenly. All meat should be checked visually to see if it is cooked thoroughly. To check for thorough cooking a fork is pierced through the thickest part of the meat and then the juices should run clear. Meat changes its colour when it is cooked. There should be is no pink meat left after it is cooked well.

Eggs

Extra care is required when preparing foods that contain raw egg, such as homemade mayonnaise. Bacteria like salmonella are present on eggshells. Inside the egg these can contaminate these types of food and cause food poisoning. It is advised to wash the eggs with clean water before storing these in refrigerator. Hands must be washed in case raw vegetables or ready to eat items are to be handled after handling egg. For example, if bread is to be toasted after making omelettes then hands must be washed in between the two processes. Cracked eggs must be discarded.

Fish

Something is fishy! This proverb is used in a situation when one can sense something is wrong. It is well said as fish and fish products are spoilt easily because of decomposition, insect, parasites infestation and poor sanitation. Fish deteriorates or loses its freshness because of autolysis which sets in after death

While procuring fish it is crucial that person should be able to identify for poisonous fish and fish parts. Sewage, bacteria and viruses (e.g. the virus of hepatitis type A) are concentrated in shellfish such as oysters, and fish may carry *Vibrio parahaemolyticus*, *Salmonella, Clostridium botulinum type E*, and other organisms. Consumption of certain fish may sometimes give rise to 'fish poisoning'. Fresh fish has following features:

- have a shiny, iridescent surface
- body is covered with a nearly transparent, uniformly spread, thin coating of slime
- eyes are bright and protruding with black pupil transparent cornea
- gills are bright free from slime

- odour is marine-like/seaweedy; fatty fish have a pleasant—margarine-like smell
- flesh is soft and flabby immediately but becomes firm after setting of rigor mortis

After freezing, good fresh fish have a delicate, pleasant odour and flavour when cooked. If the fish is a little older before freezing, it is insipid; a lack of odour or flavour is noticeable. Once procured, it is essential that the fish be unloaded as quickly as possible, minimizing bruising and rough handling, and conveyed to the initial processing area without undue delay. Fish can be cooked by grilling, frying, roasting, baking, poaching, steaming and microwaving. Ensure adequate temperature is reached while cooking. The flesh should be opaque and separate easily with a fork when fully cooked.

Cooking is surely the first art that human beings ever attempted and it is still the most universal. Cooking involves combining proper ingredients in correct proportions and sequence to get the required taste and flavor. Nevertheless, the art of safe cooking must be learnt and practiced in true spirit so that all food served is safe for consumption.

21

Refrigerator Maintenance and Food Safety

Ishwarpreet Kaur, Amarjeet Singh

INTRODUCTION

Refrigerator, commonly called a fridge, is an appliance used for food storage. Its basic application is to keep food cold. Perishable foods like cooked, ready to eat and high risk foods are generally refrigerated or are frozen to preserve and keep them safe for longer time. It has become an indispensable item of every urban household. The need of it is felt 24 × 7; 365 days in a year. The hot and humid weather conditions in the country have further increased its demand.

Refrigerator market is one of the fastest growing segments of the consumer durable industry. According to "Refrigerator Market Forecast to 2015", the refrigerator market is estimated to grow at a CAGR (Compound Annual Growth Rate) of 25.7% during 2012-2015. The efforts are to offer affordable and eco-friendly variants of refrigerators to the population of smaller towns.

History: In prehistoric times, man found that food would last longer if stored in the coolness of a cave or packed in snow. With industrialisation and mechanisation man started storing and transporting frozen ice water from colder to warmer regions.

In the intermediate stage in the history of cooling foods insulating materials like sawdust or wood shavings, cork were used in the icehouses. Even chemicals like salt, sodium nitrate or potassium nitrate were added to water causing the temperature to fall. Salts were used to preserve meats. This was later replaced by ice boxes.

Refrigeration as it is known these days is produced by artificial means. The history of artificial refrigeration began in the year 1755, when the Scottish professor William Cullen made the first refrigerating machine, which could produce a small quantity of ice in the laboratory. The evolution to mechanical refrigeration, a compressor with refrigerant, was introduced in the last quarter of the 19th century. The technology of refrigerator still continues to evolve. The refrigerants used today are less injurious to the environment than that were used earlier. There are four components of a refrigerator—compressor, heat-exchanging pipes (inside or outside the unit), expansion valve, and refrigerant. The compartments of a common refrigerator are shown in Fig. 21.1.

In India, refrigerators are increasingly finding their way in homes, departmental stores, and various business establishments. It is no longer a luxury but has become a necessity in homes and shops/industries. The total Indian refrigerator market was 8.4 million

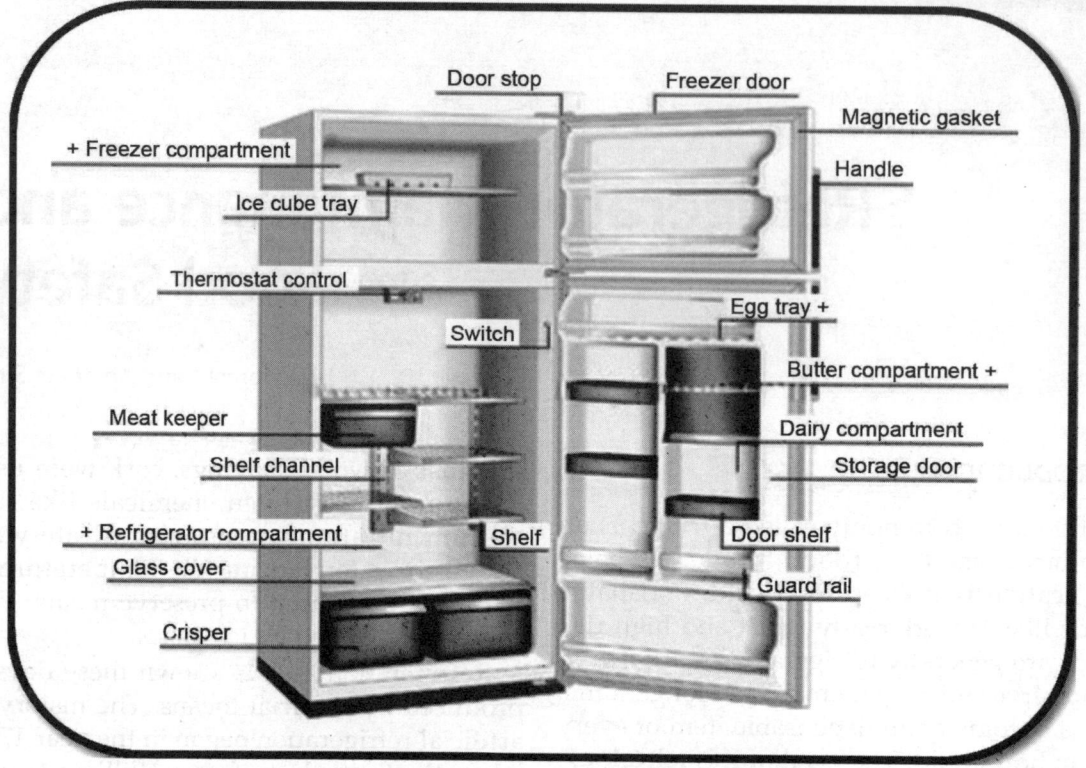

Fig. 21.1: Compartments of a common household refrigerator

units in 2010–11, registering a 15% increase from 7.3 million units in 2009–10.

Importance of refrigeration: All foods contain bacteria. If they are provided favourable conditions they multiply fast and spoil the food. Bacteria grow most rapidly in the range of temperatures between 5 and 60°C, the "Danger Zone," some doubling in number in as little as 20 minutes. Thus, bacteria can spoil the food left at room temperatures in few hours. Such foods on consumption by human beings can lead to foodborne diseases. By just reducing the temperature of food storage, depending on the type of food the shelf life can be increased from few days to months.

Cold temperatures help food stay fresh longer. Lower temperatures lower the reproduction rate of bacteria and enzymatic activity of foods. The same principle works for refrigerators and reduces the rate of spoilage. A refrigerator maintains a temperature a few degrees above the freezing point of water. The temperature ranges from 3 to 5°C (37 to 41°F). However, the freezer maintains a temperature below the freezing point of water. Freezing can stop the bacteria altogether. Refrigerator also helps maintain potency and chemical composition of medication.

Refrigerator is one of the miracles of modern living that has completely changed the life of people. Without them we would be throwing out leftovers instead of saving them for another meal. For this reason, earlier people would cook food in amounts that would be consumed in a single meal.

Types of bacteria in refrigerator: It is a myth among people that no bacteria can grow in a refrigerator. The environment inside a refrigerator usually creates an inhospitable environment for many bacteria. But some bacteria are able to grow at cold temperatures, these are called psychrophiles. This explains why food products still go bad in refrigerators. Bacteria such as some Coliforms, *Pseudomonas* spp., *Vibrio* spp., *Listeria* spp. and moulds such as *Penicillium* spp and *Cladosporium* spp. are all known to survive low temperatures and can cause illness.

Maintenance and care in operating refrigerator to keep food safe—Refrigerator and freezers are used at almost all places from homes, restaurants, shops, etc. Generally in commercial set-ups separate freezers are used for storing purposes. The maintenance of these is very important for their efficient and optimum functioning. The care for freezers is also almost the same as that of the refrigerator. The key points to keep in mind are:

a. *Safe temperature*: For food safety it is important to maintain the temperature below 5°C. Verifying of temperature on routinely basis can prevent any lapse in food quality. Some refrigerators have built-in thermometers to measure their internal temperature. For those without this feature can keep an appliance thermometer within refrigerator to monitor the temperature. This holds special importance at the time of power outage. The 'down time' period and temperature fluctuation is very crucial from food safety point of view. Once the electricity is back the temperature of refrigerator must be checked. The food is safe if the refrigerator is still 5°C. Foods held above 5°C for more than 2 hours should not be consumed and leftover of this must be discarded. In the event of a power outage foods can be kept cool for several hours by refraining from adding foods to the appliance and opening the doors. If one knows there will be a power outage, produce more ice cubes and place them in the top section of the refrigerator.

In order to monitor for mechanical failure or electrical outages that may occur overnight, or on weekends and holidays, place a coin on top of a small container of ice and keep it in the freezer. In case of any disruption of power, the ice will melt and the coin will no longer be on top of the ice, discard the food items.

Even otherwise ensure right temperature by ensuring that the doors of refrigerator are closed tightly at all times. Don't keep the refrigerator door open unnecessarily and close them as soon as possible. Never overload the refrigerator.

The temperature requirement varies with type of food and other items. The frozen food or highly perishable food items are kept best in freezer at –20° to –10°C. Other food stuff requires temperature between 1° to 5°C. It is important to keep in mind that some medicines kept at home or chemist shops require cool temperature to remain potent and must be stored at 2° to 8°C.

b. **Safe handling of foods:** It is again a misconception that hot food cannot be placed directly in the refrigerator. For this reason people keep hot foods at room temperature for long to lower their temperature. This habit should be disowned and instead food must be rapidly chilled in an ice or cold water bath before refrigerating. Always remember to place it in refrigerator within 2 hours of heating/cooking.

Food can be cooled in the following ways:

• Use shallow pans (it is recommended that the food be uncovered and no more than 2 inches deep)

- Cutting large portions of food into smaller portions (it is recommended that the food be uncovered and no more than 4 inches thick)

Food must cool from 60°C to 20°C in 2 hours and from 20°C to 5°C or lower in 4 hours—adding up to a total of 6 hours. This is called the Two-Step Cooling Process. Make sure to cover foods to prevent them from picking up odours from other foods. For even cooling a large cut of meat or whole poultry should be divided into smaller pieces or placed in shallow containers before refrigerating. A general rule of thumb for refrigerator storage for cooked leftovers is 4 days; raw poultry and ground meats, 1 to 2 days. Throw expired foods that should no longer be eaten. Raw meat, poultry, and seafood should be in a sealed container or wrapped securely to prevent raw juices from contaminating other foods. Keep raw and cooked food separated. Store ready to eat food above raw foods.

c. **Placement of foods:** Some refrigerators come with features such as adjustable shelves, crispers, and meat/cheese drawers. These are designed to make storage of foods more convenient and to provide an optimal storage environment for different variety of foods. Vegetables require higher humidity conditions while fruits require lower humidity. Sealed crisper drawers with adjustable humidity levels provide an optimal storage environment for fruits and vegetables.

Do not store perishable foods in the door like eggs. These should be stored in the carton on a shelf. This is because the temperature in the door fluctuate more than the temperature in the cabinet.

Always keep cooked food above raw food to prevent contamination. The meat and vegetables must be stored separately. Never over stock the food. Be sure to leave space between foods and between the foods and the interior compartment walls for proper air circulation/cooling. Remove the leaves of root vegetables. Wash and dry off fresh foods before storing them. Always write the content and freezing date on the packing. Remove only the required amount to be thawed. Thawed foods must never be frozen again.

d. **Care while defrosting a refrigerator-freezer**: Most refrigerators-freezers sold today don't require manual defrosting by the consumer. However, still there are some units in the market and in homes that require periodic defrosting. Keep removed refrigerated foods cold and prevent frozen foods from thawing while defrosting. To do this, place the food in a cooler with a cold source. Never use any type of electrical heating device, knife, or other sharp object to remove frost, as this could damage the inner lining.

Manual defrost when frost has accumulated on the inside walls of the freezer to a thickness of ½ inch or so. Remove the food from the refrigerator/freezer, turn off the thermostat or unplug the unit, and allow all of the frost to melt. Once the frost has melted completely, turn the unit back on, wait for it to reach its operating temperature, and restock it with food.

e. **Refrigerator care and cleaning:** To keep food safe cleaning is very important. All refrigerators/freezers should be cleaned with an approved disinfectant weekly and as necessary for spills. Discard any expired food or drink. To keep the refrigerator smelling fresh and removing odours, place an opened box of baking soda on a shelf. Avoid using solvent cleaning agents, abrasives, and all cleansers that may impart a chemical taste to food / ice cubes, or cause physical damage to the interior finish. The exterior may be cleaned with a soft cloth and mild dishwashing detergent. The front grill should be kept free of dust and lint to permit free air flow to the condenser. Several times a year the condenser coil should be cleaned with a brush or vacuum cleaner to remove dirt, lint, or other accumulations. This will ensure efficiency and top performance (Tables 21.1)

Table 21.1: Care and cleaning chart

Part	What to use
Interior/door	Soap, baking soda and water
Liner	Use 2 tablespoons of baking soda in 1 quart of warm water. Be sure to wring excess water out of sponge or cloth before cleaning around controls, light bulb or any electrical part.
Door gaskets	Soap and water wipe gaskets with a clean soft cloth.
Drawers/bins	Soap and water do not wash any removable items (bins, drawers, etc.) in dishwasher.
Glass	Soap and water; glass cleaner; mild liquid sprays
Shelves	Allow glass to warm to room temperature before immersing in warm water.
Exterior	Soap and water; non-abrasive glass cleaner
Handles	Do not use commercial household cleaners, ammonia, or alcohol to clean handles. Use a soft cloth to clean smooth handles. Do not use a dry cloth to clean smooth doors.

f. **Placement:** Position the refrigerator away from sunny windows, heaters, warm air from heating ducts, radiators, stoves and other heat sources. The heat makes cooling harder for your refrigerator. Position the refrigerator to allow good airflow on all sides. Allow the following clearances: Sides 3/4" minimum, back and top 2" minimum.

Proper levelling of refrigerator is essential for proper closure of door. Adjust front legs higher than the rear to allow doors to self-close.

For a good seal, close the door on a piece of paper and try to remove the paper. If it is not held snugly in place, adjust the door or replace the seal.

g. **Power plug:** The power plug and cord must be in good working condition. If damaged, the power cord of this appliance must be replaced by the maker or an equally qualified person so as to prevent any hazards. Pull out the power plug in the case of a power outage, repair or cleaning. For short vacations, leave refrigerator on but use up or discard perishable food. Use of extension cords must be avoided because of potential safety hazards under certain conditions. If it is necessary to use an extension cord, use only 3 wire extension cord that has a 3 blade grounding plug and a 3 slot outlet that will accept the plug (Fig. 21.2).

Many unsafe refrigeration practices are followed in India especially at shops, stores and eating establishment. Business operators rarely give any importance to cleaning and maintenance of refrigerators. Often due to space constraints these are kept in direct sunlight and also to attract customers or advertise the food brands. They are usually overloaded and not even cleaned regularly. Temperature is never monitored. No system is there for power back up. The foods items that stay in temperature danger zone for more than 2 hours, generally melt or thaw, are still not discarded.

To achieve safety of foods during refrigeration it is imperative to follow the basic cleaning and maintenance steps. Routine inspection by food safety officers for temperature and training in this aspect can help in maintaining safety of food requiring cold temperature conditions.

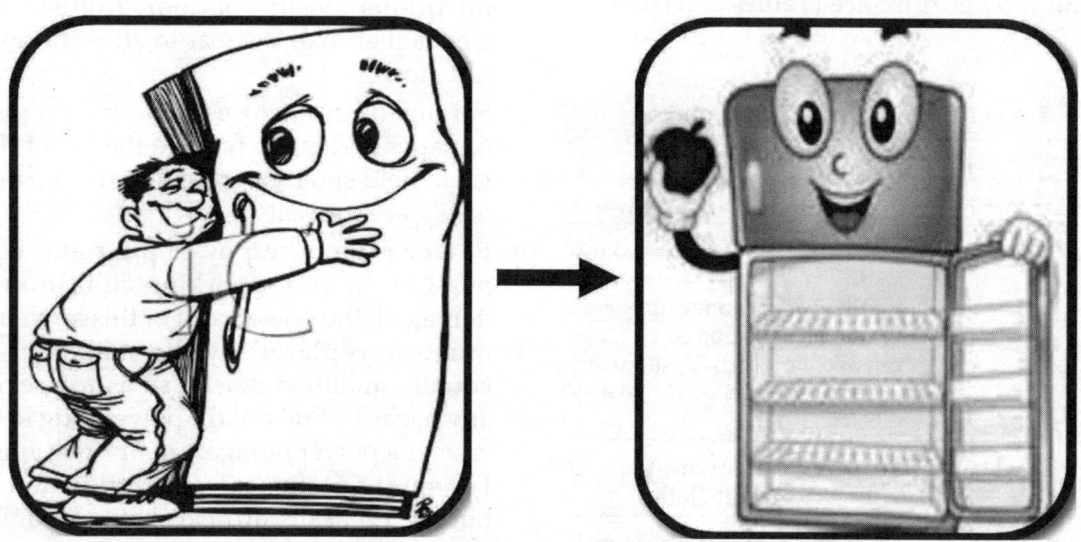

Fig. 21.2: Maintain your refrigerator to keep food safe

22

Cleaning of Kitchen Utensils

Ruchi Sharma, Puja Dudeja

It is not new to find food debris on kitchen utensils on dining table. Also, we commonly see lipstick marks on tip of washed mugs. The importance of properly cleaning the utensils can be appreciated when one realizes that these are one of the major causes of foodborne illness (FBI) outbreaks. Unless we wash our hands, utensils, and surfaces in the *right* way, we could contaminate our food. Cleaning utensils comprises many operations in the food establishment. The process is usually specific to the type of cleaning necessary. No cleaning task in the kitchen or food establishment is as important as the cleaning and sanitization of food contact surfaces of equipment and utensils. All food handlers, like doctors, have a legal and a moral responsibility to provide clean and safe food to their customers. It is up to them to prevent outbreaks of FBI.

Some incidents of FBI have been reported every now and then because of contaminated equipment.

On 1 April 2014, around 500 students at Al-Azhar University in Cairo were hospitalised after suffering from food poisoning at the dorm's cafeteria. The food poisoning was a result of the poor hygiene in the dorm kitchen because of dirty plates and utensils.

On 16 July 2013, at least 23 students died and dozens more fell ill at a primary school in the village of Dharmashati Gandaman in the Saran district of the Indian state of Bihar after eating a Midday Meal contaminated with pesticide. Investigations stated that the food was contaminated by an organophosphate. The cooking oil used in preparing food had been placed in a container formerly used to store insecticides.

Therefore, not only utensils but also surfaces that food may come in contact with should be clean and sanitised. Here it needs to be clarified that cleaning and sanitising are two separate procedures. 'Cleaning' is the removal of all food residues and any dirt or grease that may have become attached to work surfaces, equipment and utensils during the preparation and cooking of food. Thus, cleaning removes any food debris on which microbes may grow. It helps in the following:

- To reduce the risk of FBI.

- To remove any food which may attract food pests, e.g. insects, rodents, birds and domestic pets.

- To reduce the contamination of food by foreign matter, e.g. dust, flaking paint, grease from mechanical equipment.

Here cleaning generally means getting rid of clutters, getting rid of dust, making things shiny, or making it pleasant to look at, while sanitizing more often use in hospitals, public health places to get rid of germs that might cause health problems. Cleaning and sanitising are usually to be done as separate processes. A surface needs to be thoroughly cleaned before it is sanitised. This is because sanitizers are unlikely to be effective in the presence of food residues and detergents.

Cleaning of utensils may take place in a kitchen, utility room, scullery or elsewhere. Different implements are also used for this purpose, e.g. cloth, sponges, brushes or even steel wool. In hand-washing, plastic brushes with nylon bristles are preferred to wash instead of clothes or sponges, which can spread microorganisms.

If dishcloth is used it is best to change at least every other day. We need to change it immediately if it been used to wipe down surfaces contaminated with raw foods including raw meat and root vegetables. Otherwise any germs present will almost certainly spread to the dishcloth. Boiling dishcloth for 15 minutes or washing in a washing machine on a standard cycle are the best ways to kill any germs that might be present. If dishcloth is used to wipe off 'high risk' food residues such as raw eggs, raw meat/fish/poultry, raw vegetables, change it straight away. Do not use a cloth that smells bad or if it looks dirty as it is very likely it will have high numbers of bacteria on it.

Using sponge can spread contamination as bacteria live inside the sponge and spread to utensils and kitchen surfaces. Since steel wool can be abrasive, do not use it on dishes made from delicate materials, like bone china.

Person who washes kitchen utensils should also be clean and disease free. These days

we have trend of hiring maid for washing utensils. We need to assure that the person washing utensils follows proper hygiene measures.

Washing utensils may seem like a self-evident task: soak, foam up, scrub, and rinse. But there are myriads of variations in people's dishwashing habits and technique. Some standard steps followed in cleaning utensils are:

1. **Prepare the dishes:** Cleaning of utensils is either achieved by hand in a sink or using dishwasher. To avoid polluting your sink of water, begin by scraping the dishes of excess food. Do not throw the residue in the sink. Stack the dishes in preparation for washing.

2. **Prepare the water:** In principle all that is required to clean utensils is water. Water used for cleaning utensils should be clean and free from contamination. Also, make sure to use the correct water temperature. Here a major variation in method is the temperature and state of the water. Indians usually prefer running water. It is considered more hygienic as the water is not being reused. Usually cold water is used here. This is practical in environments where hot water is rarely available from the tap, and sinks are perceived as dirty surfaces (essentially a convenient drain). Westerners usually prefer standing hot water. This is practical in environments where hot water is easily available, and sinks are perceived as clean surfaces (essentially a bowl with a convenient drainage device).

3. **Soak the utensils:** Tough stuck on foods may need to be soaked first before washing. Soak dishes in warm water for 10–15 minutes. It will soften hard to get off residue and rid them of any leftover food.

4. **Choice of detergent:** Mostly detergents are used for dishwashing. Dishwasher detergents are strongly alkaline. Different kinds of dishwashing detergent contain different combinations of the items in the list below. Not all of the listed ingredients are used in some detergent.

- **Phosphates:** These bind calcium and magnesium ions to prevent 'hard-water' type lime scale deposits. They can cause ecological damage, so their use is gradually being phased out. Phosphate-free detergents are sold as ecofriendly detergents.

- Oxygen-based bleaching agents (older-style powders and liquids contain chlorine-based bleaching agents). They break up and bleach organic deposits.

- **Non-ionic surfactants:** They lower the surface tension of the water, emulsify oil, lipid and fat food deposits and prevent droplet spotting on drying.

- **Alkaline salts:** These are a primary component, in older and original-style dishwasher detergent powders. Highly alkaline salts attack and dissolve grease, but are extremely corrosive (fatal) if swallowed.

Dishwashing detergent may also contain anti-foaming agents, additives to slow down the removal of glaze and patterns from glazed ceramics, perfumes, anti-caking agents (in granular detergent), starches (in tablet based detergents), gelling agents (in liquid/gel based detergents) and sand (inexpensive powdered detergents).

Inexpensive powders may contain sand. Such detergents may harm the dishes and the dishwasher. Powdered detergents are more likely to cause fading on bone china items.

Traditionally, dishwashing is done by scrubbing the utensils with a wet cloth piece dipped in scrub ash to scrub away the dirt. Scrub ash is specially made by burning wood (may be in *"chulhas"*) for dishwashing.

5. **Washing technique:** Begin by washing glassware first to prevent spots or streaks. Then, move on to plates, bowls and other china. Silverware should be next in line for washing, followed by pots and pans and other cooking utensils. There are two ways to wash your dishes: by hand or in the dishwasher. Both ways have their devotees. There seems to be a disagreement about which is better. Some prefer the dishwasher for its convenience, and others prefer hand washing because you can control how much water you use and ensure that your dishes are clean the first time.

There are many methods you can use to wash your dishes by hand. Commonly used are single sink method, double sink method and three sink method in the kitchen. Use of soap or sanitizer is mandatory in all these methods. In these methods scrubber is used to remove tough stains. As fingernails are often more effective than soft implements like clothes at dislodging hard particles, washing simply with the hands is mostly done in India and can be effective as well. Where manual dishwashing is employed, equipment and utensils are thoroughly washed in a warm detergent solution which is kept reasonably clean, and rinsed free of such solution.

Salt is a natural abrasive that can be rubbed into the pans to help remove dried-on foods or grease. Another option is to add baking soda to the water in the pan and place it on the stove to boil. Let it soak, and then scrape off the food that remains with a spatula or knife. Then the dishes are rinsed in cool water to ensure all soap is removed.

In one sink method utensils are pre-cleaned, washed, rinsed and sanitized in the same sink.

Steps of Cleaning in two Sink Method

In two sink method, one sink is usually first filled with dirty dishes (which may have already been rinsed and scraped to remove most food) and hot, soapy water. The detergent is added while this sink is filling with water, so a layer of foam forms at the top. Then the dishes are washed one by one and thoroughly rinsed in other sink filled with hot water to remove the grease dislodged by

soap as well as the soap itself, and act as sanitizer. Then utensils are placed on a rack to begin drying, or dried and put away immediately (Fig. 22.1).

When the sink is empty, if there are more dishes to be washed they may be added to the same dishwater, or the sink may be drained and refilled if clean, hot dishwater is desired.

Public eating or drinking places should preferably conduct manual washing and sanitization of utensils only in three-compartment sinks. Most health and sanitation standards require restaurants to have a sink with three or more compartments to manually clean utensils and smaller equipment. Here too same procedure of soak, scrub, rinse and sanitize is given in Fig. 22.2.

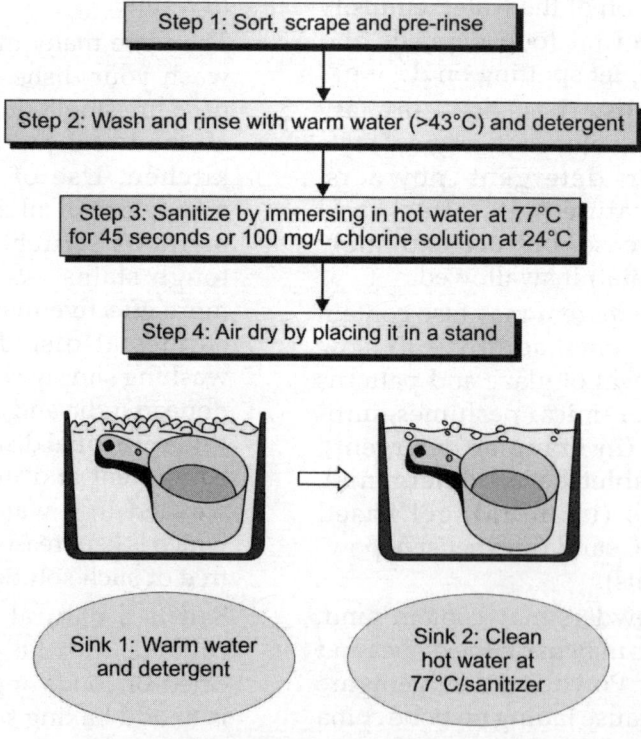

Step 1: Sort, scrape and pre-rinse

Step 2: Wash and rinse with warm water (>43°C) and detergent

Step 3: Sanitize by immersing in hot water at 77°C for 45 seconds or 100 mg/L chlorine solution at 24°C

Step 4: Air dry by placing it in a stand

Sink 1: Warm water and detergent

Sink 2: Clean hot water at 77°C/sanitizer

Fig. 22.1: Two-sink method

Steps of Cleaning Utensils (3-sink method)

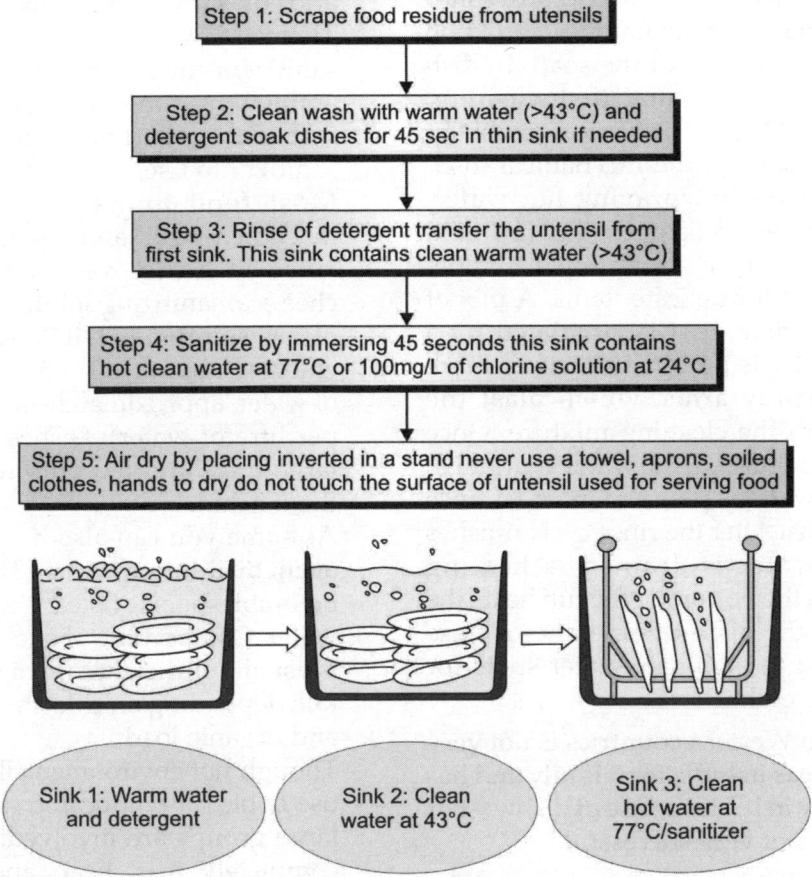

Step 1: Scrape food residue from utensils

Step 2: Clean wash with warm water (>43°C) and detergent soak dishes for 45 sec in thin sink if needed

Step 3: Rinse of detergent transfer the untensil from first sink. This sink contains clean warm water (>43°C)

Step 4: Sanitize by immersing 45 seconds this sink contains hot clean water at 77°C or 100mg/L of chlorine solution at 24°C

Step 5: Air dry by placing inverted in a stand never use a towel, aprons, soiled clothes, hands to dry do not touch the surface of untensil used for serving food

Sink 1: Warm water and detergent

Sink 2: Clean water at 43°C

Sink 3: Clean hot water at 77°C/sanitizer

Fig. 22.2: Three-sink method

6. **Rinse the dish washing suds and residue from the dishes:** As you finish with scrubbing each dish, you can rinse it off under the tap, or in the second sink if using the double-sink method. For this method, you can simply fill the second sink with lukewarm water and dunk the dishes in to rinse them, replacing the water as needed. You can also use a dish pan filled with hot water to rinse/dip your dishes clean. After rinsing in clean water utensils are hung to drip dry.

7. **Dry dishes:** If right water temperature is used, the dishes dry quickly on their own. Otherwise, dish rack may be used for drying. Be sure to put bowls and glasses in the rack upside down so that the water does not pool. The most acceptable method of drying equipment and utensils is air drying. The use of towels for drying, polishing may re-contaminate equipment and utensils with bacteria. In instance, if dish towel is also used, make sure the towel is clean. Change the towel when it becomes damp.

8. **Use of dishwasher:** Cookware may also be safely washed in an automatic dish-water although continued dishwasher may use may dull the handles and knob due to the hardness of the soap. In dish washer unlike manual dishwashing, which relies largely on physical scrubbing to remove soiling, the mechanical dish-washer cleans by spraying hot water, typically between 55 and 75°C (130 and 170°F) at the dishes, with lower tempe-ratures used for delicate items. A mix of water and detergent is circulated by a pump. Water is pumped to one or more rotating spray arms, which blast the dishes with the cleaning mixture. Once the wash is finished, the water is drained, more hot water is pumped in and a rinse cycle begins. After the rinse cycle finishes and the water is drained, a heating element in the bottom of the tub heats the air to dry the dishes. Sometimes a rinse aid is used to eliminate water spots for streak-free dishes.

Food in Western countries is not very oily, whereas in India food is oily and has lot of gravy in it. Hence use of dishwasher in India is not very successful.

If necessary dishwasher may be used for Indian cooking- exceptions are over burnt *"kadais"*, milk vessels and large vessels.

Where dishes are to be shared among many, such as in restaurants, sanitization is necessary and desirable in order to prevent spread of microorganisms.

9. **Sanitization:** Sanitize means to apply heat and/or chemicals (or other proc-esses) to a utensil so the number of micro-organisms on the surface is reduced to a level that is safe for food contact and does not permit the transmission of infectious disease. Sanitizing destroys the invisible germs and reduces their number to safe levels. Hot water (recommended tempe-rature of 77°C for at least 30 seconds) and/or sanitizers are used to sanitise. There are two common types of chemical sanitizers-chlorine bleach and quaternary ammonium compounds. Chlorine bleach is the easiest and most effective chemical sanitizer to use.

Most food joints use dish-washing machine which sanitizes dishes by a final rinse in either very hot water or a chemical sanitizing solution such as dilute bleach solution (50–100 parts per million chlorine; about 2 ml of 5% bleach per litre of water, approximately one capful bleach per litre of water). Dishes are placed on large trays and fed onto rollers through the machine.

At home you can also make a sanitizing agent by diluting bleach. The ratio would be 1 tablespoon of bleach to one gallon of water. Contact time of one to five minutes is usually sufficient to achieve a thorough kill, depending on chlorine concentration and organic load.

Though not environmentally friendly, the use of bleach is critical to sanitation when large groups are involved: it evaporates completely, it is cheap, and it kills most germs. Cabinets, refrigerators, countertops, and anything else touched by people in a large group setting should be periodically wiped or sprayed with a dilute bleach solution after being washed with soapy water and rinsed in clean water.

However, bleach is less effective in the presence of organic debris, so a small amount of food residue can be enough to permit survival of bacteria, e.g. Salmonella bacteria. Scrubbing followed by soaking in bleach is effective at reducing Salm-onella contamination, but even this method does not completely eliminate them.

Sanitizing agents may include a solution of one part vinegar added to one part water. It will also clean and sanitize your utensils. (It will also remove tough stains and grease on plastic and metal utensils.) Other agents may include 50 parts per million of available chlorine at a temperature not less than 75 degree Fahrenheit. Iodine may also be used as sanitizing agent. For proper sanitization utensils need to be immersed in this solution for at least 1 minute.

10. **Proper Storage and Handling:** Proper storage and handling of cleaned and sanitized equipment and utensils is very important to prevent recontamination prior to use. Cleaned and sanitized equipment and utensils must be stored on clean surfaces and handled to minimize contamination of food contact surfaces.

According to HACCP food safety standards cleaning and sanitizing in kitchen involves:
- Cleaning and sanitizing multi-service eating utensils after each use
- Cleaning and sanitizing utensils or equipment other than multi-service articles as often as is necessary
- Equipment and facilities for the cleaning and sanitizing of utensils consists of either:
  1. Mechanical equipment
  2. Manual equipment consisting of: three sinks of corrosion-resistant material large enough for thorough cleaning and sanitizing of utensils, and draining racks of material that is corrosion-resistant

Utensils shall be:
- Pre-rinsed or pre-scraped to remove gross food particles and solids
- Washed in a detergent solution, that is capable of removing grease
- Sanitized

In Mechanical Dishwashing

Ideal temperatures should be reached

Wash: Between 60°C (140°F) and 71°C (160°F)

Sanitizing Rinse: 82°C (180°F) for at least 10 seconds or using an approved chemical solution. Following precautions need to be taken
- Be sure the unit is working properly
- Sort, scrape and pre-rinse dishes and utensils
- Wash, follow the manufacturer's instructions
- Rinse, maintain the proper temperature
- Allow the dishes and utensils to air dry

A Little Piece of Advice

If you cannot wash utensils immediately, instead of leaving dirty dishes out on the counter for hours, at least rinse them right away so the food debris does not have a chance to set. This will make washing much quicker later on.

Processing fresh food on or with dirty equipment will contaminate the food. Food utensils and equipment must be cleaned and sanitised before each use and between being used for raw food and ready to eat food. Where utensils or equipment have been used continuously over an extensive period to prepare, process or serve the same food, they will also need to be cleaned and sanitised at regular intervals. For example, serving utensils that are provided and used for one type of salad in a salad bar, or a meat slicer used to slice fresh fish.

Also, public eating or drinking places which do not have adequate and effective facilities for cleaning and sanitizing utensils should use single-service articles or disposable which may be used only once.

23

Hazard Analysis and Critical Control Point

Meenakshi Sharma, Amarjeet Singh

To visit mall on weekend is the most sought after way to chill oneself from hectic schedule of the whole week. One buys household goods as well as enjoys the shopping spree.

Last Sunday I went to a famous mall in Chandigarh city along with my six-year-old son. We both took a trolley each and started piling things in it. After shopping for merchandise, we both finally reached grocery area. As usual I knew that my son will add all sugary things in trolley. He added three packs of juices, all of different brands. I glared at him and told him to pick the juice carton of the

brand which we generally take. But he was adamant to buy the juice carton which had some logo. I decided to be firm with him. As I was taking away the pack from his hand from nowhere a salesgirl emerged and said Ma'm, your child is very wise, unlike you, he is picking up the brand which is hazard analysis and critical control point (HACCP) certified.

I just said thanks to her; but I was perplexed. I wondered what was this HAACP? I curiously moved in the store and inspected some products for HACCP certified (Fig. 23.1). I was surprised to see so many such products.

Fig. 23.1: Dry mango powder showing HACCP certification

One can say that I was not able to believe that a salesperson had better knowledge than a doctorate degree holder. Now I was so curious to know about the HACCP word like a little child who is very eager to know the answers to her queries. Thank God! we have smart phones. I dropped my son to 'play zone' and instantly started browsing Google.

HACCP is a concept that Indian women apply to kitchens. Does it sound illogical? You will agree with me once you finished the chapter. Let us first go through the technical aspect of the HACCP (**Hazard Analysis and Critical Control Points**).

Conception of the HACCP: In 1960, National Aeronautics and Space Administration (NASA) asked Pillsbury to design and manufacture the first foods for space flights. It was imperative to provide 100% safe food to the astronauts during expedition to avoid any foodborne illness. HACCP was developed as a logical tool for adapting traditional inspection methods to a modern, science-based, food safety system. Afterward, this NASA risk prevention system has been called "the most revolutionary institutional innovation to ensure food safety of the twentieth century". HACCP regulations have been implemented by the US food and Drug Administration (FDA) to maintain the integrity of seafood and juice in the United States. The US department of Agriculture (USDA) also relies on HACCP systems in the nation's meat and poultry plants and slaughterhouses. In 1963, the World Health Organization (WHO) issued HACCP principles in Codex Alimentations. In 1973 NASA (USA national air space Foundation), Natick American Army laboratory and Pillsbury group company's common project for astronauts in food production with zero error, i.e. HACCP entered to literature. In 1985 USA national science academy suggested that HACCP should be applied in food operations for food safety. On 14 june 1993, HACCP entered to regulations of Europe Community Countries with directives "Hygiene of food matters". In 1994, the organization of *International HACCP Alliance* was also established for the US meat and poultry industries to assist them with implementing HACCP. Now, its membership has been spread over other professional/industrial areas also.

HACCP in India: In India, there has been a slow take up of HACCP. It began to change with the liberalization of Indian economy in early 1990s. The HACCP implementation has made large progress largely in response to strict requirements on safety and quality enforced by major trading partners such as USA and EU. As a result of early consignments being blocked as unsafe, the export industry has undergone fundamental changes.

Definition of HACCP: It is a management system in which food safety is addressed through the analysis and control of biological, chemical, and physical hazards from raw material production, procurement and handling, to manufacturing, distribution and consumption of the finished product.

Based on risk-assessment, HACCP plans allow both industry and government to allocate their resources efficiently in establishing and auditing safe food production practices. HACCP is a method which food businesses can use to ensure that their products do not put consumers at risk. The details of a HACCP system will vary in different establishments as no two businesses are exactly alike—but the principles are the same.

Hazard: A hazard is a biological, chemical or physical agent that is reasonably likely to cause illness or injury in the absence of its control (here FBI).

Type of hazard: The types of hazards which a HACCP plan can focus on include:

1. Biological hazards, e.g. harmful micro-organisms.
2. Chemical hazards, e.g. those chemicals which are either naturally occurring, intentionally added or unintentionally added.
3. Physical hazards, e.g. glass, stones or metal.
4. Compromised packaging quality.
5. Compromised equipment reliability.

CCP (critical control points) is an identifiable point in the food production chain where a hazard may occur. Action is taken at CCP to prevent the hazard from occurring. This can either be a point, step or procedure at which control can be applied and is essential to prevent or eliminate a hazard or reduce it to an acceptable level. A CCP can be used to control more than one hazard–Alternatively, several CCPs may be needed to control one hazard. Points may be identified as CCP when hazards can be prevented, for example:

1. Introduction of chemical residue can be prevented by control at the receiving stage.
2. A chemical hazard can be prevented by control at the formulation or ingredient-addition stage (Fig. 23.2).

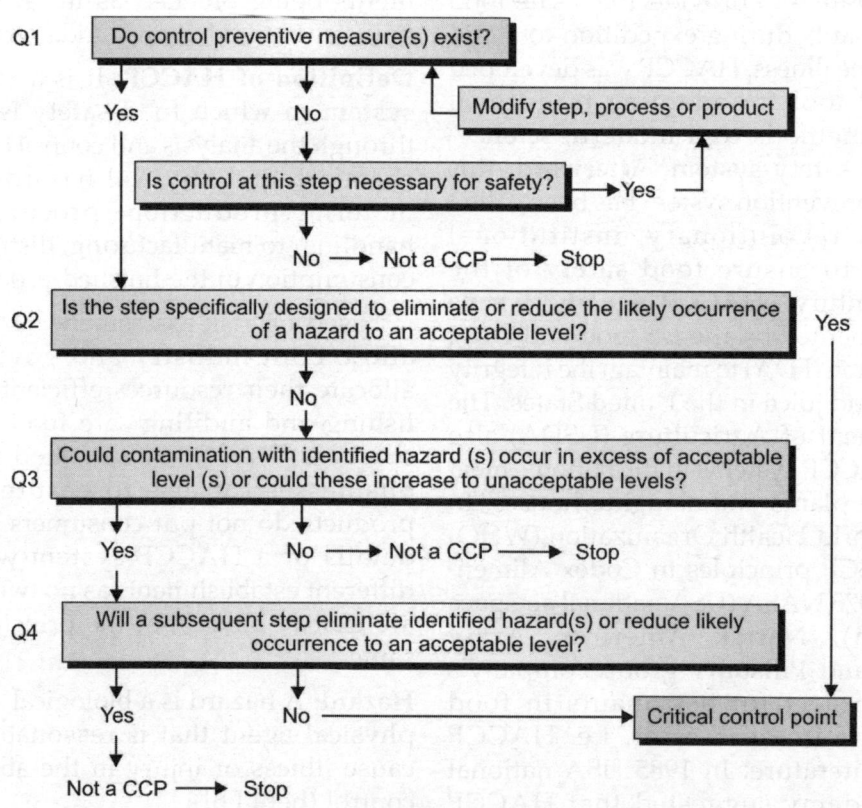

Fig. 23.2: Critical control points and decision tree

3. Pathogenic bacteria growth can be controlled by refrigerated storage or chilling.

CCP may be identified where hazards can be eliminated; for example:

1. Pathogenic bacteria can be killed during cooking
2. Metal fragments can be detected by a metal detector and eliminated by removing the contaminated product from the processing line
3. Parasites can be killed by freezing.

Points may be identified as CCPs when hazards are reduced to acceptable levels, for example the occurrence of foreign objects can be minimised by manual sorting and automatic collectors.

The critical limit is the criterion that separates acceptability from unacceptability. Critical limits must be specified and validated if possible for each critical point. In some cases more than one critical limit is elaborated at a particular step. Criteria often used include measurements of temperature, time, moisture level, pH, sensory parameters such as visual appearance and texture (Fig. 23.2).

PRINCIPLES OF THE HACCP SYSTEM

The HACCP system consists of the following seven principles:

PRINCIPLE 1

Conduct a hazard analysis.

PRINCIPLE 2

Determine the CCPs.

PRINCIPLE 3

Establish critical limit(s).

PRINCIPLE 4

Establish a system to monitor control of the CCP.

PRINCIPLE 5

Establish the corrective action to be taken when monitoring indicates that a particular CCP is not under control.

PRINCIPLE 6

Establish procedures for verification to confirm that the HACCP system is working effectively.

PRINCIPLE 7

Establish documentation concerning all procedures and records appropriate to these principles and their application.

Qualities of the HACCP System

1. Systematic—All the potential hazards are identified before there is a problem.
2. Efficient—It concentrates the control effort at the stages where the risk is potentially the highest
3. On the spot—The processes can be controlled immediately by the food business.

Prior to the application of HACCP to any sector of food chain general principles of food hygiene and appropriate food safety legislation should be ensured. Management commitment is necessary for implementation of an effective HACCP system. During hazard identification, evaluation, and subsequent operations in designing and applying HACCP systems, consideration must be given to the impact of raw materials, ingredients, food manufacturing practices, role of manufacturing processes to control hazards, likely end use of the product, categories of consumers of concern, and epidemiological evidence relative to food safety.

So far in this chapter enough technical jargon on HACCP has been used. Let us compare the sequence of the application of HACCP in both manufacturing unit and in our household kitchen (Table 23.1). HACCP principles are applied in commercial kitchen also. It is essential to ensure that the food coming off the kitchen is safe for customers to eat. Here to explain the concept, example from household kitchen has been taken.

Most important is regular cleaning of kitchen. In our literature we had a practice of not taking slippers in kitchen. Even now many households do practice that. Cleaning of kitchen ensures daily, taking out trash, cleaning the gas stove and dusters, etc. Failing to do these petty jobs on time can result in bacteria build up and potential food safety problems.

Then is the most powerful tools in public health, i.e. proper hand washing. Proper hand washing prevents the spread of everything from common cold to H1N1 hepatitis A. In our culture practice of not only washing hands but bathing before entering the kitchen is prevalent. It has been seen that mother-in-law usally tells the newlywed daughter-in-law *"rasoi main jane se pehle chaapal bahar uttarni hai, aur nahane ke baad jana hai"* (Do keep your slippers outside and bath before entering the kitchen)

Hazard description should be done and critical limits are set by expert team on the basis of food and drug regulations. Accordingly, monitoring procedures are drafted. Deviation procedures and verification procedures are also documented in the HACCP sheet. Finally the utmost important is to maintain the HACCP records (Tables 23.2 and 23.3).

Let us take another example from commercial company "XYZ" producing. We will try to illustrate the procedures used in plan development for HACCP in this fictitious company.

Company: XYZ

Final product: Refrigerated tomato juice

Procedures/steps:

1. **Incoming materials:** Locally grown fresh tomatoes are procured directly from farms. Tomatoes are received in bulk in wooden boxes and upon receipt are visually exam-

Table 23.1: Application of HACCP in manufacturing unit and in household kitchen

	In a manufacturing unit	*Household kitchen*
Assemble HACCP team	A multidisciplinary team generally comprising experts from corporate quality assurance, manager, site in charge quality assurance (QA) officer and chemists	Generally home maker in addition to mother-in-law and maids
Describe product	Manufacturing of pasteurized honey	Cake
Identify intended use	Ready-to-eat	Ready-to-eat
Construct flow diagram	Figure 23.1	In the mind of homemaker
On-site confirmation of flow diagram	Done by multidisciplinary team	In the mind of homemaker
List all potential hazards conduct a hazard analysis at all levels*	Physical hazards in raw honey viz metal and non-metal particles such as wood, stone, or glass	Physical hazard like dead insects in flour, sodium bicarbonate, cocoa powder.

* Hazards are classified at all levels, i.e. incoming material, honey testing, in-process step and packaging. Here to explain the concept, I have illustrated only physical hazards.

Table 23.2: Physical hazard CCPs present in the received raw honey

Hazard Description	Critical Limits	Monitoring Procedures	Deviation Procedures	Verification Procedure	HACCP Records
Reception of possibly unsafe materials (not listed) or reception of lids, raw honey and seed honey from non-contract suppliers could result in honey contaminated with harmful chemical residues	Meets food and drug regulations Listed suppliers and types of barrels	Receiver to ensure raw honey is from producers with contractual specification regarding antibiotics Lids are from approved supplier Barrels and seed pails are correct type	Receiver will reject lot notify management and record rejected supplier.	Quality control (QC) will review records review contract specifications and carry out periodic random antibiotic residue testing	Supplier lists Contract specifications Lot receipt record Lot reject record Antibiotic test results Deviation record

Table 23.3: Physical hazard CCPs present in the raw material for cake

Hazard Description	Critical Limits	Monitoring Procedures	Deviation Procedures	Verification Procedure	HACCP Records
Contamination of flour, sugar, cocoa powder, baking powder with harmful dead insect wings, stones or any other kind of adulteration	Meets the standards set by homemaker	To ensure specification of the product by naked eyes	Homemaker will reject and notify the grocery store.	Done by sieving/manual sorting of the flour the requisite sieve	Mind of the homemaker

ined for gross filth and weight of the boxes. Following acceptance, the tomatoes are assigned a lot number, and placed in refrigerated storage.

Packaging materials are delivered in clean, well maintained and covered vehicles. All materials are checked for integrity and other specifications. They are assigned lot numbers and placed into dry-storage warehouse.

2. **Processing:** Tomatoes are transferred from refrigerated storage to processing area in the slotted hopper. From there, the tomatoes go into flume tank containing treated

water. Tomatoes are elevated, dewatered and moved to the processing facility over inspection rollers where visually defective tomatoes are removed. Finally they are elevated, rinsed in potable water, drained, and dropped into a grinder. Juice is collected and pumped to a balance tank where juice is held until it goes to the pasteurizer. At the end, the juice is pumped into a refrigerated bulk storage tank and from there pumped to the filter.

3. **Packaging:** Plastic containers are cleaned using compressed air. Each primary container is identified by the production date, code and lot number.

Juice is pumped into reservoir on the filler and gravity fed into 1 gallon plastic containers that are pre labeled. After filling, immediately caps are mechanically applied to the plastic containers. Filled, dried containers are checked weighed and packed into shipping cartons as required by the customer.

4. **Storage shipping:** All finished product is placed into cooler storage without delay. All product is stored and shipped on a first in, first out basis.

Hazard Identification and Evaluation, and Justification for Decisions for Tomato Juice

Biological hazards such as vegetative pathogens and cryptosporidium and chemical hazards including aflatoxin and pesticides are identified as potential hazards. No physical hazards are identified. The cryptosporidium contamination can occur even if the potable water is used after ensuring sanitation procedures. The aflatoxin is also a significant hazard in the juice and its level could increase further during cold storage of tomatoes. Pesticide residues may be found on incoming tomatoes. However, the occurrence of unapproved pesticide residues in the food is likely to be infrequent.

Control Measures

Control measures are actions and activities that can be used to prevent or eliminate a food safety hazard or reduce it to an acceptable level.

For tomato juice it can be noted cryptosporidium contamination cannot be controlled only at the supplier agreement specifying the use of plant-picked and undamaged tomatoes along with the subsequent culling and washing steps.Since levels of aflatoxin could increase a more effective control measure need to followed after the cold storage step in the process. Table 23.4 shows the identification of hazard and measures to be applied for controlling the significant hazards.

Similarly we can apply the concept to other hazards, i.e. biological, chemical, etc. Foodborne illness can be prevented in home after following certain processes or handling practices. Practices which can prevent or control the "dinner plate" microbial con-

Table 23.4: Hazard identification and measures to control significant hazards

Ingredient	Identify potential hazards introduced, controlled or enhanced in this step	Are any potential food safety hazards significant (Yes/No)	Justify your decision	Measures applied to control the significant hazards
Receiving (raw tomatoes)	Biological (B) 1. Vegetative pathogens 2. Protozoan pathogens	1. Yes 2. Yes	History of outbreaks In India, unapproved pesticides levels	Pasteurization

Contd.

Table 23.4: Hazard Identification and measures to control significant hazards (Contd.)

Ingredient	Identify potential hazards introduced, controlled or enhanced in this step	Are any potential food safety hazards significant (Yes/No)	Justify your decision	Measures applied to control the significant hazards
	Chemical (C) 1. Pesticides 2. Aflatoxin	1. No 2. Yes	rarely occur Cuses illness or injury. May exceed regulatory specifications if not controlled	1. NA 2. Culling
Receiving (packaging)	None			
Dry storage Cold storage	None B –None C–Aflatoxin P–None	C–yes	Aflatoxin levels may increase during cold storage due to fungus	C–Culling
Remove debris (slotted hopper)	None			
Pateurizer/Cooler	B 1. Vegetative pathogens 2. Protozoan pathogens C–None P–None			
Holding tank Fill Case/code/ palletize Cold storage Ship	None			

tamination associated with FBI, are under the direct control of the homemaker. Our homemakers are very alert while they are **purchasing, storing, pre-preparation, cooking, serving, and handling leftovers**. It would not be wrong to say that they all are HACCP certified. That is the reason why it is universal that *"ghar ke khane ki kya baat hai"*(Home food is the best).

Conclusion

As, most of the food industry in India is disorganized, hence establishing effective systems for assuring food safety is an uphill task. Focus in India is mostly on the export products. How HACCP or some similar kind of principles can be taught to our *pani puri wala* or road side *dhabas*? This is a question that needs to be pondered and answered.

24

ISO Certification for Food Safety: Quality Management System (ISO 9001:2008)

Arun Gupta

INTRODUCTION

ISO (the International Organization for Standardization) is a worldwide federation of national standards bodies (ISO member bodies). The work of preparing International Standards is normally carried out through ISO technical committees. Each member body interested in a subject for which a technical committee has been established has the right to be represented on that committee. International organizations, governmental and non-governmental, in liaison with ISO, also take part in the work. ISO collaborates closely with the International Electrotechnical Commission (IEC) on all matters of electrotechnical standardization.

The main task of technical committees is to prepare International Standards. Draft International Standards adopted by the technical committees are circulated to the member bodies for voting. Publication as an International Standard requires approval by at least 75% of the member bodies casting a vote. Attention is drawn to the possibility that some of the elements of this document may be the subject of patent rights. ISO shall not be held responsible for identifying any or all such patent rights. ISO 9001 was prepared by Technical Committee ISO/TC 176, Quality management and quality assurance, subcommittee SC 2, quality systems.

This fourth edition cancels and replaces the third edition (ISO 9001:2000), which has been amended to clarify points in the text and to enhance compatibility with ISO 14001:2004.

History: ISO 9001

Formalised quality assurance originally came from the Defence Industry's need for standards. For example, to supply the Ministry of Defence (MoD) a company had to write up its procedure for making its product, have the procedure inspected by the MoD and then ensure that its workers followed the published procedures.

The idea of quality assurance spread beyond the military and in 1966, the UK Government led the first national campaign for quality and reliability with the slogan "Quality is everyone's business." However, by this time, suppliers were being assessed by number of their customers and it was widely recognised that such duplication of effort was a chronic waste of time and money. Progress was finally made in 1969, when a UK Government committee report on the subject recommended that suppliers' methods should be assessed against a generic standard of quality assurance.

In 1971, the British Standards Institute (BSI) published the first UK standard for quality assurance (BS 9000), which was developed for the electronics industry. Then, in 1974, the BSI published BS 5179; Guidelines for Quality Assurance. This led to a shift in the burden of inspection from the customer to the supplier, as quality assurance could be guaranteed by the supplier to the customer through third-party inspection.

Through the 1970s, the BSI organised meetings with industry to set a common standard, which culminated in the BS 5750 standard in 1979. Key industry bodies agreed to drop their own standards and use BS 5750 instead. The purpose of BS 5750 was to provide a common contractual document, demonstrating that industrial production was controlled.

The ISO 9000 certification standard has evolved over several revisions. The initial 1987 version (ISO 9000:1987) had the same structure as the UK Standard BS 5750, with three 'models' for quality management systems, the selection of which was based on the scope of activities of the organisation. The language of this first version of the Standard was influenced by existing US and other Defence Military Standards, so it was more accessible to manufacturing and was well suited to the demands of a rigorous, stable, factory-floor manufacturing process. With its structure of twenty 'elements' or requirements, the emphasis tended to be overly placed on conformity with procedures rather than the overall process of management; which was the original intent.

The 1994 version (ISO 9000:1994) was an attempt to break from the practices which had somewhat clouded the use of the 1987 standard. It also emphasised quality assurance via preventive actions and continued to require evidence of compliance with documented procedures. Unfortunately, as with the first edition, companies tended to implement its requirements by creating shelf-loads of procedure manuals and become burdened with ISO bureaucracy. Adapting and improving processes could be particularly difficult in such an environment.

The 2000 version of the standard (ISO 9001:2000) sought to make a radical change in thinking. It placed the concept of process management at the heart of the standard, making it clear that the essential goals of the standard, which had always been about 'a documented system' not a 'system of documents', were reinforced. The goal was always to have management system effectiveness via process performance measures. This third edition makes this more visible and so reduced the emphasis on having documented procedures if clear evidence could be presented to show that the process was working well. Expectations of continual process improvement and tracking customer satisfaction were also made explicit in this revision. A new set of eight core quality management principles, designed to act as a common foundation for all standards relating to quality management, were also introduced.

The fourth and current edition of the standard (ISO 9001:2008) arrived on 14 November, 14th 2008. This revision contains minor amendments only. The aim of this revision is to clarify existing requirements and to improve consistency of approach with other management standards, like ISO 14001:2004.

It has been announced that ISO are working on the ISO 9001:2015 revision of the standard to bring it up-to-date and reflect latest quality management good practice. However, fundamentally the standard will stay the same and, as such, migration to ISO 9001:2015 will be straightforward.

ISO 9001:2008 Background

A quality management system (QMS) is a collection of business processes focused on achieving quality policy and quality objectives to meet customer requirements. It is expressed as the organizational structure, policies, procedures, processes and resources needed to implement quality management. Early systems emphasized predictable outcomes of an industrial product production line, using simple statistics and random sampling. By the 20th century, labour inputs were typically the most costly inputs in most industrialized societies, so focus shifted to team cooperation and dynamics, especially the early signalling of problems via a continuous improvement cycle. In the 21st century, QMS has tended to converge with sustainability and transparency initiatives, as both investor and customer satisfaction and perceived quality is increasingly tied to these factors. Of all QMS regimes, the ISO 9000 family of standards is probably the most widely implemented worldwide.

ISO 9001:2008 is an international standard. It is one of the most widely known and the basis of many highly effective quality systems. The standard can be applied to any size of company or organization, any industry and any country, for both services and products. The standard itself consists of a set of requirements. These requirements specify what a company must do, but not how to do it. The ISO 9001:2008 Quality Management System certification enables to demonstrate commitment to quality and customer satisfaction, as well as continuously improving quality systems and integrating the realities of a changing world.

The international organization for standardization has reviewed ISO 9000 standards of version 1994 for a group of reasons which include:

- International standards should be reviewed each 5 years.
- Increasing focus and demand on meeting customer needs and other stakeholders (owners, employees, suppliers, etc.) and the increasing demand towards performance excellence
- Need to have consistency and compatibility with other management standards such as ISO 14000 standards

Besides problems with the old version of ISO 9000 (1994) standards such as wording, not process oriented, not easy to use and difficulty in implementation by small enterprises. Accordingly the ISO issued new standards of ISO 9000 in December 2000, and a transition period was given till 15 December 2003 for previously certified companies according to the old standards (ISO 9001, ISO 9002. ISO 9003:1994)

The changes to the standard have taken into account the above findings and allow organizations to focus on the processes that are essential to their operations.

ISO 9000:2008 Series

ISO 9000 series was established in accordance with the eight quality management principles[4] which are:

A. Principle 1: Customer focus—This standard relates to customer needs and customer service: a business should understand their customers and seek to meet their requirements. Where possible, they should aim to exceed customer expectations. The benefits of this are increased customer loyalty, increased revenue due to the ability to spot new customer opportunities and increased effectiveness of processes related to customer satisfaction.

B. Principle 2: Leadership—This standard relates to the direction of the organization:

a business should have clear objectives and employees should be actively involved in achieving this. The benefits of this are primarily employee engagement and increased motivation. Research has shown that if employees are kept 'in the loop' with regards to business vision they are likely to be more productive. One of the most comments employee complaints is lack of communication; this principle seeks to rectify that.

C. **Principle 3:** Involvement of people—This principle recognizes that an organisation is nothing without its staff and that their abilities should be used to full effect for business success. The benefits of this principle are employee motivation and increased innovation. When people feel that their skills are being used well they are more likely to work to their maximum potential and contribute ideas. This principle also emphasises the importance of making employees accountable for their actions, leading to a greater feeling of responsibility.

D. **Principle 4:** Process approach—The process approach relates to efficiency and the understanding that appropriate processes will speed up activities. The main benefits of this, aside from efficiency, are reduced costs due to effective use of resources, improved and consistent results and focussed improvements.

E. **Principle 5:** System approach to management—Closely related to system 4, ISO defines this principle as: 'Identifying, understanding and managing interrelated processes as a system contributes to the organisation's effectiveness and efficiency in achieving its objectives.' This means that multiple processes are managed together as a system which should lead to greater efficiency. When implemented, this prin-ciple allows a business to focus their efforts on the processes that are key to their success as well as aligning complementary processes for improved efficiency. This process fosters a greater understanding of the interrelation of various business elements.

F. **Principle 6:** Continual improvement—This principle is very straight forward: Continual improvement should be an active business objective. The benefits of this are clear: increased ability to embrace new opportunities, organisational flexibility and improved performance. Especially in difficult economic times, the businesses that thrive are those that can adapt to new market situations.

G. **Principle 7:** Factual approach to decision making—A logical approach, based on data and analysis, is good business sense. Unfortunately, in a fast paced workplace, decisions can often be made rashly, without proper thought. The efficiency that will have been imbued in the organisation after the implementation of prior principles will allow decisions to be made with clarity. Informed decisions lead to improved understanding of the marketplace as data is collated and analysed, and the ability to defend past decisions.

H. **Principle 8:** Mutually beneficial supplier relations—This principle relates to supply chains and acknowledges that the relationship between an organisation and its suppliers is interdependent. A strong relationship between the two will enhance productivity and encourage seamless working practices. ISO state that the benefits of this principle are optimisation of costs and resources, fostering long term relationships and the 'flexibility of joint responses to changing market or customer needs and expectations'.

The series comprises of:

- ISO 9000 Quality Management Systems— Fundamentals and vocabulary (this supersedes ISO 8402 and ISO 9000–1).
- ISO 9001 Quality Management Systems— Requirements (supersedes the 1994 year version of ISO 9001, 9002, 9003.)
- ISO 9004 Quality Management Systems— Guidelines for performance improvement (this supersedes ISO 9004–1)
- ISO 19011 Guidelines on Auditing Quality and Environmental Management Systems (supersedes the ISO10011–1/2/3, ISO 14010, ISO14011.ISO14012)

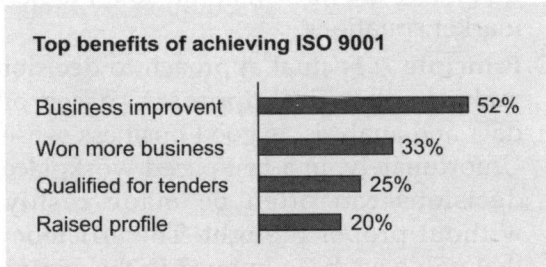

Top benefits of achieving ISO 9001

Business improvent	52%
Won more business	33%
Qualified for tenders	25%
Raised profile	20%

Intotal, 81% cited they were more competitive or had won business as a result of ISO 9001 certification.
(Source: Lake market research, 2014)

The ISO 9001: 2008 Standard

It is the requirement standard which was issued to assess an organization ability to meet customers and applicable regulatory requirements and accordingly it is the only standard in ISO 9000 series that certification from 3rd party can be obtained for conformance with.

The requirements contain little that any well run, successful organization is not already doing either formally or informally. The revisions contained in the standard are therefore attempting to ensure approval really means something and is of recognizable value.

Terminology and definitions used in ISO 9001:2008 are contained in ISO 9000:2008 which also includes the concepts on which the standard has been developed.

ISO 9001:2008 standard is organized in five main clauses which include:

- Clause 4: Quality management system
- Clause 5: Management responsibility
- Clause 6: Resources management
- Clause 7: Product realization
- Clause 8: Measurement and analysis and improvement

ISO 9001

ISO 9001 is a quality management standard. It suits all organisations large or small and covers all sectors, including charities and the voluntary sector. It aims to help organisations become structured and efficient.

A. Benefits

1. The wide-ranging benefits of ISO 9001 are why the quality management standard has been implemented by millions of organisations across the globe.

2. Designed to create a more disciplined work environment, ISO 9001 will help your workforce know exactly who does what, when and how. This saves time and cost by reducing mistakes, consequently helping to improve customer service.

B. A Business Winner

1. Maintaining a quality-led philosophy is exactly why ISO 9001 is stipulated within tenders. With buying authorities seeking to establish that their suppliers are reliable, asking for an internationally recognised standard eliminates the need to assess every single potential supplier. Having an ISO 9001 certificate saves work too, allowing organizations to skip many lengthy pre-qualification questions.

C. Improved Internal Processes

1. The standard's intention is to improve internal working in order to help you achieve greater consistency and quality of service. Together with improving what you already have in place, this is done by replacing bad or even non-existing processes with ones that are relevant, functional and documented. The results are greater efficiency and productivity, linking to increased profitability.

2. ISO 9001 is also aimed at achieving customer satisfaction by setting out what needs to be in place in order to consistently meet customer requirements. In a tough economic climate, retaining clients is vital; ISO 9001 will help you to do so.

3. In addition, ISO 9001 is designed to be compatible with other ISO management system standards such as ISO 14001 (Environmental), OHSAS 18001 (Health and Safety) and ISO 27001 (Information Security). All or any combination of these complementary standards can be integrated seamlessly. By sharing many principles, choosing an integrated management system can reduce cost considerably.

ISO 9001 Certification: Process Approach

This International Standard promotes the adoption of a process approach when developing, implementing and improving the effectiveness of a quality management system, to enhance customer satisfaction by meeting customer requirements. For an organization to function effectively, it has to determine and manage numerous linked activities. An activity or set of activities using resources, and managed in order to enable the transformation of inputs into outputs, can be considered as a process. Often the output from one process directly forms the input to the next.

The application of a system of processes within an organization, together with the identification and interactions of these processes, and their management to produce the desired outcome, can be referred to as the "process approach". An advantage of the process approach is the ongoing control that it provides over the linkage between the individual processes within the system of processes, as well as over their combination and interaction. When used within a quality management system, such an approach emphasizes the importance of:

a. understanding and meeting requirements,

b. the need to consider processes in terms of added value,

c. obtaining results of process performance and effectiveness, and

d. continual improvement of processes based on objective measurement.

The model of a process-based quality management system shown in Fig. 24.1 illustrates the process linkages presented in clauses 4 to 8. This illustration shows that customers play a significant role in defining requirements as inputs. Monitoring of customer satisfaction requires the evaluation of information relating to customer perception as to whether the organization has met the customer requirements. The model shown in Fig. 25.1 covers all the requirements of this International Standard, but does not show processes at a detailed level.

In addition, the methodology known as "Plan-Do-Check-Act" (PDCA) can be applied to all processes. PDCA can be briefly described as follows.

Plan: establish the objectives and processes necessary to deliver results in accordance with customer requirements and the organization's policies.

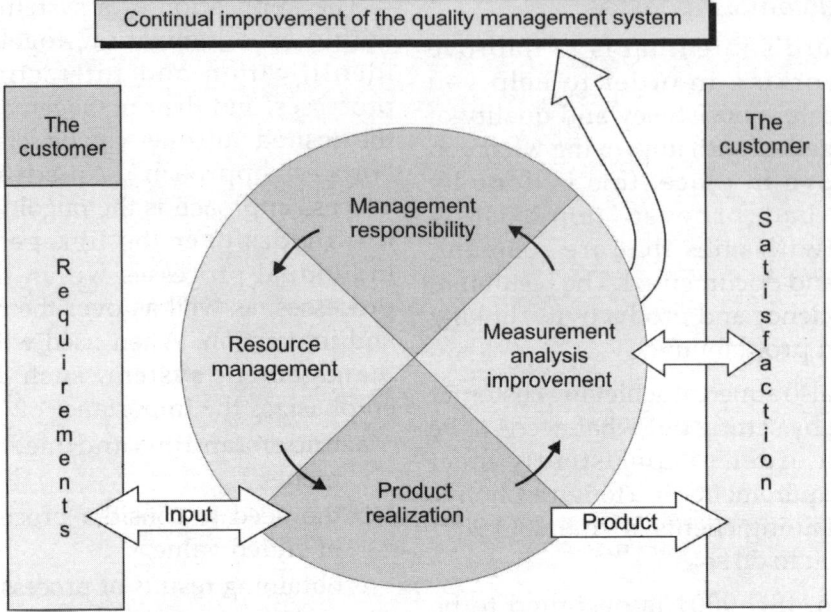

Fig. 24.1: Model of a process-based quality management system

Do: implement the processes.

Check: monitor and measure processes and product against policies, objectives and requirements for the product and report the results.

Act: take actions to continually improve process performance.

Food Safety Management: ISO 22000

Many standards have been developed within food safety: specific company standards, legal regulations, as well as standards developed by retailers. Their proliferation made their application difficult. At the same time, food safety principals have spread across the world. The consequences of unsafe food can be serious. As many of today's food products repeatedly cross national boundaries, International Standards are needed to ensure the safety of the global food supply chain. The safety of foodstuffs has become a major issue for all players in the food chain. A single standard with worldwide reach, and in-cluding the food safety principals, thus become need of the hour.

ISO 22000 is a standard developed by the International Organization for Standardization dealing with food safety. This is a general derivative of ISO 22000, a technical specification on food safety requirements and guidelines. Food safety is linked to the presence of foodborne hazards in food at the point of consumption. Since food safety hazards can occur at any stage in the food chain it is essential that adequate control be in place. Therefore, a combined effort of all parties through the food chain is required.

The ISO 22000 international standard specifies the requirement for a food safety management that involves the following elements:

• Interactive communication
• System management
• Prerequisite programs
• HACCP principals

Critical review of the above elements has been conducted by many experts in the field. Communication along the food chain is essential to ensure that all relevant food safety hazards are identified and adequately and identified and adequately controlled at each step within the food chain. This implies communication between organizations both upstream and downstream in the food chain. Communication with consumers and suppliers about identified hazards and control measures will assist in clarifying consumer and supplier requirements. Recognition of the organisation's role and position within the food chain is essential to ensure effective interactive communication throughout the chain in order to deliver safe food products to the final consumer.

The most effective food safety systems are established, operated and updated within the framework of a structured management system and incorporated into the overall management activities of the organisation. This provides maximum benefits for the organization and interested parties. ISO 22000 can be applied independently of other management system standards or integrated with existing management system requirements.

ISO 22000 integrates the principles of the HACCP system and application steps developed by the Codex Alimentarius Commission. By means of audible requirements, it combines the HACCP plan with prerequisite programmes (PRPs). Figure 24.2 shows ISO 22000 pyramid, the base is made of good practices, that include PRPs and OPRPs; in the middle is the HACCP program and on the top is management system elements like ISO 22000.

PRPs are the conditions that must be established throughout the food chain and the activities and practices that must be performed in order to establish and maintain a hygienic environment. PRPs must be suitable and be capable of providing food that is safe for human consumption. PRPs are also referred to as good hygienic practices, good

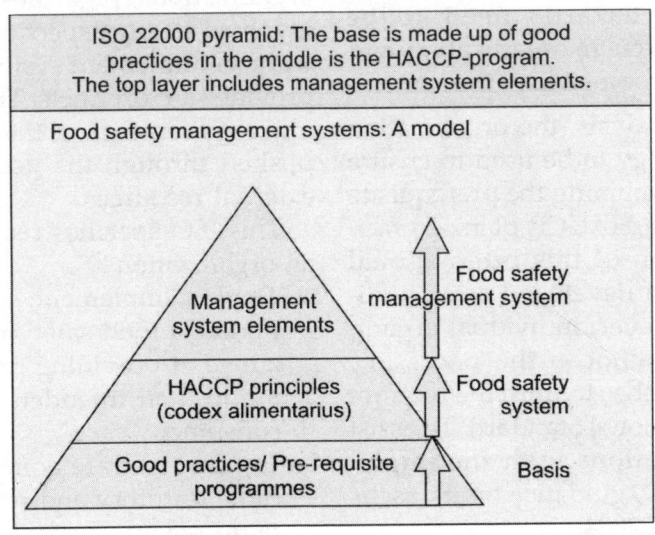

Fig. 24.2: ISO 22000 pyramid

agricultural practices, good production practices, good manufacturing practices, good distribution practices and good trading practices.

Operational prerequisite programmes (OPRPs) are prerequisite programmes that are essential. They are essential because a hazard analysis has shown that they are necessary in order to control specific food safety hazards. OPRPs are used to reduce the likelihood that products will be exposed to hazards, that they will be contaminated, and that hazard will proliferate. OPRPs are also used to reduce the likelihood that the processing environment will be exposed to hazards.

Hazard analysis is the key to an effective food safety management system, since conducting a hazard analysis in organising the knowledge required to establish an effective combination of control measures. ISO 22000 requires that all hazards that may be reasonably expected to occur in the food chain, including hazards that may be associated with the type of process and facilities used, are identified and assessed. Thus it provides the means to determine and document why certain identified hazards need to be controlled by a particular organization and why others need to comply with the same.

During hazard analysis, the organization determines the strategy to be used to ensure hazard control by combining the prerequisite programmes and the HACCP plan. To facilitate the application of this International Standard, it has been developed as an auditable standard. However, individual organizations are free to choose the necessary methods and approaches to fulfil the requirements of this International Standard. To assist individual organizations with the implementation of this ISO, guidance on its use is provided in ISO/TS 22004.

This ISO is intended to address aspects of food safety concerns only. The same approach as provided by this International Standard can be used to organize and respond to other food specific aspects (e.g. ethical issues and consumer awareness). It allows an organization (such as a small and/or less developed organization) to implement an externally developed combination of control measures.

The aim of this ISO is to harmonize on a global level the requirements for food safety management for businesses within the food chain. It is particularly intended for application by organizations that seek a more focused, coherent and integrated food safety management system than is normally required by law. It requires an organization to meet any applicable food safety related statutory and regulatory requirements through its food safety management system.

This ISO specifies requirements for a food safety management system where an organization in the food chain needs to demonstrate its ability to control food safety hazards in order to ensure that food is safe at the time of human consumption. It is applicable to all organizations, regardless of size, which are involved in any aspect of the food chain and want to implement systems that consistently provide safe products. The means of meeting any requirements of this ISO can be accomplished through the use of internal and/or external resources.

This ISO specifies requirements to enable an organization

a. To plan, implement, operate, maintain and update a food safety management system aimed at providing products that, according to their intended use, are safe for the consumer

b. To demonstrate compliance with applicable statutory and regulatory food safety requirements

c. To evaluate and assess customer requirements and demonstrate conformity with those mutually agreed customer requirements that relate to food safety, in order to enhance customer satisfaction

d. To effectively communicate food safety issues to their suppliers, customers and relevant interested parties in the food chain

e. To ensure that the organization conforms to its stated food safety policy

f. To demonstrate such conformity to relevant interested parties

g. To seek certification or registration of its food safety management system by an external organization, or make a self-assessment or self-declaration of conformity to this International Standard.

All requirements of this ISO are generic and are intended to be applicable to all organizations in the food chain regardless of size and complexity. This includes organizations directly or indirectly involved in one or more steps of the food chain. Organizations that are directly involved include, but are not limited to, feed producers, harvesters, farmers, producers of ingredients, food manufacturers, retailers, food services, catering services, organizations providing cleaning and sanitation services, transportation, storage and distribution services. Other organizations that are indirectly involved include, but are not limited to, suppliers of equipment, cleaning and sanitizing agents, packaging material, and other food contact materials (Fig. 24.3).

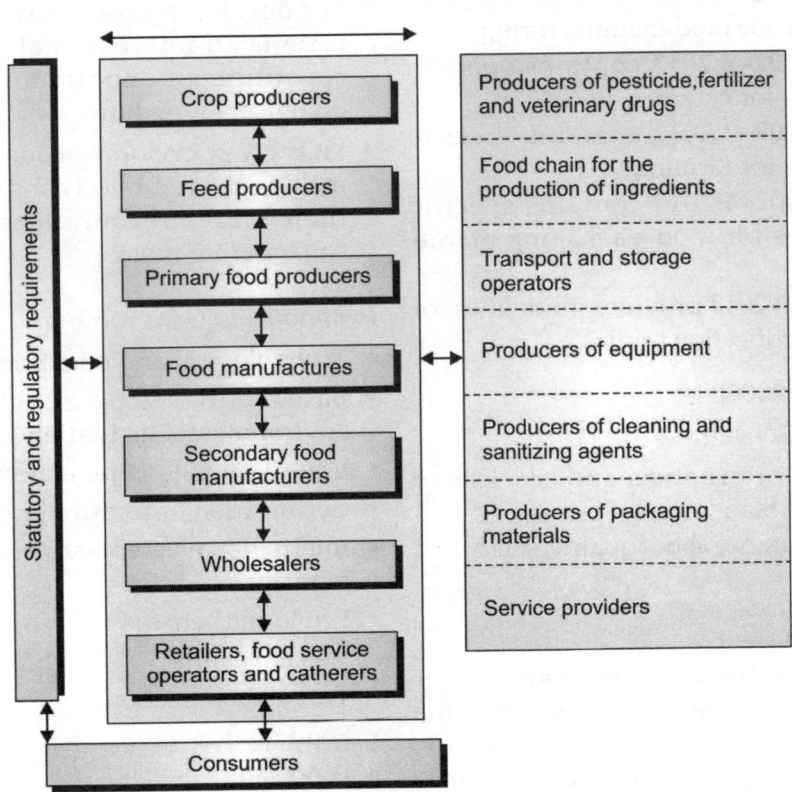

Fig. 24.3: Communication with the food chain

This ISO allows an organization, such as a small and/or less developed organization (e.g. a small farm, a small packer-distributor, a small retail or food service outlet), to implement an externally developed combination of control measures.

ISO has additional standards that are related to ISO 22000. These standards are known as ISO family of standards. The ISO 22000 family contains standards each focusing on different aspects of food safety management.

- ISO 22000:2005 contains the overall guidelines for food safety management.
- ISO 22004:2014 provides generic advice on the application of ISO 22000
- ISO 22005:2007 focuses on traceability in the feed and food chain
- ISO/TS 22002–1:2009 contains specific prerequisites for food manufacturing
- ISO/TS 22002–2:2013 contains specific prerequisites for catering
- ISO/TS 22002–3:2011 contains specific prerequisites for farming
- ISO/TS 22002–4:2013 contains specific prerequisites for food packaging manufacturing
- ISO/TS 22003:2013 provides guidelines for audit and certification bodies

Benefits of ISO 22000

A. Benefits to Consumers

- Conformity of products and services to International Standards
- Provides assurance about quality, safety and reliability

B. Benefits to Society

- For businesses, the widespread adoption of ISO 22000 means that suppliers can base the development of their products and services on specifications that have wide acceptance in their sectors.

- This, in turn, means that businesses using ISO 22000 are increasingly free to compete on many more markets around the world.

C. Benefits to Trade Officials

- For trade officials negotiating the emergence of regional and global markets, ISO 22000 creates "a level playing field" for all competitors on those markets.
- The existence of divergent national or regional standards can create technical barriers to trade, even when there is political agreement to do away with restrictive import quotas and the like; ISO are the technical means by which political trade agreements can be put into practice.

D. Benefits to Developing Countries

- For developing countries, ISO 22000 represents an international consensus and constitute an important source of technological know-how.
- ISOs give developing countries a basis for making the right decisions when investing their scarce resources and thus avoid squandering them.

E. Benefits to Government

- Technological and scientific know-how
- Bases for developing health, safety and environmental legislation
- Education of food regulatory personnel
- Certification or registration
- International acceptance of standards used globally
- Economic benefits
- Social benefits
- Trade liberalization
- Food quality
- Food safety
- Food security

F. Miscellaneous Benefits

1. ISO 22000 can contribute to the quality of life in general by:
   - Ensuring safe food
   - Reducing foodborne diseases
   - Better quality and safer jobs in the food industry
   - Better utilization of resources
   - More efficient validation and documentation of
     - Techniques, methods and procedures
     - Increased profits
     - Increased potential for economic growth and development
2. The benefits of ISO 22000 for other stakeholders:
   - Confidence that organizations implementing ISO 22000 have the ability to identify and control food safety hazards.
   - International in scope.
   - Provides potential for harmonization of national standards.
   - Provides a reference for the whole food chain.
   - Provides a framework for third party certification.
   - Fills a gap between ISO 9001:2000 and HACCP.
   - Contributes to a better understanding and further development of Codex HACCP.
   - Auditable standards with clear requirements.
   - System approach rather than product approach.
   - Suitable for regulators.

Conclusion

Food safety is related to the presence of and levels of foodborne hazards in food at the point of consumption. As food safety hazards may be introduced at any stage of the food chain, adequate control throughout the food chain is essential. Thus, food safety is a joint responsibility of all parties participating in the food chain. Failures in food supply can cause human suffering, death, poor reputation, violations, poor nutrition, poor quality products and decreased profits. ISO 22000 ensures integrity of food supply chain by minimizing foodborne hazards throughout the food chain by ensuring that there are no weak links.

It therefore makes good scientific sense for those who are involved in food processing, manufacturing, storage, distribution of food and food products to adopt and implement ISO 22000. ISO 22000 for food safety management systems is intended to provide food safety and security. It can be applied on its own or in combination with other management system standards such as ISO 9001:2000.

25

Safe Quality Foods

Ishwarpreet Kaur, Shalini Dwivedi, Puja Dudeja

INTRODUCTION

Safe Quality Food (SQF) is a food safety and quality management program. It is a comprehensive system that ensures safety and quality of food products by covering the entire food supply chain. It is the only program that is applied from farm to fork. It is tailored to the needs of food suppliers and primary producers. It empowers them to meet the product needs, regulatory, food safety and quality criteria in a structured and cost effective way. This program was developed in Australia in early 1990s. It is now owned and managed by the Food Marketing Institute (FMI) in the United States of America. The activities of the program are managed by the Safe Quality Food Institute (SQFI), a division of the Food Marketing Institute. This program was designed with the help of primary producers and experts of quality management, food safety, food retailing, food distribution, agriculture production system, and the HACCP guidelines.

SQF Codes

The SQF system provides both management system certification and product certification. There are two codes, SQF 1000 code for primary products (farming) and SQF 2000 code for food manufacturing or food sector services. Both codes are HACCP-based and address food safety controls, management system requirements and the control of quality. They are internationally recognized by the Global Food Safety Initiative (GFSI).

1. **SQF 1000 Code:** This code is applicable for primary production. It includes all farming/animal husbandry activities including crops, meat, and poultry, as well as the provision of crop spray and field harvest services.

2. **SQF 2000 Code:** This code applies to businesses manufacturing food products or services to food industry. The SQFI classifies these businesses within twenty-eight food sector categories. These include businesses like meat and poultry processing, dairy foods processing, manufacture of food packaging materials, etc.

SQFI provides guidance documents for a variety of food sectors in order to facilitate the implementation of SQF systems. These are available for download from the SQFI website. The SQF Codes can be implemented and certified at three distinct levels (Fig. 25.1). These are:

1. **Level 1—food safety fundamentals:** This is an entry level for new businesses covering only pre-requisite programs (req-

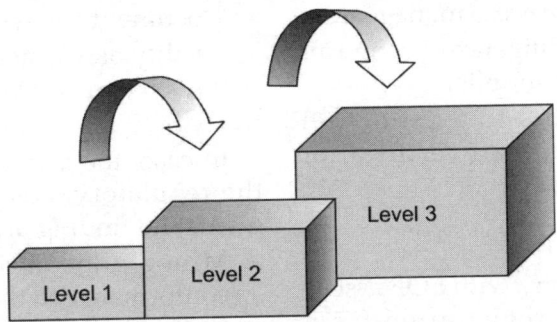

Fig. 25.1: SQF codes certification at three levels

uirements for assuring a product meets regulatory requirements and is wholesome and unadulterated) and basic food safety requirements.

2. **Level 2—certified HACCP-based food safety plans:** This incorporates implementation of HACCP for food safety in addition to the Level 1 requirements.

3. **Level 3—comprehensive food safety and quality management system:** This level requires the implementation of a HACCP-based food quality plan to enhance control over the quality of the product along with the incorporation of Level 1 and Level 2 requirements.

Note: It is advised that businesses that undertake high risk processes or manufacture high risk products begin with the Level 2 certification. Customers (e.g. retail, foodservice, etc.) may suggest the business start at a particular level, but ultimately it is the business who decides at which level they will be audited and certified to.

SQF Certification: To acquire certification, business houses use applicable SQF code to develop and implement program. Later a third party audit company (a certification body) verifies the extent to which the safety measures are implemented as documented in the SQF code.

A business cannot simply implement the requirements and "self-declare" they comply with the requirements. External verification through the certification process is required under the SQF program. Certification lasts one year and the business must be re-certified on an annual basis. Once the business unit achieves SQF certification the details of the same are listed on the SQFI website.

Benefits of SQF Certification for a Business Establishment

• Provides proof of due diligence.
• Provides increased brand protection, consumer confidence.
• Promotes food safety and quality.
• Allows consistent improvement in process quality and safety.
• Increases yield by reducing material waste.
• Streamlines risk and process management.
• Provides global recognition and access to new markets through the SQF website.

Benefits for a company purchasing products manufactured by SQF certificated business establishments:

• Helps protect the brand by focusing on hazard analysis, risk assessment and proactive prevention strategies.
• Increases consumer confidence and loyalty.

- Streamlines risk and process management.
- Provides proof of due diligence by requiring SQF certification from suppliers.
- Provides online access to a list of SQF suppliers, along with their certification status and audit results.

Steps to SQF Certification

1. Registration of company with SQF assessment database: The registration fee is based on the gross sales of a company.
2. Designate an employee as the SQF practitioner: This individual will serve as your company's internal expert on SQF. The SQF practitioner must have knowledge of HACCP (including proof of HACCP training), the applicable SQF Code and product and process knowledge. He undertakes specific communication, validation and verification activities to for the implementation and maintenance of the SQF code.
3. Choose the required level of certification–Out of the three levels of certification the business operator chooses their level of SQF standard. Level 1 is mainly for low risk products. Level 2 is a certified HACCP food safety plan that is benchmarked by GFSI. Level 3 is a comprehensive implementation of safety and quality management systems which includes level 2.
4. Obtain proposals from SQF licensed certification bodies.
5. Conduct a pre-assessment (optional)-Either an SQF auditor or company SQF practitioner can identify the "gaps" between the program and the desired level of SQF certification.
6. Choose a licensed certification body and schedule an audit.
7. Certification body conducts initial Certification audits.

- Document review
- Facility assessment

If the system has been implemented effectively SQF certification is confirmed.

In case, the activities do not comply with the regulatory, customer and SQF requirements the finding are then categorized as:

a. Minor nonconformity: This is a nonconformance to the documented system or to the SQF code that does not cause an immediate food safety risk or indicate ineffective implementation of a system element. This must be corrected within 30 days and will be verified at the next audit.

b. Major nonconformity: It is a nonconformance to the documented system or to the SQF Code that could cause an immediate food safety risk or indicate ineffective implementation of a system element. This must be corrected and verified within 14 days of the finding.

c. Critical nonconformity—results in a suspension of certification

This consists of a loss of control at a critical control point, prerequisite program or other process step that could cause a significant risk and product safety is compromised or any tampering of records

8. Surveillance—Depending on the audit result, the business site may need to be visited in 6 months to ensure that system continues to meet the requirements of the SQF Code. The need for a surveillance audit is defined in the SQF Code.

9. Re-certification—After certification, an SQF auditor visits the organization annually to ensure that system continues to meet the requirements of the SQF Code. Businesses that fail to meet the requirements of the code upon re-certification will have their certificate of registration suspended or withdrawn.

26

Safe Meat

Puja Dudeja, Surjinder Singh, Amarjeet Singh

Meat has always comprised an important part of the human diet for a large part of history. It is still the chief component of most meals in developed countries. In many developing countries, non-animal-based sources of protein are still dominant. There has been a considerable increase (62%) in the available food consumption of meat worldwide, with the biggest increases in the developing countries (a threefold increase since 1963) especially in Asia.

India has predominantly been a land of vegetarians. For a very long time, many people in Indian respected the religious taboos on avoidance eating meat. Centuries-old Hindu texts extol the virtues of forgoing meat. With globalization meat eating habit is on the increase all over the world. Even in India, people are turning to non-vegetarian diet. With influence of the west, the Indian middle class dietary consumption patterns have also changed. Over the past decade, the amount of meat eaten in India has more than doubled. In 2009, it reached around 5.5 kilograms per head per year, according to the World Food Program, though it is still less than many other countries in the world.

Fast-food restaurants and their meat based recipes are increasingly becoming popular both in the metro and level II cities. Inter-national brands such as Kentucky Fried Chicken (KFC) are attracting many consumers with their new choices of non-vegetarian dishes. Meat eating by the present generation is also considered a part of the cosmopolitan attitude. It is also seen as a sign of affluence, as meat dishes are more expensive than vegetarian food. Nevertheless, the growing consumption of meat brings with it new problems. Figure 26.1 describes the main source of food.

According to Food Safety and Standards Regulations (FSSR) 2011 meat means the flesh and other edible part of a carcass (dead body of slaughtered animal). Meat food products are articles derived from meat by means of drying, currying, smoking, cooking, seasoning, flavouring or freezing. Various meat base receipes are mutton/pork/beef/chicken curry, biryanis, sausages, nuggets, patties, puffs, rolls, samosas, koftas, chops, tikkas and soups from meat. Hot dogs and hamburgers are also included in this.

Meat has been implicated as a cause of many foodborne illnesses. These are caused by following organisms:

- *Salmonella*, by consuming raw and under-cooked meat, poultry, dairy products, and seafood. *Salmonella* may also be present on egg shells and inside eggs.

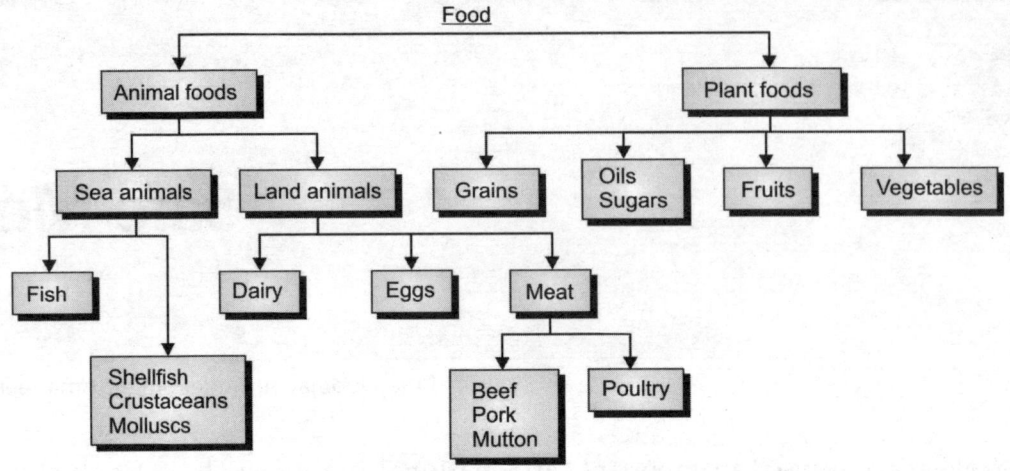

Fig. 26.1: Sources of food

- *Campylobacter jejuni (C. jejuni),* found in raw or undercooked chicken and unpasteurized milk.
- *Escherichia coli (E. coli). E. coli O157:H7* is the strain that causes the most severe illness. Common sources of *E. coli* include raw or undercooked hamburger and milk.
- *Listeria monocytogenes (L. monocytogenes),* which has been found in raw and undercooked meats, unpasteurized milk, soft cheeses, and ready-to-eat hot dogs.
- *Vibrio,* a bacterium that may contaminate fish or shellfish.
- *Clostridium botulinum (C. botulinum),* a bacterium that may contaminate improperly canned foods and smoked and salted fish.
- *Trichinella spiralis* by consuming raw or undercooked pork.

Apart *from this* harmful chemicals that cause illness may contaminate foods such as fish or shellfish, which may feed on algae that produce toxins. Consumption of such fish leads to high concentrations of toxins in their bodies. Some types of fish, including tuna may

be contaminated with bacteria that produce toxins if the fish are not properly refrigerated before they are cooked or served.

This chapter deals mainly with the safety of meat as food. Journey of safe meat can be traced in the form of farm to fork or stable to table (Fig. 26.2). Good animal husbandry practices has been dealt in a separate chapter. In this chapter we shall learn about food safety practices from butchery to fork. Chicken is one of the worst offenders for food poisoning.

Safety Issues of Meat

Meat for consumption is available to people as raw meat (butchery, supermarkets), frozen meat/snacks, ready to eat dishes. Irrespective of the form of meat product it is imperative for ensuring supply of safe meat that the examination of animals before slaughter (antemortem) is carried out or supervised by qualified veterinarians; properly trained inspectors who can look for questionable conditions that can then be referred to veterinary judgement. The same is true of examination of animal organs and carcasses after slaughter (postmortem).

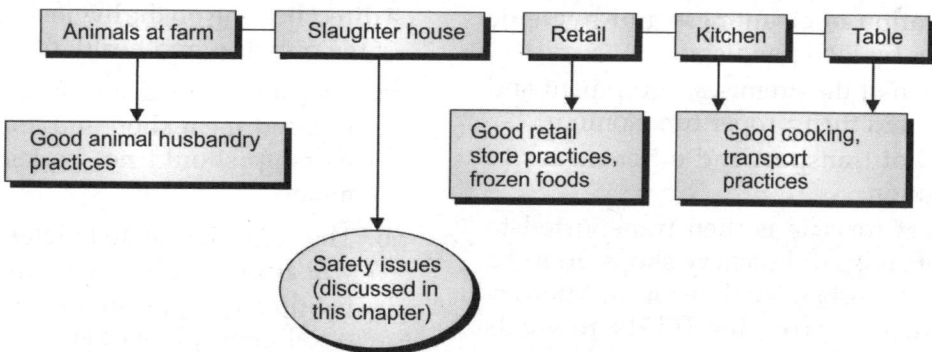

Fig. 26.2: Food safety concerns in stable to table journey of meat

Antemortem Inspection

This examination is designed to determine the fitness of the animal to provide meat suitable for human consumption. The inspector should check for evidence of animal abuse. Animals should be free from illness, infection, tiredness, etc. reasonably clean; and transported to slaughter in a manner that will protect them from accidents in a conveyance which is clean and in a sanitary condition. Conveyance cleaning should include disinfection to prevent transmission of disease. On arrival in the slaughterhouse area, the animals should rest for 24 to 48 hours in clean, airy stalls before slaughter. Care should be taken to prevent injuries during the waiting period.

Postmortem Inspection

In abattoirs, postmortem inspection should be carried out immediately after slaughter. Its purpose is to ensure the detection of diseases and abnormalities, so that only meat fit for human consumption is passed as such. The inspection should be carried out by veterinarians or by specially trained inspectors under veterinarian supervision. Routine postmortem inspection should include viewing, palpation, and where necessary incision and sample-taking for laboratory examination. The inspection should be carried out in a hygienic and systematic manner.

Prior to the final examination, all parts required for the inspection of the animal should remain identifiable with the carcass.

Meat that has been passed by the inspector as fit for human consumption should be branded in accordance with local regulations. Brands and stamps bearing the marks of inspection should be kept clean while in use, and should be held in the custody of the inspector and used only under his supervision.

Meat Processing Unit

It is essential that each meat establishment have access to a laboratory for control and diagnostic work. Following aspects are checked in meat processing units during inspection of meat products to ensure safety:

a. The inner temperature and pH of meat should be recorded

b. Cleanliness, temperature and other storage conditions

c. Control of the manufacturing process and conformity to prescribed standards for the means of preservation (heat treatment, salting, sugaring, etc.); (*refer* Chapter 16 on good manufacturing practices)

d. Control of packing, packaging and storage conditions, and observance of prescribed holding times, validity periods, etc.

e. Observation of cleanliness and hygienic practices by food handlers;

f. Sanitation of the premises, equipment and fittings, and their proper functioning;

g. Control of transport and other means of distribution.

The meat for sale is then transported to retail meat shops or butchery shops. In India The Animal Husbandry division of Ministry of Agriculture provides funds towards expansion and modernization activities of the meat shops. The respective local bodies in towns and cities are mainly responsible for day-to-day operation/maintenance of the slaughter houses. Most of the slaughter houses in our country are service-oriented and, as such, perform only the killing and dressing of animals without an onsite rendering operations. Most of the slaughter houses are more than 50 years old without adequate basic amenities viz. proper flooring, ventilation, water supply, transport, etc. In addition to these deficiencies, slaughter houses suffer from very low hygiene standard posing a major public health and environmental hazards due to discrete disposal of waste and highly polluted effluent discharge. Unauthorized and illicit slaughtering has also increased manifold and thus the related problems. For ensuring the hygiene and safety of meat being sold at retail meat shops, the following requirements need be ascertained under the supervision of the qualified Veterinary staff.

1. Location of Meat Shop

i. The meat shop/sale outlet should be a unit of meat market located away from vegetable, fish or other food markets and shall be free from undesirable odour, smoke, dust or other contaminants. Wherever a meat markets is not available, individual meat shop can be set up considering the above factors, which have a direct bearing on the hygiene conditions of the premises and health of consumers.

a. The minimum distance between the licensed meat shop and any place of worship should not be less than 50 meters;

b. The condition of 100 meters distance will apply in case the premises situated directly opposite to the entry gate of religious place of any community.

ii. All the meat shops located in the vicinity of religious places shall be fitted with black glass doors, which must be kept, closed all times except in case of entry or exit. It must be the responsibility of the meat shop owners to maintain a high standard of hygiene not only inside the shops, but also in the way leading to the shops road pavements or other adjoining place, particularly for insanitary materials originating from the meat business for example, blood, part of offal, meat scraps.

2. Size of Meat Shops

i. Considering the constraints of commercial space in residential areas in concerned Panchayats/Municipalities the size of meat shops may vary according to the size of business and activities being carried out there in the meat shops. However it will be desirable that shops are not less than 4 sq m of floor area.

ii. The height of shop in all above categories of meat shops should be not less than 3 meters, while in case of air-conditioned meat shops; it should not be less than 2.5 meters.

3. Premises

i. The premises shall be structurally sound. The walls up to the height of minimum 5 feet from the floor level shall be made of impervious concrete material (e.g.

glazed tiles or hygienic panels, etc.) for easy washing and cleaning purposes.

ii. The floor should be made of impervious and non-slippery materials with a slope for easy cleaning and removal of filth, waste and dirty water. The slope of the floor shall not be less than 5 cm for a floor of 3 meters.

iii. All the fittings in the stall should be of non-corroding and non-rusting type.

iv. All processing tables, racks, shelves, boards, etc. shall have zinc/aluminium/stainless steel/marble-granite to facilitate proper cleaning.

v. A sign board indicating the type of meat sold shall be displayed prominently. Nothing else but meat should be sold at the premises.

vi. The premises should have provision of sewer connection for drainage of waste-water.

vii. There should be provision of continuous supply of potable water inside the pre-mises. In case the water supply is from bore well the arrangement for softening of water for making the same potable shall be made in the premises and intermittent adequate store arrangement should be made.

viii. The door of the shop should be of self-closing type, and the sale counter should have a provision for small window with wire glass sliding. The door of the shop should be of dark glass top and be kept closed. No carcasses should be kept in a manner so as to be seen by the public view from outside.

4. Equipment and Accessories

i. The meat shop should have suitable arrangement for fly proofing in the form of air-curtains, flytraps, etc.

ii. It should have display cabinet type refrigerator of adequate size for main-taining a temperature of 4 to 8°C. or freezing cabinet if the meat is to be stored for more than 48 hours.

iii. The weighing scales used shall be of a type which obviates unnecessary hand-ling and contamination and the sketch of the scale shall be made of stainless steel or nickel coated.

iv. The knives, tools and hooks used shall be made of stainless steel. Sufficient cupboards or racks should be there for storing knives, hooks, clothes and other equipment.

v. The chopping equipment should be cleaned with hot water at a temperature of 82°C.

vi. There should be a provision of geysers in all the meat shops to have hot water at a temperature not less than 82°C to clean the premises and equipment used in meat shop.

vii. Washbasin made of stainless steel/porcelain shall be provided with liquid soap dispenser or other soap and nail brush for thorough cleaning of hands.

viii. The chopping block should be of food-grade synthetic material, which does not contaminate the meat. If the block is wooden it should be of hardwood trunk, which is solid enough and should not contaminate the meat.

ix. The rails and hanging hooks, if provided for hanging carcasses, should be of non-corrosive metal. The non-corrosive hanging hooks for carcasses shall be 30 cm apart and the distance between rails shall be 60 to 70 cm depending upon the size of animals slaughtered and carcasses hanged.

x. A waste bin with a pedal operated cover shall be provided in the premises for collection of waste material.

5. Sanitary Practices

i. The chopping block should be sanitized daily by covering its top with sea-salt, after cleaning it with hot water at close of business activity.

ii. The refrigerated/freezing cabinet should be regularly cleaned and well maintained.

iii. Slaughtering of animal/birds inside the shop premises should be strictly prohibited.

iv. The carcasses shall not be allowed to be covered with wet-clothes.

v. Wholesome meat obtained from the authorized slaughter house shall only be sold at the meat shops and a record thereof shall be kept in the premises to be shown to any officer of the concerned Panchayats/Municipalities responsible for local administration corporation at the time of inspection.

vi. Waste bins should be emptied, transported for disposal as per the arrangements made by the concerned Panchayats/Municipalities and waste bin/dhalau (burial pits) shall be treated daily with a disinfectant.

vii. The premises shall not be used for residential purposes nor shall it communicate with any residential quarter. No personal belonging like clothing, bedding, shoes, etc. shall be kept in the premises. Only dressed carcasses of clean meat shall be stored at the premises.

viii. Hides, skins, hoofs, heads and unclean gut should not be allowed to be stored in the premises at any time.

ix. The preparation of food of any type inside the meat sale outlet should be strictly prohibited.

x. The meat obtained from unauthorized sources or unstamped meat is liable to be confiscated and destroyed.

xi. Waste of the meat shop to be disposed off packed in heavy polythene bags in dhalaos (burial pits).

6. Transportation

i. The transportation of carcasses from the slaughter house to the premises shall be done under hygienic conditions in boxes of adequate size linked with zinc/aluminium/stainless steel or wire gauze meat safes, which must be washed daily.

ii. The transportation of carcasses from the slaughter house to the meat shops should be done in insulated vans which are refrigerated. Under no circumstances, carcasses shall be transported in vehicles used for commuting of human beings, or in an exposed condition.

27

Milk and its Safety

Ishwarpreet Kaur, Surjinder Singh, Satinder Pal Singh, Sumeet Kaur

Milk is nature's ideal and perfect single food for the newborn. It is also one of the best food products for adult human beings as it contains almost all the nutrients in a fairly balanced proportion. It has always been a part of Indian diet since the rise of civilization. It is an exceptionally versatile raw product. World-wide some 8,000 to 10,000 different milk products are available. As per the annual report (2010–11) of National Dairy Develop-ment Board (NDDB), India is the largest milk producer in the world contributing about 17% of the world's total milk production. Even this quantity of milk is insufficient for our massive population. The Indian dairy industry is facing a tremendous stress to meet the needs of ever increasing population. The wide gap between demand and supply has not only led to increased price of milk but also greater incidences of its adulteration. Milk based products like *ghee*, butter, cheese, sweets like *kalakand, ras-malai*, etc. are also frequently adulterated. This has become a serious health concern in modern society.

According to a media report in Times of India newspaper dated 10th January 2012, about 68.4% of the milk samples collected was found to be contaminated. In the national capital alone, about 70% of the collected samples failed the qualitative tests. The problem of adulteration is far more serious in urban areas as compared with rural. Many a times, the milk sold by even the most reputed companies fails the standards set by Food Safety and Standard Authority of India (FSSAI). In a national survey conducted by FSSAI, 2011, on milk adulteration found alarmingly high level of nonconformity in the states of Bihar, Chhattisgarh, Jharkhand, Mizoram, Daman and Diu and Odisha. Most other parts of the country also presented a gloomy picture. The non-conformity of samples in rural areas were 381(31%) out of which 64 (16.7%) were packet samples and 317 (83.2%) were loose samples respectively and in urban area the total non-confirming samples were 845 (68.9%) out of which 282 (33.4%) were packed and 563 (66.6%) were loose samples.

Quality of milk: The term 'quality' refers to a combination of characteristics that enhance the acceptability of a product. Milk quality is a broad term, which includes the chemical, physical, technological, bacteriological and aesthetic characteristics of milk and milk products. Milk from a healthy animal is pure, clean, safe, unadulterated and wholesome. Good quality milk is essential for preparing good quality milk products. The desired characteristics of safe milk are:

a. Free from debris and sediment.

b. Free from off-flavours.

c. Low in bacterial counts.

d. Normal composition and acidity.

e. Free of antibiotics and chemical residues.

Importance of milk quality: Supply of quality milk to general public is a complicated process. It is the concern of so many stakeholders, which include:

1. Production sector (milk producers)—small holder, large holders, periurban dairy farms, commercial dairy farms, institutional dairy farms.

2. Processing sector (milk processor)—cooperative sector, private sector, government milk scheme, joint sector, informal sector (*halwais*).

3. Marketing sector (marketing channel)—private company, state co-operative, milk marketing federation, wholesalers and retailers, informal sector (milkmen)

Good quality milk is safe for human consumption and free from disease producing microorganisms. It has a high nutritive value and keeping quality. The quality of milk is important to everyone in the milk chain. These stakeholders are:

a. **Dairy farm operators**—the dairy farm operators expect a fair price in accordance with the quality of milk produced.

b. **Milk processor**—the milk processor who pays the producer must assure that the milk received for processing is of normal composition and is suitable for processing into various dairy products.

c. **Consumer**—the consumers expect to get a fair deal for the money they spend for buying milk and milk products, i.e these need to be safe and of acceptable quality.

Quality control agencies ensure that the health and nutritional status of the people are protected from the consumption of contaminated and sub-standard foodstuffs and that prices paid are fair to the dairy farm owners, the milk processor and to the final consumer.

Knowing about the issues of milk safety strict following of hygiene and quality standards can only ensure safe milk.

Milk contamination and adulteration: Food contamination means the deterioration in quality of the foodstuff. Contamination in milk occurs accidentally due to poor hygienic conditions, or negligence of people milking milch animals. It occurs due to physical, chemical, biological and environmental factors. However, adulteration means the addition of ingredients which are not permitted in food. They are added because of business profit only. Adulterated foods are harmful for human health as they contain the unauthorized food ingredients. For example, addition of melamine in milk powder for business profit. As per FSSAI, an adulterant is defined as any material which is or would be employed for making the food unsafe or sub-standard or mis-branded or containing extraneous matter. Making profit via adulteration is very appealing to some unscrupulous people who totally ignore the health risks to public and indulge in such practices. Adulteration not only puts a financial burden on the consumers but may also play havoc for their health.

Contamination and adulteration can occur at any level in the dairy chain from farm, collection and storage to processing centres (Fig. 27.1). Both of these adversely affect the quality of milk. Some of the sources of contamination are given in Fig. 27.2.

The residues left in milk from various contamination sources is given in Table 27.1.

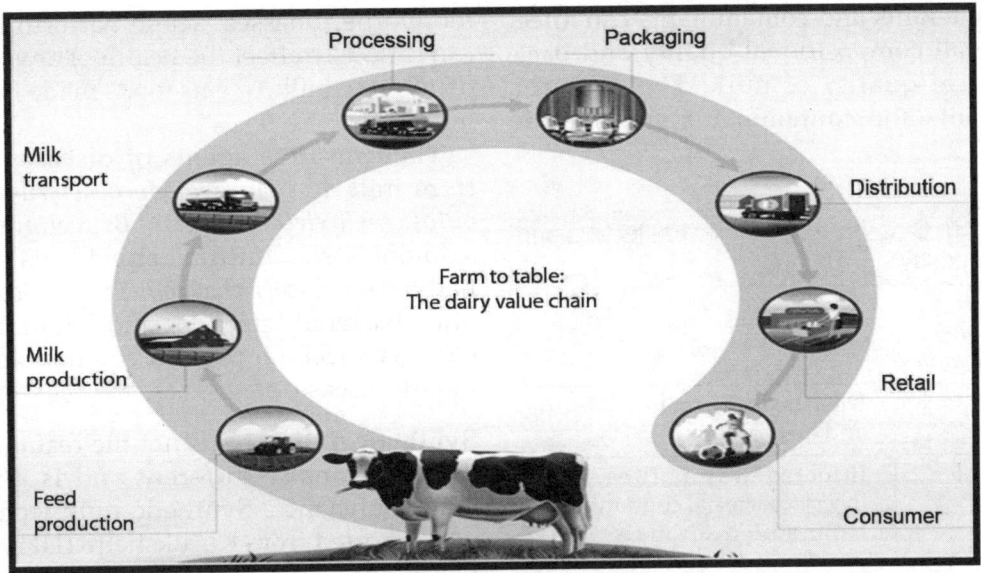

Fig. 27.1: The dairy chain

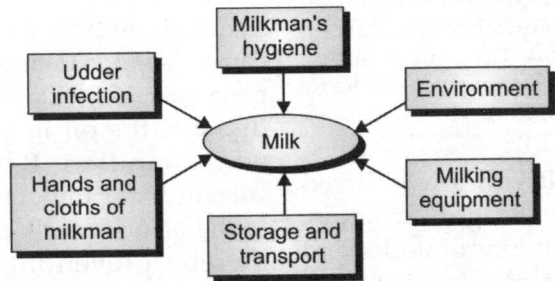

Fig. 27.2: Some of the sources of milk contamination

Table 27.1: Residues of different types left in milk	
Source	*Residue*
Farm animal	Veterinary drugs e.g. antibiotics, hormones, etc. Feed additives, e.g. trace elements, antioxidants, feed drugs, etc.
Environment	Agrochemicals, e.g. pesticides and growth promoters. Emissions e.g aerosols, fumes and dusts. Minerals of the soil, e.g. lead and cadmium.Environmental organics, e.g. mycotoxins and radionuclides
Milking and processing	Hygiene formulations, e.g. cleaning and disinfecting agents. Insecticides surfaces, e.g metals, plasticisers from non-food-grade equipment.

The adulterants and contaminants can affect the overall compositional quality and bacteriological quality of milk. The common adulterants and contaminants are given in Table 27.2.

Table 27.2: Classification of milk adulterants and contaminants

Type of Adulterants/ Contaminants	Names
Physical	Dust, bedding materials, dung, and animal hair
Biological	Infected water, flies, other insects, bacterial contamination from soiled animals, soiled hands and equipment, contamination from abnormal milk (mastitis pathogens, blood and clots)
Chemical	Glucose, skimmed milk powder, salt/detergents, starch, refined oil, urea/melamine, formalin/hydrogen peroxide/ boric acid, flour

Compositional quality: It varies with breed of the milch animal, species, stage of lactation, feed, season, disease conditions of udder and variation in milking. It also varies with accidental contamination and deliberate adulteration.

Bacteriological quality—milk serves as an ideal medium for the growth of microorganisms due to its high nutritive value and moisture content. There are ample opportunities for the contamination of milk by transmission of pathogens. The further growth of the microbes in milk is dependent upon both intrinsic and extrinsic factors. Thus raw, unpasteurized milk can carry dangerous bacteria such as *Salmonella, E. coli*, and *Listeria*, which are responsible for causing numerous foodborne illnesses. These harmful bacteria can seriously affect the health of anyone who drinks raw milk, or eats foods made from raw milk.

The causative agents of diseases spread from milk to man are *Mycobacterium tuberculosis, Coxiellaburnetti, Brucella abortus*, salmonellosis, anthrax, shigellosis, enteropathogenic *Escherichia coli, Streptococcus*, and other bacterial infections and viral infections such as vaccinia, pseudo cowpox, foot and mouth disease, etc.

Synthetic milk: This is not the real milk. It is compositionally different and is a factory made substance. Synthetic milk technology was reported from Kurukshetra (Haryana) 15 years ago. This unscrupulous technology spread to Rajasthan, Himachal Pradesh and Uttar Pradesh, and later to rest of the country.

Synthetic milk formula ingredients - Synthetic milk is prepared by mixing urea, caustic soda, cooking oil and common detergents (Fig. 27.3). Detergents help emulsify and dissolve the oil in water giving the frothy white solution. Refined oil is used as a substitute for milk fat. Caustic soda is added to the blended milk to neutralize the acidity, thereby preventing it from turning sour during transport. Urea is added for Solid-Not-Fat (SNF). This milk is blended with natural milk as its taste is disgusting.

Procedure—vegetable refined oil (butyrorefractometer reading <42) is put into a wide mouthed container along with a suitable emulsifier and thoroughly mixed till it is a thick white paste (Fig. 27.4). Then water is slowly added to the paste until the density of the liquid is similar to that of milk. Urea or sodium sulphate, glucose, maltose and any commonly available fertilizer are dissolved in hot water and then added to the solution. There are variations in the ingredients of synthetic milk depending in different parts of the country: blotting paper, hydrogen peroxide, formalin can be present in the sample. It is basically a type of milk adulteration. Making synthetic milk is essentially a crime (Fig. 27.3).

This milk easily passes the basic tests carried out at the village level dairy cooperative society (fat and lactometer reading). It is carcinogenic, and is also harmful for heart, liver and kidney. Urea content above 700 ppm can cause serious health concerns like indigestion, acidity and ulcers. Caustic soda also hinders lysine utilization in the body. Consumption of such milk can lead to gastrointestinal disorders

Fig. 27.3: Mixing and preparing of synthetic milk

Common Adulterants of Milk and its Products

Water: It is the commonest and oldest adulterant of milk. Though adding clean water to milk is not detrimental to health, it causes a financial loss to the consumer who has paid more for a defined quantity of milk and is getting less instead. *Clean water is probably the 'safest' adulterant.* However, the water that is added to dilute milk might not always be clean as the *dudhwalas*/milkmen or dairy workers rarely wash their hands or animal udder before milking. It is also not uncommon for milkmen to use pond water for this purpose. This leads not only to contamination of milk but also to gastroenteritis or other infections/infestations (Fig. 27.4).

At many places in our country, the widespread usage of pesticides has polluted the water table. Adulteration of milk with such water brings in pesticides to milk that accumulate in body tissues over a period of time. The water may also be contaminated with heavy metals that are known to interfere with cellular enzymatic machinery.

Fig. 27.4: Procedure of preparing synthetic milk (India TV report)

Glucose and skim milk powder (SMP): Skimmed milk powder is not allowed to be added to fresh milk. However, it is added to enhance to volume of milk in lean seasons. Glucose is added to provide a natural taste to milk.

Starch: It is added to prevent curdling of milk.

Refined oil: It is added as an alternative to fat of milk and to provide an oily appearance.

Sodium bicarbonate, formalin, hydrogen peroxide and boric acid: The natural raw milk has a short shelf life of only a few hours. The above mentioned adulterants are added to milk to increase its shelf life (preservation) and to maximize the profits.

Salts, detergents and glucose: These agents have been found to enhance the thickness and viscosity of milk. The detergents are added for the dissolution and emulsification of oil in water to give the typical foamy appearance of milk. An ICMR report states that detergents cause gastrointestinal problems and food poisoning while other man made substances that are added to milk interfere with cardiac functions and sometimes may prove fatal. The glucose is added most likely to enhance the SNF (Solid Not Fat).

Melamine and urea: These compounds are toxic and are added to the milk to enhance its nitrogen concentration and mimic a high protein content of milk. Urea is also believed to increase to whiteness of milk. It, however, overburdens the kidneys may even lead to renal failure. In addition, it can adversely affect the normal functioning of heart and liver.

A human tragedy struck China in 2008 when melamine levels up to 2500 ppm were detected in many infant formulas feeds and powdered milk products. The Chinese Ministry of Health reported that more than 290,000 infants were affected by this contaminated formula feed. Hospitalization was required in more than 50,000 cases and six infants were confirmed dead. High melamine concentration is responsible for producing crystals in urinary systems. Many of the infants in the Chinese tragedy had developed stones in their renal systems. Many nations have now set limits for melamine in infant formula feeds. As per WHO, a TDI (Tolerable Daily Intake) limit of 1 mg/kg of body weight in powder infant formula feeds and 2.5 mg/kg of body weight in other foods would offer an adequate safety margin for dietary exposure to melamine.

The protein content in milk/food is generally estimated by total nitrogen quantification by means of the Kjeldahl and Duma tests. One of the major drawbacks of these tests is that they cannot differentiate between nitrogen originating from protein and non-protein sources. Consequently, melamine cannot be detected and hence used as adulterant. But nowadays, several methods based on ELISA, HPLC-MS, solid phase extraction, ultraviolet spectrum, Raman spectroscopy and ion chromatography are available that can detect melamine and urea in milk/food.

Flour: Low quality flour is added to increase the bulkiness of milk to a level that is acceptable to consumers.

Tests to Detect Adulteration in Milk and it Products

I. Adulteration of Milk

Water: Place a drop of milk sample on a polished slanting surface. A trail is left behind when pure milk flows. No such trail is detected in case of milk that is adulterated with water.

Urea: Take a small quantity of milk sample in a test tube an add ½ teaspoon of soybean or arhar powder. Shake the test tube so that both are properly mixed up. After a gap of 5 minutes, dip a red litmus paper in it and remove it after ½ a minute. If the color changes from red to blue, urea is present in the milk.

Starch: A few drops of iodine solution or tincture of iodine are added to milk sample. Blue colour develops if starch is present.

Formalin: Add concentrated sulphuric acid (5 ml) to milk sample (10 ml) in a test tube. A violet or blue ring is formed at the junction of the two layers if formalin is present.

Vanaspati: Add 10 drops of hydrochloric acid to 3 ml of milk sample in a test tube. Further add a teaspoon of sugar. Mix well and examine the mixture after 5 minutes. The development of red color indicates the presence of Vanaspati in milk.

Detergent: An equal amount of milk sample and water are mixed and shaken. Formation of lather indicates the presence of a detergent.

Synthetic milk: It leaves bitter taste in mouth after consumption. It gives a soapy feeling on rubbing between the fingers and turns yellowish on heating.

Glucose/inverted sugar: The presence of glucose/inverted sugar is tested with the help of urease strip. A positive test indicates the presence of glucose or inverted sugar.

II. Adulteration of Milk Products

Rabdi: It is usually adulterated with blotting paper. For testing purpose, add 3 ml of hydrochloric acid and 3 ml of distilled acid to a sample of *rabdi* in a test tube. Stir well the contents with a glass rod. Remove the glass rod and examine it. If there is a presence of fine fibres on the glass rod, blotting paper is present.

Khoa/Paneer: Add water to a small quantity of food article. Let the mixture cool and add a few drops of iodine solution. If blue color develops, starch is present.

Vanaspati in ghee: To test the adulteration, mix a teaspoon of fully melted ghee with equal quantity of hydrochloric acid (concentrated) in a test tube with stopper and add a pinch of sugar. It is then shaken for a minute and left for 5 minutes. Formation of crimson color indicates the presence of Vanaspati.

Mashed potato/Sweet potato in ghee: Another adulterant of ghee is mashed potato or sweet potato. To test for its presence, add a few drops of iodine to the sample of ghee. A change of color to blue indicates that ghee is adulterated.

Vanaspati in butter: Butter is also sometimes adulterated with *Vanaspati*. The adulteration is checked in a similar way as the adulteration of ghee with *Vanaspati*.

Adulteration and the law: The agricultural Product Standards Act, 1990, regulations relating to dairy products and imitation dairy products (Regulation 2581 of 20 November 1987) states that 'milk' represents the normal secretion of the mammary glands of bovines, goats or sheep.

The sale of spurious/adulterated food items including milk is punishable under the FSSA, 2006. It is the duty of states and union territories governments to implement FSSA, 2006 and FSSR, 2011 and take stringent action against the offenders.

The Supreme Court of India, in December 2013, urged the state governments to amend the laws and make production and marketing of adulterated milk, an offence punishable with life imprisonment. Such necessary amendments have been made in the states of Uttar Pradesh, Odisha and West Bengal.

In a historic judgment to protect the health of common man, a bench of Jammu and Kashmir High Court, in December 2013, penalized three industries to the tune of Rs 30 crore as the foods manufactured by them were adulterated an unsafe for human consumption. It included milk, turmeric and saunf powder manufacturing units.

Milk-Marketing System in India: Milk production in India is dominated by small and marginal landholding farmers and by landless labourers. Eighty percent of milk is marketed through the highly fragmented un-organized or informal sector, which includes local milk vendors, wholesalers, retailers, and producers themselves. On the other hand, the organized dairy industry or formal channels, which accounts for about 20 percent of total milk production, comprises two sectors: government and co-operatives. A schematic diagram of milk marketing channels in India is presented in Fig. 27.5.

CUSTOMERS

Within the organized sector, the co-operative sector is by far the largest in terms of volumes of milk handled, installed processing capacities, and marketing infrastructure. The 82 thousand Dairy Cooperative Societies (DCSs)

across the countries have a strong membership of nearly 10 million landless, marginal, and smallholder milk-producer families.

Milk sector growth in India: With the growth of the population in urban areas, consumers had to depend on milk vendors who kept cattle in these areas and sold their milk, often door-to-door. As a result, several cattle sheds came into existence in different cities. In the cities of Mumbai, Kolkata, Chennai, New Delhi and even in some large townships, processed milk, table butter and ice-cream were available by 1940s, though not on a large scale. Polsons, Keventers and the Express Dairy were some of the pioneer urban processing dairies.

With the initiation of India's first Five-Year Plan in 1951, modernization of the dairy industry became a priority for the government. The goal was to provide hygienic milk to the country's growing urban population.

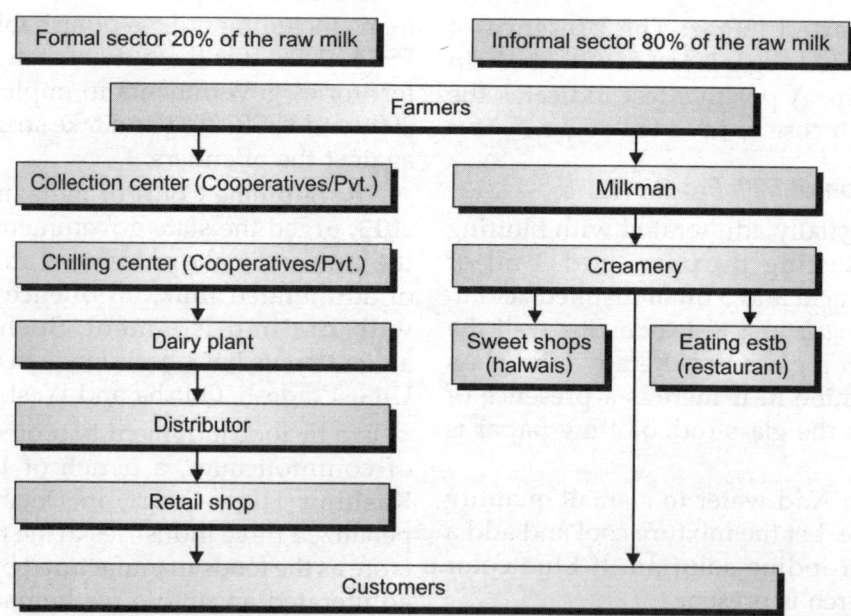

Fig. 27.5: Milk supply channels in India

Initial government action in this regard consisted of organizing "milk schemes" in large cities. To stimulate milk production, the government implemented the Integrated Cattle Development Project (ICDP) and the Key Village Scheme (KVS), among other similar programmes. In the absence of a stable and remunerative market for milk producers, milk production remained more or less stagnant.

During the 1960s, various state governments tried out different strategies to develop dairying, including establishing dairies run by their own departments, setting up cattle colonies in urban areas and organizing milk schemes. Almost invariably, dairy processing plants were built in cities rather than in the milksheds where milk was produced. This urban orientation to milk production led to the establishment of cattle colonies in Mumbai, Kolkata, Chennai. These government projects had extreme difficulties in organizing rural milk procurement and running milk schemes economically, yet none concentrated on creating an organized system for procurement of milk, which was left to contractors and middlemen. Milk's perishable nature and relative scarcity gave the milk vendors leverage, which they used to considerable advantage. This left government-run dairy plants to use large quantities of relatively cheap, commercially imported milk powder. High-fat buffalo milk was extended with imported milk powder to bring down the milk price, which resulted in a decline in domestic milk production. As the government dairies were meeting barely one-third of the urban demand, the queues of consumers became longer while the rural milk producer was left in the clutches of the trader and the money-lender.

All these factors combined left Indian dairying in a most unsatisfactory low-level equilibrium. The establishment and prevalence of cattle colonies emerged as a curse for dairying in the rural hinterland as it resulted in a major genetic drain on the rural milch animal population, which would never be replaced. City dairy colonies also contributed to environmental degradation, while the rural producer saw little reason to increase production.

Operation flood, launched in 1970, introduced co-operatives into the dairy sector with the objectives of increasing milk production, augmenting rural income, and providing fair prices for consumers. It was started to effectively utilize donated milk products from abroad for domestic dairy development. These surpluses were used to speed up Indian dairy development in two ways. First, the donated milk products were used to reconstitute milk and therefore provide the major cities' liquid-milk schemes with enough milk to obtain a commanding share of their markets. Secondly, the funds realized from the reconstitution and sale of donated products were used to resettle city-kept milk animals and permit their progeny to multiply; to increase organized milk production, procurement, and processing; and to stabilize the major liquid-milk schemes' position in their markets

The three phases of operation flood were:

1. During its first phase, operation flood linked 18 of India's premier milk sheds with consumers in India's four major metropolitan cities: Delhi, Mumbai, Calcutta, and Chennai.

2. Operation Floods Phase II (1981–1985) increased the milk sheds (collection centers) from 18 to 136; 290 urban markets expanded the outlets for milk. By the end of 1985 there was a self-sustaining system of 43,000 village co-operatives covering 4.25 million milk producers. Domestic milk-powder production increased from 22,000

tons in the pre-project year to 140,000 tons by 1985, all of the increase coming from dairies set up under Operation Flood. Producers' co-operatives increased direct marketing of milk by several million liters a day.

3. Phase III (1985–1996) enabled dairy co-operatives to expand and strengthen the infrastructure required to procure and market increasing volumes of milk. Veterinary healthcare services, feed, and artificial-insemination services for co-operative members were extended, and member education intensified. Phase III consolidated India's dairy cooperative movement, adding 30,000 new dairy co-operatives to the 42,000 existing societies organized during Phase II. Milk sheds peaked to 173 in 1988–89 with the numbers of women members and Women's Dairy Cooperative Societies increasing significantly.

Today there are 22 state federations in India, with 170 district-level unions, more than 76,000 village-level cooperative societies, and 11 million milk-producer members in the different states. These co-operatives collect an average of 15 million liters of milk each day. Fresh liquid milk, packed and branded, is marketed in over 1000 cities and towns in India by these co-operatives.

Most of the dairy co-operatives in India are based on the principle of maximization of farmer profit and productivity through cooperative effort. This pattern, known as the Anand Pattern, is an integrated cooperative structure that procures, processes, and markets produce.

Even the packaging and retailing of milk has seen a tremendous change. Earlier milkmen distributed milk on the doorsteps of the customers or people themselves brought milk from the dairy farmer in steel containers. Later, the trends changed and milk was marketed in glass bottles. Difficult handling, transport and breakage of glass bottles resulted in shift to convenient milk packets. Another reason for the switch was loss of riboflavin from milk on exposure to sunlight. Still newer trend of tetra packs and pyramid packs have arrived which assure higher shelf life and quality (Fig. 27.6).

Constraints in milk marketing: The dairy sector is characterized by small-scale, scattered, and unorganized milk-animal holders; low productivity; inadequate and inappropriate animal feeding and health care; lack of an assured year-round remunerative producer price for milk; an inadequate basic infrastructure for provision of production inputs and services; an inadequate basic infrastructure for procurement, transpor-

Fig. 27.6: Different packagings of milk

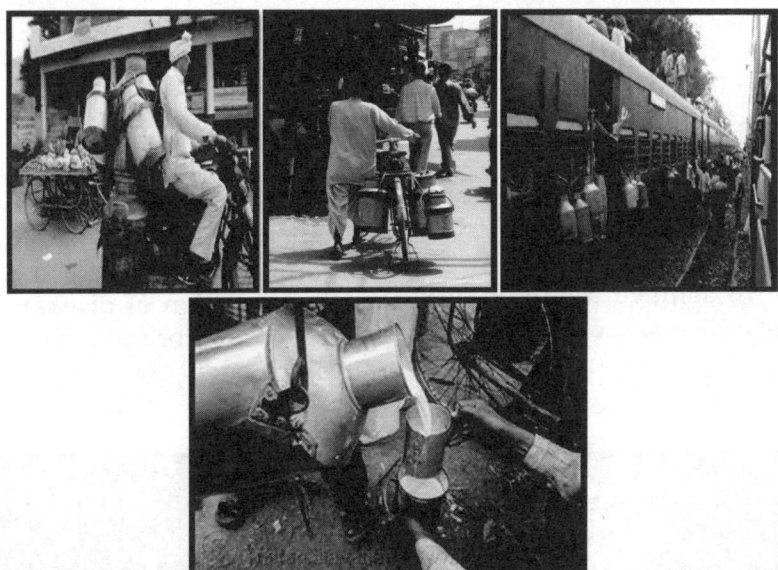

Fig. 27.7: Existing system of deliver of milk in India through milkmen

tation, processing and marketing of milk; and lack of professional management (Fig. 27.7). Additionally, the dairy-development policies and programs that are followed, including those relating to foreign trade, are not congenial to the promotion of sustainable and equitable dairy development.

Milk hygiene practices in India and other countries—In India, animals are generally milked twice a day. Prevailing conditions like tropical climate, inadequate cooling facilities, widespread adulteration, lack of quality consciousness and small-scale scattered production can influence hygienic quality of milk considerably.

The highly organized method is carried out in few states like Gujarat (ANAND pattern co-operative system), Verka in Punjab, Parag in U.P., Nandini milk dairy in Karnataka, etc. In this system the individual milk producer supplies the milk within 1–3 hr of production to a village level society. This is transported twice a day in cans within 3–5 hr to the district level dairy plant under ambient conditions. As dairy plant is far off, the milk from village society goes to a chilling centre, cooled below 5°C and then transported to the district level dairy plant. Here, milk is pasteurized and supplied to the consumer.

However, in developed countries the milk distribution system is quite different. It is a common practice to cool the milk immediately at the farm, and the same is collected by the dairy plants every day or alternate day or twice a week.

Even though such ambient and hygienic conditions are not strictly followed in our country, the habit of boiling the milk before consumption has probably saved customers from serious milk borne illnesses.

Effect of poor quality milk on humans—Milk borne diseases and illnesses from adulterated milk are the major problems resulting from poor quality milk. Flavoured milk *(badam milk, kesar milk)* sold in local bottles prepared and packed under non-standardised quality conditions can also pose a threat to health of people (Figs 27.8 and 27.9). During festival season (like Diwali) usually, maximum instances of adulteration of milk products are reported. This links to the fact that the demand for sweets prepared from milk increases in most Indian cities during this period. These unscrupulous activities result in many milk borne diseases.

Quality control to achieve hygienic milk production in the milk marketing chain: Milk can be contaminated at any point in the milk production process. It is the responsibility of the milk producer to identify these points and implement control measures to protect milk from contamination.

Steps should be taken as and when necessary to ensure supply of safe milk to people. An efficient hygiene program should begin at the farm. Proper hygiene practice can prevent the transmission of disease from animals to man. In order for milk to reach the processor and ultimately the consumer in good con-

Fig. 27.8: Flavoured milk prepared at a sweet shop

Fig. 27.9: Bottles flavoured milk of local brands

dition, a number of points must be observed right from the farm level to the processing factory, and ultimately from retailers and the consumer.

A. **At the dairy farm:** Quality control and assurance must begin at the farm. Farmers must use approved practices of milk production and handling. Also all regulations regarding the use of veterinary drugs on lactating animals and against adulteration of milk, etc. must be observed. The following must be kept in mind:

1. **Milch animals:** These must be adequately protected from diseases as the pathogens may be either excrete into milk or may contaminate milk through environmental contamination. Mastitis control is a must. To achieve this government agency, cooperatives or the milk collecting dairies can provide technical and laboratory assistance to dairy farmers. Also ensure that good milking equipment, disinfectants and antibiotics are available. Bathing and cleaning of the animal must be done daily. Teats of animals can be cleaned with towel soaked in bleaching powder (10 mg/litre) or potassium permanganate (1%) should be adopted (Fig. 27.10). The foremilk may contain the microorganisms and it should be collected separately and removed from the cowshed.

2. **Environment:** The contaminating materials in dairy farms are dung, mud, bedding materials, and straws. To prevent the entry of these contaminants into milk, routine brushing and washing must be carried out. The cow shed must be kept clean, ventilated and well lighted. Proper system for manure disposal must be assured. Clean potable water supply should always be available.

3. **Milker's hygiene:** The hands and clothes of milker must be clean. He must be disease free and in good health. He should not be suffering from diseases, should wash hands properly and cut nails periodically before milking. Milker should avoid the wrong milking practice like knuckling and incomplete milking, which leads to multiplication of organisms in the left over milk. The buckets in which milking is done must be clean (Fig. 27.11).

4. **Proper sanitation of milk cans:** Use clean hygienic milking machine and storage equipment. Stainless steel or

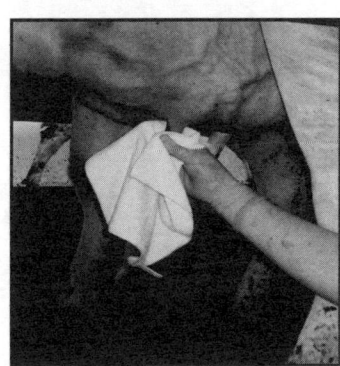

Fig. 27.10: Cleaning of teat before and after milking

Fig. 27.11: Clean milking bucket

aluminium cans should be used for milk transportation. The equipment surface should be free from dents, pits, rough spots, crevices and also it should be non-toxic, and noncorrosive. Cans should be cleaned properly and dried (Fig. 27.12). Immediately after milking store the milk at 4°C. The cans should be washed, rinsed, sanitized and dried properly. Exposure to sunlight will enhance killing off bacteria during drip drying of cans. For milking machines CIP (cleaning in place) must be followed.

5. **Milk transport vessels:** All milk transport vessels should be cleaned routinely. Milk cooled on the farm or cooling centre may be transported in milk cans or in bulk tankers. Bulk tankers are insulated, so the milk will remain cold until it reaches the processing factory. In the case of farmers delivering milk via cans it is advisable to place them in a shaded and cool area while awaiting pick-up by a milk transport vehicle.

6. **Packaging:** Packaging materials should be stored in a dry place away from manufacturing areas. It must be used in a clean and sanitary manner to avoid contamination of processed products.

B. **At milk collection centres:** All milk from different farmers must be checked for quantity, wholesomeness, acidity and hygienic quality.

C. **At reception in dairy factories:** Milk from individual farmers or bulked milk from various milk collection centres must be checked for quantity plus bacteriological and compositional quality. Also tests for the presence of antibiotics need to be carried out regularly.

D. **Within the dairy factory:** Once the dairy factory has accepted the farmer's milk it has the responsibility to ensure that the milk is handled hygienically during processing. At the dairy plant quality assurance tests must be carried out to ensure that the products are processed and they conform to specified standards. These relate to the adequacy of the processes applied and to the keeping quality of the manufactured products.

E. **During marketing of processed products:** Public health authorities are employed by law to check the quality of food stuffs sold for public consumption. Their inspectors may impound substandard or contaminated foodstuffs and prosecute possible culprits.

Fig. 27.12: Draining and drying of milk can and milk measuring sets

Apart from taking hygienic precautions, the growth of microbes can be restricted by cooling or other methods such as activation of natural inhibitory systems like lacto-peroxidase-thiocyanate-hydrogen peroxidase system and other food grade bio-preservatives, which can temporarily restrict the multiplication of microorganisms in milk. However, pasteurization is the universally adopted method to enhance the shelf life of milk and make it safe for human consumption. At room temperature milk can be stored only for 3 hours immediately after milking. The shelf life of milk can be extended to 24 hours by cooling to 5°C. Its shelf life is further extended to 4 to 7 days by pasteurization. By UHT (ultra high temperature) treatments the shelf life is extended to few months with lacto-peroxidase system milk can be preserved for 6–12 hours without the need of equipment and electricity.

Milk and milk products quality can be maintained by examination of samples at frequent intervals. The assessment of microbial load at various stages of manufacture or processing may serve as a useful tool for quality assessment and improvement.

Conclusion: The milk we drink today is no longer nature's pure food. Adulterated milk has put health and safety of people at stake. Mushrooming of small dairies with no regulatory supervision the quality of milk is deteriorating day by day. The offenders who freely adulterate the milk can only be stopped from such horrendous activities by imple-

The steps of maintaining safety of milk are summarized below

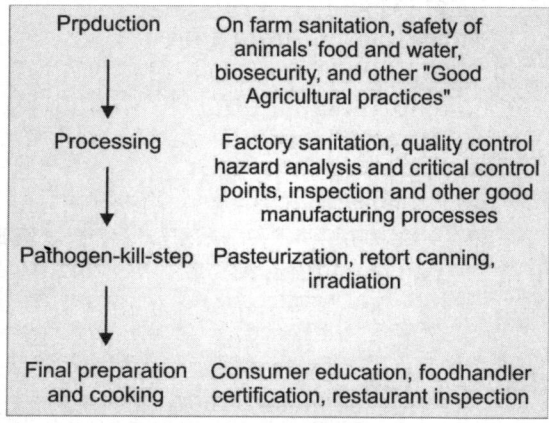

menting strict laws. Such people should be heavily punished and fined. It is important to sensitize people involved with milk business (small or large scale) regarding the harmful effect of adulteration. Strict licensing policies and routine checks even for small scale dairy farmers is a must. A milk hygiene training program needs to be evolved. It should be compulsory for milkmen to attend this before issuing licence. This can help maintain milk hygiene. Government must extend support in this endeavour by proving basic infrastructure and equipment to poor and marginal dairy farmers. Designing easy to use and cheap milk adulteration testing kits for homes is the best solution to overcome this problem. To achieve purity of milk and thereby health, people and the government must join hands against milk adulteration menace.

28

Safety of Beverages

Puja Dudeja, Amarjeet Singh, Surjinder Singh

INTRODUCTION

Any drink usually other than water is defined as a beverage. Most beverages consist of water as their main ingredient. Beverages have been a part of India culture since Indus valley civilization. *Sura*, a brewed drink made from rice meal, wheat, sugarcane, grapes and other fruits was quite popular. People used to ferment a portion of their crops to make alcoholic drinks. In Ayurveda also, uses of alcohol have been described both as a medicine and as poison.

Water is the chief constituent in all drinks. The term "soft drink" specifies the absence of alcohol in contrast to "hard drink" and "drink". The term "drink" is theoretically neutral, but often is used in a way that suggests alcoholic content. Beverages such as soda pop, lemonade, fruit punch, tea, coffee, milk, hot chocolate, milkshakes and tap water and energy drinks are all soft drinks. People consume hot tea and coffee in the winter months and shift to cool refreshing drinks in hot summers. Among the hot beverages manufactured in India, tea is the most dominant beverage that is ruling both the domestic and international market even today. Among cold drinks some like *Lassi* (curd with water and spices) common in north while others like coconut water is prevalent in south India. It is a common sight to see roadside vendors selling ice cold *nimboo pani* (Indian lemonade), fresh fruit juices, sugarcane juice, rose drink (*rooafza*) at crowded places, outside a market, hospital, railway station, bus stand, cinema hall, colleges, etc. These attract customers who want to quench their thirst at an affordable price. In Indian culture a guest is often aksed '*kya lengae app: thanda ya garam*'? 'what will you have: something hot or cold?. '*Chai-paani*' is a common term for any meeting, ceremony or any other occasion. Tea is ubiquitious. Even during travel there is an easy access to these beverages.

Indian beverages market is one of the fastest growing sectors in food industry. It occupies USD 230 million market. The rising middle class who can spend on beverages like wine, imported whiskey, fancy energy drinks has contributed largely in the growth of this industry. For the rural section, where the purchasing power is restricted the beverage companies offer small packs of Rs 5/10 (*panch ka coca cola; thanda matlab coca cola, toofani uthao bus thus mae*, etc.). Beverages are manufactured both by the organized and unorganized sector. They market their products in PET and glass bottles, aluminium cans, tetra packs and fountains in disposable glasses. These are sold both at retail stores and by street vendors (Fig. 28.1). On the other hand, fresh juices and drinks by a street vendor are served in disposable or glass tumblers. The consumption of cold beverages during summer months provides relief from the scorching heat. There is a strong temptation towards the wonderful array of drinks served on the roadside.

Classification of Beverages

Beverages are broadly classified as alcoholic and non-alcoholic drinks. Non-alcoholic drinks are further classified as milk based, soy based, fresh juices, tea, coffee, energy drinks, sports drinks and soft drinks (Fig. 28.2).

Food Safety Aspects of Beverages

In view of the fact that beverages consumption is an integral part of daily routine it is vital that food safety issues are given due importance. Beverages can be contaminated during its journey from manufacturing unit to the consumer. They hygiene of the bottling plant is important for the safety of drink (refer to Chapter 16 on good manufacturing practices).

Fig. 28.1: Beverages displayed at a roadside stall

There is shortage of water in India and the situation is only going to worsen in times to come (*refer* to Chapter 35 on water supply in eating establishments). For all the companies manufacturing beverages in our country due emphasis is given to the purification process of water in the plant. They have filters, Reverse Osmosis (RO) plant, chloronomes, etc. to ensure potability of water. Apart from this they do get water tested from time to time and maintain proper records of the same. Beverages can also get contaminated during transportation (in case of leakage) or storage

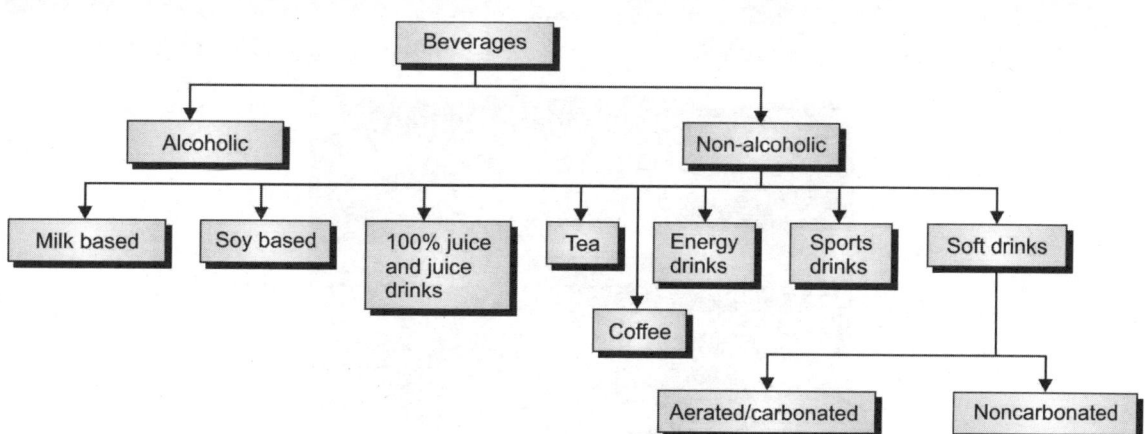

Fig. 28.2: Classification of beverages

(external surface of cans can get contaminated with rat urine). A safe beverage can become unsafe at the time of serving if contaminated ice is mixed with it to cool it.

There are many safety issues related to consumption of these beverages depending upon the type of Eating Establishment (EE) serving it. While most restaurants and café serve beverages in apparently hygienic conditions than *dhabhas* or roadside vendors. These days cafes and restaurants either have vending machines or serve packaged drinks in their crockery. They essentially follow the cleaning and maintenance procedure of vending machines (*refer* to Chapter 34 on maintenance of equipment). For packaged drinks the best before and use by date is seen

(*refer* to Chapter 48 on food labeling). In case fountain drinks are prepared potable/packaged water is used. They also use potable water along with permitted/colours/flavouring agents/syrups is used in making special beverages. Some of these EEs use only bottled/packaged drinking water for preparation of drinks. Even if they use tap water, it is treated and is safe for consumption.

Safety of beverages is an issue with roadside vendors that serve and attract more number of people. A variety of common man beverages sold by street vendors is given in Figs 28.3 to 28.7.

In India, especially in the metropolitan and other cities a huge section of the population of all income and age groups consume

Fig. 28.3: Sugar cane juice vendor

Fig. 28.4: *Jal jeera* and lemon juice stall

Fig. 28.5: Mango shakes for sale

Fig. 28.6: Lemon soda vendor

Fig. 28.7: A typical juice stall

beverages sold by street vendors. Lack of potable water supply is the main constraint faced by these vendors for mainitaing hygiene (*refer* to Chapter 35 on safe water supply). These stalls are generally placed in markets, outside school, colleges, hospitals, bus stops, railway stations, parks, cinemas where they can have maximum clientele. They do not have adequate water and space for washing of fruits, cleaning of hands and utensils leading to contamination of the beverage served.

They generally have 10–12 glasses with them for serving the beverage. In a day they often scores of customers ranging from 50–500. These vendors prefer to use glass tumblers, which have a peculiar shape. This shape exaggerates the amount it serves to the consumer. These glasses are washed and reused for different customers. Hands and utensils washing is usually done with the help of stored water in one or more buckets, and sometimes without soap. Unclean glasses often become a source of contamination and compromise the safety of beverage served in these.

Wastewaters and garbages are discarded nearby, providing nutrients for insects and rodents. They generally store solid waste in a carton or plastic bag and dump it in a nearby place at the end of the day. Improper washing of utensils and glassware used in preparation and serving of juices add to the risk of bacterial contamination. The unhygienic ways of preservation without refrigeration, unhygienic surroundings like swarming flies and dust act as source of pollution. Some beverages have shown to be potential sources of bacterial pathogens notably $E.\ coli$ 0157:H7. Fresh beverages prepared in an unhygienic way are recognised as an emerging cause of foodborne illness. Over and above this the food handlers managing these stalls have poor personal hygiene. This adds to the chances of contamination. They are untrained in food hygiene and they work under unsanitary conditions with little or no knowledge about the causes of foodborne illnesses (FBI).

To make more profit fruit juices are served by street vendors after diluting them with water and ice. Such addition of unsafe water or ice can cause outbreaks of gastroenteritis by $E.coli$, *Salmonella* and *Shigella*. The vendors use locally made ice, which is often transported in a vehicle. Here it comes directly in contact with the floor of vehicle. On unloading, it is covered with a dirty gunny bag and then handled directly. Then it is pierced by a dirty ice pick and added to the drink.

The microbiological quality of fresh juices available on streets is questionable. Some of the juices are not efficiently protected against flies, which may carry foodborne pathogens. During peak time of sale, which is generally in the hotter part of the day they prepare, juice in advance. Often peeled or deseeded cut fruits are kept unrefrigerated. These are directly used for juice extraction. The vendor does not apply safe food storage temperatures here. Thus, chances of FBI are quite high.

However, the practice of consuming fresh fruit and juice in the prevailing scenario from a street vendor by general public cannot be stopped on unhygienic grounds. Even it is difficult to image that the street vendors can be prohibited from selling such items. This is usually their only source of livelihood. However, it is the responsibility of the government to ensure food safety. So, the authorities must adopt measures to educate the vendors on food safety and hygienic practices and enforce adequate food safety guidelines for street food vending. Health education of the vendors and implementation of standard hygienic protocols may improve the quality of these beverages. Regular monitoring of the quality of these refreshing drinks for human consumption must be introduced to avoid outbreaks of FBI.

In a study by Chirag Gadi in Allahabad city 66.6% of lime juice and 83.3% of sugarcane juice samples showed the presence of Salmonella. The study also documented the fact that samples from crowded sites were more contaminated (69%) as compared to less crowded sites (31%). On stalls and *dhabas* with only one servant or only the owner, the degree of juice and food contamination was high (43%) as compared to having two servants (31%) or three servants (26%). This has been be explained by the fact that a single servant does all the work right from the peeling, preparation and serving, while performing these multiple tasks one person does not wash or clean the hands frequently and thus is likely to contaminate juices and foods. Similar findings have been documented by Titarmare et al., 2009 who evaluated presence of Salmonella in pineapple, orange and carrot juices.

Reddy et al in their study on street vended fruit juices and its safety evaluation in Bellary city of our country showed high levels of contamination fruit juices as well as water samples. This was mainly because of the poor quality of water used in juice preparation.

Sale of Duplicate (Fake) Aerated Drinks

During summer months some FBOs involve in unscrupulous practice of making duplicate soft drinks and sell them at low cost. They target consumers of low socioeconomic status residing in the urban slums or rural parts of the country. These aerated drinks look and taste quite similar to their original drink. Even their name sounds alike the original one except for a minor change in the spelling, which is not easy to be perceived by the consumer. For example, 'Fanda' for 'Fanta', 'Limna' for 'Limca'. Such manufacturing units usually do not practice adequate hygienic measures. Quite often they bottle fake beverage in a grossly unhygienic way in a dingy room in slum colonies. Consumption of such duplicate beverages can cause FBI. They are neither licensed or registered units and at times they run from the homes of these money minded FBOs.

Sale of Illicit Liquor

'Dozens of people are recovering in hospitals and officials say they expect the number of dead to rise further'. A news item herd commonly where poor villagers die after consumption of liquor spiked with excessive chemicals. Deaths from contaminated alcohol are a regular occurrence in our country especially in rural areas. In Oct 2013/42 people lost their lives in West Bengal. In December 2011, 169 people in West Bengal died from consuming bootleg liquor, while more than 100 perished in Gujarat in July 2009 from the same affliction. In 2008, more than 100 were killed in Karnataka, while 41 expired in Tamil Nadu. In 1992, in a particularly egregious case, almost 200 people in Orissa died after drinking a concoction of methyl and ethyl alcohol.

The homemade alcohol-commonly called *desi daroo* usually costs less. Chemicals and pesticides are often added to the alcohol to increase the strength or improve the flavour . Sometimes the antifreeze 'methanol' is added. This liquor is poisonous as it contains toxic chemicals. Ammonium nitrate can cause headache, dizziness, abdominal pain, vomiting, heart irregularities, convulsions, collapse, and death. Methanol is highly toxic to humans, and ingestion of just 10 ml can result in blindness, and 30 ml or more is usually fatal. This locally brewed is a popular poor man's drink. It is manufactured and produced in people's homes, also in large industrial units. Monitoring and supervising the safety standards in these units is quite difficult. Strict implementation of law is required to deal with this menace of unsafe liquor.

Conclusion

Beverages are well well-known for their nutritive value and give a refreshing feel to the consumer. In tropical countries like India, they are consumed by all sections of society. The consumption pattern is even increasing with increasing in paying capacity of people, lowering cost and ubiquitous availability. However, consumption of contaminated beverages can lead to outbreaks of FBI. Strict implementation of law for manufacturers and education and training of those serving beverages can ensure their safety. Last but not the least the consumer should also play their part by discouraging (by not patronizing the) sale of unhygienically prepared beverages. For this, consumer awareness needs to be imparted by various concerned agencies.

Section IV

Food Safety in Eating Establishments

29

Location and Layout of Eating Establishments

Puja Dudeja, Amarjeet Singh

Eating out Culture in Contemporary Indian Society

With changing times, our eating practices have changed. My daughter who is 16 years old was planning her birthday party. The most important decision to be taken was the choice of EE where the party would be held and menu was secondary. It was a difficult task as the underlying condition was that no birthday party should have been previously held there in her peer group. Times have really changed! I was remembering my days when my mother would cook *halwa, puri, sabzi*, etc. on my birthday and my friends would enjoy the same at home. Be it a birthday, anniversary, promotion, birth of a child we all prefer to treat our near and dear ones out in an eating establishment (EE) (e.g. restaurant/hotels/eating joint, etc.). Even when there is no reason to celebrate, it has become a ritual for the urban nuclear families to eat outside home at least once in a week, preferably on weekends. They try all new eating joints in the city even ignoring the travel distance from place of residence. The kitty parties which initially used to be conducted at homes where the guest would cook her best and serve it to her friends have shifted to restaurants with special halls for kiity parties. Eating out culture has increased both in the upper and lower income class equally.

Classification of EE

An EE includes the premises where public is admitted for repose or consumption of any food or drink or any place where cooked food is sold or prepared for sale. It includes:

a. Eating houses

b. Restaurants and hotels

c. Snack bars

d. Canteens (school, colleges, office, institutions)

e. Food service at religious places

f. Neighbourhood tiffin service/dabba walas

g. Rail and airline catering

h. Hospital catering

Location of EE

There are varying perspectives regarding location of an EE. A FBO would always want a location which would attract maximum number of customers. He would like to run his business from the main road rather than in the interior where customers would find it difficult to reach. He would prefer a place that has adequate parking facility for the clients who come. Also if the EE is co-located with other shopping outlets it attracts more customers. In fact most of the EEs are co-located in shopping malls/shopping complexes/markets. Even a street vendor places his cart in these busy locations. A street vendor also places himself next to a park, sazi mandi, outside a school/college, bank, cinema, hospital and other public places. Many canteens/dhabhas/tea stalls are placed in close proximity of busy offices who cater for the food requirements of people working in these buildings. There location is convenient for the office worker/wage earners who have a limited lunch break to eat. Similarly many EEs are located on the highway for the convenience of customers travelling long distances.

The sole purpose of their location is to have maximum number of customers and maximum business. In such a scenario often location related factors which affect food safety like general surroundings, air pollution, water supply, garbage disposal, etc. take a back seat.

As far as the customer is concerned, he would prefer an eating joint based on convenience. For example, imagine a patient visiting a hospital with a family member accompanying him. If there is long waiting in the OPD they would prefer a EE for tea and snacks which is closest to the OPD. Similarly, the relatives of patient waiting outside an OT/wards, etc. will prefer the nearest location. Most of the times if a family has gone for shopping they prefer to eat in the EE located in the same area example during an exhibition usually there are stalls of edible substances. However if it is a planned eating out visit people select the EE based on taste, past experience, menu variety, affordability, etc. For some food safety is the only criteria for selecting an EE.

As far as food safety is concerned location of an EE should be such that it is away from open drains, garbage dumps, water logging and excessive dust. There should be no overhead sewer pipes. It may not be possible

to relocate all EEs but the endeavor should be to make the surrounding clean and suitable as required for running an EE. Some of the important aspects of a good location are given below:

- The surrounding area should be free from air pollutants like smoke, dust, fumes, etc.
- The area should be clean and free from breeding places of flies and mosquitoes, rodents, etc.
- There should be adequate supply of potable water
- Proper sewage disposal facility
- Away from public urinals, garbage collection bins, open drains, etc.
- The premise should not be used for residential purpose.

If the surrouding area of an EE is not clean and hygienic then it can affect its functioning. I would like to share a personal experience here. There was a small restaurant in a market inside a hospital premises. During summer season, there were lot of flies in the market. This nuisance was progressively increasing each day. The problem worsened in magnitude to such an extent that it started affecting the business of all shopkeepers especially the restaurant owner. Every one blamed the restaurant owner for this situation. On detailed inspection of this EE, it was found that all hygienic practices were being followed in the kitchen. On inspecting the surrounding area it was realized that people who were bringing home cooked food were eating in open and throwing the leftover food in open. The garbage bin co-located to the market was also open and serving as a breeding ground for the flies. Here it was clear that any amount of maintenance and rigorous hygiene inside the premises of EE did not have any effect on fly nuisance. It was the surrounding (general hygiene) which mattered.

Layout of EEs

When we shift into a new house, we decide its layout in terms of drawing room, bedrooms, study room, guest room, entrance, location of dustbin. These decisions are based on the construction of building, convenience to people staying in house, sometimes *vastu shastra*, security aspects, source of natural lighting, etc. A well designed layout of an EE will contribute in the hygiene and sanitation of building and hence safety of food. Premises should be designed so that cleaning procedures are performed effectively and there is reduced chance for accumulation of soil and contamination of food. An ideal layout would be such that there is unidirectional flow from receiving area to preparation and service area. The shorter the distance of travel of food the lesser are the chances of contamination and cross contamination. Before finalizing the layout plan of an EE it is important to have a thorough understanding of the working of kitchen in respect of different menus being offered, and details of functioning of equipment . The layout of food premises shall be such that:

i. Food flow is in one direction as far as possible (i.e. receiving–storage–preparation–packaging/serving–transportation–retailing). In EEs, this chain may end at serving with collection of dirty utensils back to the kitchen. This unidirectional flow will reduce chances of microbial contamination. The flow of food and dirty utensils shall be decided at the start of business. Many a times due to paucity of space food flow is haphazard. However, small modifications in the process (no cost) or structure (low cost) can make the process safe. For example, while inspecting an extremely busy canteen we observed that there was a small window between the kitchen and outside area through which the orders would be transferred. The same window was also used to receive dirty used utensils back to the kitchen. At times during rush hours the cooked order was kept next to dirty utensils awaiting pick up by the waiter on one side and dirty utensils by dishwasher on the other side. The manager was suggested to use the window only for one purpose or construction of a new adjacent window. So next time we went the orders were sent from the door to outside.

ii. Adequate spaces are provided for food preparation, food storage, scullery, storage of equipment/utensils and installation of sanitary fitments. In case, the kitchen of the EE is small then separate arrangements should be made for cleaning of utensils or use of disposables(plates/spoons/bowls, etc.) with proper waste disposal can be adopted. Storage of raw and cooked items can be kept to minimum by changing the procurement policy. Area where food is processed/cooked should preferably be visible or easily accessible on demand to the consumers.

iii. Incompatible areas such as toilets should be completely segregated from food rooms. This not only adds to the aesthetics of the EEs but also prevents contamination. Customers also should not have to pass through a food room while going to the toilet. Such practices are unacceptable to us both in our homes and while eating outside.

iv. Layout of equipment shall be such that it permits adequate maintenance and cleaning. The functioning should be in accordance with its intended use and shall allow good hygiene practices, including monitoring. For example, it is recommended that the refrigerator should be kept at least few inches away from the wall and above the ground. This will allow the heat to escape and ensure cleaning in the space between wall and the equipment.

v. Walls of EE shall be of a design and construction that they are easy to clean and prevent harbourage for pests. Internal surfaces of walls and partitions in kitchens and food rooms shall be surfaced with smooth, light coloured, durable, non-absorbent and easily cleaned materials (e.g. glazed tiles or stainless steel) to a height of not less than 2 m. The rest may be lime washed or painted in light-colour. Junctions between walls, partitions and floors should be coved (rounded). Ceilings and false ceilings shall be of continuous construction so that there are no empty spaces or joints. False ceilings in food rooms shall have smooth, easily cleanable and impervious surfaces. Floors in kitchens and food rooms shall be surfaced with non-slippery, light coloured, non-absorbent and easily cleaned and durable materials (e.g. mosaic tiles). It should be coved at the junctions with walls and be sloped towards a floor drain and shall allow adequate drainage and cleaning. Floors, ceilings and walls must be maintained in sound condition to prevent accumulation of dirt, condensation and growth of undesirable moulds.

Floor drains in kitchens and food rooms shall be so constructed as to prevent accumulation of waste water, easily accessible for cleaning and clearing if choked; and be properly trapped, vented and connected to a proper drainage system. Placement of grease traps in the floor drains can prevent choking. The staff handling cleaning of utensils shall be instructed to first remove the leftover food from the used utensil before cleaning. Care at this step will prevent choking of drain in the long run.

Doors shall have smooth, non-absorbent surfaces, and be easy to clean and, where necessary, disinfect. Windows shall be made of material which is easy to clean, minimize the built up of dirt. Where necessary, windows should be fixed.

The layout of main areas of an EE are discussed below.

Receiving Area

It is the area where raw materials (both dry and fresh) are received. It is imperative that the receiving area should be colocated to the storage area so that the distance for storage is reduced. It should have a separate door opening to the exterior. There should be a proper road so that the items can be directly unloaded from the vehicle to the receiving area. It should preferably have washing and sanitizing facilities for raw fruits and vegetables. There should be a weighing scale to record the actual weight of raw material received after sorting. There should be a garbage bin. An air curtain at the entrance will prevent entry of flies along with the raw materials.

Storage Area

It can be dry food store commonly called ration store, cold store (refrigerator), frozen stores (deep freezers). There should be adequate lighting in the dry food storage area. Slabs should be constructed at least 6" above the ground so that no items are kept on the floor. It should not be a thoroughfare and staff should enter only when required.

Kitchen Area

In designing the kitchen, care should be taken to avoid placing them at locations where there are manholes or foul-waste/sanitary fitments (water closets, urinals and toilets). Resiting of the manholes and/or drainage alterations may be required. Kitchen should not be so designed as to result in the conveyance of food or clean eating utensils from the kitchen to the customers' seating area through an open space.

Planning needs to be done before placing various equipment. This will depend not only on the size of the equipment but also upon the process involved and the stage of food preparation. There shall be unidirectional flow of items. Prevent cross contamination during food preparation/manufacturing processes from viz. pre-processing, packaging, dishing/portioning of ready-to-eat food. This can be achieved by compartmentalizing and strict measures should be taken so that material movement happens only in one direction without any backward flow and any mixing up of various activities. Area occupied by machinery shall not be more than 50% of the manufacturing area.

If the kitchen area is small then shelves, racks and cupboards can be placed. Cooking ranges should be kept in the centre of kitchen away from doors to keep away from dust. Provision of exhaust should be made above the cooking range to reduce fumes, smoke and odour from the kitchen.

Sinks should be placed close to the drain. All work tables should be movable for the ease of cleaning and maintenance.

Racks for pots and pans should be placed on the wall and stored upside down so that the water is drained. Hooks can be placed to hang pots and pans with handles.

Refrigerator should be placed at an area which is away from direct heat of sun. Separate sinks should be available for utensils and hand washing.

Staff rooms: It should have a connection with the kitchen. Facilities of lockers should be available.

Washrooms: These should be adequate in number based on the load of EE and staff working there. It should be readily assessable at a distance of 3 m. These should not open directly into the food area. Toilets should be well lit, ventilated and labeled for males and females. Hand washing facilities with soap and disposable tissues should be present. A warm air dryer can be installed.

Seating arrangement: The number of customers to be accommodated is calculated at 1.5 m^2 per person of the seating area provided. The customers should not pass through the kitchen when going to toilets.

It is a common practice in our country to start food business by renting a shop in a busy locality which has not been originally designed as an EE. Such areas need to be extensively remodeled with suitable renovations that will change the structural and equipment layout and/or will involve demolition work to walls, ceilings and floors anywhere on the premises. However, this work is not undertaken to cut down the initial investment cost of the business. Later, these disorganized and compromised arrangements can impair food safety.

30

Lighting in an Eating Establishment

Puja Dudeja, Amarjeet Singh

We all have heard about candle light dinners for creating a romantic aura from the dim light in the eating establishments. The thought of it only takes us to a dream world of fine dining experience with our beloved ones! However, have you ever wondered how they can compromise food safety of the customer eating food in such a dim light? A pest in the food like fly/cockroach can be mistaken for any Indian spice. Isn't it true?

Lighting is a very important feature as far as the cooking area/kitchen and storage area of an EE is considered. Proper lighting at receiving area is required for sorting of raw material. It helps to identify spoiled items and dirt. It facilitates cleaning of the area and improves the quality of work.

It is important that at the time of construction adequate planning is done for construction of windows for natural light and placement of electricity points for artificial light. Dark gloomy corners accumulate dirt and become difficult to clean.

Well planned windows can be a good source of natural sun light. However if the glass pane of window is broken or screening of windows is not done properly the same window can be a source of dist and foreign particles in the kitchen. It is recommended that the window area should be one fifth of the floor area to provide sufficient lighting in the room. Natural light of sun that comes through the windows also has germicidal properties.

Artificial lighting can be placed in the form of tungsten filament bulbs, florescent tubes and sodium vapour lamps. Bulbs produce a glare and shadow where as tubes produce an even spread of shadow less light and little glare. This gives a comfortable lighting and increases work efficiency. It is preferred to use white light as colored or fluorescent lights can change the perception of original colour of food items.

Food preparation and sink area need special attention, as adequate lighting is a must here. There shall be no glare or flicker. There shall be a constant continuous source of glare free light in these areas. Lights shall be so placed that the interior of various equipment like cooking range, ovens, etc. is easily visible. The standards for amount of light falling in different areas are:

- Food storage area: 150–200 Lux
- Food preparation, serving, washing area: 200–500 Lux
- Service area: As desired

Inadequate lighting not only compromises food safety but also leads to headache, eye strain, irritability in the workers.

Light Fittings

Light fittings should be easy to clean and corrosion resistant. These should be maintained well and kept in good state of repairs. They should have a shatter-proof covers to prevent broken glass from contaminating food or food contact surfaces. Electric wires should not be hanging loose in the kitchen. These should be fitted in such a way to avoid dust or dirt accumulation.

31

Ventilation in an Eating Establishment

Puja Dudeja, Amarjeet Singh

In traditional Indian culture, it was a common practice that when men would eat food the wives would make the surrounding environment comfortable by using hand held fans. In the palaces, the people employed for the job would perform this action. The underlying aim was to provide comfort to the person eating food. With the invention of electricity and availability of mechanical ventilation devices the practice was stopped. Ventilation in eating establishments of our country is of concern both to the operator and customer as humidity and temperature levels are high in most parts of our country during summer months. People go to an Eating Establishment (EE) not only for eating food but also for relaxation and comfort. Even the *dhabha* owners feel that the customers will enjoy their food if they are made comfortable. They have also gone to the extent of placing air conditioners in *dhabhas!*

Adequate ventilation is equally important in the seating area and kitchen. Lot of heat is generated in kitchen during cooking along with fumes and smoke. Inadequate ventilation can cause stress, contributing to unsafe systems of work and high staff turnover, Adequate ventilation can make the work place comfortable for food handlers by cooling it. They will sweat less and are less likely to touch body parts while handling food. They are also more likely to comply with the recommended guidelines of wearing their aprons and cap if surroundings are cool. Good ventilation also supports reduced accumulation of grease or toxic fumes and gases. An ill-ventilated work place affects the work place and work efficiency of the worker. It also promotes the multiplication of harmful microorganisms in food, leading to food spoilage and food poisoning.

The objectives of an effective kitchen ventilation system are to:

- remove cooking fumes at source, i.e. at the appliance

- remove excess hot air and bring in cool, clean air, so the working environment is comfortable

- make sure that the air movement in the kitchen does not cause discomfort, eg from strong draughts

- provide enough air for complete combustion at fired appliances, and prevent the risk of carbon monoxide accumulating

- be easy to clean, avoiding build-up of fat residues and blocked air inlets, which lead to loss of efficiency and increased risk of fire

Systems of Ventilation

There are mainly two types of ventilation systems

1. Natural ventilation: Windows, doors and ventilators
2. Artificial ventilation: Exhaust/extraction system, propulsion system, air conditioners

Natural ventilation: We all know that when hot air is lighter, it rises, cool air comes and takes its place and causes cooling. The combination of windows and ventilators work together based on this principle. However, many a times these issues are neglected at the planning and designing stage of a kitchen. Sometimes a room not originally designed as a kitchen is used for the purpose. Such rooms lack adequate ventilation. In case of windows and ventilators, it is imperative that they are screened properly by mesh. In case it is not done they will serve as entry points for pests and dusts. The mesh should be cleaned during the general maintenance of the kitchen to prevent accumulation of dust and grease. In case there is damage to the screen it should be repaired. As a temporary measure, it can be sealed with a cardboard or piece of cloth but in no case it should be left open as pests can enter from here.

Efficacy of natural ventilation depends upon the location of kitchen. Natural ventilation is not available if the kitchen is located in the basement. Even if the kitchen is located at ground floor or upstairs ventilation will depend upon the surrounding area. The windows and ventilator can only provide natural lighting if the outside area is busy and overcrowded. These work alone if the surround area is open and there is cross ventilation. These days focus is on eco friendly buildings which conserve and save energy. Even otherwise electricity supply in India is quite erratic and not dependable. In such a scenario it is imperative that due emphasis is given to natural ventilation in kitchen and eating area. Design of the building should cater for cross ventilation.

Artificial ventilation: This is the only way to ensure adequate ventilation where natural methods do not work. At times the kitchen area of an EE is small and there are a number of people working inside for long hours continuously. In such situations if the working environment is made comfortable by adequate ventilation and temperature control there are less chances of errors. The ventilation requirements of kitchen are different from rest of the areas of EE.

The commonest form of mechanical ventilation device is an exhaust fan. This most of us have commonly seen in our home kitchens too. The principle behind its working is that it propels the hot air outside by mechanical suction and creating a vacuum in the kitchen area. Along with this, an injector fan along with a filter can be placed which allows pure air to enter. The inlet fan should not be placed near a refuse dump or a garbage bin else the waste can come and spread inside the kitchen along with air flow. These fans should not create loud noise and hence nuisance. The blades of the fans need to be regularly cleaned so that the dust and grease accumulated can be cleaned and prevented from falling in food items.

In case the kitchen area is large then separate exhaust fans can be placed above a tandoor or a cooking range or near the frying area. These days we come across various advertisements of chimneys in home kitchen. These are nothing but an exhaust system with a hood and a duct. The hood is placed at a height of approximately 1.2 m above the cooking equipment. There is an exhaust fan and a filter in it. This system works for specific equipment only. The hoods are made of corrosion resistant material like steel. The filter captures the dust and grease from the fumes and prevents them from dropping back. The size of the hood varies with the size of the equipment. The filters or grease extractors are detachable and should be cleaned every quarterly. Documentation of this cleaning should be made. If the grease is not removed and allowed to accumulate it may start falling into the food over the cooking range and make it unsafe. In case the cooking range is being used heavily for frying or heating oil purposes then the frequency of cleaning is shortened. The duct above the hood propels the air and fumes to the exterior.

Along with this propulsion of cool air is required to be done with the help of fans. Direct fans have an effect on the heating efficacy of cooking range and are not preferred. They can be placed in the staff room of the workers or relaxing area and away from the cooking range. The best option for maintaining ideal temperature and humidity in a kitchen is installing an air conditioner. However, this system is expensive and consumes a lot of electricity. Eating establishments like reputed hotels with star rating do have centralized air conditioning in the kitchen to maintain a comfortable environment for the food handlers and hence get quality work from them.

Factors which need to be considered before planning the ventilation system of a kitchen are:
- Cooking load of the kitchen
- Amount of cooking equipment used
- Layout and shape of the kitchen
- Number of staff
- Need for easy cleaning and maintenance

Apart from looking at the comfort of the food handlers, occupational and food safety, the food business operator has to work the expenditure incurred in installing ventilation equipment in the kitchen. Many a times if the EE is offering *tandoori rotis*, the tandoor is placed outside in open. EEs with a star rating do have a centralized air conditioning system for the comfort of workers. As a contrast in average EEs' kitchens have ceiling/cabin fans for propulsion and exhaust fans for expulsion of hot air and fumes. The cleanliness and maintenance of these fans is not taken seriously. Usually these are coated with grease and dust. Also, these are placed at a height. So cleaning them requires suspension of all other operations in the kitchen. Since their cleaning does not affect the business of FBO their maintenance takes a back seat. At times, in case of a fault they remain non-functional for a long time for want of repairs. Last but not the least these systems of ventilation are at the mercy of electricity failures and power cuts. Hence, a FBO needs to ensure that:
- Adequate means of natural or mechanical ventilation are in place
- Direction of flow should not be from contaminated areas to clean areas
- Cooking range inside kitchens and food rooms shall be equipped with an exhaust system that can efficiently and effectively remove all fumes, smoke, steam or any vapour arising during cooking
- The exhaust system is installed with a metal hood properly connected to an air-duct fitted with an extraction fan of sufficient capacity.
- Fresh air supply system fitted with propulsion fans with adequate capacity is installed in food rooms and kitchens.

Personal experience: While visiting various EEs there was a dhabha which had a huge exhaust fan. The working of fan produced so much noise that it was causing headache and irritability to the workers in kitchen. It was also creating nuisance for the shops in neighborhood. When the exhaust was turned off one day everyone felt relieved. All the food handlers preferred poor ventilation over noise nuisance!

32

Pest Control in Eating Establishment

Puja Dudeja, Amarjeet Singh

Food pest can be defined as "an animal or an insect which lives in or near man's food and is destructive, noxious and troublesome". Pests are the uninvited guests of any eating establishment (EE). All EEs typically dread the thought of pests, which they should. Food, water and shelter, the three things any pest needs to survive, are all present in EEs. The nuisance of pests has been vividly illustrated in the famous old story 'The Pied Piper of Hamlein' where rats made the life of people in Hamlein difficult. At times pest nuisance can force the closing of food businesses. Controlling the pests is, therefore, an essential step in the maintenance of hygiene and sanitation in every society. Some pests carry and transmit pathogens, while others can cause a shock and death through their stings or cause allergic reactions in consumers. In fact their mere presence in food or food preparation areas causes distress and makes consumers unhappy and above all, makes food unsafe!

Common pests include rodents: rats and mice; insects: flies, cockroaches, wasps, moths, silverfish and ants; birds: pigeons; animals mainly stray cats are found near EEs.

All EE are at a risk from the damage caused by pests and by the time their, presence is noticed in the premises, some damage has already have taken place. These tiny creatures are a definite enemy of a food business operator. Potential human pathogens have been isolated from a wide variety of pests. For example, rodents are capable of transmitting food poisoning bacteria like salmonella, listeria, etc. and a range of viral, bacterial, protozoan and endoparasitic diseases, either by direct contact with the food from their contaminated bodies or legs, faecal deposits or by urine. They also cause physical damages. Cockroaches are ubiquitous pests, which feed on faeces, human and animal wastes and then on human foodstuffs and so transmit bacteria such as Salmonella and Staphylococcus. Flies can also cause extensive problems. Contamination of food article occurs through their vibrating wings/legs and bodies. They transmit pathogens, including food poisoning organisms.

Effective pest control is therefore required in all EE in order to avoid food contamination. This helps to prevent wastage of food and to avoid damages. Good pest control is also mandated for all FBOs and EEs. To comply with the law and to avoid losing business, staff and the profit it is necessary that pest detection is done early in all EEs and appropriate control measures are taken.

Actually pests nuisance is secondary to poor hygiene and waste disposal both inside and outside the EE. Still despite good hygiene in our home kitchens we do face the menace of cockroaches and rats. All the same it is vital to maintain hygiene and tackle pest nuisance. It is important to dispose off the waste correctly. Availability of adequate number of pedal operated bins can ensure that waste is not littered in EEs. These bins then need to be emptied into bags for further disposal. This alone is not sufficient.

I would like to share my personal experience here. A FBO came with loss of business due to fly nuisance in his EE. He reported that all the waste was collected in bags, sealed and collected by the municipal authorities. On inspecting the place, I found that there was a park nearby. People were eating food items in the park and leaving the leftover in the open. Also, there was a garbage bin which was overflowing and was not regularly emptied by the authorities. This was serving as a breeding ground for flies!!! Hence, not only the kitchen but also the premises and surroundings of EEs should be kept clean to prevent pest nuisance. Besides above mentioned general hygiene measures some pest specific actions may also be required.

Cockroaches

They generally hide in cracks and crevices and warm places like heating pipes. The transmit diseases mechanically by polluting food with infective material carried on their legs and bodies. To prevent their occurrence good housekeeping is the mantra. All cracks and crevices should be properly sealed. There shall be no left over food particles to support and nourish them. It will help to keep a check on their density. Switch on the light late in the night and count for a fixed period of time like 2 minutes. Once their presence is confirmed, they need to be controlled by use of insecti-

cides. These can be in the form of insecticidal sprays, dusts or baits. The insecticide should be applied thoroughly to runways, cracks, crevices, undersides of tables and even under the table spreads, rear of sinks, meat keeping and other harbourage areas.

Newer insecticides like Fipronil and Imidacloprid Gel have been found to be very effective in controlling cockroaches. Fipronil has a cascade effect and cockroaches exhibit necrophagy amongst themselves, whereby they consume Fipronil killed cockroaches and get killed in turn. An important note at this stage is that chemical control with insecticides is a temporary solution for this problem. There is no substitute for maintaining hygiene inside and outside the premises of EEs. Continuous use of these chemicals can make the cockroaches resistant also. Examples of chemicals which can be used against cockroaches are 100% boric acid, as baits/sprinkle along corners, Imidacloprid Gel (1.85–2.15%)/ Fipronil Gel (0.01–0.03%), 2% Fenitrothion/ 3% Malathion/0.03% Deltamehtrin as spray.

Flushing agents within the spray can cause the cockroaches to spread to previously uninfected areas. Additionally, spraying does not penetrate egg cases to kill eggs. When those eggs hatch, the baby cockroaches eat anything they can, grow, reproduce and the cycle begins all over again. It is important to do a thorough cleaning before spray. It will always be helpful to remove food sources for the cockroaches from EEs. This includes dead cockroaches and egg cases as well as feces and body parts which look a little like pepper sprinkled around cracks and crevices and is also a food source, particularly for baby cockroaches. Bait works very well in cockroach infestations because baits do not disperse by themselves like sprays do. The cockroaches actually do part of your job for you by eating the bait and then carrying it back to where the other cockroaches have gathered. When they defecate, the feces are poisonous, and the baby cockroaches will eat them and die. One major concern with baits is that they can be contaminated by sprays, so it is very important not to use any sprays around them.

Rodents

Another type of pest that commonly infests EE is rodents. Before we discuss their control it is important to make ourselves clear that rodents and mice are different and require different strategies for control. Mice are *not* baby rats. They are two completely different species. The commonest rodents which are pests to man are called 'commensals', a Latin word for "sharing the same table." This name is appropriately given to them as they do share our table! It is estimated that rats and mice consume or contaminate up to one-third of all the food produced in the world annually. In addition to eating food, they constantly urinate and defecate, and spread disease causing organism whatever they've been walking through.

Let's read a short real story to understand the damage they can cause to human health.

On Sunday, a family picnic, brought with them few drinks in tin. However, on Monday, two family members (who joined the picnic) were admitted to hospital and placed in the Intensive Care Unit. One died on Wednesday. Autopsy results concluded it was *leptospirosis*. The bacteria, known as *Leptospira interrogans*, was stuck to the tin cans, which were drunk, without the use of glasses, cups or sip straws. Test results showed that the soda tin was infected from mice urine, and that had dried. The mice' urine contained leptospirosis. That is why it is highly recommended to rinse the parts evenly on all soda cans before drinking it. Cans are usually stored in the warehouse and delivered direct to retail stores without cleaning. Similarly, Hantavirus can be transmitted from dried rat urine.

The problem here is that, particularly in India cans are considered a status symbol for kids and youngsters. They often consume canned drinks without washing them and can easily fall prey to such diseases.

Detection of Rodents

Before we discuss rodent control, let's first understand how to confirm the presence of rodents in the EE. This can be ascertained by a thorough inspection. Look for the presence of **gnawing marks**: Wood scrapings around doors, windows and frames, **droppings** which are visible along rat runs, near rodent burrows and at the feeding sites. Another way of detection is rat **runways**, which can be seen along walls and behind stored objects.

In case there are droppings present, then cleaning of the area needs to be done very carefully by performing following steps:

1. Wear rubber or plastic gloves.
2. Spray urine and droppings with a disinfectant or a mixture of bleach and water.
3. Make sure urine and droppings get wet.
4. Let it soak for 5 minutes.
5. Use a paper towel to wipe up the urine or droppings.
6. Throw the paper towel in the garbage.
7. Spray disinfectant.
8. Mop or sponge the area with a disinfectant or bleach solution.
9. Wash gloved hands with soap and water or spray a disinfectant or bleach solution on gloves before taking them off.

10. Wash hands with soap and warm water after taking off gloves.

Do not sweep or vacuum up mouse or rat urine, droppings, or nests. This will cause virus particles to go into the air, where they can be breathed in.

Prevention of Rodent Infestation

The common breeding places for rodents are debris/garbage/clutter of house/surroundings, which are dark and undisturbed areas. Good housekeeping and maintenance of general hygiene will prevent their infestation. It is imperative that surroundings of EE should be cleaned up and made free from debris. Displacement of furniture routinely will prevent dark undisturbed areas and hence rodent breeding. Rats can also come up through toilets and broken sewer pipes. Rats do require a source of water, so eliminating any free standing water in a facility will do a great deal to resolve a rat problem.

All openings/holes in the walls call for strict closure using steel wool and copper mesh/mortar. Doors are required to be made rodent proof by placing metal sheet at bottom. The drainage points must be made rodent proof by covering with sieves and securing with cement. Use of rodent proof utility wires limit access of rodents to buildings as the tube rolls when the rodent tries to walk over it. It is essential to make chimneys and air vents rodent proof air vents and chimneys using 1/4" hardware cloth.

Another way of preventing rodent infestation is to clean all food spillages immediately. Metal bins or heavy duty plastic bins with tight-fitting lids should be preferred over plastic bags for storing garbage. Avoid placing food scraps in compost piles. Storage of food grains must be done in rodent proof containers.

If the rodents still gain access to EE then certainly they need to be controlled using traps, baits with rodenticides or fumigation. Selected rodenticide should be tasteless and odorless with a delayed effect. Baiting should be done along the rat runs. Since rats are nervous about new things in their environment, it is often the best strategy to pre-bait using nonpoisonous grains at first, then gradually introducing lethal, slow-acting poisons. Poisons must be both painless and slow acting, because rats will observe their companions to see whether they have adverse reactions. Rats are so intelligent that in large colonies, the smallest and weakest rats are used as tasters. They are the ones that eat the new food, and the rest of the colony watches to see what happens.

Rats are also hierarchical. One successful strategy for baiting or trapping is to search out the best hiding place for rats and bait exclusively at that location. The top rat in the hierarchy will be eliminated; the others that replace him will die in turn until all the rats are gone.

Mice are much smaller than rats. Mice, like rats, can collapse their rib cages to squeeze through any opening. Unlike rats and cockroaches, mice can obtain their water from the food they eat. Mouse droppings are about the size of a grain of rice. Usually, droppings are one of the first signs of an infestation to be noticed. Some other common signs are gnawed food and a musky smell. When baiting for mice, it is best to place either bait or traps where droppings are seen. Leave them in place for a day or two and monitor the area for additional activity. If more droppings are spotted, replace the bait after a week to 10 days. Mice are curious, but that curiosity wears off after a day or two, so the baits and/or traps become ineffective. They also have limited memories, so after they forget about the bait and/or traps that have been placed out, they will come back for another visit.

Since both rats and mice are prey animals, they tend to avoid open spaces where they can be seen. A good way to prevent an infestation before it can start is to create an open space around building. Both rats and mice will be reluctant to cross any space where they could be spotted by one of their predators. Keeping mouse and/or rat-infested areas clear of clutter will also do a great deal in restricting their ability to hide.

Flies

Flies are another major pest problem and disease carriers. They are capable of carrying over 100 pathogens, such as those causing typhoid, cholera, salmonellosis, bacillary dysentery, anthrax and parasitic worms. It is known that presence of one fly in an operating room is cause for closing down a surgery. The best method of control of houseflies is to eliminate their breeding places and to maintain hygiene and sanitation. This can be done by proper disposal of excreta, swill, garbage and all other decaying organic rubbish, offal and carcasses. Kitchens should be made fly proof by putting screens on windows and doors. The doors of all entrances and windows should open outwards and preferably should have vacuum levers especially in kitchen. If a fly gains access to fly-proofed rooms, it can become large fly-traps. Clean cups and bowls should be kept in inverted position when not in use.

In case there is a fly nuisance then spraying with pyrethrum should be done . Other group of chemicas that can be used are organo phosphorous and synthetic pyrethroids.

Ultra low volume spray can be done in large areas to control houseflies. Baiting can be done with Propoxur, Imidacloprid (with pheromone muscalure), which helps in attracting the flies to the bait. During the night, houseflies prefer to rest on strings and hanging wires or any object. Use of insecticide treated cords and strips which are hung from ceilings in kitchens can prove to be effective. Use of light traps (electrocutors) is very useful in the eating places. These shall be placed at least 1.5 m (preferably 4.5 – 6 m) away from a food handling area.

Most FBO operators become concerned about flies when they see them flying around as adults. However, it is much easier to control them as larvae. Flies hatch from eggs as maggots, feed on wet organic matter and then leave that matter to pupate, emerging from the pupae as adult flies. Wet, organic matter occurs in many places in a EE, such as garbage cans, floors and in floor drains. It takes only a day or so for the fly to hatch from the egg, and about a week or more to pupate. Twice-weekly thorough cleaning of wet organic matter from the areas mentioned can completely disrupt the fly's life cycle.

Pests have always been a problem in the food industry. Many restaurants and dining establishments have suffered huge losses in business because of presence of pests. It is recommended that all precautions must be taken to prevent their occurrence as controlling them after occurrence is a difficult task. Moreover, if due precautions are not taken use of chemicals as such can make the food unsafe.

A new product, which has been endorsed by HACCP, is called Bait Safe. This pest control is designed to bait and catch rodents. However, unlike other pest control products, Bait Safe has been designed specifically for rooms that include food preparation or storage.Bait Safe is a locking device that does not allow pests to leave once they have been caught. This keeps the rodents from eating the poison and then going elsewhere, creating a hazard. The devices can be placed in eaves, floors, cupboards, or other spaces.

According to FSSR 2011 regular inspections shall be conducted at least once in a fortnight for early detection of pest and to apprehend pest situations at the premises. Proper records of pest control activities and inspections, in respect of the premises shall be maintained. Air curtains at main entrances shall be affixed. Rodenticides and insecticides shall be applied in such a manner as not to contaminate foods. They shall not be applied while food preparation is taking place, and all open foods should be well covered and protected.

General Tips for Keeping Pests Away

A few tips for pest contol in premises are:

1. Seal cracks and keep screens closed.
2. Keep food covered and clean all spills immediately.
3. Goods should be stored 15 cm away from walls and floor.
4. Close unwanted openings around drains pipes, wiring, vents to make them rodents and insects proof.
5. Dispose garbage and trash promptly.
6. Wash waste bins daily with hot, soapy water.
7. Food products such as flour, sugar, etc. should be removed from their original containers and placed in approved sealed tight containers that are properly labeled and more impermeable to pests.

The first way is that pest control is handled by one of the staff working in EE. It is imperative that the person given this responsibility has an in-depth knowledge about various pest control methods, chemicals used and the method of application. The other way out is to outsource it to a pest management company on payment.

A few criteria for choosing a pest control contactor are:

1. The pest control contractor should be a licensed company
2. The contractor should inspect the premises on regular basis to ensure the complete absence of pests from the immediate surrounding area; only the destruction of pests that are observed in the premises is not sufficient.
3. The contractor should provide the EE with a written contract indicating services, methods and frequency of visits.
4. The pesticides used should comply with existing municipality regulations.
5. The experience of pest control contractor in the food industry and provision of appropriate references from current clients.
6. Must have sufficient resources in terms of trained/qualified staff and the necessary equipment to carry out proper services.
7. The contractor should be able to undertake a complete survey and provide a clear report of recommendations and actions required.
8. Environmental criteria should also be taken into consideration; the method should be least hazardous to humans, least possible to come in contact with humans, and also most readily biodegradable.

To conclude a simple systematic effort can help control pest nuisance. Pest control should always be a priority for FBOs of all EEs.

33

Cleaning and Maintenance of Eating Establishment

Puja Dudeja, Amarjeet Singh

Last Sunday I went with my family to a popular franchise outlet of international food chain. To my surprise I was offered something different which is not routinely in the menu: A free kitchen tour!!!. We gladly accepted the proposal. All of us were given shoe covers, masks and cap for visiting the kitchen. We were glad to see the state of art equipment and impressed by the standard of hygiene and cleanliness inside the kitchen.

Clearly the idea was that the Food Business Operator (FBO)/Eating Establishment (EE) wished to flaunt its good hygiene to promote itself.

Another experience I would like to share at this moment is that of a restaurant were I went for lunch last month. The kitchen was next to the serving area with a glass door where the customer could see the food handlers working inside! This is a total contrast to other EE where a board stating 'Entry of outsiders is not allowed' is placed to ensure, no visitor barges inside kitchen premises for a surprise check on hygiene and cleanliness of cooking area.

It is well said that cleanliness is next to Godliness. Whether it is a cloth/shoe/bag shop or an eating outlet a customer prefers to visit a shop which is tidy and clean. Consumers expect that premises are germ-free when they visit a restaurant. FBO generally ensure that the service area or the place where people sit and eat is clean. However, it is the hygiene of cooking area, which is cost cutting and often neglected. Reasons are many. Small time FBOs wish to spend less and their endeavor is to employ same people for cooking and cleaning activities. Alongside they want to spend minimum time in maintenance. In addition, this is generally a hidden area from customer's eyes.

However, with changing times beside the pricing and taste the consumers demand food safety along with aesthetics and ambience of the EE.

Cleaning of EE involve removal of soil, food residue, dirt, grease or other objectionable matter. Separate cleaning materials, including cloth, sponges and mops should be used for the designated clean area. Use of disposable, single-use cloths is recommended wherever possible. Effective cleaning is essential to get rid of harmful bacteria and stop them spreading to food. Water alone is not a very efficient cleaning agent because of its high surface tension. Adding of detergent to water facilitates the contact between water and surface soil. The detergents enable water to penetrate soil by lowering the surface tension. A good detergent should be able to soften water completely, non-corrosive to surfaces (metals and buildings), non-toxic and biodegradable and economical in use. It should have good wetting or penetrating ability, emulsifying ability on fat, dissolving ability on food solids, deflocculating, dispersing or suspending ability, good rinsing properties and scale and rust removing properties.

The types of chemical compounds used to achieve the functions of cleaning described above are:

1. Alkalis and alkaline salts: Sodium hydroxide, sodium carbonate (soda ash), sodium metasilicate
2. Surface active agents: Aryl sulphonates
3. Sequestering agents: EDTA (ethylenediaminetetra-acetic acid), NTA (nitrilotriacetic acid)
4. Inhibitors (anti-corrosive agents): Silicates
5. Acids: Phosphoric acid, nitric acid, sulphuric acid and hydrochloric acid. The organic acids used are gluconic acid, tartaric acid, citric acid, acetic acid and sulphamic acid.
6. Fillers: Sodium chloride or sodium sulphate

Disinfection is the killing of infectious agents outside the body by direct exposure to chemical or physical agents. Disinfection reduces the number of microorganisms in the environment, to a level that does not compromise food safety or suitability. Disinfection can be used to destroy bacteria from surfaces. However, chemical disinfectants only work if surfaces have been thoroughly cleaned first to remove grease and other dirt. For effective disinfection, it is important to first clean the surface and remove visible dirt, food particles and debris, and then rinse to remove any residue. After this step, application of a disinfectant is done using the correct dilution and contact time, according to the manufacturer's instructions, and then rinsing with

drinking water. Sanitisers, have both cleaning and disinfection properties in a single product, but cleaning and disinfecting process must still be carried out as above to ensure the sanitiser works effectively, that is, to first provide a clean surface and then again to disinfect. Disinfection may be necessary after cleaning in high hygiene areas. Cleaning and disinfection programmes shall be continually and effectively monitored for their suitability and effectiveness once in six months and records maintained. Cleaning and disinfection schedule in an EE as given by FSSAI are given in Tables 33.1 to 33.5 (*Source: FSSAS*). Disinfection methods are of two types:

1. Non-chemical disinfection methods like heat/steam: Expensive, impractical
2. Chemical disinfection methods: Commonly used disinfectants. There are a large number of disinfectants, each claimed to be the best on the market. Nevertheless, the only ones suitable for the food industry contain chemicals of one of the following groups:

- Chlorine and chlorine-releasing compounds
- Quarternary ammonium compounds
- Amphoteric (ampholytic) compounds
- Phenolic compounds
- Peracetic acid

Chlorine is the most effective disinfectant available and sodium (or calcium) hypochlorite is a cheap disinfectant commonly in use. The hypochlorites have a characteristic smell produced by free hypochlorous acid which is considered to be the germicidally active form of chlorine. A practical disadvantage of sodium hypochlorite is the risk of corrosion to all common metals (especially aluminium and galvanized iron), except perhaps high quality stainless steel. Other chlorine-releasing compounds are chloramines (chloramine T) and chloroisocyanurates.

Table 33.1: Cleaning of structure

Component	Least frequency	Equipment and chemicals	Method
Floors except washroom	End of each day or as required	Brooms, damp mops, brushes, detergents, sanitizers	1. Sweep the area and remove debris 2. Apply detergent and mop the area 3. Use scrub for extra soil 4. Rinse thoroughly with water 5. Remove water with mop
Walls, doors, ceiling, ventilators fans and exhaust fans	Fortnightly or as required	Clean wiping clothes (one time use) brushes and detergents	1. Remove dry soil 2. Rub with wet cloth or rinse with water 3. Apply detergent and wash 4. Wipe with wet cloth or rinse with water 5. Air dry
Air conditioners	As per manufacturers maintenance manual		
Desert coolers	Fortnightly or as required	Water, mop	1. Remove water 2. Rub with cloth or rinse with water 3. When not in use remove water and keep dry
Washroom	Once every 4 hours	Brooms, damp mops, brushes, detergents, sanitizers	1. Sweep the area 2. Apply detergent and mop the area 3. Use scrub for extra soil 4. Rinse thoroughly with water 5. Remove water with mop
Store	End of each day or as required	Brooms and camp mops	1. Sweep the area 2. Mop the area 3. Use scrub for extra soil 4. Air dry
Water storage	Once in six month	Clean wiping clothes (one time use), detergents, sanitizers	1. Remove foreign matter and tank 2. Rub with wet cloth or rinse with water 3. Apply detergent and wash 4. Rinse with water and sterilizer 5. Air dry
Insect Electrocuting devices	Once in a week or as required	Clean wiping clothes (one time use)	1. Remove insects and other foreign matter 2. Rub with wet cloth 3. Reinstall insectocutors
Waste bins and waste areas	End of each day or as required	Water, clean wiping clothes (one time use), detergents	1. Remove foreign material and soil 2. Rub with wet cloth and rinse with water 3. Apply detergent and wash 4. Air dry

Contd.

Table 33.1: Cleaning of structure (Contd.)

Component	Least frequency	Equipment and chemicals	Method
Parking and open spaces with water	End of each day or as required	Water	1. Sweep the area and remove debris 2. Wash parking space thoroughly
Street lanes and other public places or the common part of building which are adhering and or nearby food premises	End of each day or as required	Water or mop	1. Sweep the area and remove debris 2. Wash thoroughly with water

Table 33.2: Cleaning of food contact surfaces

Component	Least frequency	Equipment and chemicals	Method
Work tables	After use	Clean wiping clothes (one time use), detergents, sanitizers	1. Remove food debris and soil 2. Rub with wet cloth or rinse with water 3. Apply detergent and wash 4. Wipe with wet cloth or rinse with water 5. Apply sanitizer 6. Air dry
Sinks	After each use	Running water, detergents,	1. Remove food debris and soil 2. Rinse with water and or detergent

Table 33.3: Cleaning of equipment

Component	Least frequency	Equipment and chemicals	Method
Utensils, cutting boards, knives, other cooking-equipment, serviceware, crockery, and cutlery	After use	Clean wiping cloths (one time use), brushes, detergents and sanitizers	1. Remove food debris and soil 2. Rinse with water 3. Apply detergent and wash 4. Rinse with water 5. Apply sanitizer 6. Air dry

Contd.

Table 33.3: Cleaning of equipment (Contd.)

Component	Least frequency	Equipment and chemicals	Method
Food processing-equipment, vending machines	As per manufacturers cleaning and maintenance manual		
Refrigerators, freezers and storage areas, refrigerated display counters	Weekly or as required	Clean wiping cloths (one time use), brushes and detergents	1. Remove food debris and soil 2. Rub with wet cloth or rinse with water 3. Apply detergent and wash 4. Wipe with wet cloth or rinse with water 5. Dry with clean cloths/air dry

Table 33.4: Cleaning of hand contact surfaces

Component	Least frequency	Equipment and chemicals	Method
Doors and door knobs	Daily	Damp cloths and detergents	1. Remove debris 2. Apply detergent 3. Rinse or wipe with damp cloths 4. Dry with paper towels/air dry
Upholstery	Daily	Clean wiping cloths (one time use), steam/chemicals	1. Remove food debris and soil 2. Wipe with dry cloth
	Fortnightly or as and when required		1. Remove debris 2. Apply chemicals 3. Vacuum dry

Table 33.5: Cleaning of furniture and decorative items

Component	Least frequency	Equipment and chemicals	Method
Chairs and tables, reception and cashcounters, display counters held at ambient temperature	Fortnightly or as required	Clean wiping cloths (One time use), brushes and detergents	1. Remove dry soil 2. Rub with wet cloth or rinse with water 3. Apply detergent and wash 4. Wipe with wet cloth or rinse with water 5. Air dry
Paintings, artificial plants and decorations	Fortnightly or as required	Clean wiping cloths (One time use), brushes and detergents	1. Remove dry soil 2. Wipe with wet cloth 3. Air dry
Plants	Daily	Washing	Water

Quarternary ammonium compounds are: surface active agents which are free of odour and colour, are highly stable and have little corrosive action on metals when used in recommended concentrations. They are more expensive than hypochlorites. Utensils and equipment should be thoroughly rinsed after applying these compounds as disinfectants because of possible toxic effects.

Amphoteric compounds are essentially alkyl or acyl amino acids. They combine detergent and disinfectant properties. They are of low toxicity, are non-corrosive and are expensive.

Phenolic compounds are not generally suitable for use in the food industry. Some halogenated phenol derivatives can be used in the meat industry. They are effective against spores, viruses, moulds and gram-positive and gram-negative bacteria. They are corrosive and can irritate the skin of personnel.

A quite new disinfectant is a mixture of peracetic acid, hydrogen peroxide and acetic acid, which is stable and is effective against bacteria, spores, yeasts, moulds and viruses. The active agent is peracetic acid. The mixture is non-corrosive.

The choice of disinfectants will normally depend on several factors, one often being the supply situation when the more specific disinfectants are marketed by only a single or a few companies. Expenditure on disinfectants will also be important and the cost must be compared to the characteristics of the disinfectant before a choice is made. The previous cleaning programme should also be considered. A disinfectant will never assure a demanded hygienic level without previous cleaning.

The choice will often be hypochlorites that are inexpensive disinfectants with a good germicidal effect. Precautions to prevent corrosion of the surfaces and to prevent development of toxic gases, i.e. to prevent the possibility of mixing hypochlorites and acids, should be taken.

As per the FSSA 2006, food premises, their fixtures, fittings, equipment and utensils shall be maintained clean, and in a good state of repair and working condition. A well-planned, well-executed and controlled cleaning and sanitation programme for eating establishments (service area, kitchen, equipment, utensils) is very important to achieve a hygienic standard. Cleaning and sanitation alone, however, will not assure a hygienic standard in production as process hygiene as well as personal hygiene of food handlers are also equally important factors

34

Cleanliness and Maintenance of Equipment for Food Safety in an Eating Establishment

Puja Dudeja, Ishwarpreet Kaur, Amarjeet Singh

An eating establishment (EE) is a place where food is served. It is the endeavour of all Food Business Operators (FBO) to provide a comfortable seating to the customers while food is being served. In all EEs many types of equipment are required both in the seating and cooking area. These may also affect food safety aspect. The seating area may have coolers, air conditioners, exhaust fans, music system, television, lighting system, etc. to provide comfort to the clients. These are common to all EEs except for the quality and type of services, which are provided in consonance of the standards of EE.

Requirement of size and number of kitchen equipments in an EE is based on type of food production, convenience to user and ease of cleaning and maintenance. Small FBOs often tend to purchase substandard (local brand) equipment to cut down on costs. For example, it is a common practice in our country to get utensils cleaned manually rather than using a dishwasher. For storage and sanitation equipment refer chapter on good cooking practices.

Equipment in a typical kitchen of an EE is classified in Figs 34.1 and 34.2.

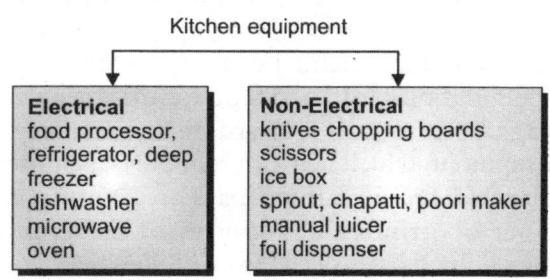

Fig. 34.2: Classification of kitchen equipment based on requirement of electricity

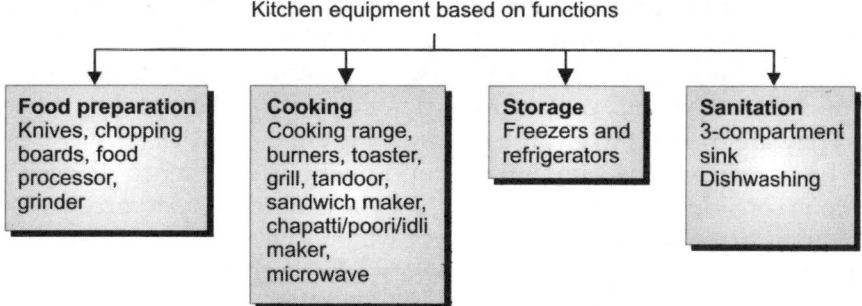

Fig. 34.1: Classification of kitchen equipment

Equipment as a Food Safety Hazard

Equipment, fixtures, utensils and food contact surfaces are a source of trouble if there surfaces are not smooth, easy to clean, durable, nontoxic, nonabsorbent, and corrosion resistant. Accumulation of dust and dirt, harborage of pests or microorganisms can contaminate food when they come in contact with food. If there is a dent, buckle, chip similar material can be logged here and hence may become a source of contamination. Many a times food particles get lodged in the blades, corners, internal edges of equipment. The bottom tray of toasters, grills, *tandoors* is often contains old bread crumbs or residues of burnt *chapattis*, etc. Cleaning and maintenance are easily put off until tomorrow in the busy work hours of the EE. But if they are not cleaned properly tomorrow may be too late! Such contaminated equipment can cause foodborne illnesses if used.

There is a marked reduction in efficiency, also in case the pieces of equipment are not cleaned and maintained well. Apart from this accumulation of dirt and food residues can get lodged which can jeopardize food safety. Equipment, which has different parts, is more subjected for such deposits. For example, a mixer or grinder has blades of different shapes. Such parts needs to be dissembled and then reassembled for maintenance and cleaning.

Sturdy pieces of equipment are able to withstand heat, water and cleaning agents. However, over time equipment may develop cracks, crevices, holes. Such spaces are also a place to hide for microorganisms, pests, dirt, food particles. They are difficult to clean too. Here it should be noted that any opening of less than 0.8 mm is considered closed.

Build up of dirt is also dependent on the shape of equipment. Many equipment with internal angels of 90° can facilitate accumulation of dirt more than those with mulation of dirt more than those with

rounded edges. Another source of contamination from equipment is the material used. For example, if lead is used as coating material it can chip or crack during handling and may become a source of contamination.

Installation of Equipment

The first and foremost requirement before installation of any equipment is to create enough space for it. The equipment can be a large one which is either permanently placed on the floor or is mobile, e.g. deep freezers, refrigerators. Mobile equipment with casters are preferable as they are easy to move which allows cleaning of walls and floor. Adequate toe space (6") from floor and gap between wall and equipment is maintained to do cleaning and prevent pest nuisance. Small equipment like food processors, microwave, toaster are placed on the shelves/counter tops. Certain small pieces of equipment which are used occasionally are kept in racks. These equipment like toaster, sandwich maker, etc. are light in weight and can be lifted easily to clean the area between rack and equipment and between rack and wall.

Worktables and shelves used for preparation of food can also accumulate dust, dirt and food particles in case they are not smooth and have developed cracks and crevices. This applies to chipped crockery also. Wooden surfaces are difficult to maintain and hence are not recommended. Wood also absorbs water, get chipped easily and encourages the growth of microorganisms. Aluminum tops are also difficult to maintain as they can get dented easily and durability is poor. Stainless steel, granite, marble tops are easy to clean and maintain and hence advised to be used . They are impervious to grease, food particles or water. The edges and supporting stands of these shelves/tops need to be designed in a way that there is no place for accumulation of dirt and dust and these are easy to clean.

Cleanliness and Maintenance of Kitchen Equipment

The word maintenance means to keep the equipments in good operating order, in clean and usable state. Neglecting this may effect the efficiency of equipment and jeopardize food safety also. Cleanliness means to keep equipment and kitchenware that come in direct contact with food safe and bacteria free. Neglecting cleanliness of equipments can lead to foodborne illnesses. Soiled equipments and facilities have a negative effect on the health of the customers.

There are three general ways to establish a maintenance and sanitation program: corrective, regular and preventive.

- **Corrective maintenance** is the type used by most FBOs who wait until something breaks down and then summon an employee to fix it. If he fails, then to cut down on cost of repair from the manufacturing company he often calls a local repair service as the next step. As is often the case, the repair man is not always familiar with the equipment and has to struggle for the repair of parts. It is obvious that this corrective maintenance approach is not the best approach to be taken.

- **Regular maintenance** requires that major pieces of equipment be inspected and serviced at regular fixed intervals. When repairs are necessary, they can be scheduled and anticipated before there is a total break down, and generally not disrupt normal operations. Some operators employ their own staff for this purpose while others may prefer to establish a regular maintenance contract with a service company to keep their equipment running at peak operating efficiency.

- **Preventive maintenance,** often referred to as "PM", is more of a systematic approach and is based upon schedules of inspection and replacement that reflect the requirements of each piece of equipment. While failures can and do occur, they happen with less frequency. While this approach may initially be more expensive than the other two systems mentioned, PM will more than pay for itself by increasing overall equipment life, maintaining top operating efficiency and have an overall lower frequency of failure.

PM is a systematic maintenance, and encompasses cleaning of equipment also. As a routine all pieces of equipment which are used need to be cleaned daily as it prevents dirt and food residues from building up. For proper maintenance procedures owner's manual, that comes with the equipment should be referred to. In case they are lost, instructions specific to the equipment can be downloaded using internet. All food handlers must be educated regarding proper use of the equipment.

It is advisable to nominate one person from the staff who is responsible for implementing the preventive maintenance program. Having a maintenance staff earmarked reduces the likelihood that an equipment breakdown would occur. This individual/s would gradually become very familiar with all the pieces of equipment in the operation. Not only this there is also a need to have a contingency plans. For example, what would be done if there were a power cut? How would food be prepared? Is there a generator back up? What would be the alternative menu? If a contingency plan is in place from the outset, when a need arises, things will go much more smoothly. Maintenance for equipment is an important ongoing task.

In addition to implementing a PM plan for equipment, there needs to be ongoing upkeep of facilities. An equipment wise record of maintenance activities will ensure better compliance.

Thorough cleaning of all equipment on regular basis will ensure removal of food scraps from hard to reach parts of the equipment. Cleaning of equipment depends upon nature of material used in construction, general construction features and ease of dismantling and reassembling. Although each equipment in kitchen will require specific instructions for operating and cleaning there are general guidelines which are applicable to most of them. Availability of a Standard Operating Procedure (SOP) regarding cleaning and maintenance in the local language will be of great help to the person performing the task. He is less likely to miss any important step during this operation. For example, in cleaning of blender, following steps need to be followed:

1. **Take the blender apart:** Remove the blender from the base. Remove the lid. Unscrew the bottom component, being careful with the blade. Remove the gasket seal and the blade (Fig. 34.3).

2. **Wash the jar:** The jar of the blender is the main portion and where most of the cleaning will need to happen. Use warm soapy water and a dish rag or sponge. Rinse thoroughly. Dry carefully. The jar of the blender can also be washed in the dishwasher on the top rack, although hand washing is faster and reduces the risk of breakage and scratching.

3. **Wash the other small pieces:** The gasket seal, cutting blade, and locking ring all need to be washed as well. Use warm soapy water and take care to watch the sharp edges of the blade. Rinse thoroughly and dry.

4. **Run the blender for stuck on items:** For easier cleaning of dried stuck on food mix a 1:1 ratio of baking soda and water, and run it through the assembled blender. Dump out the solution. Take apart the blender. Follow the steps above for washing the components.

5. **Clean the motor housing:** Wipe down the motor/base of the blender with a warm, damp cloth. Never submerge the motor/base in water. It is not safe to put in the dishwasher, either. A wipe-down is all it needs. Remember to wipe down the cord periodically and to check it for damage.

6. **Reassemble the blender and store:** If blender is used often enough, store it on your countertop or in another easily accessible location. If blender is only used for special occasions, store it in a less accessible place.

Some more smelly foods can seem to linger in the blenders. Avoid this by not storing items in the blender. For a smelly blender, blend a 1:1 ratio of baking soda and water together in the blender. Then let the solution sit in the blender for 5–10 minutes, before washing normally.

For Easier Cleaning of Blender

Blender clean-up can be very quick, depending on whether pre-cleaning has been done right after using it. Immediately after finishing one recipe, place two cups of warm water and a drop of dish detergent in the blender jar (for eliminating odors half a lemon with rind can be added). Apply the lid and pulse a few seconds. Rinse thoroughly. Repeat if necessary to remove all food residues from jar and blade. Also wash the lid completely.

Continue with your next recipe, pre-cleaning in between each batch. Once finished, disassemble the blender jar and blade unit. Wash carefully the jar, lid, blade unit and rubber gasket in warm soapy water. Rinse well, dry and reassemble.

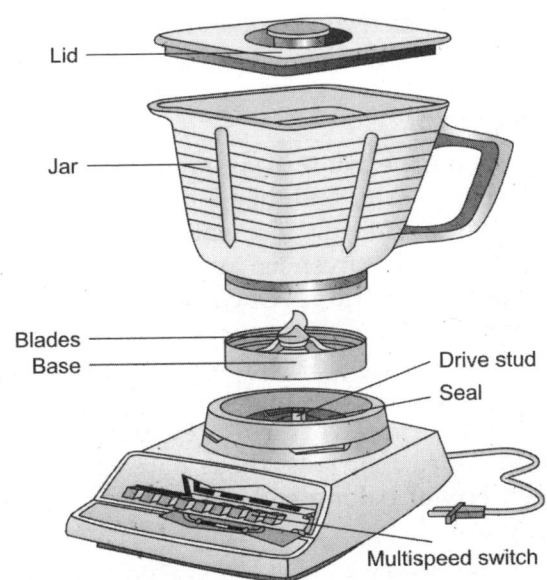

Lid

Jar

Blades
Base

Drive stud

Seal

Multispeed switch

Fig. 34.3: Parts of blender

Special Procedures for Specific Equipment

Cleaning in place equipment (CIP): This practice is used for cleaning equipment that is difficult to dismantle and reassemble. In this method detergent and sanitizer are circulated for specific period of time at specific speed and in a specific sequence so that the food contact surfaces are cleaned and sanitized. These pieces of equipment need to be self draining and leave no residue of cleaning material. This is time saving, labour saving, cost effective and minimizes the damage to the equipment. For example, the tea/coffee/juice/soup/crush dispensers/dairy milking machines are cleaned this way at the end of the day.

Chopping boards and meat blocks: Earlier these used to be made of wood which would easily develop cracks and crevices. These would then be a source of contamination. Even particles of wood may get into food. Presently these are made up of marble or non-absorbent food grade plastic or galvanized zinc. Irrespective of the material they need to

be cleaned immediately after use. The surface should be wiped, scrubbed, and cleaned with a cleaning solution. The sides and under surface should be cleaned with a wet cloth. Sanitizer solution should be poured over the board and kept for drying. Rubbing common salt on meat blocks helps in keeping them dry.

Tandoors: For *tandoor* maintenance, it is important to clear the soot from collecting inside. All accumulated ash should also be removed daily. The cracks and crevices develop need to be filled with clay. For those using electric *tandoor*, the tray is removed cleaned, scrubbed, sanitized and dried for next use.

Water purifiers: These are maintained by the service staff form the manufacturing company. They should be called for maintenance as per the schedule of the equipment.

Microwave: Splatters of food from microwaving look bad, smell bad, and decrease the efficiency of microwave. There are four methods used for cleaning microwave.

i. Using Vinegar

a. Place a microwave-safe bowl half filled with water and a tablespoon of white vinegar inside the microwave

b. Turn on for 5 minutes. This will steam up the walls of microwave and loosen the dried-on gunk.

c. Remove the container. Wipe the inside of microwave with paper towel or a clean rag.

ii. Using Lemon

a. Cut a lemon in half. Place both halves cut-side down on microwave plate with a tablespoon of water.

b. Microwave for 1 minute or until the lemon is hot and inside of the microwave is steamy.

c. Wipe the inside of the microwave with kitchen paper and wash the plate.

iii. Using Dish Soap

a. Add dish washing liquid in a microwave-safe bowl and fill it with warm water. Place the bowl in microwave and turn it on for 1 minute or until it starts to steam.

b. Take the bowl out. Wipe the inside of microwave with damp sponge.

c. Baking soda can also be added to the bowl to serve as a deodorizer.

iv. Using Window Cleaner

a. Mix cleaning solution in a bowl (two parts window cleaner with one part warm water).

b. Dip a sponge in the cleaning solution and use it to wipe the inside of the microwave. Remove the spin tray and wipe the microwave until all spots and stains have been removed. Also wipe the vents of the microwave oven.

– Make sure microwave is unplugged.

– Soak tough stains and spots in the window cleaner solution for 5 minutes before scrubbing them away.

c. Wipe it down with a clean rag. Finally clean with a rag soaked in fresh water. If you still smell the window cleaner, wipe again with a clean rag soaked in fresh water.

– If tough spots remain, clean with a cloth soaked in olive oil.

– Never use unsafe chemicals that can cause the unit to catch fire or cause safety hazards.

d. Let the microwave dry.

Tips

• After use leave the microwave open for a few minutes afterward in order to let it dry and air out a bit inside.

• Clean microwave once or twice in a month.

• Do not use abrasive cleaners on microwave.

Refrigerator

Try wiping the inside of your refrigerator with hot, soapy water and then a rinse wipe with: solution of vinegar and water to help fight mildew. It also freshens and deodorizes the inside of your fridge. Once every 7–10 days, remove all the food from your refrigerator, wipe down the inside with warm soapy water, clean all the shelves and trays, then replace the food. As you work, check expiration dates and discard food past those dates. Place an open box of baking soda in the back of the fridge.

Do not forget to pull out your refrigerator regularly to vacuum off the back coils or pull the front kick plate off to get at the coils if they are at the bottom. The coils are a major dust collector and this means your fridge has to work harder to keep the inside cool. Unplug the appliance first.

Fixtures

All lighting and light fixtures should be designed to avoid accumulation of dirt and be easily cleaned. Structures within food establishments shall be soundly built of durable materials and be easy to maintain, clean and where appropriate, able to be disinfected. Other fixtures as door knobs, handles, switches should also be cleaned to prevent accumulation of dust and dirt.

Use of Mops and Issue of Food Safety

In our set up often mops are used to wipe dirt from surfaces of equipment. It is extremely important to note here that dirty mops should never be used on a food contact surface. For example, a mop which has been used to clean the microwave should not be used to clean the work table which is to be used for chopping of vegetables. Adequate number of clean dried mops should be available in the kitchen to prevent cross contamination. All used mops should be washed well and dried before reuse. In case fresh clean mops are not available disposable mops may be used. A separate mop for each worktable and equipment per day can ensure safety.

Quick Fixes for Kitchen Equipment

- To quickly clean burned food on a pan, add some dish soap and 1/2" of water. Bring to a boil, then let the liquid cool in the pan. The burnt food will be easy to remove.

- When food spills over and burns on the oven floor, sprinkle a handful of salt on the mess. The smoke will be reduced and the spill easier to clean after the oven cools. Add some cinnamon to the salt to help reduce odors. This is useful for egg spills also.

- In case of a non self-cleaning oven scrape up any large spills then spray cleaner inside the oven, close the door and let it sit overnight so the cleaner has time to work.

- One can also place racks of oven in the bathtub with about 1/2 cup dishwasher detergent and cover them with several inches of warm water. Let the racks soak for 45 minutes, then rinse and dry.

- To clean sluggish drains, pour 1/2 cup baking soda down the drain. Add 1/2 cup white vinegar and cover the drain. Let this mixture foam for a few minutes, then pour 8 cups of boiling water down the drain to flush it. Do not use this combination after using any commercial drain opener or cleaner.

- If plastic from the bread wrapper melts onto your toaster, use a little nail polish remover to get it off. Let the toaster cool before you try this.

- Clean your coffee maker every few weeks by filling the water reservoir with equal parts white vinegar and water and putting it through the brew cycle. Then use clean fresh water and repeat the brew cycle to rinse the machine. Repeat with fresh water two more times.

- For glass cook tops, there are special commercial cleaners that work well. Use them with a hard plastic scraper to remove burned-on food.

- For lime and mineral deposits on your kitchen sink faucet, wrap vinegar-soaked paper towels around faucets for about an hour. This breaks down the mineral scale, and the chrome will be clean and shiny after buffing with a dry paper towel.

Cleaning Solutions for Various Purposes

- For your own window washing solution, mix 1/3 cup vinegar and 1/4 cup rubbing alcohol in a 1 quart spray bottle. Fill up with water.

- Dry baking soda cleans chrome perfectly.

- Cream of tartar and water mixed to a paste will clean porcelain.

- A paste of baking soda and water will clean coffee stains.

- Use a cut lemon half sprinkled with salt to clean copper.

- Dissolve 1/4 cup baking soda in 1 quart of warm water for a good general cleaner.

Make sure to clean up spills as they occur so you will not be faced with one huge cleaning session. Following these maintenance tips will ensure much easier cleanup, smoother working and a longer life to equipment.

35

Water Supply in Eating Establishments

Sonia Puri, Maninder Kaur, Puja Dudeja

Water is the most precious natural resource required for sustenance of life. It is necessary for survival of all living organisms. A lady of the house will be jittery if her kitchen does not have a regular fresh water supply. This is because water is not only used for cooking meals, but also used for washing utensils and for ensuring cleanliness of kitchen. Bulk of the daily requirement of water for human beings is used in maintenance of hygiene. Even today, in a country like India, quantity (availability) of water is the primary requirement. Quantity takes precedence over quality. If adequate quantity is not there quality is of no use. Even WHO evolved a concept of water washed diseases which implied that diseases like gastroenteritis, scabies can be 'washed away' even if adequate quantity of water is available(though of doubtful quality). After all you need sufficient water to rinse kitchen utensils first (lesser quantity of double distilled water is of no use!).

In today's era, water scarcity is being faced globally. But, sadly in India, the world's 2nd most populous country, the problem still remains unresolved. This facilitates continued episodes of outbreaks of food poisoning, diarrhea, Hepatitis A and E, giardiasis, etc. Hence, having safe and adequate water supply should be the priority of every Food Business Operator (FBO).

Water supply is the lifeline of any eating establishment. Both potable and non-potable water is needed at an EE. **Potable water** is defined as water that is safe, clear and wholesome for consumption. The term 'potable' is derived from the latin word *portabilis* which means 'drinkable'. In an EE, water is required for cooking, washing of raw fruits and vegetables, cleaning utensils and washing hands. Water, that contains waste, poisonous substances, pathogenic organisms, odour, and harmful chemicals, is termed **non-potable** water. But that does not mean it should be discarded; it can still be used for cleaning toilets, floors, for fire control, washing purposes, maintaning garden, air conditioning, etc.

Any place which has been designed and constructed with an aim of establishing an EE must have a supply of safe water in the master plan. This water usually comes from muni-cipal sources. The responsibility of its purification, treatment and supply rests with the water supply board. However, over time number of EEs, unauthorized market stalls, colonies have increased due to unplanned expansion and urbanization of cities. Often unauthorized water pipe lines are laid and new water points are created. These pipes are of lower quality and cross sewage pipe lines at various places. If water pipes, are not maintained in good condition and in order at all times, leakage or defects would result in contamination of water and thereby spoilage of food. In case there is a reduction in water pressure or cut in water supply then adequate amounts of potable water should be hygienically transported for the EE.

It is a common practice for FBOs to start a food business/EE without that place having a safe water supply system with some make shift temporary arrangement. Recently I got an invitation card from a friend of mine to attend inauguration ceremony of his new EE on a highway somewhere in North India. I distinctly remember receiving a similar invitation from him last year for his cloth store. On enquiring he confirmed the change of business as the cloth shop was not doing well.

My immediate concern, which was the availability of a water supply point in that shop was least of his. Isn't it shocking? Most of the *dhabhas*, roadside restaurants, food outlets in shopping complexes, do not have a running water supply. They try to manage the entire water requirement through stored water in buckets or drums or other vessels. Even the hygiene of such containers is questionable.

In some EEs taps are connected to a water cooler. In such cases the coolers need to be cleaned at regular intervals, well covered and date of cleaning should be endorsed on it. Water can be contaminated inside the cooler. Hence, it is important to take a water sample and send it for testing at least once a month and a proper record of such testing along with the water report must be maintained.

As per the **Food Safety and Standards Regulations, 2011**, source of water must be displayed in an EE. This becomes even more important in case of a dual water supply system. This will help in recognition of different taps supplying potable and non potable water. It is a common observation in star-rated EEs to have 'HOT'/'COLD' water displayed. Along with this, it is recommended to display drinking water/source of water in case of potable supply or otherwise. It is commonly seen that the taps are fitted with a water purifier (UV/ RO system) in these EEs to ensure safety of water. If such equipment is used then it must be maintained well, with timely service and a record. Otherwise it will give a false sense of security.

Another issue related to water supply is the type of water supply **continuous or intermittent**. If free water supply is not available throughout then water is stored in tanks. Water storage tanks for potable water shall be such that they prevent contamination of water. They should be accessible for cleaning purposes. In case of underground storage tanks, these should be waterproof from inside to prevent seepage. All storage containers should be kept covered to prevent contamination from pests, insects and animal and vegetable matter. They should be regularly inspected by removing the cover. They should be cleaned and disinfected at least once in six months to prevent contamination. A **record** of all such maintenance activities must be maintained. Non-potable water systems should be identified and not allowed to be connected with or refluxed into potable water systems.

Quite often water is stored in tanks, buckets or drums in the EEs. Potability of this water depends on the cleanliness of these storage containers. Hygiene aspect of drawing water from the container is also an important issue. A ladle should be used to take out water and later hung or kept in a safe place. Bare hands should not be used to draw water (Figs 35.1 and 35.2). All water storage containers except for tanks must be dried and cleaned before refilling.

Many customers demand packaged water while consuming food outside. It is the responsibility of EE staff to check the date of expiry and seal before it is served.

In those EEs, where alcohol or beverages are served, usage of **ice** is another area that has to be given due emphasis. Freezing of water for ice does not remove chemical hazards/biological hazards. When contaminated ice is added to foods and beverages chances FBI increase. Hence, all ice to be used in food and drinks should be made from potable water. Even in guidelines so issued for all types of medium to small food outlets by **"Food Safety and Standard Authority of India" (FSSAI)**, safe ice usage has been emphasized upon. It should be manufactured, stored, transported and handled in a sanitary way. Also, ice so used to cool open foods in buffet displays, shall also be made from potable water. Ice for drinks shall not be handled with bare hands. An ice scoop should be used and, except for the handle, the other parts of scoop should not be touched. Unused ice should never be returned back to the ice storage container. Ice and steam should be

produced from potable water, handled and stored to protect them from contamination. If procured from outside, it should be procured from licensed vendors.

In our country temporary vendors sell tea, fresh fruit juice, sugarcane juice, roadside food stall (thali, pakoras, bhel, chaat, etc.) are commonly seen. Some of them are registered while others are not. These vendor earn their livelihood by selling food which suits the taste buds of the local population. These foods are relished by the students and working community. Often such food suits the pocket of the customers also. However, the main drawback is that their food is unhygienically prepared and is unsafe. The main problem that these vendors face is availability of potable water with pressure. They make their own arrangements for storage of water and cleaning of utensils. Various studies done on these street food and the water used by them have documented presence of *E. coli* and other coliforms in water suggesting sewage contamination. Such water is a cause of FBI. Hence these vendors must be educated about use of potable water. Water should be stored

Fig. 35.1: Water stored in an uncovered bucket — In a fast food outlet

Fig. 35.2: Use of bare hands to draw water from an open bucket by a roadside vendor

in clean covered tank or drum fitted with a tap. Hands or any utensils should not be dipped in source of water. A long handled ladle must be used. For disinfection of water on small scale chlorination (bleaching powder/chlorine, iodine tablets can be used). Raw ingredidents and utensils should be washed with potable water.

Availability of adequate quantities of potable water in any food premises is of prime importance. Water is required for cooking and maintenance of hygiene and sanitation. In case, continuous supply of water is there then adequate storage arrangements should be made. Water storage tanks should be covered all the time. They should also be regularly cleaned and disinfected to prevent contamination (once a month for big tanks). Small utensils holding water in the kitchen, should be cleaned daily and completely drained before refilling. Non-potable water systems shall be identified and shall not connect with, or allow reflux into, potable water systems. Water pipes, either hot or cold, should be maintained in good condition. A good plumbing system of an eating establishment with identification of potable and nonpotable supply will ensure safety of food.

Waste Management in Eating Establishments

Do not just throw your garbage. Get rid of it!

Sonia Puri, Maninder Kaur, Puja Dudeja

It is often said that amount of waste produced by a society is directly proportional to its state of development, i.e. more developed a society, more is its waste generation. This is easily illustrated by the available state in USA where solid waste generation per Kg/capita/day is much more than that in India. India's population has been growing exponentially over the last few decades. With this rapid increase in population, urbanization, industrial growth the generation rate and volume of solid waste has also accelerated. Proper disposal of these solid wastes, has become highly challenging.

Waste is generated in households, markets, and commercial establishments such as hotels, restaurants, shops, etc. Apart from this waste is also generated from industries, hospitals, slaughter a house, sanitary drains, etc. In most of the cities of India, there is gross misman-agement in handling waste, which poses serious health and environmental problems. The plague outbreak in Surat is a good example of a city suffering due to the callous attitude of citizens as well as authorities in maintaining cleanliness in the city. In India, the urban local bodies, popularly known as the municipal corporations/councils, are responsible for solid waste management. These bodies try their best to manage with their meager resources. However, they have

not been able to successfully handle the increasing amount of waste. So the problem of waste disposal still persists. This is particularly troublesome for putrescible waste, e.g. from kitchen.

In India, hospitality industry is growing at a fast pace. Consequently, it also has signi-ficant environmental impact due to huge generation of solid waste. With changing times, there has been an increase in number of restaurants, hotels and street food vendors. Often they do not have a proper waste disposal system. They dump the waste generated to next collection point in the locality. It has been estimated that out of the total municipal solid waste generated, the hotels and restaurants solely contribute 25–30%. Except for a few star rated EEs who have a waste management plan of their own majority of others dump their waste in an unplanned way, often on open dumping site.

Waste from an EE is in the form of leftover food, disposable crockery, organic waste during peeling of fruits and vegetables, wrappers of raw material, tea leaves, fruit pulp (juice bar), cans, unused oils, etc. Waste from an EE can be classified as solid (garbage, refuse); liquid (sewage); gaseous (smoke, fumes). The liquid waste is normally disposed off through sewer pipes laid by municipal bodies. The gaseous wastes are expelled into atmosphere where they get diluted in the air. The main issue with waste management is that of solid wastes. Refuse is the dry waste in the form of cans, bottles, paper bags, polythene bags, lids, cartons, paper napkin, straws, toothpicks, etc. Garbage is the wet waste resulting from preparation, cooking and consumption of food. It also includes spoilt, leftover food, peels of fruits and vegetables, bones and skins, egg shells, etc.

In most cities, it is commonly seen that the EEs use bins to collect their waste without any segregation into dry (refuse) and wet (garbage) and throw it in nearby vacant areas, government vacant land, drains, streets, etc. Rarely it is disposed off into community waste collection bin. Needless to say, very few EEs have a door collection system of waste as they try to avoid the expenses involved in a proper waste disposal system. In India we see that most of the FBOs running small EEs do not have any civic sense and throw the waste in discriminately as it does not cost a penny. Even when they throw waste in community bins, there is often spillage of lot of waste outside the bin. The representatives of EEs are least concerned about this spillage of waste while emptying their waste bins. The managers too are reluctant in paying for waste collection services in case they are offered. Often the municipal collection and disposal services are not up to the mark. In case the waste is collected as scheduled, the spilled waste is left as such. It keeps lying there and later acts as a breeding ground for flies.

Broken Pipe of Sink

There is often backflow of water from blocked drains inside the EEs. The major problem area is near the kitchen sink. Though leftover food from the utensils is scrapped as the first step of cleaning utensils, still small particles are left behind. Grease and oil that is allowed to enter the sewer system causes serious damage. The greasy water separates from the wastewater and accumulates on the inside of sewer pipes. Over time, these deposits get larger as more grease and other solid material builds up. These deposits reduce the capacity of sewer pipes. Subsequent blockage causes sewage overflows, offensive odour and an unhealthy environment. A clogged drain/sewer causes backflow of waste water and emits bad odour, posing hazard to food safety and environmental hygiene (Fig. 36.1).

It is a common practice in our EE to use empty cardboard cartons or paint buckets as

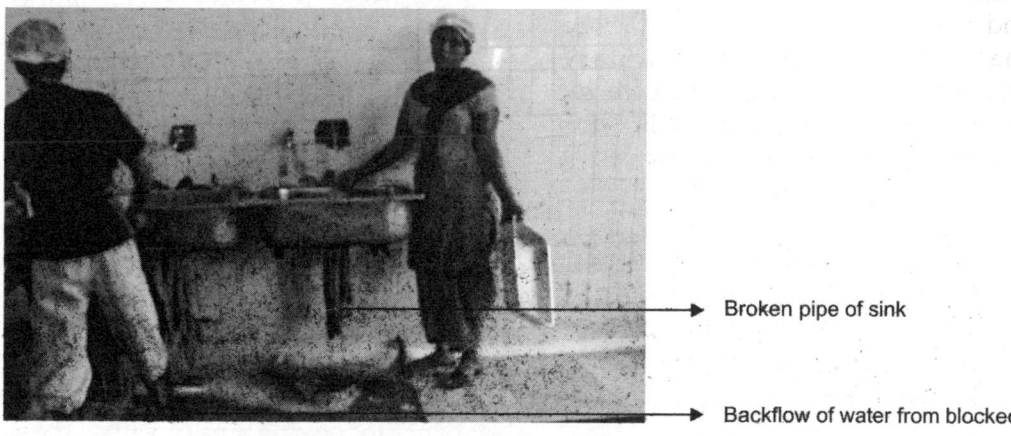

Broken pipe of sink

Backflow of water from blocked drain

Fig. 36.1: Poor waste management near kitchen sink

Fig. 36.2: Overflowing garbage bin

Fig. 36.3: Uncovered waste bin next to kitchen sink

waste containers which are often without lids. They are inadequate in number too leading to spillage of waste (Figs 36.2 and 36.3). In most commercial EEs kitchen waste collection and disposal system is pathetic. Bins usually overflow and smell a lot. Waste collection frequency is inadequate.

Street food vendors usually congregate in overcrowded areas where there are high numbers of potential customers. Such areas usually provide limited access to basic sanitary facilities. It is a common sight to see street food vendor co-located to a garbage bin/urinal/manhole or in insanitary conditions. They add to this nuisance by adding the waste from their own stall. Such food prepared in unhygienic conditions is predisposed to contamination. A common reason for this is the fact that most of these vendors are unauthorized and unlicensed. They always work through temporary structures. So hygiene usually remains compromised.

Another disturbing practice by small EEs and street vendors is to prepare food or wash dishes behind restaurants near a public tap (Fig. 36.4). It is damaging to both the public drainage system and the environment. It is incorrect to dump wastewater or waste food into storm water drains in the street. These drains are meant to carry rainwater. If clogged with grease and other waste the result is breeding of rats, cockroaches and it overflows. During heavy rain storms, serious flooding with damage to life and property can occur if storm water drains are clogged with waste.

As discussed in chapter on cleanliness it is of vital to maintain hygiene of an EE at all times. However to maintain hygiene, it is equally important to have a proper waste disposal system. Accumulation of waste in kitchen of EE, dumping of waste outside the EE, overflowing and uncovered garbage bins, blocked drainage pipes with backflow in kitchen, etc. are bound to affect safety of food. Often due to space constraints the sink, washing and cooking area are in close proximity. Backflow of water from blocked drains, uncovered dustbin with waste in close distance to cooking area can affect the safety

Utensils for washing behind the EE.

Fig. 36.4: Washing utensils kept in proximity to public tap

of food. This will also undo the effect of cleaning and maintaining the EE.

Waste buildup in kitchen of EE/vicinity of EE can give foul odours if kept too long. It serves as a breeding ground for flies, rodents and other insects. These pests then transmit pathogens to humans. They may also pollute water supply of the EE. In case such stored refuse and garbage is burnt, it adds to the air pollution. Over and above, it is aesthetically bad to store waste for longer duration.

Proper solid waste management involves collection, storage and disposal of waste. It is best to collect waste at the point of generation. Appropriate location of garbage bins is a must for proper collection of waste. They are required at the point of cutting and chopping raw food articles and near kitchen sinks. In case it is not feasible to place a bin there waste can be collected in small bowls. These days tables fitted with garbage bins are available in the market where waste can be immediately thrown into the bin. All small bins need to be emptied into the large bin when full else, overflow will compromise on cleanliness of the EE. Waste containers with cracks should immediately be replaced.

Waste is stored before disposal. Under all situations, it should not be left overnight as then it will serve as a breeding place for cockroaches and pests. These storage bins are preferably placed in a cool area to delay the decay process. Waste storage rooms if any are there should be clean and well maintained. Waste and refuse should be removed at a frequency that will minimize the development of objectionable odour and other risk of attracting or harbouring pests or animals, and should at least be once daily. In case the waste is not cleared due to some reason the FBOs often have it thrown outside the EE. Such practices create fly nuisance and pollute the environment. In these situations, the waste should be taken and thrown in the nearest municipal bin in the locality without any spillage.

An ideal garbage bin should be sturdy, durable, rustproof, covered with tight fitting lids, leak and pest proof, operable with a foot paddle and easy to clean. They must be in adequate number to prevent overflow of waste. The bins are lined with appropriate size garbage bags which keeps the bins dry. They are often placed at a raised platform (6") above the ground. Once the bins are emptied they must be cleaned with soap/detergent, scrubbed, rinsed, dried and then reused. It is better to have spare set of bins so that when one set is getting cleaned and dried the other can be used. In case the EE has adequate space organic waste can be disposed by composting or vermiculture.

Liquid waste comprise of waste water from kitchen sinks and drains and toilets. This is done through a public sewerage system or through septic tanks. Sewage is treated before disposal. A good plumbing system is essential to prevent contamination from sewage in food preparation area. Leak in sewer pipes, cross connection with potable water system, back flow and blockage can create possible contamination. It is advisable to place grease traps in the kitchen drains to prevent clogging due to food waste and oil.

According to Food Safety and Standards Regulations (FSSR) 2011 the drainage and waste disposal systems and facilities in an EE shall prevent the contamination of food and/or the potable water supply. All areas of food premises that generates refuse shall have a waste container for temporary storage of solid waste on the premises. Waste storage areas/rooms shall be away from food rooms/kitchens and be well ventilated. The walls, floors and ceilings shall be such that they enable easy cleaning.

All sanitary fitments and hand washing facilities shall be connected to a proper sewage or waste water disposal system. No manhole shall be situated inside any kitchen or food room. All soil /waste/rainwater pipes inside any kitchen, food room or seating accommodation shall be enclosed in impervious rustproof pipe ducts.

All restaurants and food processing factories are required to install grease traps so that greasy materials will be separated from wastewaters before passing to communal sewers. These are the main sources of greasy waste and therefore it is very important that the grease traps used at these establishments are effective in removing grease from wastewater before it passes to the sewer system.

Waste is a potential source of pathogens and food contaminants. Proper disposal of waste is important for preventing the spread of pathogens inside food premises and contamination of food. Properly maintained waste containers can discourage the access of pests and animals.

Some key points for waste management in EE are:

- Scrape dirty serving dishes and cooking utensils into a garbage bin before washing.
- Dispose of floor sweepings and food scraps to a garbage bin before washing floors and food preparation surfaces.
- Use metal strainers or baskets in all drains. If this slows things down, keep two strainers on hand; quickly place one over the drain while the other is emptied.
- Never pour waste cooking oil down a drain or toilet. Waste oil and grease should be poured into a storage container which is discarded with other solid kitchen waste.

Section V

Food Handlers: An Important Link in Food Safety

37

Role of Food Handlers in Food Safety

Puja Dudeja, Shalini Shachdeva, Amarjeet Singh

Foodborne illnesses (FBI) are an important public health problem. Food safety is a critical issue in our country with reports of outbreaks of FBI every year resulting in substantial costs to individuals, health care system and the country. There are many hurdles on the road to food safety. Food contamination can occur at any stage from production to consumption, i.e. from ' Farm to Fork'. Food handlers are a important link in the chain from kitchen to fork.

Food handler is any person who handles either **food or surfaces** that are likely to be in contact with food such as cutlery, plates and bowls. They work at place where food is cooked and served, i.e. our kitchens, *dhabhas*, canteens, messes, restaurants, juice bars, street food vendors, snack bar, take away joints, etc. Even the housemaids working as help in the kitchen for chopping vegetables, kneading flour, etc. are food handlers. They are involved in preparation of raw material, cooking, packing, storing, displaying and serving of food. Examples of food handlers are waiter staff/service staff, chefs, head cooks, dishwashers, receiving and food storeroom staff, bartenders, host/hostesses that handle food, street vendors who sell food items and housemaids.

Food handlers on one hand can provide us with tasteful and safe food whereas on other hand they can also be source of contamination and compromise food safety. They are an important source for the transfer of micro-organisms to the food. They can transmit pathogens passively from a contaminated source. For example, bacteria may be transferred from raw poultry to food such as cold cooked meat that is to be eaten without further heating. Many a times food handlers are themselves the sources of organisms either during the course of gastrointestinal illness or during and after convalescence, when they no longer have symptoms. During the acute stages of gastroenteritis, the patients excrete large numbers of organisms. If food handlers have such a disease and they continue to work in kitchen, disease may spread. Food handlers who are asymptomatic may present a real hazard. They can transfer pathogens from their body/body secretions/fluids while working in kitchen.

Good hygiene, both personal and in cooking is the basis for preventing the transmission of pathogens from food handling personnel to consumer. If a food handler does not observe food safety precautions pathogenic organisms present in or on his body can be transferred to food during handling. They multiply to an infective dose in contact with food. Such food can be a potential source of food poisoning to the person consuming it.

Apart from this a food handler may commit common food handling mistakes which include inadequate cooking, heating, or reheating of foods, cooling food inappropriately, etc. Many studies have documented that there is a wide gap between knowledge about various food safety practices and their practical implication by these workers. Reasons are many. These involve time constraints, poor motivation levels, low wages, poor working conditions with high temperature and humidity levels, long working hours, ill-treatment by FBOs, lack of respect in the profession, poor supervision by the FBOs, unavailability of items as water and soap for hand washing, inadequate toilet facilities, nonconduct of medical examination of handlers, lack of training, etc.

A food handler has certain set of responsibilities while working in kitchen for ensuring the safety of food. These are ensuring and observing

a. Handwashing practices
b. Good personal hygiene
c. Clean work attire
d. Management of illnesses
e. Safe food practices
f. Regular training

Hand washing: Unclean hands of food handlers are often the culprit in transmission of microorganisms to the food. The hand hygiene of a food handler at work place is of utmost importance. Hence, it is the duty of the FBO to ensure availability of soap and clean water for washing hands all times. It is a common practice of food handlers to use the kitchen sink for washing hands. This practice is not advisable. A separate sink exclusively for hand washing with clean water source is required to maintain hygiene. In food manu-

facturing units with HACCP and ISO certifications sanitizers are also provided near the sink. This liquid is rubbed on hands after washing. It ensures cleanliness of hands. However, use of sanitizer without washing hands should not be done. Such units also have a checklist where a food handlers has to log entry of washing hands. This ensures an appropriate frequency of washing hands. It also acts as a tool for check by the supervisor.

It is often seen that small EEs like dhabhas, small restaurants do not have appropriate hand washing facility for customers and food handlers. Most often kitchen sink is used for this purpose. It is not uncommon for food handlers to wash hands without using soap. The food handlers just rinse their hands with plain water and then dry them by wiping with their dirty clothes or apron. This gives only a false sense of security to them. They may or may not inform the FBO regarding non availability of soap. The onus for supplying soap lies with the FBO. These FBOs economize on soap use since their main motive is business and profit. They hardly supervise washing of hands by food handlers. They do not punish or warn them frequently to ensure compliance of good hand washing practices.

The other groups of food handlers are the street vendors who do not have any provision of washing hands. Usually they just wipe hands with the mop they are using to clean the food contact surface. These days some of them have starting using gloves to attract customers. They feel as if wearing disposable gloves is a symbol of hygiene and cleanliness. It has been well said that little knowledge is a dangerous thing. Similarly these food handlers do not use the gloves in the correct way due to lack of awareness and cost cutting. They tend to reuse these gloves. Rather they keep gloves 'on' for hours together. They keep touching dirty surfaces of their cart, utensils, etc. with gloves on. A dirty glove is as bad as or even worse than an unclean hand.

Unclean hands can transmit germs from the hands to the food. Hands can get contaminated after handling raw food for example, after touching raw potatoes, carrot, onions, meat, etc. coughing, sneezing, visiting toilets, eating, drinking, smoking, handling money, using mobile phone, touching hair, scratching body, handling dustbin, and soiled equipment and utensils. Hands must be washed properly with soap and water after these activities. Hand washing is a must before commencement of work in the kitchen.

The correct steps of hand washing are shown in Fig. 37.1. These steps are:

1. Wet the hands with running water

2. Apply soap

3. Rub hands for 20 seconds (wash all surfaces thoroughly, including forearms, wrists, palms, back of hands, fingers and under fingernails).

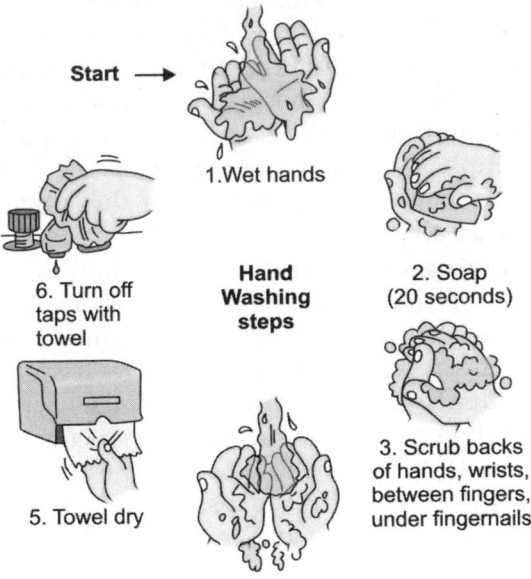

Fig. 37.1: Steps of hand washing

4. Rinse hands thoroughly

5. Dry hands with a clean paper towel/drier

Good Personal Hygiene and Work Attire

Personal hygiene and cleanliness is essential for food preparation, not only in eating establishments but also in the domestic setting also, to avoid foodborne illness. It is the moral and professional responsibility of a food handler to maintain high standards of personal hygiene and cleanliness to protect the consumer from becoming ill following the consumption of their products. It includes the following practices:-

Hair: Hair has also been known to cross-contaminate food. Bacteria cling to the hair and scalp, living off dandruff, dust and perspiration are collected in the hair. The hair of a food handler should be short and clean. Long hair is not only difficult to maintain but also their possibility of falling into food while cooking/handling increases many fold. Hence, food handlers should preferably keep their hair short. Some people's hair grows a little faster and some grows a little slower. Hair of a food handler should preferably be short and should be trimmed every 4 to 8 weeks to keep them in shape.

In the course of handling food, hair should be covered with a clean cap or hair net. Long hair should be tied back. It should be ensured that combing of hair is not done in food handling areas. Using a cap prevents food handlers from directly touching hair and scalp with fingers to move it out of their face and passing bacteria from the hair onto food. Hair can be partly covered by staff who are serving or handling food that is protected, e.g. a person employed in a takeaway front counter, working in soft drink dispenser counter, ice cream shop, etc. Irrespective of the style, design, cost, material of cap all hair must be properly tucked inside the cap. No hair should come out of the cap. Also, a torn cap should not be worn.

Face and neck: A food handler should not wear ear rings or any necklace at the time of work. The reason behind this is that if the jewelry parts are loose then they can fall into the food and contaminate it.

Beards, if any, must be trimmed and tidy. The use of beard nets is strongly recommended for bearded food handlers working in all food preparation establishments. This will prevent also prevent loose hair or broken hair fragments from falling onto food and food preparation areas.

Clothes: The clothes of a food handler should be neat and clean. They should not be dirty. Clothing must be sufficient to cover the entire body including arms if necessary to block body hair from getting into the food. Also, it is suggested that there should be no outer pockets on the shirts. In case there is any it should be empty. Items like pen/pencils/mobiles/medicine/paper, etc. if kept into the outer pocket can fall into food while working. This may compromise food safety.

Apron: The next characteristic identifying feature of a food handler is the apron. It should be changed at least once during the day. Preferably it should be done in the middle of an eight-hour shift. Apart from this cooks should change aprons whenever these get soiled. Whenever a food handler changes workstations from raw food preparation activities to ready-to-eat food preparation activities, the apron should be changed.

Food handlers should not wear aprons outside food preparation areas like while going to washroom, etc. This procedure minimizes possible contamination of aprons by airborne pathogens, dirt, dust, and possible soiling by washroom fixtures and other unsanitary articles. So, they should remove apron before going to toilet. It is advisable not to have pockets in apron above the waist. Buttons should preferably be avoided on the clothing as they may come off and fall on the food.

Hands and wrists: Wearing jewelry or other cosmetic enhancing items during food handling activities is discouraged. These include, but are not limited to rings, nail polish, wrist watches, bracelets, clip-on earrings, false nails, false eye lashes, etc. Jewelry can hide microorganisms that cause foodborne illness and make it hard to wash hands. Jewelry can also fall into food. In cases where rings are difficult to remove, clean gloves should be worn by the food handler while handling food.

Nail cutting: Long nails are a strict 'no no' for food handlers. Nails of a food handler should be short and clean. Long nails accumulate dirt and bacteria, which can enter food while handling. One must fix a day in the week for trimming nails. Blades should not be used to pare nails as these can cause injury.

Foot wear: These must be clean and free of dirt and accumulated food particles on both the top and bottom. Street wear should preferably be avoided inside the kitchen Accumulation of food particles and dirt on footwear may allow microorganisms to multiply. This may consequently affect the general sanitary conditions of the kitchen premises.

Dressing on wounds: A food handler is exposed to various kinds of injuries through use of sharp knives, grater, and other items while preparing food. There are frequent incidents of cuts, abrasions, burns while handling food. It is imperative that cuts and wounds should not be left open. These should be covered with a dressing to prevent contamination of food. The dressing/bandage should be coloured. This is because in case it falls into the food it can be easily identified. A food handler may work if the cut has been bandaged and a disposable glove is worn.

Management during Illnesses

A food handler who is suffering from infectious disease may transmit infection from his body to the food and make it unsafe. Persons consuming such infected food may fall ill and suffer from FBI. A food handler must report to his senior about his illness in case he is suffering from diarrhea, vomiting, fever, cough, skin lesions (including boils/cuts), eye or nose discharge.

Such a person should not be engaged in handling of food. In case it is unavoidable then all measures must be taken to prevent food from being contaminated as a result of the disease. For example, an infected sore must be completely covered by bandage and clothing, or by a waterproof covering if it is on an area of bare skin. In case of cold a disposable tissue or a handkerchief must be used to handle the secretions. Various restrictions for food handlers during illnesses are given in Table 37.1 (*Source*: Training manual, FSSAI).

Table 37.1: Restrictions for food handlers during illnesses

Disease	Work status	Duration of work Restriction/ comments
Abscess, boils, etc.	Relieve from direct contact and food handling	Until drainage stops and lesion has healed or employee has negative culture
AIDS or ARC (AIDS related complex)	May work (as per CDC guidelines) No open lesions, upper respiratory diseases or communicable diseases	Employee will be counseled and educated
Diarrhea		
Acute stage (etiology known)	Relieve from direct food handling	Until symptoms resolve and infection with Salmonella, Shigella or Campylobacter are ruled out
Campylobacter	— do —	Until symptoms resolve or after appropriate antibiotic therapy for 48 hrs
Salmonella	— do —	Until stool is free of the infecting organism in two consecutive cultures not less than 24 hours apart
Shigella	— do —	Until stool is free of the infecting organism in two consecutive cultures not less than 24 hours apart
Hepatitis A	— do —	Until seven days after onset of jaundice. Must bring note from physician on return.
Staphylococcus aureus	— do —	Until lesions have resolved and the employee has negative culture

Good Hygienic Practices at Work

Food handlers need to know how the work they do can affect the safety of the food they handle. Following good hygiene practices should be observed by a food handler to ensure food safety.

- Wash and dry your hands whenever you think they are contaminated.
- Do not sit on food preparation shelf.
- Do not leave your personal belongings on the food preparation shelf/cooking area.
- Cover exposed sores with a waterproof dressing or disposable gloves.
- Wear clean outer clothing. Change aprons or other clothing if they are soiled.
- When sneezing or coughing inside food preparation area is unavoidable, food handlers should turn away from food and cover their noses and mouths with tissue paper or handkerchiefs. Hands should then be thoroughly cleaned at once.
- Never blow into a bag to open it to put in food.
- Never blow on food for any reason.
- Do not spit, smoke or use tobacco in areas where food is handled.
- Only eat when you are outside the food preparation area.
- Do not touch ready-to-eat food with bare hands.
- Do not taste food with fingers.
- Do not reuse a sampling spoon without washing.
- While cooking in kitchen do not touch hair or other parts of bodies such as noses, eyes or ears.

Food handlers must tell their senior if they know or think they may have made any food unsafe or unsuitable to eat. For example, jewellery or a Band-Aid worn by a food handler may have fallen into food, or glass may have broken into or near exposed food.

Training: *Please Refer to Chapter on Training of Food Handlers.*

Food Safety Requirements for Food Handlers under Food Safety and Standards Regulations (FSSR) 2011

According to FSSR 2011 any food handler believed to be suffering from or to be a carrier of a disease or illness likely to be transmitted through food shall not be allowed to enter into food handling area. A food handler can transmit staphylococcal, Salmonella, Shigella, *E. coli, Entameoba histolytica, Campylobacter,* Hepatitis A, influenza, threadworm, and giardia infection. A system should be developed in all EEs whereby any affected person shall immediately report illness or symptoms illness to the management or FBO and medical examination is carried out apart from periodic checkups, if clinically or epidemiologically indicate. All arrangements should be made to get food handlers examined at least once in a year to ensure that they are free from any infectious, contagious or communicable diseases. A record of these illness signed by a qualified doctor should be maintained for inspection purposes. In case of a food manufacturing unit the staff shall be inoculate against enteric group of diseases and a record kept for inspection. In case of an epidemic, all workers are vaccinated irrespective of the scheduled vaccination. At the time of recruitment also recent history of illness along with a medical checkup must be done.

In nutshell food handlers are an important cause of FBI. They have a definite role in prevention of FBI. Their food handling practices can affect the well being of many people. They are a crucial link in food safety.

38

Occupational Safety of Food Handlers

Puja Dudeja, Amarjeet Singh

INTRODUCTION

At present, India is the world's second largest producer of food and have the potential of being the biggest. We are backed by a powerful and strong food and agricultural sector. The food processing industry is one of the largest industries in India. Globally it is ranked fifth in terms of production, consumption, export and expected growth. The total food production in India is likely to double in the next 10 years with the country's domestic food market estimated to reach US$258 billion by 2015. Currently, the Indian food processing industry accounts for 32 percent of the country's total food market. Owing to lucrative opportunities and profits in this sector, the food industry in India has been attracting a investors. At the same time, the work force involved in this food industry has also been constantly on the rise. Like other professions, the workers employed in various sections of food industry namely production, storage, packaging, transport, retailers, and EEs are also exposed to certain hazards by virtue of their being in these jobs.

The job of a food handler is of low status and poorly paid which leads to poor motivation. Keeping occupational health of the these employees at high priority is the key to improve manager and handler relationship. Looking after the health, welfare and safety of his employees is of immense benefit to the employers/managers of these food industries/EEs. A satisfied and motivated food handler will give a high quality work. On the other hand, employees who feel unmotivated to work are likely to be casual in their approach regarding food safety practices like hand washing, personal hygiene, etc.

As per (ILO/WHO 1950) occupational health is the promotion and maintenance of the highest degree of physical, mental and social well-being of all kind of workers in their occupation. It seeks to prevent departures from health among the workers as well as control of job related health risks. Adaptation of work to people, and people to their jobs is one of the main strategy of occupational health. A safe and healthy work environment is essential for control of risks arising from physical, chemical and other work place hazards.

The main focus in occupational health is on three different objectives:

i. Maintenance and promotion of workers' health and working capacity

ii. Improvement of working environment

iii. Optimizing the main work itself to make it conducive to safety and health of the worker

303

iv. Through development of work organizations and working cultures in a direction which supports health and safety at work.

v. Promotion of a positive social climate and smooth operation at work place

All these measures will and may enhance productivity of the undertakings (Joint ILO/WHO Committee on Occupational Health).

Classification of Occupational Hazards for Food Handlers

Occupational hazards are present at all places where food handlers are deployed be it a food processing unit, EE, *sabzi mandi*, food packaging industry, retailer shop, etc. These hazards in food handlers can be classified as under:

1. Physical   4. Mechanical
2. Chemical   5. Social/Psychosocial
3. Biological

Physical Hazards

Temperature

Food handlers often have to work in environment with high temperatures. This is especially when they are employed near a cooking range or tandoor. Lot of heat is generated in process like deep frying and cooking. The problem becomes worse in summer season. Often food handlers have to work in such areas for prolonged hours. Some EEs with star ratings do have a centralized cooling system to provide comfort to their workers from heat. However, such kitchens in any city are very few in number. In low budgeted food processing industries and EEs least importance is given to creating a comfortable working zone for the food handlers. Working for long hours in high temperature areas leads to lot of sweating. This can get not only dropped into the food from the handlers but also can lead to prickly heat, dehydration, heat exhaustion, heat cramps, heat stroke and burns in food

handlers. For better productivity it is essential that temperature is maintained (corrected effective temperature 20–27°C). High humidity along with temperature can make the working environment uncomfortable for workers.

Lighting

The workers may be exposed to the risk of low illumination or excessive brightness. This seen at work places where either the owner's intentions are to save money by providing less number of lighting points or by not replacing the old one. Poor illumination can lead to eye strain, headache, eye pain, Iachrymal congestion and eye fatigue. There should be sufficient and suitable lighting, natural or artificial, wherever persons are working. Inadequate lighting can compromise food safety. (*See* Chapter 30 on Lighting in Eating Establishment.)

Noise

Noise is also a health hazard in many food industries. An old advertisement on Doordarshan '*Hawkins ki seeti baji* and *khushboo hi khushboo udi*' may not hold true for food handlers. I am reminded of an incident here. One day my 2 years old daughter was suddenly got up from her sleep and started crying. I rushed from kitchen where I was working, looked around to check for any reason for her sleep disturbance. I put her back to sleep and came back to kitchen. But this was repeated. I realized it was the noise of my old mixer and grinder which was the culprit!!

Various equipment in the kitchen produce noise like chimney exhaust, exhaust fan, *chapatti maker*, food processor, etc. The effects of noise can be auditory or nonauditory leading to nervousness, fatigue, decreased efficiency and annoyance. The degree of injury from exposure to noise depends upon a number of factors such as duration of exposure and frequency range along with individual susceptibility. Among food handlers annoyance due to noise can also affect food safety.

Fire

Fire hazards/explosion of gas cylinders/stoves/pressure cookers can also take place in the kitchen. These can lead to burns and may be fatal also.

Lifting Loads

At times food handlers have to lift heavy loads in carrying raw food items like vegetables, fruits, grains, etc. This can lead to musculoskeletal problems like back ache, sprain, strain and pain in limbs, etc. They also have to stand for long hours and can develop varicose veins. Quite often ergonomics is not applied while designing the working shelves in the kitchens. Discordance between heights of food handlers and kitchen shelves may lead to low back pain and easy fatigability. This can also affect the working of food handler in a negative way as he may not be able to mop/clean/cut as per required instructions of food safety.

Chemical Hazards

The use of dishwashers for washing utensils is restricted to high budget EEs only. FBOs prefer to employ staff for cleaning of dirty utensils and use low cost detergants. In this way food handlers come in contact with chemicals in the form of detergents and disinfectants. Prolonged exposure or increased duration of exposure can cause allergic contact dermatitis especially in hands. Such hands can be super infected with bacterial infections and jeopardize food safety.

Biological Hazards

Food handlers come in contact with organisms like Salmonella while handling raw foods particularly of animal origin like eggs. Training and education on safe handling of raw foods can prevent infections in them. However, most of them do not get an opportunity or exposure to any kind of formal training on food safety issues. Most of the times they are not aware of simple steps like washing of hands thoroughly with soap and warm water while handling raw poultry and eggs. Apart from this persons handling plates and utensils that have been in contact with such foods are also at risk. These can be fatal in case of avian influenza.

Food handlers may be carrying pathogens in/on their bodies and can be a source of infection to other co-workers in the same working area. Cases and carriers can also transfer pathogens to various foods compromise food safety.

They need to taste the food prepared by them before it is served to the clients. In case the food has become unsafe they themselves fall a prey to FBI. Most of the low budget EEs, messes, canteens provide food to its workers after the lunch or dinner timings are over. This makes them the last ones to eat the meal, which most often has been lying in temperature danger zone (5–60° C) for more than 4 hours exposing them to risk of FBI.

Mechanical Hazards

The commonest mechanical hazard is due to cuts with knives and other sharp equipment. Protruding parts of various machinery can also lead to injuries. It is known that 10% of accidents in any industry are said to be due to mechanical causes.

Psychosocial Hazards

Food handlers do face monotony in their work environment. Most of the times they continue to work at same level for years together. There is very little scope of career progression. These people are always behind the curtain do not get recognition and appreciation for good work very often. Even the lady of the house waits for an appreciation from the family members after cooking meals. However, times are changing. Various live TV shows hosted by famous chefs have given due recognition to the skills and progression. The emerging hotel management industry, catering industry, eating out culture in changing society, globalization has improved the social status of the occupation.

The food handlers are often exploited with low wages and restricted leave. Most of them work in the unorganized sector. At times they are paid less for the skill they possess.

Easy availability of food and alcohol, results in obesity and alcoholism. They work hard to earn their livelihood but the job does not require strenuous activity. Often they are required to stand for long hours. Access to tasty and rich food at all times makes consume more calories than required and hence risk of non-communicable diseases.

They also work under lot of stress as even ignorance on their part can dissatisfy the consumer/client and harm the reputation of EE.

Prevention of Occupational Hazards in Food Handlers

Engineering Measures

Measures for the prevention of occupational diseases need to be emphasized during designing the building. Once the building is constructed, it is extremely difficult to make alterations. Due care should be given to the ventilation and lighting area of work place. Principles of ergonomics should be applied while designing the height of working shelves and food flow. Good housekeeping of the building will help in controlling occupational hazards. It also contributes to efficiency and morale of workers in industry. There should be good general ventilation. It has been recommended that in every room of a factory,

ventilating openings shall be provided in the proportion of 5 sq. feet for each worker employed in such room, and the openings shall be such as to admit a continued supply of fresh air. Chimneys should be installed to prevent accumulation of fumes.

Trolleys with wheels (like those in super markets) should be available to carry heavy loads. Equipment should be regularly serviced as per manufacturer's instructions.

The quality of chemicals used should be checked before exposing food handlers to them.

Medical Measures

Regular medical examination of food handlers will ensure good health. In case a worker is unwell he should be given leave and rest. However, this is done as a lip service only because of work force constraints. In case complete leave is not possible, then the manager should employ a sick workers in an alternate job not involving direct contact with food.

It should be noted that negative stool samples, for employees recovering from gastroenteritis, are not a necessary condition of their employment or return to work with the exceptions of enteric fever and E. coli infection.

Periodical medical examinations for early detection and treatment should be done. Ordinarily the workers are examined once a year. A record of such examination needs to be maintained.

Health education about the processes, handling of raw material, correct use of gloves, aprons, caps has an impact on both safety of handler and food. It also includes guiding the worker about various legislations available and social security schemes for the benefit of the workers.

Pre-employment Screening of Food Handlers

The most important infections attributed to transmission from infected food handlers are norovirus, *Salmonella enteritidis* and *Salmonella typhimurium*, which together account for the largest numbers of outbreaks and individual infections. The most common routes of transmission are faecal–oral, and via aerosol formation from vomit. Food handlers can be symptomatic or asymptomatic carriers of foodborne infections—both the transmission of norovirus and *Salmonella enteritidis* have been attributed to asymptomatic food handlers. All food handlers before given employment should undergo a medical examination by a registered doctor and a stool test. Only those found fit should be allowed to handle food.

Legal Measures

Society has a responsibility to protect the health of the workers engaged in various occupations. Laws serve an important purpose as an institution in all socities. Citizen friendly laws are a must for any society to advance. Safety and health of the workers in all professions occupy a significant place in Constitution of India. It is the policy of the State to make provisions to secure just and humane conditions at work in all occupations. The Constitution provides a broad framework under which policies and programmes for occupational health and safety are established. There are many laws covering different occupations in our country especially hazardous ones like mines, dockyard, foundary, etc. The government is entrusted with the responsibility of implementation of these laws in true letter and spirit. The onus of implementation of the same rests with the employer in each occupation. Some responsibility is also shouldered upon the employees to observe safety precautions while working.

Hence, occupational safety and health is a shared responsibility (Fig. 38.1).

Legislations on occupational health and safety have been in existence in India for over 50 years e.g. The Factories Act, 1948 and the Workmen's Compensation Law in which a worker can claim compensation have been made to support. Workers in different occupaticops. However, both these are not applicable to hotels and restaurants. The workers in hotels and restaurants are covered under the Employees Sate Insurance Act (ESI Act), which is a contributory social insurance scheme that protects the interests of workers in contingencies such as sickness, maternity, employment injury causing temporary or permanent physical disability or death, loss of wages or loss of earning capacity.

In spite of a good framework of this law, the implementation of this welfare act is far from satisfactory in case of food handlers working in EEs. Indian government has expanded the coverage of Child Labour Prohibition and Regulation Act and banned the employment of children less than 14 years of age as domestic workers and as workers in restaurants, dhabas, hotels, spas and resorts with effect from 10 October 2006. The official estimates for child labour working as domestic labour and in restaurants is more than 2,500,000 while NGOs have estimated the figure to be around 20 million. Strict implementation of the law would ensure that these children come out of the viscious cycle of poverty and hard working conditions at a younger age. The changes were necessitated after the Right to Education Act came into effect which promised free and compulsory education to all children aged between 6 and 14 years.

However, it is a often seen that children work in roadside restaurants, tea or sweet

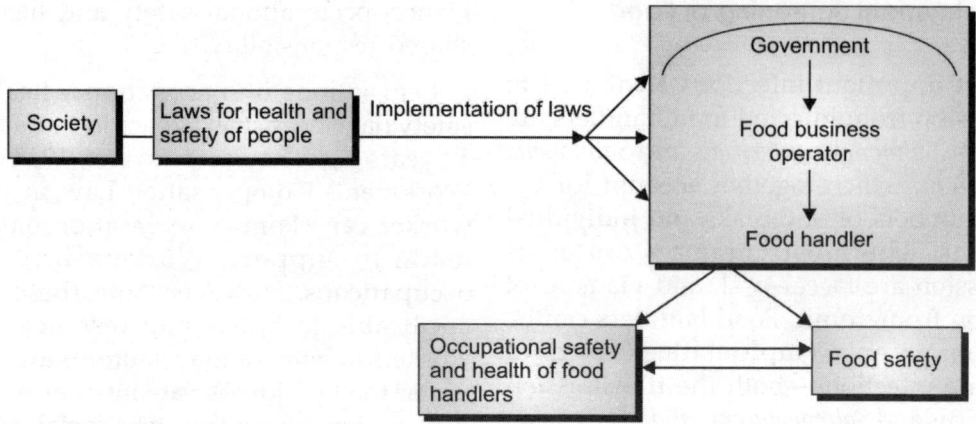

Fig. 38.1: Relationship between legal measures for occupational health and food safety

shops to earn money. Poverty stricken families compel their children to move out of their villages to nearby towns or cities in search of work to sustain themselves and to support their families back home. These parents do not give priority to the education of their children as they assume that education does not have immediate results. These children find refuge working in restaurants as food and shelter are generally provided in addition to paltry minimal wages for their efforts. In return, they carry out menial tasks for their employers, responsible for shopping, lifting loads during transportation of raw materials, counter service, dishwashing, cleaning or any other type of work required in the restaurants. The life of a child restaurant worker normally starts before sunrise performing a long list of chores determined by the employer. They include demands such as cutting vegetables with dangerous tools, cooking in smoky kitchens without proper facilities for ventilation or exhaust fans, lighting kerosene/wood/gas stoves or hauling water from distant taps. A child worker continues these tedious chores throughout a day's work, without any rest, exhausting them physically and mentally. At times they are punished with

verbal abuse in case they break a glass item or make a mistake.

Personal hygiene, healthcare, education, rest, breaks or any form of entertainment are least expected by children working in restaurants. All the working children have to sleep on the floor or on the tables which is possible only after closing time sometimes late into the night. They sleep late at night when customers have gone. Most of the restaurant workers wear dirty clothes and lack sufficient water for hygiene and cleanliness. Fatigue from continuous working hours does not provide them time for washing clothes or taking a bath.

A large number of food handlers mostly women in India work as maids and nannies in the unorganized sector. They are often involved in cooking along with other household chores. In February 2014, the International Labour Organization (ILO) stated: "Millions of maids working in middle class Indian homes are part of an informal and "invisible" workforce where they are abused and exploited due to a lack of legislation to protect them". According to the National Sample Survey of 2004-05, there are around 47.50 lakh domestic workers in the country.

Out of these, 30 lakh are women working in urban areas. An ILO report suggests that the number of maids has surged by close to 70 percent from 2001 to 2010 in India. They number in at least 10 million.

There is a cyclical pattern of their daughters becoming maids to the same families has also been observed. The employers often hesitate in giving them timely leave, decent salary and often suspect them in case of a theft. The rule of minimum wages is not maintained most of the time. In 2008 the government drafted a National Policy on Domestic Workers, the main points included were minimum wages, working hours and conditions, social security protection and the right to form trade unions and develop their skills. The policy however has not been approved by the cabinet yet. Other laws which are applicable to domestic workers are the Minimum Wages Act, 1948, the Employees Compensation Act, 1923, the Equal Remuneration Act, 1976 and Inter-State Migrant Workmen Act, 1976, they are often denied the benefits because they are generally illiterate and do not hold mutually agreed contracts, unlike say those working in factories.

Every occupation has its own set of problems and hazards associated with it. What makes food industry different from the rest is that maintaining occupational safety and looking after the health and welfare of employees will ensure safety of food. It has a direct bearing on the quality of product.

39

Food Safety Training of Food Handlers: An Experience from a Tertiary Care Hospital

Puja Dudeja, Shalini Sachdeva, Amarjeet Singh

Maintenance of hygiene in an eating establishment (EE) is the first prerequisite to minimize the occurrence of foodborne illnesses (FBI). Improper handling of food has been implicated in 97% of outbreaks of FBI. Knowledge, awareness and practices of food handlers have a direct bearing on the safety of food. Hence, food hygiene training of food handlers is of vital importance for food safety. The status of food hygiene training in EEs in India depends upon the level of EE. The star rated EEs often have well trained staff to handle the kitchen. Majority of the remaining food handlers working in small EEs, *dhabhas*, canteens, messes, street vendors lack adequate knowledge about food safety.

In this context in India, Food Safety and Standards Regulations (FSSR), 2011 have now come into force which makes all EE to put into place, implement and maintain a permanent procedure based on these guidelines which includes training of food handlers on food safety issues. Eventually all EEs in India will have to comply with the prescribed standards of FSSR, 2011. When this is properly implemented the customers would be the greatest beneficiaries, as it would result in better quality of food and its safety.

FBI result when food is improperly prepared, or is mishandled. The case of Mary Mallon (Typhoid Mary) who as a food handler caused seven outbreaks, 57 cases and 3 deaths is well known example to the food industry. Poor standards of hygiene during food preparation and the lack of training in food safety by the food handlers are probably the most common causes of FBI. Mishandling of food and disregard of hygienic measures on the part of food handlers may enable pathogens to come into contact with food, and in some cases, to survive and multiply in sufficient numbers to cause illness in the consumer. To decrease the burden of FBI, WHO Department of Food Safety and Zoonoses (FOS) actively promotes training as the means to improve the practices of food handlers for food safety.

Food handlers in small EEs in cities are often migrants from rural areas/suburbs who are school dropouts. They come to city in search of livelihood. These EEs offer them food, place to live in and a minimum salary. They neither have any knowledge about cooking per se and nor about food safety. Often they have poor personal hygiene. Many of them suffer from gastrointestinal infections. Over time, they learn their job and keep the FBO happy for continuation of work. If either is not satisfied, they shift to another EE of similar standards. So as of now almost all food handlers in small scale EEs lack any formal training in food safety.

However, the onus of training them lies with the FBOs and the authorities. No initiative is usually taken by the FBO to teach them the basics of food safety and hygiene. They continue to wipe utensils with dirty mop, wearing of wrist bands, unclean nails and long hair. Sometimes they take bath once in three days for want of clean dry clothes especially in winters. In case soap is not available for washing hands, they hesitate to inform the FBO. They do not consider it important because the focus of most FBOs is on cost cutting. The FBOs main concern is their availability during peak hours of business. Moreover, these food handlers are reluctant to take leave when they are not feeling well to avoid any salary loss. Personal hygiene efforts are seen as a wasteful expenditure.

Pajot and Aubin, 2011 have reviewed studies on food handler training and brought out in their report that there is insufficient research evidence on the fact that food handler training improves food safety practices. Also there is limited evidence that it may enhance knowledge and behaviour. Implementation of various training courses like 'Hygienic Minimum' course to the food handlers has been studied and it was found that there was improvement in the most of examined parameters.

This chapter shares the experiences of the authors in training 280 food handlers working in various EEs in a tetiary care hospital. Initially, baseline knowledge, awareness, and practices of food handlers was done. Consent to participate in the training was taken from all the participants. Ethical clearance was obtained from institute ethical committee. To assess the need and later evaluate the effect of training "In-Depth Interviews and Focus"

group discussion were also held with the food handlers.

Based on the baseline knowledge an intervention package was devised. The package comprised of the following
• Distribution of self instructional manual in Hindi on food safety (Fig. 39.1)
• Computer based training through screening of short films on foods safety in Hindi

Training of food handlers was done both at the place of work and centrally. Their training involved distribution of a booklet in Hindi which contained the correct procedures to handle raw food items, cleaning of raw items, heating and reheating practices, etc. in a simple and easy to understand language. Posters in Hindi highlighting Do's and Don'ts for food safety were displayed at the place of work. A short comedy film in Hindi titled 'Gravy Extra' was, screened to sensitize the subjects to the issue of food safety (Fig. 39.2). Another documentary in Hindi titled 'Food Safety: Farm To Fork' was screened to teach best practices in food safety at place of work. Food safety practices were taught during interactions by the authors with the food handlers. Certificate of training on food safety were given to all participants. A logo was also devised for training of food handlers (Fig. 39.3).

We found that the baseline awareness of food handlers with respect to food safety aspects of FSSR 2011 was dismal. They lamented lack of scope for personal and professional growth. They also rued about the poor image of their profession. Verbatim responses before training were:

'Madam hamae seekna hai'

'hum jab es kaam me aatae hai koi bhi hamae nahi sikhata hai, hamare senoors ki bhi kabhi koi training nahi hui haa

'mae to night duty me aata hoo

Fig. 39.1: Booklet in Hindi given to food handlers for training

Directed by
Ravinder Kumar

Concept
Dr Amerjeet Singh

Script writing
Dr Puja Dudeja

Artists
Abhimanyu Badhwan
Anurag Shamdilya
Shivangi Vats
Soumya Singh
Dr Tejinder Kaur
Dr Harmeet Kaur
Dr Meenakshi Sharma
Dr Raman Sharma
Dr Shalini Dwiwedi
Geetu

Fig. 39.2: CD cover of movie on food safety 'Gravy Extra'

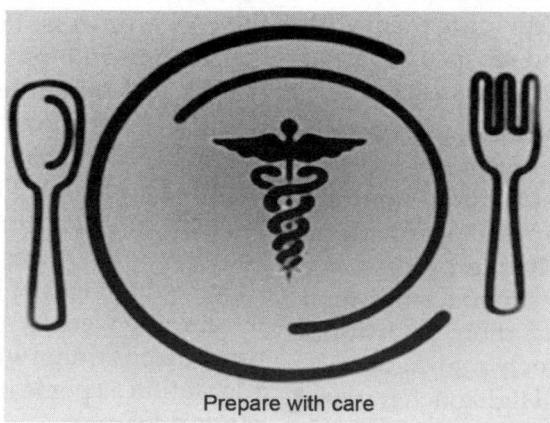

Prepare with care

Fig. 39.3: Logo for certificate for training in food safety

'please mujhe phone karkae training ke samay bula lena'

'hume to garmi me aur lagatar khade ho kar apna kaam karna padta hai'

They main reasons for dissatisfaction in their profession were poor social status, lack of recognition of the talent, high level of commitment and difficulty in getting leave, poor personal and professional growth, lack of training, poor work environment, etc. They felt that the working conditions like high temperature and humidity provoked irritation along with mental and physical fatigue. Lack of training for them made them feel that they were not of sufficient importance to deserve training courses.

There was a striking change in their morale after the training. There was visible improvement in their personal hygiene. They felt valued and felt important as reflected by their verbatim response after training which were as given below

'hamae bahut accha laga'

'hamare jeevan me bahut badlav aaya hai'

'aab hamara shareer bhi sawach hai'

'ye training aage bhi honi chaiyae'

'iss se pahale kabi kisi ne nahi poocha aur na hi seekha ya,

'hame ab lagta hai ki hum bhi important hai'.

The use of IT, a technological marvel and its associated gadgets have facilitated trainings in various fields. But often this approach has been under attack for lack of humane component during sessions. Whatever strategies we may employ to, promote High-Tech E learning, it cannot handle issues like social justice and dignity of food handlers. We have to realize the power of humane/personal touch even when imparting high tech training. The same was demonstrated in our training. The food handlers emotional quotient was awakened with high touch training which has a lot to do with the safety of food served to clients. The impact of personal touch cannot be underestimated. Government has done its role by issuing guidelines for its officials for training of food handlers. The task of imparting them in true spirit is where the researchers have a role to play.

Food handlers are the most important human resources for the burgeoning food industry of our country. Providing them with soap for hand washing and an apron to wear alone cannot ensure their dignity.

Research has demonstrated that people make behavior changes in response to those trainings that contain emotional content. Emotions drive behaviors making them central to the behavior change process. We also used this principle in training our food handlers in transforming their traditional knowledge with high-tech methods containing latest information. High touch training method embedded with relevant emotional content helped in building bond with the participants. Our training with a personal touch was successful in honest and respectful exchange of information. The emotional component in our training had the power to encourage, provoke, evoke and evolve people in action. As a result the food handlers felt liked, accepted, respected and valued. It reduced professional distance for better communication. They listened fully through the entire process. People are likely to listen consider and adopt the perspectives of others if they feel others listened and understood them. It was acknowledged that their opinions are important. Hence, data collected was truthful and real. This was so because the participants felt confident that what they share would be respected and honoured. Other advantages of high touch training over 'E' learning were face-to-face interactions, real time discussions with real time feedback, adaptation as per learner's level including the slowest learners.

Strategies for implementation of FSSR 2011 will encompass training of food handlers as an important intervention after which food safety can be enhanced and ensured. Training of managers can also be effective in reducing food safety problem by implementation of realistic food safety practices within the workplace. If managers were trained to advanced levels, they would then provide basic training for food handlers in-house.

Section VI

Government Efforts to Ensure Food Safety

40

Food Sampling and Analysis

Ishwarpreet Kaur, Amarjeet Singh

INTRODUCTION

Food includes a wide range of edible solid and liquid materials. Each one has a unique nutritional composition. These nutrients in foods support human survival and maintain health. However, being a part of nature any food item is vulnerable to biological, chemical and physical spoilage. Besides this, consumption of spoiled food may be detrimental to the health of people. The presence of toxins, contaminants, excessive and harmful additives in food items can affect consumers' health. Due to ulterior profit motive involved food manufactures/retailers indulge in many unscrupulous activities regarding food production, distribution and marketing. They also indulge in mislabelling. This has further worsened the situation. Therefore, ensuring food safety and quality is an extremely important public health concern. Stringent food safety laws are framed by most countries for this purpose. Non-conformance by FBOs to the established standards of quality of food products necessitates enforcement of food laws. These laws are essential to protect the health and safety of consumers. Evidence for violation of these standards is obtained through various tests. Thus, conviction of culprits is contingent upon positive test results showing compromised food safety standards.

In this context, the Government of India enacted **Food Safety and Standards Act, 2006. This act consolidated various prevailing laws related to food. The government also** established the FSSAI which is authorised to lay down science based standards for articles of food; to regulate manufacture, storage, distribution, sale and import of food; and to facilitate food safety. In 2011 the FSSAI devised Food Safety and Standards (Laboratory and Sample Analysis) Regulations. This new mechanism helps in identifying discrepancy in quality of food products and pinpoints where enforcement attention is required. It not only ensures compliance to the product standards but also promotes fair trade practices.

Thus, food sampling and analysis is a vital tool required for implementation of these regulations. The analysis of raw or processed samples helps achieve food safety, quality and better nutrition. Food samples can be sent for analysis by any of the stakeholder viz. Food Safety Officer (FSO)/Authority Officer, FBOs or customers. The food samples are investigated for quality control. Any consumer can approach FSO and register their complaint. Food contents are also analysed for correct product labelling. Proper labelling helps consumers to make informed choices about the food items they are buying. It also facilitates enlightenment of the consumer that they are getting the exact food content and quality as claimed by the manufacturers. However, the FSSR requires that food sample sent for testing must meet the laid down standard and safe limits. Analytical results are notified to food producers/retailers. Product quality advice is also provided.

There are few basic activities involved in sampling and analysis of food products, i.e. sampling plan, sample selection, sample preparation, analysis using appropriate methods and instruments, analysis of measurement, and data reporting.

Sampling of food products: A food analyst often has to determine the quality of a large quantity of food material, e.g. products stored in warehouse, content of trucks arriving at a factory, etc. The analyst can not test every part of the food material anywhere. Always a sample of the whole is tested.

Food sampling is thus defined as the selection of a representative portion, number of container or product units from a particular lot of the same food or consignment. It is a portion or sample that represents the whole.

The whole or total quantity from which a sample is obtained is called the population. A sample should represent a population as much as possible.

A sample is a set composed of one or several items (or a portion of matter) selected by different means in a population (or in an important quantity of matter). It is intended to provide information on a given characteristic of the studied population (or matter). The results obtained from the sample forms a basis for a decision concerning the population or the matter or the process, which has produced it.

A lot is definite as a quantity of some commodity manufactured or produced under uniform conditions while consignment is a quantity of some commodity delivered at one time. It may consist in either a portion of a lot, or a set of several lots.

Why sampling is needed: Food samples may be collected and sent for analysis as a part of routine inspection, or for legal reasons, and also at times for quality accreditation. Routine sampling is done for purposes like surveillance, collection of data for a monitoring, and to determine the conformity to product standards specified in the regulation. It is not just needed to monitor the characteristics of processed foods but also to get an idea about the raw materials and ingredients. Sometimes samples are collected to solve legal matters, as in the case of complaints from purchaser or consumer. Food companies or restaurants may also send samples for analysis to get/renew their quality clearance certificates or accreditation. Apart from this, it is a policy in food manufacturing units to have internal checks on products manufactured.

Importance/significance of sampling: The aim of food sampling programme is to test food for the presence of microbiological contamination, chemical hazards and natural toxicants. However, it is humanly impossible to check or analyse each grain or food product before giving clearance for consumption. This will not only lead to unnecessary wastage of food but also loss of other valuable resources (like time, effort and money). To overcome this problem sampling is done. It is a practical approach, where a portion of total product is selected assuming it to be of the same quality as the whole lot.

Even in our day to day life, we make decisions using this basic principle of sampling. For example, at grocery store while selecting grains we simply check a few grains from the whole lot or a bag. These handful grains serve as a sample and help us decide the overall quality of the food item. Thinking the quality of the whole to be same, the customer generalizes the result to entire lot and makes his purchasing decision.

Good sampling and its requirements: In modern era of scientific advancements only very small quantities of food sample are enough for analysis. Therefore, representative sample selection is must for achieving reliability of analytical data.

A good and adequate sampling helps to ensure that the sample quality measurements are accurate and precise estimate of the attribute of the population. Quality estimation from a fraction of population helps in saving time, money and human effort.

Samples are useful for their intended purpose only when they are collected following a good sampling technique/process. This requires the following:

- Inspection of the lot before sampling
- Use of suitable sampling devices as per the food commodity and type of sample desired.
- Use of suitable containers to hold sample
- Proper maintenance of the sample and records
- Use of adequate precautions in preserving, packing, and delivery of the sample
- Timely delivery of samples
- Provision of adequate and appropriate storage conditions for samples

Sampling plan: Samples should be collected following a particular plan. This will depend upon:

- Type of food product

- The size of food articles to be sampled (production units, cans, packages, etc.)
- The nature of the defect anticipated: bacterial contamination, chemical toxin or residue, insufficient heat exposure, etc.
- The capability of the laboratory to carry out analyses
- The degree of hazard to human health
- Acceptance and rejection criteria: Absence of pathogens, adulteration, tolerance limits, compositional standards, net contents
- Degree of confidence required so that the test result is valid

Selection of suitable sampling plan: It depends on various factors like:

1. **Purpose of the analysis:** Samples are analyzed for various purposes such as official, raw/processed materials in manufacturing unit, research and development, etc.

   a. *Official samples.* These are selected for official or legal requirements by government laboratories. This helps in ensuring that manufacturers are supplying safe foods that meet legal and labeling requirements. The analysis is done on the basis of officially sanctioned sampling plan and analytical protocol.

   b. *Raw materials.* Food manufacturing units/factories analyze raw materials for appropriate quality before Processing.

   c. *Process control samples.* For quality assurance food is often analyzed during processing. It helps to ensure that the process is operating in an efficient manner. Thus, if a problem arises during processing it can be quickly adjusted so that the properties of the food are not affected.

   d. *Finished products.* To achieve high and consistent quality, finished foods are tested. Tests ensure that the food is safe, meets legal and labeling requirements, and is of a high and consistent quality.

e. *Research and development.* Samples are analyzed by food scientists involved in fundamental research or in product development.

2. **Nature of measured property:** It is important to establish which all properties will be measured, e.g. color, weight, presence of extraneous matter, fat content or microbial count.

The properties of foods are classified as either attributes or variables. An attribute is something that a product either does or does not have, e.g. it is or is not spoilt. A variable on other hand is some property that can be measured on a continuous scale, such as fat or moisture content of a material. Variable sampling usually requires less samples than attribute sampling.

3. **Nature of population:** While deciding the sampling plan it is important to define the nature of population. Population can be defined in various ways:

a. Finite or infinite. A finite population is one that has a definite size, e.g. a truckload of potatoes, a tanker full of milk. An infinite population is one that has no definite size, e.g. a conveyor belt that operates continuously, from which foods are selected periodically. Analysis of a finite population provides information about the properties of the population, whereas infinite population analysis provides information about the properties of the process.

b. Continuous or compartmentalized. A continuous population is one in which there is no physical separation between the different parts of the sample, e.g. liquid milk or oil stored in a tanker. A compartmentalized population is one

that is split into a number of separate sub-units, e.g. boxes of apples in a truck, or bottles of juices moving along a conveyor belt.

c. Homogenous or heterogeneous. In a homogeneous population the properties of individual sample are same at every location within the material (e.g. a tanker of well stirred liquid oil), whereas a heterogeneous population is one in which the properties of the individual samples vary with location (e.g. a truck full of onions, some of which are bad). If the population is homogeneous then there is no problem in selecting a sampling plan because every sample would be representative of the whole. In practice, most populations are heterogeneous and so we must carefully select samples from different locations within the population to obtain a representative sample.

Who all can collect a food sample? Sampling is done at various levels. For legal or official requirement, FSO may collect samples from any place where any article of food is manufactured, stored, or sold. It can be from factory premises to retail outlets in markets. As far as work load and routine surveillance is concerned it is the duty of FSO to send at least 36 food samples/month for analysis from its designated area. However, an authorized officer is responsible for collecting samples of imported food products. FSSAI also tests the samples of food products imported into India before their consignments are cleared at the ports.

Even a food manufacturing unit/factory or eating establishments can send their samples for analysis as a routine quality control measure.

Any purchaser or consumer of food product can also send the sample for testing. The consumer can complain to the Food Safety

Officer/Designated Officer of the area or Food Safety Commissioner of the State about the food bought from a shop or restaurant.

The three situations are described in subsequent/succeeding text as per the purpose of sampling/testing: Food sampling and testing procedure by FSO/authorized officer

Food Safety Officer (FSO): It has the power and duty to inspect all food establishments licensed for manufacturing, handling, packing or selling of an article of food within the area assigned to him. He can procure and send food samples for analysis for the purposes of surveillance, survey and research. He may also collect samples to investigate any complaint which may be made to him.

Procedure for Taking Sample and Manner of Sending it for Analysis

Sample Collection: Proper sampling requires adequate sample size, suitable containers, use of appropriate preservatives and transportation to prevent any spoilage before analysis. It also involves proper sealing/packing, identification and dispatch. For objectivity and to remove bias during sampling it is important to take 'random sample'. This will ensure that the representative sample is taken from a number of locations within the population.

However, even selective or biased sampling has to be done when random sampling is unnecessary or even undesirable. As in case of visibly contaminated, adulterated, or defective food products (e.g., swollen/puffed cans, or flour containing live insects), the FSO/Authorized Officer, after examination of the lot, should select the units which will most clearly demonstrate the violation.

In case the sample is drawn from an open container, a similar sample must also be drawn from a container in original condition (if such container is available) of the same

article bearing the same declaration and inform the same to the Food Analyst.

Sampling should be done as per Section 47 defined in FSSA Act.

The generic formula used for packed foods is:
1. For number of packages <100, minimum of 10 samples has to be sampled.
2. For number of packages >100 in a batch, square root of the total number of samples is needed.

Where no specific instructions are given, the general rule is to collect samples from the square root, plus one, of the number of units in the lot.

After deciding about the sample size, the number of packages is selected from bulk consignment in a random manner so that each package in a lot has an equal chance of being selected.

For better understanding of sampling an example of cereals, pulses and milled products is given as follows.

Method of Taking Samples

a. Sampling from bags (for cereals, pulses, etc). Increments shall be taken from different part of a bag (e.g. top, middle, bottom) by means of a sack/bag spear from the number of bags specified in Table 40.1.

Table 40.1: Number of bags to be sampled	
Number of bags in consignment	*Number of bags to be sampled*
Up to 10	Each bag
10 to 100	10, taken at random
More than 100	Square root (approx.) of total number, taken according to a suitable sampling scale.

When using mechanical sampler, increments shall be taken from a minimum of three different sampling points.

b. Sampling from silos, bins or warehouses (for cereals, pulses, etc.) Increments shall be taken throughout the whole depth of the lot. Suitable instruments must be used to achieve this requirement. The grain should be sampled using a grid system, for example similar to that used for rail/road wagons, barges or ships. The number of increments to be taken shall be determined as follows. Take the square root of the tonnage in the static bulk. Divide by two and round up to the next whole number. This is the minimum number of increments that is to be obtained.

Sampling procedure (Fig. 40.1): The laboratory results depend on the condition in which the samples are received. The analysed results will be meaningless if the sample is not representative of the lot, or is mishandled during collection. For ensuring uniformity, established sampling procedures must be followed. The

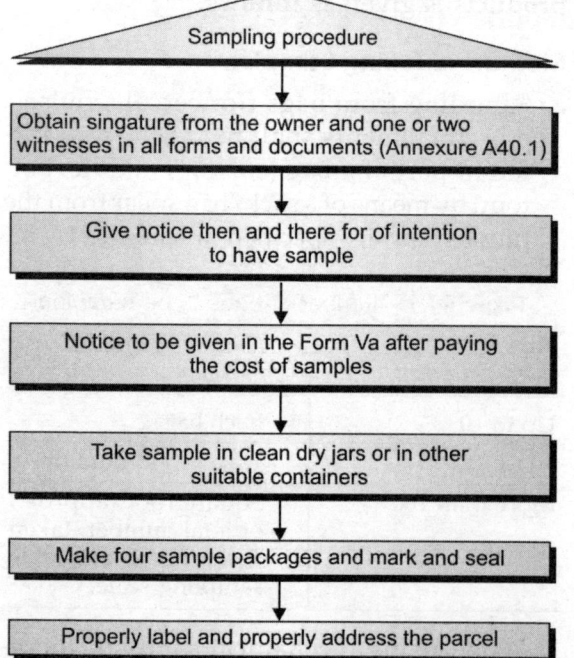

Fig. 40.1: Flowchart depicting the sampling procedure

proper statistical sampling procedure, according to whether the food is solid, semisolid, viscous, or liquid, must be determined by the FSO at the time of sampling.

The FSO must follow the specified procedure under clause A of Sub Section 1 of Section 38 and Section 47 (Except 47 (5)) of the Act while taking sample of food for analysis.

FSO must ensure that the sample lifting is done in the presence of one or more witnesses. All the forms and documents should bear the signatures of witnesses. The FBO is served the notice in Form VA. In case the FBO discloses that the product has been obtained from the manufacturer, distributor or supplier, a notice is also given to them.

Methods to store samples: All kinds of sample collected for laboratory analysis must be placed in suitable containers for storage and handling. Samples should be collected in hermetically sealed and sterile containers. Samples should be stored in containers that can protect its content from degradation by environmental factors. The containers used should be sterile and dry. Appropriate waterproof and leak proof material preferably made of stainless metal or plastic must be used. Container must have a secure closure/seal. The containers must not influence the odour, flavour, pH, or composition of the sampled products. All containers must have air-tight closures. Suitable plastic bags may also be used.

Airtight containers can protect changes in the moisture content of samples. Light sensitive samples must be stored in opaque glass or the container should be wrapped in aluminium foil. Oxygen sensitive samples should be stored under nitrogen or inert gas. Chemically unstable samples should be stored in cool places. However, unstable emulsion should not be frozen. Preservatives like mercuric chloride, potassium dichromate and chloroform can be used to stabilize certain foods.

If a sealed package marketed by the manufacturer/FBO is taken as sample, further sealing in separate containers will not be required.

Maintaining sample integrity: The FSO/ authorized officer is responsible for collecting, holding, sealing, storing and delivering the sample. The food samples must reach the laboratory in the same condition as that at the time of sampling. Food control authorities should use special tamper-proof containers. These must be sealed and bear a stamped with the FSO/authorized officer's designated identification number.

The following information should be collected while collecting the sample: name of the food, lot number, container size, product code numbers, labelling information, and condition of the lot-broken packages, evidence of rodent or insect infestation, debris, etc.

Packaging and sealing of samples: Sample packages should be secured or sealed to prove their authenticity. The stopper of containers shall be securely fastened to prevent leakage, evaporation or to avoid entrance of moisture. The person taking the sample shall divide the sample in four parts or take four already sealed packages. Each container is then completely wrapped in fairly strong thick paper.

The ends of the paper shall be neatly folded in and affixed by means of an adhesive.

A paper slip of the size that goes round completely from the bottom to top of the container must be pasted on the wrapper. This slip should bear the signature of the designated officer and code number of the sample.

The person from whom the sample has been taken shall affix signature or thumb impression in a manner that both the paper slip and the wrapper carries a part of his signature or the thumb impression. If he refuses to affix his signature or thumb impression, the signature or thumb impression of one or more witnesses can be taken in the same manner. However, it is important to note that where the purchaser or an authorized officer draws the sample no such paper slip shall be required to be affixed.

The paper cover must be further secured by means of strong twine or thread both above and across the container. Then the twine or thread is fastened on by means of sealing wax which must have clear impression of the seal of the sender. The seal must be at the top of the packet, one at the bottom and the other two on the body of the packet.

The outer covering of the packet shall also be marked with the code number of the sample. Labelling should be done with water insoluble ink. To prevent tempering official or legal samples should be sealed.

Dispatch of the sample: All bottles or jars or other containers containing the samples for analysis shall be sealed, labelled and the parcel shall be properly addressed before despatch. The sample must be prepared, handled and dispatched to prevent breakage or spoilage, and to ensure that the sample examined by the laboratory is the same as that collected and documented by the FSO/authorized officer. See Fig. 40.2 for the content details to be mentioned on label.

Content	Check list
Code number of the sample	√
Name of the sender with his official designation	√
Date and place of collection Nature of articles being sent for analysis	√
Nature and quantity of preservative, if any, added to the sample	√

Fig. 40.2: The label on any sample of food sent for analysis

If in case the sample is collected from Agmark sealed container, the label shall bear the following additional information (i) Grade (ii) Agmark label no./Batch no. (iii) Name of packing station.

The marked and sealed samples are then addressed and dispatched under appropriate conditions to retain the integrity of sample. One out of four sample parts is sent to the food analyst along with memorandum in Form VI. Two parts to the sealed sample along with and two copies of memorandum in Form VI is to be sent to the designated officer (referral labs). And the fourth, if FBO request is send to accredited lab. If the FBO does not request to send the sample to an accredited lab then the fourth part also shall be deposited with designated officer (Figs 40.3a and b).

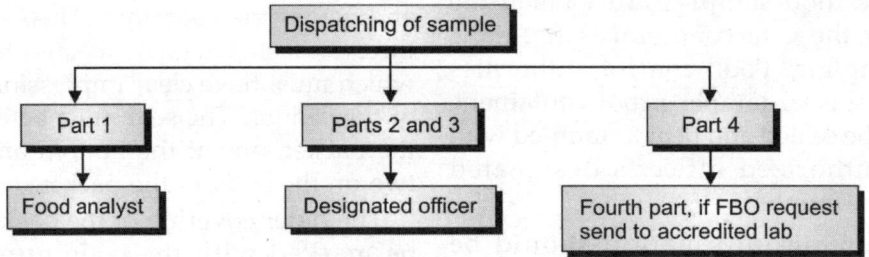

Fig. 40.3a: Dispatching of food samples

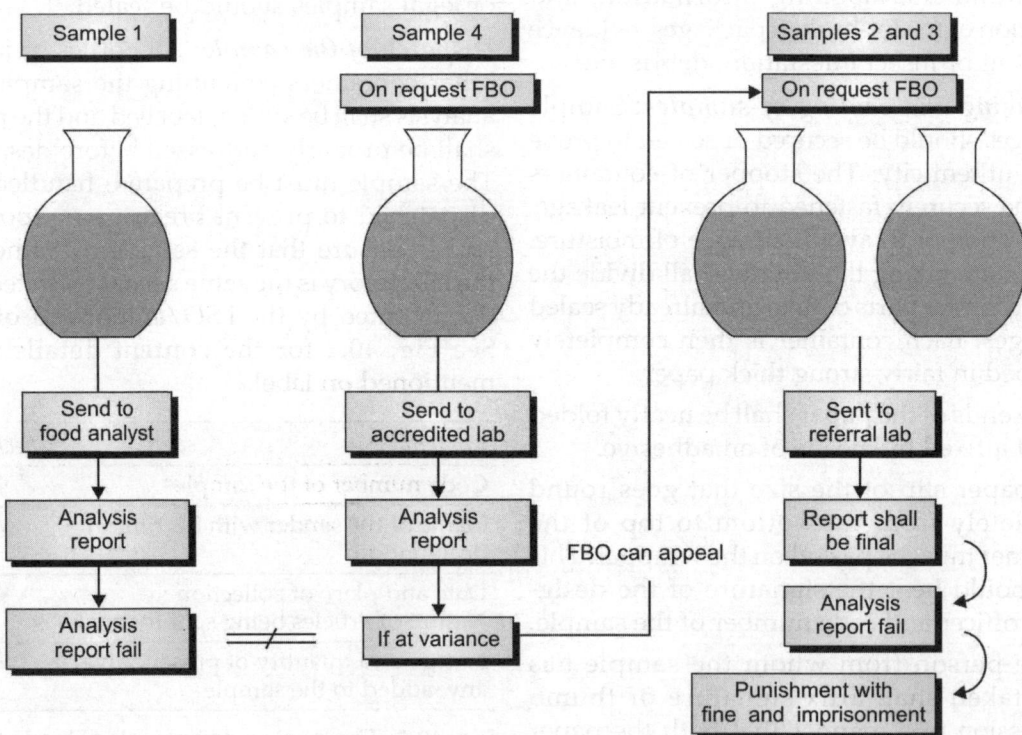

Fig. 40.3b: Process for dispatching of samples

All samples must be secured in shock-absorbing package. Frozen food samples are sent overnight in insulated cartons.

For lifting a sample for testing micro-biological parameters, the method of lifting sample, type of container, temperature to be maintained, method of transportation and any other condition to maintain the integrity of the sample shall be notified by the food authority from time to time.

The FSO or the authorized officer, while taking sample for analysis except in the case for microbiological testing, may add to the sample, a preservative as prescribed in the regulations for the purpose of maintaining it in a condition suitable for analysis.

The nature and quantity of the preservative added shall be clearly noted on the label to be affixed to the container.

The quantity of sample of food to be sent to the food analyst/referral lab for analysis shall be as specified in regulations by the food authority.

Foods sold in packaged condition (sealed container or package) shall be sent for analysis in its original condition without opening the package as far as practicable, to constitute approximate quantity along with original label. In case of bulk packages, wherever preservatives are to be added, the sample shall be taken after opening sealed container in the presence of the FBO or one or more witnesses. The contents of the original label must also be sent along with the sample.

Sampling cost: When FSO is getting food analysed the concerned food authority will bear its cost. FSO will pay the cost of sample to the person from whom the sample is taken, calculated at the rate at which the article is sold to the public.

Sampling tools: All sampling equipment must be clean and dry when used to collect a sample. The tools used by FSO/authorized officer range from pliers, spoon, screwdriver and knife. These are useful for opening containers, cutting bags of food products.

In case of a homogeneous liquid sample it is important to shake the liquid prior to collecting the sample from containers. Aeration ensures homogenous sampling in case of large volume of liquids. The liquids are sampled using pipettes, pumping or dippers. The dippers should be of smooth metal construction, preferably of stainless steel, to facilitate sterilization. Figure 40.4 depicts sampling dipper.

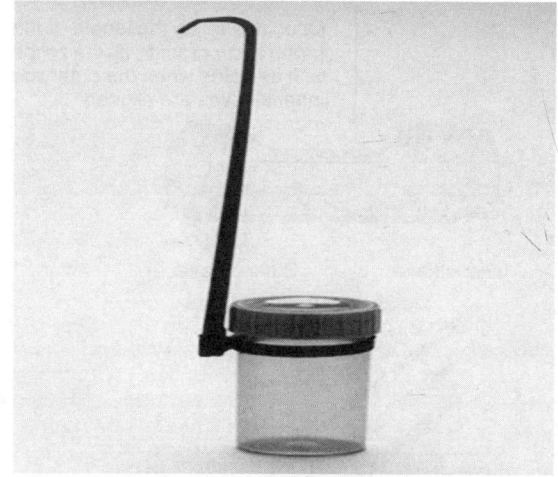

Fig. 40.4: Sampling dipper

However for solids, like grains, probes or triers are inserted in population at several locations to collect the sample. A dry borer tube may be used for flour, dried milk and dried milk products. The product will usually have to be pushed out of the trier with a spatula. A special probe is needed for sampling railway wagons or lorry loads of dried grains such as wheat and maize. A conical-shaped metal probe often referred to as a bag thief, is used for sampling bags of grain, coffee beans and spice. The probes are stored in leather or cloth sheaths when not in use. Special probes or triers are used for butter and cheese. Figure 40.5 shows images of probes and triers.

Rubber or latex surgical gloves may also be used to permit handling without adding bacteria to the sample. Isopropyl alcohol or ethyl alcohol should always be carried by the FSO/authorized officer in a plastic bottle, for disinfecting sampling tools.

Below given is the example of *instruments used for product sampling (cereals, pulses, etc) are:*

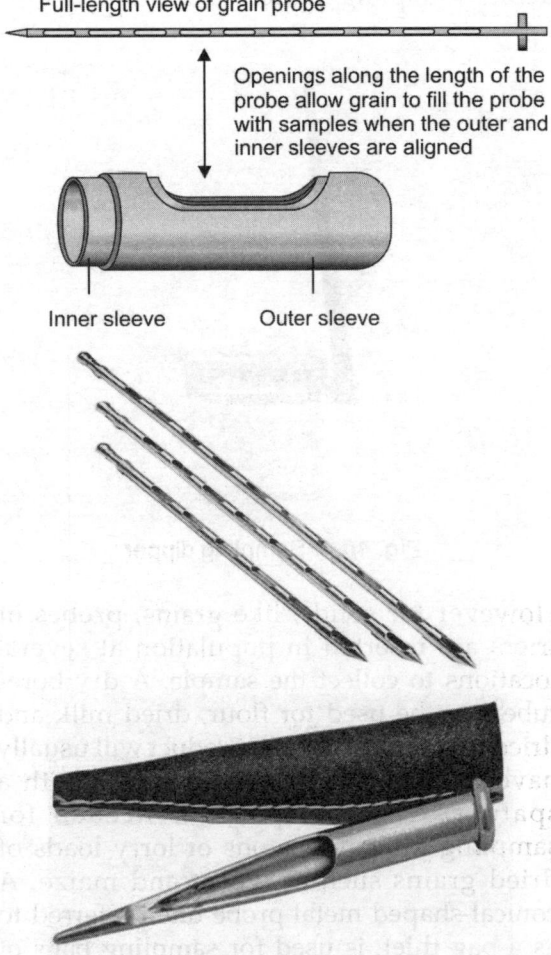

Full-length view of grain probe

Openings along the length of the probe allow grain to fill the probe with samples when the outer and inner sleeves are aligned

Inner sleeve   Outer sleeve

Fig. 40.5: Probes and triers

1. *Sampling from bulk:* Use appropriate apparatus for obtaining increments from static bulk (example, hand-held spears, mechanical or air-assisted apparatus).
2. *Sampling from bags*: Use sack type spears.
3. *Mixing and dividing*: Use shovels and dividing apparatus or automatic random dividing apparatus.

II. Food Sampling and Testing Procedure by Customer/purchaser in case of Complains

The purchaser must state and give the reason or intention to have food analysed in form V of FSS Rules, 2011, to the person from whom he has purchased. If the purchaser desires to have the food article purchased by him to be analysed by the Food Analyst, he shall give a notice in writing, then and there, his intention to have it so analysed to the person from whom he has purchased the food article.

The purchaser pays the fee for the sample analysis to food analyst. The report of the sample is to be sent by food analyst in 14 days of receipt of the sample to the purchaser. If the sample does not conform to the standards prescribed (i.e. adulterated/misbranded/contaminated) under the Act or the Regulation a copy of report is also sent to the Designated Officer of the area in which the article of food was purchased. If the sample fails the cost is to be refunded to him from the Designated Officer.

III. Food Testing Procedure by FBO

Food business operator (FBO) can also get the samples tested in same way as purchaser. The aproximate quantities of some food samples has been prescribed by FSSR, 2011.

FBO right to have the food analysed: In case FBO desires to have the fourth part of the sample analysed, he shall request the FSO to send the sample to any accredited laboratory for analysis under intimation to the designated officer.

The cost of testing by the accredited lab will be borne by the FBO. The payment shall be made through Bank draft or online transfer or treasury chalan or any other suitable means as specified by the designated officer.

The food analyst in-charge of the accredited laboratory shall analyse the sample within fourteen days from the date of the receipt of the sample. If in case the sample cannot be analysed within fourteen days from the date of its receipt, the food analyst of the accredited laboratory, shall inform the designated officer and the Commissioner of Food Safety giving reasons and specify the time to be taken for analysis. The food analyst shall send four copies of the analysis report to the designated officer indicating the method of analysis.

Appeal to the Designated Officer

On receipt of analysis report from the Food analyst, that the sample of food sent for analysis has failed the FBO can appeal before the designated officer against the findings of the food analyst.

Such appeal shall be in Form VIII and must be filed within 30 days from the date of the receipt of the copy of the analysis report from the designated officer. The appellants in the appeal, may, require the designated officer to send to the referral food laboratory one part of the sample which is with him for analysis and the report of the referral laboratory shall be final and binding on the appellants.

The designated officer shall fix a date of hearing of the appeal after giving notice of such hearing to the Appellants. If on a consideration of materials placed before him, the designated officer is of the opinion that the matter be referred to the referral laboratory for opinion, he shall forward one part of the sample with him to the referral laboratory and the report of the referral laboratory shall be final and binding.

Forms Detail

A list of various forms alone with their purpose is given in Table 40.2.

Table 40.2: Forms detail*	
Form	*Purpose*
Form II	Receipt for article of food seized by a food safety officer—Seizure memo
Form III	Form of order/bond not to dispose of the stock (*to keep any article of food in safe custody of the vendor*)
Form IV	*Vendor to execute bond*
Form V	The notice to be given by the food safety officer or an authorized officer or the purchaser to the person from whom he has taken the sample and to the person
Form VI	Memorandum to food analyst (manner of dispatching containers of samples)
Form A	Certificate of analysis by the referral food laboratory
Form VII A	Report of the food analyst
For VIII	Form of appeal before designated officer

* A copy of each form is given in Annexure A40.2

Regulation in India for Laboratory and Sample Analysis

Food testing laboratories' competence is of paramount importance to the health and safety of the public. To cater to this aspect the FSSR for laboratory and sample analysis came into force on 5th August, 2011. It is dedicated to safeguarding food and thereby enhancing the health and well-being of people. This regulation provides details on laboratories for imports; notified and referral laboratories, their function, area of jurisdiction and quality of sample sent for analysis.

Under the provision of Rule no. 13 (FSSA), the approximate quantity of different food samples to be sent for analysis is specified. Foods sold in packaged condition (sealed container/package) shall be sent for analysis in its original condition without opening the package. In case of the unspecified products, the quantity drawn shall be determined in consultation with the food analyst.

In India, the FSSA (2006) and FSRR (2011) have prescribed 140 laboratories for food testing or analysis. Sixty eight are NABL accredited private laboratories. These have been recognized and authorised for testing by FSSAI. State Governments have set up 72 food testing labs. And there are 4 referral food laboratories under the act, these works for the analysis of appeal samples collected by food inspectors of states and local bodies and imported samples (Table 40.3). The details of laboratories can be obtained from www fssai. gov.in

After test or analysis, the certificate thereof shall be supplied forthwith to the sender in Form B. The fees payable in respect of such a certificate shall be (₹ 1000) per sample of food analysed as prescribed by the food authority. Certificates issued under these regulations by the laboratory shall be signed by the director.

The test report are duly signed by the Director of referral laboratory or food analyst in standard format, i.e., Form B, certificate of analysis by referral food lab; and Form A, format of report of food analysis is attached with these regulations.

The test report duly sealed in confidential cover is sent to the officer/Food Business operator/consumer, who has sent the sample and requested the testing. The test report shall clearly indicate who has drawn the sample and the reference method.

The laboratory shall issue the test reports immediately after completion of the tests and not later than a maximum period of 14 days

Table 40.3: Referral food laboratories in India

Name of the referral Laboratories local Areas/State/UTs	Name of the referral laboratories local areas/State/UTs
Referral food Laboratory, Kolkata	Arunachal Pradesh, Assam, Chhattisgarh, Manipur, Meghalaya, Mizoram, Nagaland, Orissa, Sikkim, Tripura, Uttarakhand and Union Territories of Andaman and Nicobar Island and Lakshadweep
Referral Food Laboratory, Mysore	Gujarat, Haryana, Himachal Pradesh, Maharashtra, Uttar Pradesh and Union Territory of Chandigarh
Referral food laboratory,	Andhra Pradesh, Delhi, Jammu and Kashmir,
Pune	Karnataka, Kerala, Rajasthan and Tamil Nadu
Referral food laboratory, Ghaziabad	Bihar, Goa, Jharkhand, Madhya Pradesh, West Bengal, UnionTerritories of Dadra and Nagar Haveli, Daman and Diu and Puducherry.

Note:

"Notified laboratory" means any of the laboratories notified by the Food Authority under sub-sections (1) and (2) of section 43 of the Act. "Referral laboratory" means any of the laboratories established and/or recognized by the food authority by notification under sub-section (2) of Section 43 of the Act.

from the date of receipt of sample in case of routine samples and 5 days in case of imported food sample.

An appeal against the report of food analyst shall lie before the designated officer who shall, if he so decides, refer the matter to the referral food laboratory as notified by the food authority for opinion.

The laboratory shall keep the remaining sample after complete required analysis for a minimum period of one month in the desired storage conditions as required by the test procedure. Exceptions can be made for perishable food items. Discard time period of perishable samples can be mentioned in the test report to avoid claims.

The laboratory shall maintain the record of observations and a copy of the test report for a minimum period of three years.

Food Analysis

Food analysis is the discipline dealing with the development, application and study of analytical procedures for characterizing the properties of foods and their constituents.

According to the requirement for regulatory or monitoring purpose the food analysis is categorised into: Chemical analysis, microbial analysis, physical analysis, and sensory analysis.

Importance of food analysis: The reason for analyzing food is to ensure that they are safe. We analyse food to meet statutory and voluntary obligations. It is also done as a part of quality assurance. Scientifically the overall quality refers to technological, physical, chemical, microbiological, nutritional and sensory parameters to achieve the wholesome food. Food analysis is required for checking shelf-life or authenticity, and assuring legal compliance. Analysis of food and drink is very important part of product development.

Need for food analysis: Food Analysis serves as a unique and invaluable tool for all food scientists, technologists and regulatory authorities for quality assurance and control of food products, to study the different aspects of food products.

Food is a complex matrix consisting of different components. These components can be categorized into different categories which are listed as given below:

1. Nutrients: For example, proteins, amino acids, total cholesterol, trans fats and lipid profile, carbohydrates, dietary fiber, vitamins, minerals, etc. Depending upon the food product some of them may be present at high concentration levels while others may be present at low concentration levels of parts per million.

2. Additives: For example, colours, stabilizers, antioxidants, flavours, preservatives, etc. The additives are added to the food products for the purpose of giving the food products desired appearance, texture, flavour and extending the shelf-life. The additives are usually present at very low concentration levels.

3. Adulterants: They are added intentionally to the food products mostly for the purpose of cost benefits and they may be present at higher as well as lower amounts. They may be safe or sometimes highly toxic, such as, argemone in mustard oil, animal cholesterol in ghee, low cost vegetable oil in high cost vegetable oil, etc.

4. Contaminants and toxicants: Toxicants can be classified into:

   a. Physical toxicants, e.g. glass, wood, metal, insect matter, etc.

   b. Biological toxicants, e.g. microbes and pathogens

   c. Chemical toxicants, e.g. residual pesticides, residual antibiotics, mycotoxins and environmental pollutants like PAH (polycyclic aromatic hydrocarbons), PCB (polychlorinated biphenyls), Dioxins, toxic metals, etc.

Most of the times these contaminants are not added intentionally, but find their way into the food products from environmental pollution or if proper practices are not being followed during agriculture, animal breeding, storage or processing. The various toxicants are present at low levels of concentration and if present beyond a certain prescribed level of concentration in food products may prove to be highly toxic or carcinogenic to humans.

Tools for food analysis: Food is considered to be highly unsafe because it is easily contaminated and adulterated by toxic chemicals (pesticides), harmful micro-organisms (Salmonella) and extraneous matter (metal, wood, glass, insect and other matter). The analytical techniques must be highly sensitive to detect low levels of harmful material, microbes and various chemical substances. It is important to use the correct analytical tool in order to get meaningful answers to questions.

Functions of Food Analyst

On receipt of the package containing a sample of food for analysis, the food analyst or an officer authorized by him shall compare the seals on the container and the outer cover with specimen impression of seal received separately. A statement/record to this effect shall be made on receipt of sample and in the test report by the concerned laboratory.

If the sample container received by the food analyst is found to be in broken condition or unfit for analysis, he shall, within a period of seven days from the date of receipt of such sample, inform the designated officer/food business operator/consumer about the same

and request him to send the second part of the sample for analysis.

On receipt of the sample, the food analyst shall analyse the sample and send the analysis report mentioning the method of analysis. The analysis report shall be signed by the food analyst and such report shall be sent within fourteen days of the receipt of the sample by the food analyst. In case the sample cannot be analysed within fourteen days of its receipt, the food analyst shall inform the designated officer and the commissioner of Food Safety/ food business operator/consumer giving reasons and specifying the time to be taken for analysis.

The laboratory is liable to maintain confidentiality of samples and information thereof.

After completion of analysis of article of food, the food analyst shall send his report to the designated officer, the purchaser of article of food, as the case may be, in Form VII A.

National Accreditation Board for Testing and Calibration Laboratories (NABL):

It provides laboratory accreditation through third party assessment for formally recognizing the technical competence of labs. The accreditation services are provided in accordance with ISO.

Referral Food Laboratory: The laboratory having competence to carry out the analysis as per "The Food Safety and Standards (Food Products Standards and Food Additives) Regulations, 2011" and "Food Safety and Standards (Contaminants, Toxins and Residues) Regulations, 2011" i.e Level 1 Food laboratory and Level 2 Food laboratory.

Level 1 Food laboratory: The laboratory which is competent to carry out the complete analysis as per "The Food Safety and Standards (Food Products Standards and Food Additives, Parts-I and II) Regulations, 2011" for 18 categories of food covered in food code. The level 1 laboratory will carry out the following analysis—Physical analysis; Chemical analysis; Microbiological analysis; Rheological analysis; Functional testing; Basic nutrient analysis such as fat, protein, calorific value; Sensory analysis.

Level 2 Food laboratory: The laboratory which is competent to carry out the complete analysis as per Food Safety and Standard (Food Products Standards and Food Additives) Regulation (FSSR) 2011" and (Contaminants, Toxins and Residues) 2011" for 18 categories of food covered in food code.

The level 2 laboratory will carry out the analysis covered in Level 1 Food Laboratory as well as the following analysis—Contaminants (chemical, microbiological); Toxic substances; Pesticides residues; Antibiotics and pharmacologically active substances; Irradiation of food; Detailed nutrient analysis; Molecular analysis (genetically modified food)

Equipment: The applicant laboratory for recognition as Level 1 food laboratory, Level 2 food laboratory and referral laboratories should have all the equipment required for testing under their scope of recognition.

Advance Lab Equipment in Food Analysis

Table 40.4 highlights the advance equipment used in food analysis along with their applications.

Table 40.4: Advance lab equipment and application in food analysis

Equipment	Application
Gas chromatography	It is used for separating, analysis and determination of different compounds in food products that can be vaporized without decomposition. Such as: • Cholesterol, fatty acid profiling and trans fat analysis Analysis of residual pesticides and environmental contaminants • Antioxidants and preservatives like TBHQ, benzoic acid, sorbic acid, acetic acid, etc. • Characterization of flavours and fragrances • The GC profiling of the essential volatile oils gives a reasonable fingerprint which can be used to characterize the identity of the particular oil.
High Performance liquid chromatography	It is used for profiling and analyzing of various components in food products such as: • Amino acids profiling, peptides and proteins • Carbohydrates and carbohydrate profiling, sweeteners • Lipids and alcohols • Fat soluble and water soluble vitamins, carotenoids • Organic acids and organic bases

Contd.

Table 40.4: Advance lab equipment and application in food analysis (Contd.)

Equipment	Application
	• Residues of mycotoxin, antimicrobial and veterinary drugs, pesticides, etc.
	• Pigments, colorants and phenolic compounds • Bittering substances • Additives, preservatives, antioxidants and stabilizers in processed food products
Mass spectrometry	It is an analytical technique used for: • Identifying unknown compounds by the mass and mass fragmentation pattern • Used for elucidating the chemical structures of molecules, such as peptides and other chemical compounds. • Quantifying the amount of a compound
Gas chromatography-mass spectrometry (GC-MS)	This method combines the features of gas-liquid chromatography and mass spectrometry to identify different substances within a test sample. It is used for: • Pesticide residue analysis in all raw and processed food products • Analysis of environmental contaminants such as polychlorinated biphenyls, polyaromatic hydrocarbons, dioxins, etc. in food products • Flavours and fragrance in food products • Fatty acid profiling in oils and fats • Volatiles and other residual solvents in food packaging material
Liquid chromatography/ Mass spectroscopy	• Routinely used for detection of mycotoxins; toxins produced by different fungi, e.g. *Aspergillus* spp., *Fusarium* sp., *Penicillium* spp. etc. • Some of the mycotoxin regularly analysed in food samples include Aflatoxin B1, B2, G1,G2, ochratoxin A, etc. • Residual drugs and antibiotics in different food products • Residual pesticides in raw and processed food products • Banned dyes and colourants, e.g. Sudan dyes in different food products.
Atomic absorption spectroscopy	• Used for assessing the concentration of metals and minerals that may be present in food products such as Fe, Pb, As, Cd, Zn, etc.
Reverse transcription polymerase chain reaction (RT-PCR)	• Used for the detection and quantitation of VT 1 and VT 2 toxin genes in *E. coli* O157:H7 • Used in the GMO quantification

Conclusion

Knowledge regarding fair and valid food sampling procedure is very important for procuring accurate samples for testing. These samples are further used for measuring the compliance with set food standards. Food sampling and analysis is indeed a great tool in laying down, regulating and enforcing the food quality standards. Regular monitoring of food products can curb and control violation of food safety and standards. Not just the FSO, but the purchaser and FBO play a joint responsibility in assuring food safety and hygiene. A little vigilance and initiative by purchaser; diligence and responsible working by enforcement officers; and an honest commitment from FBO can ensure safe and quality food products, and thereby welfare of the public.

Annexure A40.1: Quantity of food samples to be collected for analysis

Article of food	Approximate quantity to be supplied
1. Milk	500 ml
2. Sterilized Milk/UHT Milk	250 ml
3. Malai/dahi	200 gms
4. Yoghurt/sweetened dahi	300 gms
5. Chhana/paneer/khoya/shrikhand	250 gms
6. Cheese/cheese spread	200 gms
7. Evaporated milk/condensed milk	200 gms
8. Ice-cream/softy/kulfi/ice candy/Ice lolly	300 gms
9. Milk powder/skimmed milk powder	250 gms
10. Infant food/weaning food	500 gms
11. Malt food/malted milk food	300 gms
12. Butter/oil/ghee/margarine/cream/bakery shortening	200 gms
13. Vanaspati, edible oils/fats	250 gms
14. Carbonated water	600 ml
15. Baking powder	100 gms
16. Arrow root/sago	250 gms
17. Corn flakes/macaroni products/corn flour/custard powder	200 gms
18. Spices, condiments and mixed masala (whole)	200 gms
19. Spices, condiments and mixed masala (powder)	250 gms
20. Nutmeg/mace	150 gms
21. Asafoetida	100 gms
22. Compounded asafoetida	150 gms
23. Saffron	20 gm
24. Gur/jaggery, icing sugar, honey, synthetic syrup, bura	250 gms
25. Cane sugar/cube sugar/refined sugar/dextrose, misri/ Dried glucose syrup.	200 gms
26. Artificial sweetener	100 gms
27. Fruit juice/fruit drink/fruit Squash	400 ml

Contd.

Annexure A40.1: Quantity of food samples to be collected for analysis (Contd.)

Article of food	Approximate quantity to be supplied
28. Tomato sauce/ketch up/tomato paste, jam/jelly/marmalade/ tomato puree/vegetable sauce	300 gms
29. Non-fruit jellies	200 gm
30. Pickles and chutneys	250 gm
31. Oilseeds/nuts/dry fruits	250 gm
32. Tea/roasted coffee/roasted chicory	200 gm
33. Instant tea/instant coffee/instant coffee chichory mixture	100 gm
34. Sugar confectionery/chewing gum/bubble gum	200 gm
35. Chocolates	200 gm
36. Edible salt	200 gm
37. Iodised salt/iron fortified salt	200 gm
38. Food grains and pulses (whole and split)	500 gm
39. Atta/maida/suji/besan/other milled product/paushtik and fortified atta/maida	500 gm
40. Biscuits and rusks	200 gm
41. Bread/cakes/pastries	250 gm
42. Gelatin	150 gm
43. Catechu	150gm
44. Vinegar/synthetic vinegar	300 gm
45. Food colour	25 gm
46. Food colour preparation (solid/liquid)	25 gm solid/100 ml liquid
47. Natural mineral water/packaged drinking water.	4000 ml in three minimum original sealed packs
48. Silver leafs	1 gm
49. Prepared food	500 gm
50. Proprietary food, (non-standardised foods)	300 gm
51. Canned foods	6 sealed cans
52. Food not specified	300 gm

Annexure A40.2

FORM II
Seizure Memo
(Refer Rule 2.3.1)

In exercise of the power delegated to me under Section 38 of the FSS Act, I hereby seize/detain the under mentioned food products/documents which contravene the provision of section _____ of this act at the premises of M/S _____

_____

_____

Sl. no.	Name of the products	Batch No.	No of units	Qty in kg
1.				
2.				
3.				
4.				
5.				

The detention/seizure has been made and the inventory has been prepared in presence of the following witnesses.

Name and address of the witness signature.

1.

2.

The products detained/seized have been duly sealed and are left in the custody of Shri.——————— ——————————————————— with the instruction not to tamper with the seals and not to dispose of the products till further order.

Signature of manufacturer/dealer signature of food safety officer

Name: _____

Place: _____

Date: _____

FORM III

(Refer Rule 3.3.2.(1))

(to keep any article of food in safe custody of the vendor)

To

(Name and address of the vendor)

_____

_____

_____

Whereas *.................................intended for food which is in your possession appears to me to be adulterated/misbranded: Now therefore under clause (c) of sub-section (1) of section 38 of the Food Safety and Standards Act, 2006 (34 of 2006), I hereby direct you to keep in your safe custody the said sealed stock subject to such orders as may be issued subsequently in relation thereto.

Food safety officer

Area: _____

Place: _____

Date: _____

FORM IV

(Refer Rule 3.3.2.(2))

(Vendor to execute bond)

SURETY BOND

Know all men by these present that we (i) _____ son of _____ resident of _____ and (ii) _____ son of _____ resident of _____ proprietors/partners of Messrs _____ hereinafter called the Vendor(s) and (iii) _____ son of _____ resident of _____ and (iv) _____ son of _____ resident of _____ hereinafter called the surety/sureties are held and firmly borne up to the President of India/Governor of _____ hereinafter called the government in the sum of _____ rupees to be paid to the government, for which payment will and truly to be made. We firmly bind ourselves jointly and severally by these presents.

Signed this _____ day of _____ whereas Shri _____ Food Safety Officer has seized _____ here, insert the description of materials together with number/quantity and total price hereinafter referred to as the said article from _____ (specify the place);

An whereas on the request of the Vendor(s) the government agreed to keep the said article in the safe custody of the vendor(s) executing a bond in the terms hereinafter contained and supported by surety/ two sureties which the vendor(s) has/have agreed to do _____ Now the condition of the above written obligation is such that if in the event of the vendor(s) failure to produce intact the said article before such court or authority and on such date(s) as may be specified by the said Food Safety Officer from time to time the vendor(s) and/or the surety/sureties forthwith pay to the government on demand and without a demur sum of _____ rupees the said bond will be void and of no effect. Otherwise the same shall be and remain in full force and virtue.

These presents further witness as follows:

i. The liability of the surety/sureties hereunder shall not be impaired or discharged by reason of time being granted by or any forbearance, act or omission of the government whether with or without the knowledge or consent of the sureties or either of them in respect of or in relation to all or any of the obligations or conditions to be performed or discharged by the Vendor(s). Nor shall it be necessary for the government to sue the Vendor(s) before suing the sureties or either of them for the amount due, hereunder.

ii. This Bond is given under the Food Safety and Standards Act, 2006 for the performance of an act in which the public are interested.

iii. The government shall bear the stamp duty payable on these presents.

In witness whereof these presents have been signed by the vendor(s) and the surety/sureties the day hereinabove mentioned and by Shri...............on behalf of the President of India on the date appearing below against his signature.

Witnesses:
1. _____ (Signature)
  (Name and address) _____
2. _____ (Signature)
  (Name and address) _____
  Signature _____ (Vendor) _____
  Signature _____ (Vendor) _____
  Signature _____ (Surety) _____
  Signature _____ (Surety) _____

for and on behalf of the President

of India/governor of _____

signature _____

(Designation) _____

FORM V
(Refer rule 3.4.1. (3))

To

_____

_____

Dear Sir/Madam:

I have this day taken from premises of _____ situate at _____ samples of food specified below to have the same analysed by the food analyst for _____.

Details of food:

Code number:

Place: (Sd/-) Food Safety Officer

Date:

Address:

FORM VI
(Refer Rule 3.4.3 (7))
Memorandum to Food Analyst

From:

_____

_____

Date: _____

To

Food analyst

_____

MEMORANDUM (Refer Rule (V)a of 3.4.1(8))

1. The sample described below is sent herewith for analysis under ____ of ___ of section ___ of food safety and Standards Act, 2006
   i. Code number
   ii. Date and place of collection
   iii. Nature of articles submitted for analysis
   iv. Nature and quantity of preservative, if any, added to the sample.

2. A copy of this memo and specimen impression, of the seal used to seal the packet of sample are being sent separately by post/courier/hand delivery (strike out whichever is not applicable)

(Sd/–) Food Analyst

Address:

   a. the sealed container of the second and third parts of the sample and two copies of memorandum in Form VII shall be sent to the Designated Officer immediately but not later than the succeeding working day by any suitable means and

   b. the sealed container of the remaining fourth part of the sample and a copy of memorandum in **Form V** shall be sent to an accredited laboratory, if so requested by the food business operator, under intimation to the designated officer.

FORM A

(Refer Regulation 2.2.2)

CERTIFICATE OF ANALYSIS BY THE REFERRAL FOOD LABORATORY

Certificate No. _____

Certificate that the sample, bearing number _____ purporting to be a sample/of _____ Was received on _____ with Memorandum No. _____ Dated _____ From _____ [Name of the Court] _____ for analysis. The condition of seals on the container and the outer covering on the receipt was as follows:

_____

_____

I _____ (Name of the Director) _____ found the sample to be _____ (Category of food sample) _____ falling under Regulation No. _____ of Food Safety and Standards(Food Products and Food Additive) Regulations, 2011. The sample was in a condition fit for analysis and has been analyzed on _____ (Give date of starting and completion of analysis) _____ and the result of its analysis is given below/*was not in a condition fit for analysis for the reasons given below:

Reason:

_____

Analysis Report:

  i. Sample Description:

_____

  ii. Physical Appearance:

_____

  iii. Label: _____

S.No.	Quality Characteristics	Name of the	Results	Prescribed Standards as per:
	Method of the test used			
				(a) As per Food Safety and Standards (Food Products and Food Additive) Regulations, 2011
				(b) As per label declaration or proprietary foods (c) As per the provisions of the Act and Regulations, for both above

Opinion **

Place:

Date:

(Signature)

Director Referral Food Laboratory

(Seal)

FORM VIIA

(Refer Rule 2.4.4(6))

REPORT OF THE FOOD ANALYST

Report No. _____

Certified that I _____ (name of the Food Analyst) duly appointed under the provisions of Food Safety and Standards Act, 2006 (34 of 2006), for _____ (name of the local area) have received from _____ * a sample of _____, bearing Code number and Serial Number _____ of Designated Officer of _____ area* on _____ (date of receipt of sample) for analysis.

The condition of seals on the container and the outer covering on receipt was as follows:

Intact/damaged/missing (delete where inapplicable)

I found the sample to be _____ (category of the sample) falling under item no. _____ of Chapter 5 of Food Safety and Standards Regulations.

The sample was in a condition fit for analysis and has been analysed on _____ (give date of starting and completion of analysis) and the result of its analysis is given below/was not in a condition fit for analysis for the reason given below:

Reasons:

_____

Analysis Report

Refer Rule 2.4.2 (5)

i. Sample Description (What it contains)

_____

ii. Physical Appearance of sample/container

_____

iii. Label declaration.

_____

Sl

S. no	Quality characteristics	Nature of method of testused	Result	Prescribed standards asper (a) provisions of the FSSAct, Rules and Regulations
1.				
2.				
3.				
4.				

Report (sample wise)
– adulterated/misbranded/within norms/violates provision of _____ (delete where not applicable)
– any other observations
Signed this _____ day of _____ 20

Address: (Sd/-) Food Analyst.

FORM VIII
(Refer Rule 2.4.6 (1))
FORM OF APPEAL BEFORE THE DESIGNATED OFFICER
(PLACE)

In the matter of appeal under Section 46 (4) of The Food Safety and Standards Act 2006 (34 of 2006)

and

In the matter of appeal against the report dated ____ from the food analyst

1. No. and date of the report of the food analyst against which the appeal is being preferred

2. Brief details of the facts and the grounds on which the report is being challenged

3. Relief being claimed

Signature of Appellant

41

From Preventive of Food Adulteration Act to Food Safety and Standards Act

Sonika Raj, Puja Dudeja, Amarjeet Singh

India is called as land of *Annapurna*, where food, water, fire and earth are not only considered as basic necessities but are worshipped. According to Vedas, one should offer food as a sacrifice to God. Food plays an important role in worship, and the food offered to God (*prasada*) is thought to bestow considerable religious merit, purifying body, mind and spirit. *Food and the eater: That is the essence of the whole world:* Brhadaranyaka Upanisad, 1.4.6.

In ayurveda, science of life, health and longevity, food plays a prominent role in promoting health and is therefore considered medicine. The kind of *ahara* (food) has been given due importance. It focuses on "We are what we eat". Food has been given prime importance in prevention as well as in cure of diseases. It mentions a great deal on what kind of food should be taken for sustenance of a healthy body. In spite of this, the evil of food adulteration is widespread, rampant and persistent in modern India. There are frequent news in media regarding adulteration of foodstuffs including milk and milk products especially during festivals. In 2012, there were approximately forty four outbreaks of FBI as recorded by Integrated Disease Surveillance Project (IDSP) in different parts of the country. Remember, a few years back pesticides in soft drinks created a huge controversy when the Centre for Science and Environment (CSE) in the Capital, compared the pesticide levels with those prescribed by the EU standards for primary raw ingredients.

Food adulteration is the addition or removal of any substances to or from food, so that the natural composition and quality is affected. Mixing, substitution, concealing the quality, putting up decomposed food for sale, misbranding or giving false labels and addition of toxic — all these come under adulteration. It results in two disadvantages for the consumer. Firstly, he has to pay more money for foodstuff of lower quality and secondly, in some form or the other, adulteration affects health. Adulterated food is impure, unsafe and not wholesome. Food adulteration can cause immediate and long term effects on human health. Diarrhea, dysentery, vomiting are such type of effects. Tamarind and date seed powder mixed with coffee powder can cause diarrhea. Adulteration on bakery items and dairy products causes increased salivation, abdominal cramp, vomiting, prostration, etc. Improperly processed milk and canned meat may cause food poisoning and abdominal pain. Vegetables and fish mixed with formalin and other type of chemicals which are used to keep the food fresh are injurious to health. Unhygienic meat and meat products can cause food infection usually with fever and chills. Long term effects include liver damage, stomach disorder, heart diseases, epidemic dropsy, etc. Copper, tin, zinc, mercury mixed with foods can cause brain damage of a person. Calcium carbide, urea, colouring dyes like Auramine, Rhodomine B and Yellow G, brunt engine oil and even some permitted preservatives when used in excessive amount affects the multiple organs of human body. Mostly it causes peptic ulcers, cancer of colon, chronic liver diseases including cirrhosis and liver failure, electrolyte imbalance and eventually kidney failure. Heart diseases, blood disorders and bone marrow abnormality are also detected. Skin problems resulting from unhealthy, unsuitable food consumption are often seen including allergic manifestation. Non-permitted colour or permitted food colour like metanil yellow, beyond the safe limits can cause allergies, hyperactivity, anemia, liver damage, infertility, cancer and even birth defects. Figure 41.1 shows the potential sources of adulterant entry into the food.

In India, 64.8% of milk-based and cereal-based sweets and savory products tested once were found to be adulterated. In 2012, a study in India conducted by the Food Safety Standards Authority of India (FSSAI) across 33 states found that milk in India is adulterated with detergent, fat and even urea, as well diluted with water. Of the 1791 random samples from 33 states, just 31.5% of the samples tested (565) conformed to the FSSAI standards while the rest 1226 (68.4%) failed the test. The study conducted by the authors in Tricity (Chandigarh, Panchkula, Mohali) found that quality of 77% of the collected raw milk samples was below the standards. The United States Economic Research Service has estimated that the cost to the US economy of outbreaks associated

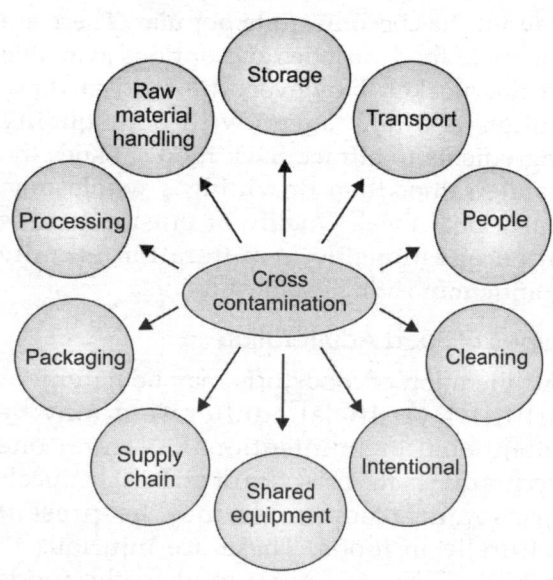

Fig. 41.1: Potential sources of adulterant entry into the food

with five major foodborne pathogens was in the region of US$6.9 billion per year. At the same time the economic burden of foodborne infections in Australia and Sweden were estimated at around AU$1.2 billion and US$123 million per year respectively (Hall et al., 2008). According to a recently published report, oils (mainly olive oil) represent 24% of reported food adulteration cases. Milk (14%), fruit juices (12%, including concentrates, jams, purees and preserves), spices (11%) and sweeteners (8%) complete the top five ingredient categories most commonly associated with fraud (Moore et al. 2012).

Traditionally, Indian families used to cook food at home with healthy ingredients and knew what was added into the meal. However, in modern times fast pace of life has changed the family dynamics. More and more women are working outside homes. With rising incomes people are spending more on food. Lack of time is forcing people to opt for ready to eat fast foods. Eating out at restaurants has become quite popular. There are plenty of food varieties and options available in the market. However, the food at these outlets is often cooked with poor quality ingredients to attract. Such food satisfies the palate rather than providing a wholesome nutritional meal. Quality of outside food is often questionable. Adulteration is quite common in eating establishments.

Types of Food Adulteration

Adulteration of foodstuffs may be natural or artificial. Artificial adulteration may be intentional or unintentional. Natural one occurs due to the presence of certain chemicals or organic compounds may be present naturally in foods. These are injurious to health. These are not added to the foods intentionally or unintentionally. For instance, toxic varieties of pulses, mushrooms, green vegetables and sea foods. About 5,000 species of marine fish are known to be poisonous and many of these are among edible varieties. Unintentional adulteration is a result of ignorance or the lack of facilities to maintain food quality. This may also be caused due to spill over effect from pesticides and fertilizers. Inappropriate food handling and packaging methods can also result in adulteration. Intentional food adulteration is usually done for financial gain. Some examples of intentional adulteration are addition of water to milk, extraneous matter to ground spices, or the removal or substitution of milk solids from the natural products. The most common form of intentional adulteration is colour adulteration. Intentional adulteration is a criminal act and punishable offense. About 25 to 30 percent of the food items in India are intentionally adulterated.

Food adulteration is a kind of slow poisoning. It is destructive to human life. It is a socio-economic crime because it is done with the purpose of gaining profit. It has tendency to erode national health, character and economy. The adulteration affects the human resource of the nation. It has direct impact on national progress and GDP of a country. Foodborne diseases impose a substantial burden on health-care systems and markedly reduce economic productivity. Poor people tend to live from day to day, and loss of income due to foodborne illness perpetuates the cycle of poverty. In recent times, consumers have become quite active for their rights to get healthy food. But not much headway has been made in this direction.

Victims of Food Adulteration

- People who have been victims of poisoning include the famous Greek philosopher Socrates(C. 469 BC–399 BC), who, after being sentenced to death by the Greek state for impious acts, drank a beverage laced with hemlock (Poisonous plants in the Apiaceae family).

- Italian witch Hieronyma Spara, who was ultimately hanged on orders from the Catholic Church, taught young Roman women how to murder their husbands using arsenic during the 1600s.

- In 1984, followers of Shree Rajneesh (1931–1990), also known as "Osho," spiked salad bars in Oregon with salmonella, resulting in more than 750 illnesses, which was perhaps the first act of bioterrorism in the United States.

Laws Related to Food Adulteration

Formal and informal laws to deal with adulteration have been always in place. There is a mention of food adulteration and the punishment given to the traders who are involved in such anti-social activities in the *Arthshashtra* written by *Chanakaya* in 375BC. During the British era, Indian Penal Code (IPC), 1860, came into force. Section 272 of IPC states that whoever adulterates any article of food or drink, so as to make such article noxious as food or drink, intending to sell such article as food or drink shall be punished with imprisonment of either description for a term which may extend to six months, or with fine which may extend to one thousand rupees, or with both. The individual state laws imposing strict liability came into force since 1912. But, there was considerable variance in rules and specifications of food which affected inter provincial trade. Until 1954, several states formulated their own food laws. Government of India appointed the Central Advisory Board and the Food Adulteration Committee in the years 1937 and 1943 respectively. But there was a considerable variance in the rules and specifications of the food, which interfered with inter-provincial trade. The Central Advisory Board appointed by the Government of India in 1937 and the Food Adulteration Committee appointed in 1943, reviewed the subject of Food Adulteration and recommended for Central legislation. The Government of India, therefore, enacted a Central Legislation called the Prevention of Food Adulteration Act (PFA) in the year 1954 which came into effect from 15th June, 1955. The Act repealed all laws, existing at that time in States concerning food adulteration.

Prevention of Food Adulteration Act (PFA)

The objective envisaged in this legislation was to protect the people from poisonous and harmful foods, to prevent the sale of sub-standard foods ensure pure and wholesome food to the consumers and also to prevent fraud or deception. The Act has been amended four times in 1964, 1971, 1976 and in 1986 with the objective of plugging the loopholes and making the punishments more stringent and empowering Consumers and Voluntary Organizations. The subject of the Prevention of Food Adulteration is in the concurrent list of the constitution. However, in general, the enforcement of the Act is done by the State/U.T Governments. The Central Government primarily plays an advisory role in its implementation besides carrying out various statutory functions/duties assigned to it under the various provisions of the Act. Various orders have been issued from time to time for regulating the licenses, permits or otherwise the production or manufacturing of essential commodities (Fig. 41.2).

FRUIT PRODUCTS ORDER (FPO) 1955

Fruit products is for non-fruit beverages, squashes, jams, jellies, tomato products, etc. Under the Central Government a Central Fruit Products Advisory Committee was con-

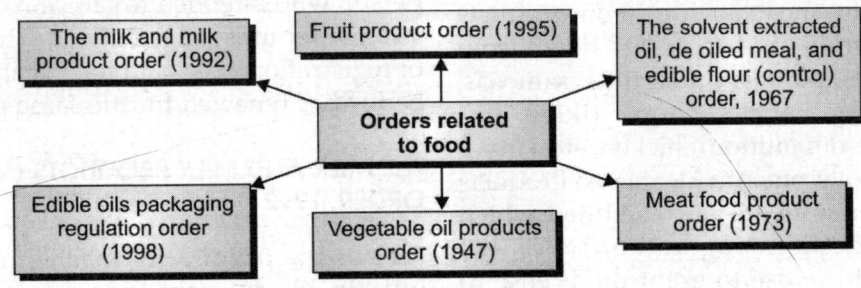

Fig. 41.2: Food related orders in India

stituted. There was appointment of a Licensing Officer (LO). Every manufacturer had to apply for a licence. The LO were given the powers of refusal of a license. The manufacturer could appeal to the Central Government within 30 days of such refusal. Under this order license number was to be displayed or embossed prominently in case of bottle, tin, barrel or any other container. It should specify the code number and date of manufacture. The code number was regulated to be given in English or Hindi numerals or alphabets or both. Label was not to be misleading or false. Any beverage not containing 25% of fruit juice was not to be described as a fruit syrup, fruit juice or syrup. Non-fruit beverages, syrups, etc. should be labeled as "Non-Fruit". This Order was not applicable to any syrup which contained fruit juices for medicinal use or are sold in bottles bearing a label with the words "For medicinal use only".

THE MEAT FOOD PRODUCTS (MFP) ORDER, 1973

Meat food products meant any article of food, being used as a food which is derived from meat by means of drying, curing, smoking, cooking, seasoning, flavouring. It shall not include the following products unless the manufacturer himself desires to be covered under the provisions of the said order: meat extracts, meat consommé and stock, meat sauces; whole, broken or crushed bones, animal gelatin, meat powder, bone extracts and similar products; fats melted down from animal tissues; patties, puffs, rolls, samosas, cutlets, koftas, kababs, chops, tikkas and soups made from mutton, chicken, etc. Under the Central Government a Meat Food Products Advisory Committee was constituted which appointed the licensing authority. It was empowered to refuse to grant the license to any applicant. Reasons were to be recorded

in writing. The manufacturer could appeal against the refusal within 30 days. The validity of license was 1 year. Rrenewal of license was mandatory.

THE VEGETABLE OIL PRODUCTS (VOP)(REGULATION) ORDER, 1998

Vegetable oil product is any product obtained for edible purposes by subjecting one or more edible oil to any combination of processes like blending, refining, etc. Under the Central Government appointment of Vegetable Oil Products Commissioner was done. Under this order no producer was to be eligible for registration unless he has his own laboratory for testing of samples and the commissioner may refuse to grant registration and such reasons for rejection are to be recorded in writing. Any person could appeal against such a refusal/cancellation of the registration within 30 days of the receipt of the said order.

EDIBLE OILS PACKAGING REGULATION (EOPR) ORDER, 1998

Edible Oils means vegetable oil and fats but does not include any margarine, vanaspati, bakery shortening and fat spread. Under the Central Government appointment of Edible Oils Commissioner was created. The registering Authority was appointed by the State Government and appointment of Inspecting Officers was created. Under the order any person who intended to carry on the business as a packer must be registered. The certificate of registration was valid for 3 years and could be further renewed for the same period.

THE MILK AND MILK PRODUCTS (MMP) ORDER, 1992

As per the order Milk means milk of cow, buffalo, sheep, goat, or a mixture thereof, either raw or processed. Milk Product means cream, curd, yogurt, cheese and cheese spread, ice cream, milk ices, condensed milk (sweetened and unsweetened), condensed skimmed milk (sweetened and unsweetened), sweets made from khoya, etc. Under the Central Government Milk and Milk Product Advisory Board was constituted and a Registering Authority was constituted. The order entailed that no person or manufacturer should set up a new plant or expands the capacity of the existing plant without obtaining registration/permission. There was provision of transfer of registration. The registering authority could suspend/cancel any certificate.

The PFA is a central legislation. Rules and standards framed under the act are uniformly applicable throughout the country. Besides, framing of rules and standards, the following related activities were undertaken by the Ministry of Health and Family Welfare.

- Keeping close liaison with state/local bodies for uniform implementation of food laws.
- Monitoring of activities of the states by collecting periodical reports on working of food laws, getting the reports of food poisoning cases and visiting the states from time to time.

- Arranging periodical training programme for senior officer/inspector/analysts.
- Creating consumer awareness about the programme by holding exhibitions/seminars/training programmes and publishing pamphlet'.
- Approving labels of infant milk substitute and infant food, so as to safeguard the health of infants.
- Coordinating with international bodies like ISO/FAO/WHO and codex.
- Carrying out survey-cum-monitoring activities on food contaminants like colors.
- Giving administrative/financial/technical support to four Central Food Laboratories situated in Kolkata, Ghaziabad, Mysore and Pune and providing technical guidance to the food laboratories set up by the states/local bodies.
- Holding activities connected with National Monitoring Agency vested with powers to decide policy issues on food irradiation.
- Formulation of a manual on food analysis method.

Penalties under PFA are described in Table 41.1. So far, there was a multiplicity of food related laws in India (Table 41.2). It created confusion

Table 41.1: Penalties under PFA	
For selling, manufacturing or distributing adulterated, misbranded food and giving of false warranty	Imprisonment not less than 6 months but which may extend to 3 years Fine not less than ₹ 1,000/–
Penalty for selling, manufacturing or distributing adulterated food or containing adulterant injurious to health	Imprisonment not less than 1 year but which may extend to 6 years Fine not less than ₹ 2,000/–
If the adulterant is likely to cause death or amounts to grievous hurt	Imprisonment not less than 3 years but which may extend to life term Fine not exceeding ₹ 5,000/–
Failure of the vendor to disclose the details of the person from whom the food article was purchased	Imprisonment not exceeding 6 months Fine not exceeding ₹ 500/–

in the minds of the consumers, traders, manufactures. Problems were faced as different products were governed by different ministries and orders. There were variations in specifications/standards in different orders. The food processing technology was getting advanced day by day and there were emerging concerns for food safety. Table 41.2 describes different laws under their respective ministries.

The provisions under PFA have been amended nearly 360 times and standards of around 250 articles of food of mass consumption have been prescribed. While making amendments, standards formulated by Codex/technological development in the food industry sector/dietary habits/nutritional status of our population and social and cultural practices were taken into consideration. Many loopholes have been noticed in the PFA. The Act did not provide for the mandatory standardization of food products. There was no requirement of training to the food inspectors. Usually they do not know how much sample is to be taken and in what quantity the preservatives are to be mixed in the sample because of which the samples are usually destroyed by the time they are analyzed in the laboratory. Minimum numbers of inspectors required for the area

Table 41.2: Role of different ministries in regulating food related laws

Ministry of Health and Family Welfare	Ministry of Agriculture	Ministry of Food and Consumer Affairs
Prevention of Food Adulteration Act, 1954 PFA Rules, 1955 Health Food Supplement Bill	Agriculture Produce Marketing Act Milk and Milk Product Order	Essential Com. Act, 1955 Standards of Weights and Measures Act,1976 Packaged Commodities Rule, 1977 Consumer Protection Act, 1986 B.I.S. Act,1986 VOP Control Order, 1947 VOP (Std. of Quality), 1975 SEO Control (Order), 1967
Ministry of Commerce Imports and Exports Regulations Export Inspection Agency Tea Board Coffee Board Coffee Act and Rules	**Ministry of Food Processing Industries** Fruit Products Order, 1955	**Ministry of Rural Development** Agricultural Produce Grading and Marketing Act, 1937 Meat Food Products Order
Ministry of Forests and Environment Trade in Endangered Species Act Ecomark	**Ministry of Science and Technology** Atomic Energy Act, 1962 Control of Irradiation of Foods Rules, 1991 G.M. and Organic Foods	**Ministry of HRD** (development of women and Child Welfare) Infant Milk Substitutes, Feeding Bottles and Infant Foods (Regulation of Production, Supply and Distribution) Act, 1992-Rules,1993

are not given in the Act. Under Section 12 of PFA the person had been given the right to get the sample tested if he thought that it contained some deleterious substance.

But for this he had to pass two hurdles. Firstly, he had to tell the seller the purpose for which he was taking the sample and secondly, he had to pay the requisite fees. As far as the first issue was concerned, no trader who was guilty would allow the consumer to take the sample. Secondly, though the fees was refundable if the test report found to be positive, but it was not possible to afford it initially. Moreover it was doubtful whether the analysis was 100% percent accurate. There was also a major problem with the procedural part of the Act. The Act failed to mark distinction between different categories of adulteration and provided the same punishment for all types, even if the type of adulteration is life threatening. Moreover, the PFA was covered under the Probation of Offenders Act, 1958. As a result of this the perpetrators of heinous socio-economic crime like this, were let loose after getting caught for the first time. Deficiency in the testing laboratories on the counts of inadequate trained manpower, inadequate testing facilities, non-availability of sophisticated equipment, inadequate budgetary provision and non-availability of reference standard material were also there.

In India, the concept of food safety is now being looked into seriously. The felt need for an integrated food law has been met with by promulgation of FSSR, 2011. The archaic Prevention of Food Adulteration Act (PFA), 1954, has been finally repealed. FSSA, 2006 consolidated the laws relating to food. The main intent of this endeavor was to ensure availability of safe and wholesome food for human consumption. The emphasis of Food Safety Act, 2006 is on highlighting the responsibility of manufacturers, recall, risk analysis, good manufacturing practices and process control, viz Hazard analysis and Critical Control Point (HACCP).

The Food Safety and Standards Act (FSSA), 2006 has been enacted to consolidate the laws related to food making it at par with the international standards. FSSA, 2006 came into effect in August, 2011, five years after it was passed in Parliament. It subsumes various central Acts like Prevention of Food Adulteration Act of 1954 , Fruit Products Order of 1955, Meat Food Products Order of 1973, Vegetable Oil Products (Control) Order of 1947, Edible Oils Packaging (Regulation) Order of 1988, Solvent Extracted Oil, De- Oiled Meal and Edible Flour (Control) Order of 1967, Milk and Milk Products Order of 1992 and also any order issued under the Essential Commodities Act, 1955 relating to food. It is meant to ensure prevention of fraudulent, deceptive or unfair trade practices which may mislead or harm the consumer, and unsafe, contaminated or sub-standard food. It will also ensure improved quality of food for the consumers and censure misleading claims and advertisement by those in food business.

The FSS Rules, 2010 contain qualifications of the enforcement agencies, sampling techniques, legal aspects and other issues enumerated under of the FSSA, 2006.

The FSS Regulations, 2011 contain labeling requirements and standards for packaged food, permitted food additives, colors, microbiological requirements, etc.

Key Provisions of FSSA

1. Effective regulation, manufacture, storage, distribution and sale of food to ensure consumer safety and promote global trade.

2. Single reference point for food safety and standards, regulations and enforcement.

3. Prevention of sale of misbranded, unsafe/contaminated or sub-standard food

4. No article of food shall contain food additive, processing aid, contaminants or heavy metals, insecticides or pesticides residue.

Important Definitions as per FSSA

Food: Any substance, whether processed, partially processed or unprocessed, which is intended for human consumption. It includes infant food, packaged drinking water, water used in food during its manufacture, chewing gum. It does not include live animals, plants prior to harvesting, cosmetics, drugs and medicinal products. Taking into account the effects of adulteration on the society as whole, FSSA defines the word 'unsafe food' instead of 'adulterated food'.

Adulterant: Any material which renders the food unsafe or sub-standard, misbranded or contains extraneous matter.

FSSAI established in 2008, has laid down the science based standards for food items and regulate their manufacture, storage, distribution, sale and import to ensure availability of safe and wholesome food for human consumption. It will collect and collate data regarding food consumption, incidence and prevalence of biological risk, identification of emerging risks and procedures for accreditation of laboratories. The data will help in the implementation of the proposed Food Security Bill and also contribute to the development of international technical standards for food.

Authorities under FSSA 2006 is described in Fig. 41.3.

Central Government

The act is a historic one and constitutes a regulatory authority that will govern the quality and standards of food right from national to village level. It has provided the mandatory standardization of food. Liability of person is civil labiality which is easier to prove. It has taken major initiative in abolition of so called 'Inspector Raaj'. If the food inspector is found to be guilty of misusing his powers the there is a provision of imposing a fine of rupees one lac. The Act has been inspired from Codex therefore the standards match the international quality. For the first

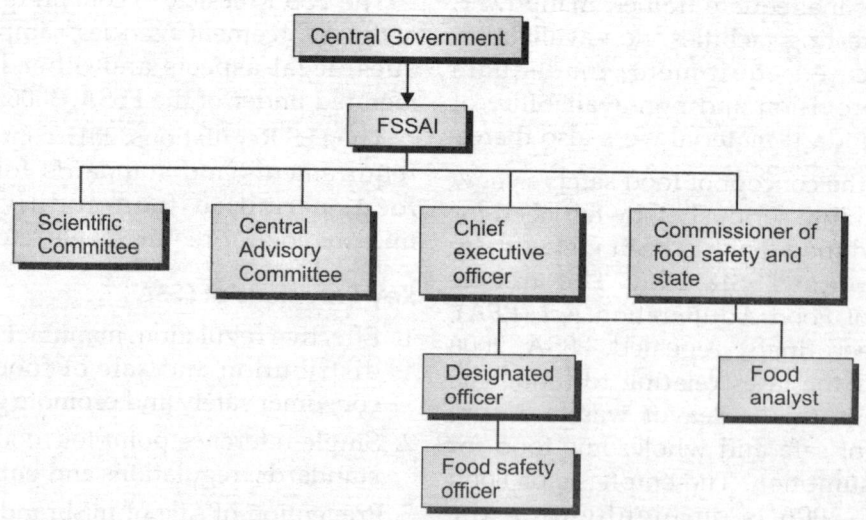

Fig. 41.3: Authorities under FSSA 2006

time in this act, there is a clause of providing compensation to the consumer who faced any health hazard along with the fine and punishment to the offender.

The Act imposes that it will be the responsibility of FBO to ensure that the articles of food satisfy the requirements of the Act at all stages of production, distribution and storage. And if he finds that they are not according to standards then it is his duty to recall those food items. Every FBO is bound to have license. Now there will be only single ministry to look into whole affairs rather than nine ministries. Now there are two types of treatment: the matters which are hazardous to health and the ones which are not. A graded penalty structure is proposed based on the severity of offences. This removes the confusion and inequality which was there in the earlier Act. The punishment imposed will be a fine for offences like manufacturing, selling, storing, or importing sub-standard or misbranded food. The guilty may be punished with imprisonment along with a fine. However, the penalties for non-compliances are very high and this only emphasizes the need for total compliance to the regulations in all respects.

Other Highlights of FSSA

FSSA 2006 has also attempted to define Nutraceuticals (Section 22) which are described in a separate chapter. It has also laid restrictions on advertisements and prohibition of unfair trade practices (Section 24). It places the liability of the manufacturers, wholesalers, distributors and sellers (Section 27) in the event any article of food supplied/sold after the date of expiry, unsafe or misbranded, unidentifiable of manufacturer, received with the knowledge of being unsafe or stored in unhygienic conditions. It also consists of Food Recall Procedures (Section 28). As per the Act, (Section 31) license is mandatory. However,

petty manufacturers, hawkers, temporary stall owners, small scale cottage industries, etc. are required to register with the concerned municipal authority. As per (Section 32) improvement notices can be issued by the designated officer. Such notices include:

- Grounds stating the failure to comply with regulations.
- Matters which constitute FBO failure to comply.
- Measures to be taken by FBO in order to secure compliance within reasonable time.

Failure to comply the notice may lead to suspension/cancellation of the license. Under section 40 rights of a purchaser to get food analyzed after paying the fees has been specified. The purchaser has to inform the FBO at the time of purchase that it is for analysis. The fee will be refunded if sample fails and failing sample will lead to prosecution. The Act brings all the FBO under a single umbrella to have regulations in control. Penalties under FSSA are given in Table 41.3.

Wide varieties of checks have been provided in the FSSA. It also has the provision of serving an improvement notice to the operator if he is not complying with the prescribed standards. But even if he is again found non-compliance then food officer can take action. The time limit for prosecutions has also been fixed. The trial has to start within a year from the date of commission of offence. The new Act has been implemented for all those in food business. Even hospital kitchens are also not excluded. Consumer safety has been given due importance. The fishermen and farmers are excluded from the purview of the Act. The Act facilitates the formation of stakeholder's forum at regional levels to channelize the voice of stakeholders. The Act also covers the food supply in the public distribution system that means it also covers the Food Security

Table 41.3: Penalties under FSSA 2006

Offences	Penalties
Selling food not of nature, substance of quality demanded	Penalty not exceeding ₹ 5 lakhs
Selling, manufacturing or distributing sub-standard food	Penalty not exceeding ₹ 5 lakhs
Selling, manufacturing or distributing misbranded food	Penalty not exceeding ₹ 3 lakhs
Misleading advertisement	Penalty not exceeding ₹ 10 lakhs
Food containing extraneous matter	Penalty not exceeding ₹ 1 lakhs
Failure to comply with the directions of food safety officer	Penalty not exceeding ₹ 2 lakhs
Unhygienic or unsanitary processing or manufacturing of food	Penalty not exceeding ₹ 1 lakhs
Processing adulterant not injurious to health	Penalty not exceeding ₹ 2 lakhs
Processing adulterant injurious to health	Penalty not exceeding ₹ 10 lakhs
Unsafe food which does not result in injury	Imprisonment up to 6 months and fine up to ₹ 1 lakhs
Unsafe food which results in non-grievous injury	Imprisonment up to 1 year and fine up to ₹ 3 lakhs
Unsafe food which results in grievous injury	Imprisonment up to 6 years and fine up to ₹ 5 lakhs
If failure results in death	Imprisonment not less than 7 years but which may extend to imprisonment for life and fine not less than ₹ 10 lakhs

Act, 2009. Table 41.4 shows the major shifts from PFA to FSSA.

Loopholes in FSSA

The FSSA also has some loopholes. The main focus is on processing industry. Unorganized sector is completely ignored. There is no mention of registration process and the registration authority. The Act specifies compulsory registration which will create problem for vendors and hawkers. Portable drinking water used for manufacture of various food articles is excluded from the Act. Certain companies avoid use of the words 'packaged' or 'mineral' drinking water to avoid meeting any standards even though they supply the same product given by those adopting these words. By using the words 'herbal' or 'flavoured' water which have been categorized as traditional food products under the FSSA and exempted from standards, the companies are able to supply drinking water cheaply. It provides for both civil and criminal procedures. So, it can create confusion which one has to be followed. No jurisdiction has been defined to the Food Safety Officer for inspection and sample seizure.

Challenges

Enormous workload of implementation of new law all across the country will be a slow and a long process. Lack of trained human resources is a definite bottle neck. The number of food safety officers in India is woefully small (approximately 2000). Manpower shortage is a main hurdle in implementation of this Act. Moreover, the definition of 'food' expressly excludes the animal feed from its

Table 41.4: Critical (major) shifts from PFA to FSSA

PFA	FSSA
It mainly focused on adulteration of food	Main focus is on consumers safety related to food
Opinion Based	Scientific evidence based
It had prescriptive standards	Standards are more or less general
Several authorities were there	Single authority system is there (FSSAI)
Based on inspection	Based on regular monitoring and surveillance
Provision of same punishment for all types of adulteration, even if hazardous to health	Grading of punishment according to type of adulteration
No standardization of food products	Mandatory standardization of food items
No provision of food recall	In order to remove unsafe food from the market and thus prevent injury to consumers food recall procedures can be initiated voluntarily by the manufacturers/distributors concerned or by the food authority
No provision of improvement notice	If the designated officer has reasonable ground for believing that any food business operator has failed to comply with any regulations to which this section applies, he may, serve an improvement notice on that food business operator.

purview. It will be a mammoth task to educate the hawkers and street food vendors about the concept of food safety.

The fact is that whatever pesticides, insecticide, etc. get into the animal feed and consumed by the animal (cow, goat, etc.) becomes a part of food chain. For example, it is present in the milk. Therefore, this should be made part of the definition of food contamination.

Food chain from farm to the products needs to be traced. But as the farmers are excluded from the purview of the Act, the tracing is possible to the *sabzi mandi* only.

As the there is lack of properly trained workforce, the Ministry of HRD can think about the role of universities, for organizing vocational training courses on food analysis and food testing.

A separate department in the concerned Ministry must look after the matter of food

adulteration since it is a serious matter that affects the health of millions of citizens. Ministry of Health as well as Ministry of Food Processing should deal with food safety and FSSA.

The Act should have a compulsory provision for black-listing of the companies or even suspension of their license when held guilty of the offence. It should be made a part of the punishment.

The organized as well as the unorganized food sectors are required to follow the same food law. The unorganized sector, such as street vendors, might have difficulty in adhering to the law, for example, with regard to specifications on ingredients, traceability and recall procedures.

The Act does not require any specific standards for potable water (which is usually provided by local authorities). It is the responsibility of the person manufacturing food to ensure that he uses water of requisite quality even when tap water does not meet the required safety standards.

The Act excludes plants prior to harvesting and animal feed from its purview. Thus, it does not control the entry of pesticides and antibiotics into the food at its source.

The Codex and the Committees have suggested Confidence Building Measures among the consumers for food safety. This can be done by attaching the logo displaying that products are safe. This logo that can be understood by literate or illiterate person should be made mandatory.

The power to suspend the license of any food operator is given to a local level officer. This offers scope for harassment and corruption.

Conclusion

An effective food safety regulatory framework is imperative to ensure safe food for consumers in a country. However, considering the size of food industry in India it will take lot of resources to implement the new food law. Laws alone cannot solve the issue of food safety. The need of the hour is not only having an integrated law but also to have an integrated approach to change the mindset of people regarding food safety. Awareness campaigns must be organized with the help of the NGOs. Consumers need to be alert about their rights, particularly regarding adulteration in food products. Government departments should publish the literature related to food safety. If awareness is built up effectively and the food inspectors carry out their duties sincerely, food adulteration is bound to come down drastically. But for all this to happen, there is need for alertness among the monitoring agencies and a proper understanding of the adverse consequences of adulterated food on human health by the authorities and the general public. More research and development is needed with regards to food safety. There are advantages as well as certain loopholes that are yet to be filled and some questions that are unanswered. The quest of pure food is still on.

Section VII

Food Safety in Special Conditions of Life

42

Food Safety Concerns Among Elderly

Gunjan Grover, Amarjeet Singh, Puja Dudeja

We all have varied food choices. We have food fads. We like some food while we avoid other items. Ever since our childhood, over the years we develop a sense of selecting food items suitable to us. We also develop skills and ways to reject the food that may harm us or is unsafe. Our natural defense through the five senses (eye, ear, nose, tongue and skin) protect us against the intake of bad food. It is easier to identify a spoiled food item using these special senses. All the senses have important role in identifying a food item potentially harmful for us. But vision, taste and smell are more important. Usually light and smell of good food produces salivation. These senses play a role both in acceptance or rejection of food. Animals also use these senses in getting appropriate food. We, human beings also rely on these senses for choosing the safe food. The colorful appearance, aromatic smell and yummy taste of food tend to tempt everyone including the old people.

Time changes everything. Our diet and food consumption pattern has changed over millennia. The way we used to grow our crops has also changed. Ploughing has been replaced by tractor based farming. Today, we are using fertilizers to grow our crops. Consequently, many chemicals enter food chain from farm to our plates. For example, fruits and vegetables we consume may have pesticides residues. Milk obtained from animals has also been shown to have traces of harmful chemicals like DDT. Eating food having such chemicals can harm human body. Also consumption of food materials that are spoiled may have an adverse effect on our health. There is also "change" in functioning of organs, color of hair, texture of skin, etc. as age advances. Power of vision, taste, teeth, smell, hearing and touch gradually deteriorates in old age.

Due to diminished power of special senses, an old person is not capable of identifying certain gross physical features which signify safe food. Impaired vision can affect the person's ability to react in physical environment. In old age, activities of daily life are also affected viz. walking, transferring, etc. Defective vision also impact the capacity of elderly to recognize unsafe materials in food. The power of smell also declines with age. Normally, we smell food to detect whether it is fit for eating or not. Generally, we check the quality of food items stored in refrigerator by smelling it. For example, checking the smell of curd and milk, cooked *dals*, vegetables and dough. Compromised immunity and decreased sensitivity of special senses make elderly people more prone to food safety related hazards. Unsafe food can cause serious illness in elderly. So, consideration of food safety is more important in this age.

Compromised food safety is not only related to natural process of aging but also to living conditions of elderly. These have also changed in modern society where joint families have broken down. Elderly either live in their homes or in old age homes. At homes, they live either alone (single/ couple) or with family (with or without working family members). When they are staying with their families, they get respect and care by their children and other family members. In these settings, they usually get freshly prepared food. Now-a-days in most families, all the younger family members go out for work. In such families, food for elderly may be prepared in advance for the whole day. This may adversely affect the food safety. For example, semisolid food materials like *sabzis, dals and dalia* prepared in the morning, if left on dining table for elderly may get spoiled by noon time. These food stuffs need to be preserved at adequate temperatures in refrigerator to avoid spoilage. Even for convenience when some food materials are preserved in refrigerator for 2–3 days, these may get spoiled on reheating (which may be cumbersome for elderly). Consumption of such food may lead to compromised food safety in elderly.

If they live alone they may have to prepare food on their own. They also have to manage housekeeping (dishwashing, maintenance of cleanliness). But the problems like joint pain, tremors in hands and poor vision may not allow them to maintain their homes and kitchens. Food safety may be adversely affected in case they hire domestic help, since they generally neglect kitchen hygiene practices. If elderly live in old age home settings, it may have its own problems of food safety.

Food safety for elderly may be affected at many stages from 'farm to fork'. Therefore, identification of unsafe food is required at different stages viz. shopping for food, cooking and consumption. Special senses play an important role in all these.

In elderly people, food safety can be compromised at various stages

I. Procurement of Raw Material

When shopping for food stuffs in departmental stores or *mandis*, elderly may have problem in finding the things they are looking for. They may also fail to notice infested grains (e.g. insects in *dals*, rice, etc.), dry fruits, fruits and vegetables. Elderly also tend to spend with care. They buy cheap fruits and vegetables from *mandis* late in the evening in the hope of getting it at lowest possible rates. After dusk time, they cannot differentiate between healthy and rotten fruits and vegetables due to darkness. Bulk buying at this time may result in buying infested fruits and vegetables. Detection of quality of fruits and vegetables by its smell in *mandis* and departmental stores is also common. It can help recognize rotten fruits, vegetables and other products. Decision to buy a dairy product like cheese, milk, curd, etc. is also affected by its smell. But an older person with diminished ability to smell cannot distinguish between healthy or spoiled food.

Usually, we taste the food material to check the quality while buying it from the market. Normally, in *mandis*, a young person buys a few fruits like cherries, chickoos, strawberries, etc. after tasting it. Dampness in dry fruits can be sensed by tasting it, flavor of cheese can be noticed for its fitness for eating. Rotten fruits can also be judged by its taste, appearance or even touch. Vegetables and fruits can be sorted from spoiled using the sense of touch. Puffed up packs of juices are also not safe for drinking. This can also be judged by appearance and touch. But reduced power of taste or touch and poor vision of an older persons does not allow them to observe this and they may buy these things as such. At sweet shops also, they are not able to spot spoiled sweets.

Due to poor vision, they also feel difficulty in checking expiry dates on food packs e.g. breads, dairy products like milk, cheese, curd, etc. They may also have age related food allergies. Such cases require reading of ingredients on food product packages which is very difficult for elderly due to small fonts size.

Procurement is the first stage in food production and consumption. It also includes purchasing of raw food materials. It needs to be in accordance with the quantity of material required and availability of storage space. It is also advisable to ensure the quality of food items. It can be done by checking its appearance, smell, taste, touch and 'use by dates'. However, it is difficult for older people because of diminished power of their special senses.

II. Pre-cooking

After buying raw food stuff from market we either store the same or cook it directly. It is the next stage in 'farm to fork ' framework where food safety may be compromised. This stage involves identification of unsafe food materials before cooking. It includes sifting, washing, cutting, chopping, etc.

While sifting grains and cutting vegetables, insects may go unnoticed by elderly people due to poor vision. A young person can detect rotten vegetables, fruits and infested grains by the appearance and smell while cutting or using these. Before making *chapatis*, smell of dough (which was stored in refrigerator) tells us whether it should be used or not. But an older person who has poor visual acuity and diminished power of smell cannot do this.

Tremors in hands affect fine movement of hands and fingers in elderly. They may get injured while cutting and chopping vegetables. Handling food stuffs with dirty bandages and dressing on hands increase the risk of contamination of food.

Poor vision in elderly also hampers their mobility. Due to this limitation, they cannot maintain adequate cleanliness in homes and kitchens. This leads to accumulation of dust on kitchen floors and shelves, spider webs on roof and walls. Dish washing is also affected. Improper disposal of kitchen wastes invites pests and vectors in kitchens. This can also contaminate food. Hygiene of preparation and consumption of food in such an environment is compromised. It has a negative impact on their health.

III. Cooking

Elderly who are able to cook for themselves may have risk of compromised food safety due to diminished sense of sight, smell and taste. Insects like cockroach or fly may drop into preparations and contaminate it. They may also use infested condiments like *jeera* or coriander seeds at the time of preparation. A person with normal vision can identify it but an older person with poor visual acuity cannot.

After cooking, food material is either consumed or stored in refrigerator for consumption later on. It may get spoiled if not stored at optimum temperature. Food material stored in refrigerator may also get spoiled, if, by mistake door of fridge remained open. Froth (which normally warns us about unsuitability of its consumption) produced on reheating any *dal* or *sabzi* also cannot be seen due to poor vision in old age. After boiling, if milk is not put in refrigerator in time, has chances of getting spoiled. An older person cannot smell its sourness too. Liquid preparations like mango shakes are also likely to get spoiled. Taste can identify unsafe food materials when the power of vision and smell fail to do so. Sense of taste has the ability to detect food materials which are unsafe for consumption. Weakened ability to taste among elders results inadvertent consumption of such materials.

Quite often, elderly are not able to prepare food for themselves. So, they hire domestic help. Even these helpers can be negligent while chopping or cooking for elderly. Thus, elderly are at their mercy. They may even not be able to correct the servants or supervise them as far as food safety is concerned.

IV. Consumption

Food items cross many stages from farm to our plates. It is not necessary that food items that cross all stages safely till cooking will remain safe for consumption later. It may also become unsafe after cooking. After preparation, if food items not stored properly, its internal temperature may enter danger zone where there is multiplication of micro-organisms. Consumption of such food result in FBI.

Aroma stimulates the desire for food. It gets satisfied only on eating food that is palatable and gives pleasure. Taste of food affects the pleasure and satisfaction obtained from it. It decides the final acceptability of food. "Taste" also has been given due importance in Hindu mythology. An old lady *Shabri* in *Ramayana* discarded sour ones and sorted sweet *bers* (berries) by taste and offered these to Lord *Rama*.

Food is not consumed only for gaining nutrients but also for pleasure. Older persons want to enjoy food of their choice and taste. Sensitivity of taste buds decrease with old age. Therefore, complaints of bad tastes in food are common in this age. It is the result of decreasing sense of taste causing abnormal taste perceptions. It results in poor appetite.

Also, due to diminished power of smell, they do not find the food pleasurable. When they find the food bland, they loss interest in eating. They may even dislike and reject some healthy foods. They may also add more salts, sugars and spices to make the food more palatable. It may result in eating less and consequently nutritional deficiencies and weight loss. Other factors like poor dental hygiene, smoking and certain medications also decrease the sensitivity of taste.

Food and dairy items which are leftover after consumption are usually stored in refrigerator. Cooked vegetables and *dals* stored in refrigerator for long have more chances of being spoiled. Consumption of such food may result in certain FBI or intoxications due to production of toxins, e.g. moldy breads. Sometimes two different things stored in refrigerator appear same and can be differentiated using power of smell only. It becomes difficult for an older person to do so. For example, vinegar in bottle may be confused with water in a bottle. Drinking of vinegar in place of water may have adverse consequences.

Due to some problems like inability to prepare food of their choice or restriction on cooking nonvegetarian food at home, elderly may prefer eating out. Some also go for tiffin service. Restaurants, cafes, food points and other food services do not maintain sufficient cleanliness in their kitchens. Pests and vectors invade their kitchens. These vectors may contaminate the food. A young person with normal vision can spot this. But dim lights in restaurants combined with poor vision of elderly may not allow them to recognize small sized insects and result in consumption of contaminated food. Staff in restaurants and hotels also do not take adequate care about the handling, preparations and preservation of food. Elder people are not able to notice unhygienic and unhealthy food by its taste. Food materials like gravy, cakes, pastries and muffins, etc. which are prepared many days in advance have greater chances of getting spoiled. Consumption of such materials may lead to diarrhea, food poisoning, etc.

Food safety issues among elderly are given in Figs 42.1 and 42.2.

In summers, food item prepared in the morning may get spoiled by evening due to high temperatures. In restaurants and hotels, food which we eat is usually partly prepared 2–3 days in advance. *Dals,* rice and potatoes are boiled in bulk quantity. As potato is called the king of vegetables, it is used in almost all preparations right from snacks to main course e.g. *samosas, pakoras, kulchas* and *sabzis,* etc. Likewise, consumption of *dals* and rice is also common. There are more chances of these food items to get spoiled. Generally, a young person detects it by its smell before eating. But an older person due to decreased sensitivity of smell would not be able to identify that something is wrong with the food item. It results in consumption of unsafe food.

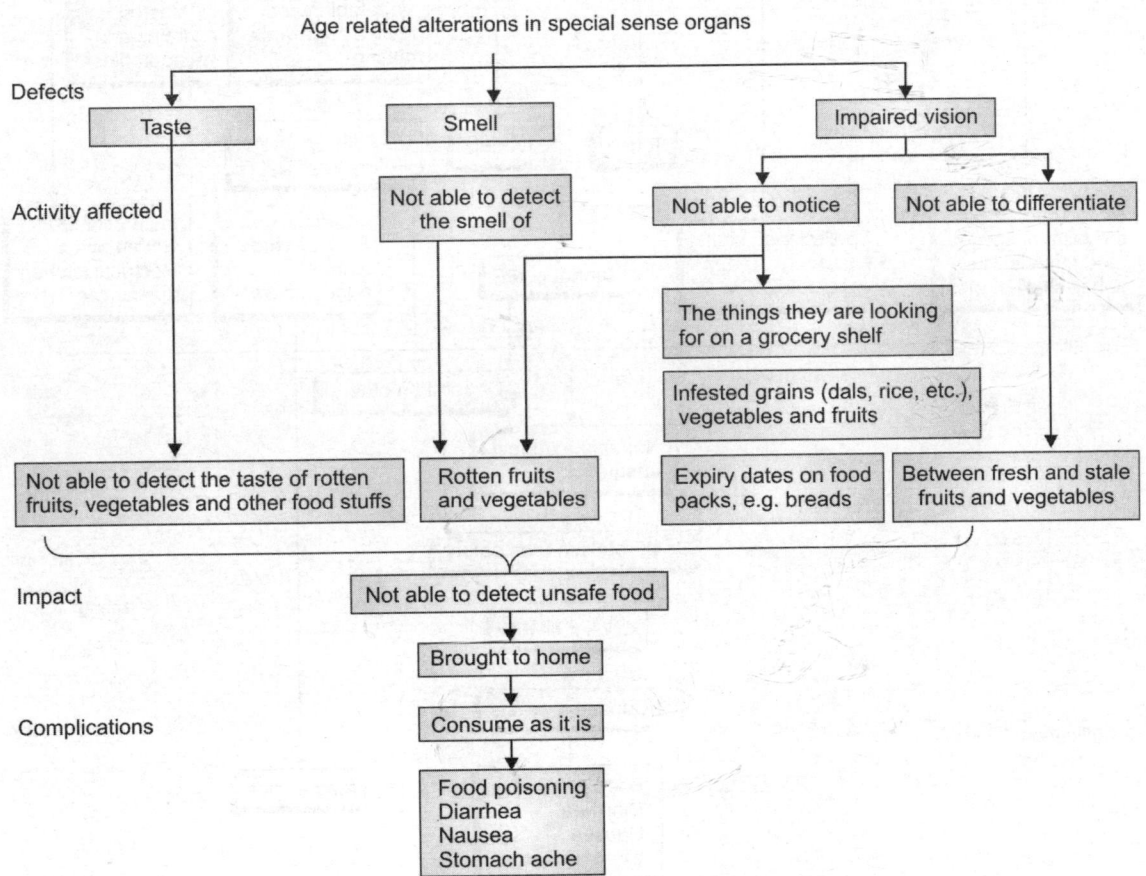

Fig. 42.1: Food safety issues among elderly

Limitation in mobility and fine movement also do not let elderly wash their hands properly before and after meals. Other problems like neglect of nail hygiene relate poor vision to food safety concerns among them. Improper hand washing along with negligence of nail hygiene make them more susceptible to FBI.

Simple solutions to improve food safety issues in elderly are given in Table 42.1

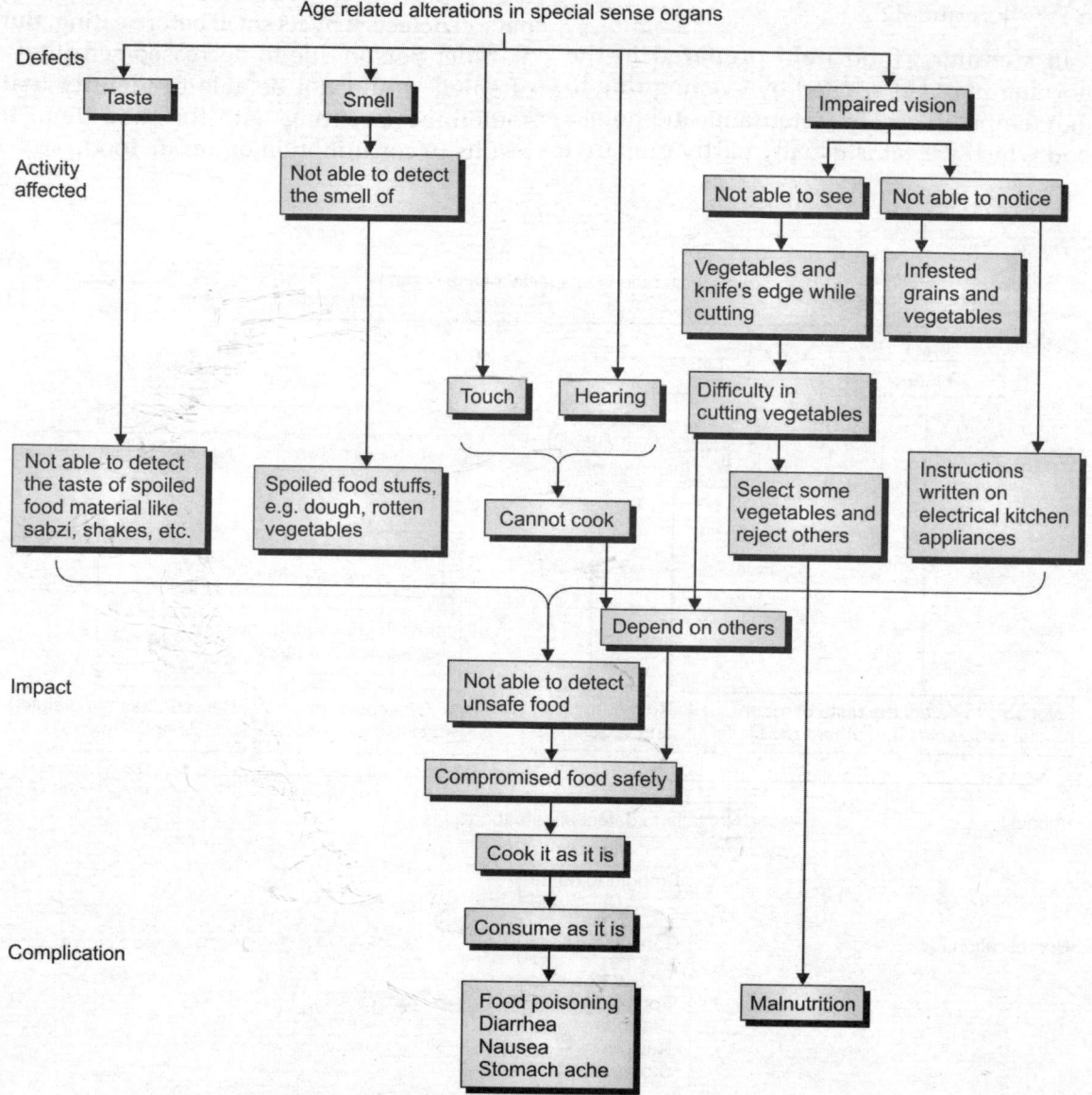

Fig. 42.2: Food safety issues among elderly during cooking

Table 42.1: Solutions to food safety issues in elderly

Lesion/Defect	Solution
Impaired vision	• Provide spectacles • Label things in grocery store on boards • Ingredients and expiry dates on food products should be highlighted in large font size • Servant/maid/or any other caregiver can accompany them in market • Lens with a string attached on shelves
Diminished power of taste and smell	• Servant/maid/or any other caregiver can accompany them in market

Good food helps us to maintain our health while bad food can harm us. Inability of elderly to prepare their own food and ignorance of hygiene practices by domestic helpers make them more prone to consumption of unsafe food. Due to diminished sensitivity of special senses and weakened immunity in this age, they cannot identify unsafe food and consume it as such. When immunity wanes with age, the reaction of body to different foods also change. It may thus, result in FBI. Therefore, it is necessary to address food safety issues among elderly.

43

Food Safety Chain in Infants from Farm to Fork

H Ravi Ramamurthy

"A chain is as strong as its weakest link"

Food is one the basic drives of humanity. The requirement of this life force is even more pronounced in the young. Among the different stages of human life, infancy is most dependent on the care giver for survival. Thus the need for sustained, appropriate and safe source of food begins with the beginning of life itself. The terms "sustained" indicates a steady supply and "appropriate" indicates suitability for consumption. But the term "safe food" encompasses a variety of factors—both natural and man-made, that determine whether the food consumed is safe or not. The factors that determine safety of food may act simultaneously or sequentially on the "food safety chain". The chain of food safety has several links beginning right from the time it is grown or manufactured. Subsequently as it is packaged and distributed over long distances, the risk increases. The final link is the method by which food is prepared and fed to infants by the end user. The need to provide safe food to infants cannot be overemphasized. However, to ensure food safety, a thorough understanding of the complexities that interplay in this food safety chain is required.

Peculiarities of infant foods: Infant foods are unique compared to food of older children or adults. The consistency is more semisolid than

solid, taste sweeter or blander than the spicier food for older children and finally in view of the small stomach size and the initial weaning foods need to contain more calories per gram of food. Based on these properties the food material of infants may be broadly classified in Fig. 43.1.

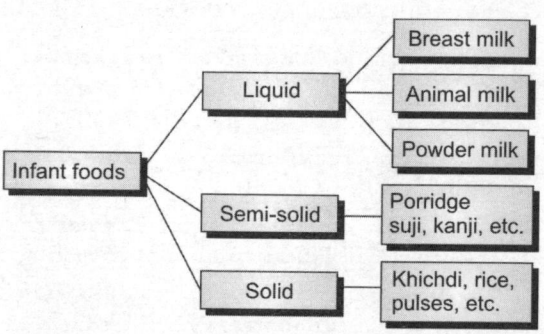

Fig. 43.1: Classification of infant foods

Peculiarities of infancy in relation to food: Infancy is a period of life with utmost dependency on the caregiver. However, it is not the only peculiarity of this age group. This period is essentially an immune challenged stage. The newborn has extremely limited defenses against infective organisms. Gradually as the infant completes the first year, he/she is exposed to a variety of antigens that build up the immunity. However, inappropriate exposure may sometimes lead to life threatening infections. Hereby, the traditional infant feeding practices assume importance. Exclusive breastfeeding of newborns for the first six months followed by gradual weaning with home based supplementary foods from six to twelve has been traditionally practiced in India and also been recommended by Indian Academy of Pediatrics (IAP). However, deviation from this practice has led to widespread use of Powdered Infant Formula (PIF) milk instead of breastfeeding. Home based supplementary feeds have taken a back seat to commercial infant cereal preparations. These preparations take a long circuitous route from manufacture to consumption and are also expensive. It is not unusual that as a cost cutting measure, preparations of sub-optimal concentration are fed in a "food insecure" nation like India leading to a subclinical chronic under nutrition. The salient features of infancy in relation to nutrition are summarized in Table 43.1.

Threats to safety of food: The chain of safe food is highly vulnerable to threats from a variety of agents. These threats are discussed as follows and described in Table 43.2.

Biological: These agents may degrade the food at any level beginning from the manufacture/packing to transport and finally at the consumers. The earliest and commonest contaminated packaged food that an infant may be exposed is PIF. It is worthwhile to note that *Enterobacter sakazakii* and *Salmonella enterica* are most commonly implicated in contamination of PIF right from manufacture onwards. This is because, using current manufacturing technology, it is not feasible to produce sterile PIF. During the reconstitution of PIF at consumer end prior to feeding, inappropriate handling practices can compound the contamination. In addition, one of the harmful traditions in India as well as other countries is the administration of prelacteal feeds to newborns like honey, *janam ghutti*, etc. in a silver spoon and avoid colostrum. These agents have no demonstrated benefits but have been implicated in life threatening infections like infantile botulism. *Clostridium botulinum* spores thrive in honey; either packaged or fresh and produce preformed toxins if allowed sufficient time. Pre-lacteals also interfere with breast milk production. Another harmful practice apparently to prevent colic certain concoctions of various spices, sweeteners and gripe water are rampantly used in Indian households.

Table 43.1: The peculiarities of infants in relation to nutrition and feeding

Key Features	Birth to Six months	Sixth month to twelve months
Immune System	• Immature, compromised	• Evolving and a little more mature
Dependency on caregiver	• Totally dependant	• Significantly dependant
Predominant food foods along with breast milk	• Breast milk or powdered infant formula (PIF) milk	• Supplementary semi solid
Ideal feeding practices	• Exclusive breastfeeding	• Home based supplementary feeds with cereals, vegetables and fruits
Current feeding practices in India	• Exclusive breastfeeding in majority	• PIF in certain groups of infants
	• Home based cereals	• Commercial infant cereals

Table 43.2: Threats to food safety: A food safety chain concept

Biological	Chemical	Food safety chain	Mechanical	Lifestyle
• Bacteria • Fungi	• Pesticides • Fertilizer • Industrial chemicals • Heavy metals in soil	Production	• Food adulterants • Weeds	• Heavy dependency on pre-cooked meals or meals ready to eat (MRE) • Huge market for MRE and packaging industry
• Contamination in packages – Bacterial – Fungal	• Harmful chemicals from packing material, e.g. Bisphenol A	Transport	• Difficult to eat (choking) – Food adulterants – Large non-edible material in packaged food. – Pins, plastic pieces, etc.	• "All-in-one" containers for packing, heating and eating • "Organic natural products" with reduced shelf life
• Infantile botulism – Honey • Poor hand hygiene • Unclean utensils • Unclean water	• Household chemicals – Pesticides, paints, cosmetics from hands, etc. • Leaching from plastic container	Consumption	• Small hard foods (nuts) • Slippery foods (candy, berries) • Sticky foods (peanut butter)	• Poorly supervised feeding • Refrigerating and reheating of food • Feeding left overs • Crèches/day care center for infants of working mothers

These concoctions are more susceptible to contamination than any other food prepared at home.

At this juncture it is worthwhile to ponder over another major source of infection—*bottle feeding of milk*; either expressed breast milk (EBM), PIF, or animal milk. Feeding bottles and their poor hygiene have been identified as a major risk factor for diarrhea and respiratory infections. Despite the available information, several mothers opt for bottle feeding of infants. The very same ignorance continues to be a hazard when the child shifts to semi solid weaning foods even when it is prepared at home. The common agents implicated in deteriorating the quality of milk as well as food prepared at home are *Staphylococcus aureus*, Bacillus cereus, *Clostridium perfringens*, *Coliforms* and *Clostridium botulinum*. These agents are also implicated in outbreaks of food poisoning. Though most foodborne diseases are sporadic and often not reported in India, a nation-wide study carried out recently, reported an alarming 13.2% prevalence at household level. More often than not the magnitude of these outbreaks decide whether it is reported or goes unnoticed.

Chemical: The presence of chemicals in our day to day consumption of food; non-intentional or intentional (adulteration) is known and probably accepted as "normal" by consumers. Unfortunately, contamination of infant food begins right from breastfeeding; there have been several reports of pesticides especially 1,1,1-trichloro 2, 2-bis (chlorodiphenyl) ethane (DDT) in breast milk have been reported. Other pesticides are known to accumulate in food stuffs and some workers have described extremely high levels of pesticides in Indian food products compared to western products. Similarly intentional food adulteration too occurs at varying levels in a variety of foods, especially milk. It is found that during milk handling and packaging, detergents (used during cleaning operations) are not washed properly and find their way into the milk. Other contaminants like urea, starch, glucose, formalin along with detergent are used as intentional adulterants. These adulterants are used to artificially increase the viscosity of milk as well as to make it appear fresh for a longer period. Unfortunately, adulteration usually goes unreported unless it occurs at a large scale, like the outbreak of epidemic dropsy which occurred in Delhi, India, due to consumption of contaminated mustard oil. Further down the chain, contamination by chemicals occurs through contaminated packing materials used. The common chemicals that can contaminant packed foodstuffs are:

- Perchlorate especially in PIF.
- Lead from paints, used batteries, etc.
- Melamine, Bisphenol A (BPA) from plastic packaging of food stuffs.

Finally, at the consumer level the water used to prepare the food and the hygiene of utensils used also contribute to food safety. It is again worthwhile to note that plastic/melamine containers like cups, bottles and plates tend to leach harmful chemicals like BPA into the food, whereas traditional steel utensils are inert and do not.

Physical: Food materials are at risk from accidental as well as intentional macro-contamination beginning from production to consumption. The presence of macro impurities like weeds, gravel, glass pieces may cause physical injury to the infants' gastrointestinal tract (GIT) or behave like foreign bodies in the airway as well as GIT. Infants have restricted ability to consume foods that are hard (candy, nuts), smooth (grapes, berries) or sticky (toffee, peanut butter). These foodstuffs increase the risk of choking and may be life threatening.

Lifestyle as a Threat to Food Safety

As mankind progresses in science and technology, a faster pace of life with a "readymade culture" has overtaken the time tested traditional, deliberate way of life. The advent of industrialization, fast pace of life and both working parents; has exploded the culture of bottle feeding to newborns. This culture evidently originated in the west, so much so that there is no traditional Indian word for a "bottle"! Moreover, before the advent of plastics, bottles were made of steel or glass which was much easier to clean. The current trend is of plastic bottles which not only is difficult to sterilize but also are known to leach harmful chemicals into milk; especially when heated. This alarming trend of bottle feeding has percolated to every socioeconomic stratum; both urban and rural. In view of these hazards of bottle feeding, it is only mentioned to be condemned. As the infant grows beyond six months, the changing practices become more evident. Among the urban affluent class, the availability of "ready to feed" baby foods in disposable plastic containers has largely replaced the traditional cooking of fresh foods and serving in clean, washed utensils. On the other hand, among the underprivileged population poor hygiene remains the major threat. The potential lifestyle risk factors identified in this group are feeding leftover overnight food; not washing hands prior to cooking and feeding; consumption of the spilled food from the floor; use of dirty cloth for wiping hands and utensils; and the use of unsterilized dirty feeding bottles for the infants. However, somewhere in between, a grey area that has been poorly studied, that of recycling the so called disposable containers is emerging. The plastic containers cannot be washed like the steel utensils; secondly the harmful effects of reheating, reusing and recycling plastics carry the risk of release of harmful chemicals into food. One compound that has deserves special mention is Bisphenol A (BPA). It is a chemical for use primarily in the production of polycarbonate plastics and epoxy resins. It can leach into food from the protective internal epoxy resin coatings of canned foods and from polycarbonate tableware, food storage containers, water bottles, and baby bottles. The degree to which BPA leaches from polycarbonate bottles into liquid depends on the temperature of the contents, container and also the age of the container. BPA has a variety of toxic effects on the human body ranging multiple organ damage to carcinogenicity. Ready to eat infant foods are available in plastic containers; heated by microwaving and fed to infants in the same container. Microwaving of plastics has been found to increase the leaching of BPA into foods.

Impact of Food Safety and its breach: Food safety has a long reaching impact. The impact is seen at varying levels in the community, it may be at an international, national and at the household level. The impact of breach of food safety and of the efforts to maintain food safety on the community is summarized in Table 43.3.

Making Food Safe for Infants

The safety of food is a politico-social problem, fuelled by the commercial interests of the manufactures; lifestyle of the consumers and pressures of mass media. Thus the technological advances in food and agricultural sectors have made food safety programs increasingly necessary. In the past, food was consumed by those who produced it or by their immediate neighbors. The changing food habits owing to sociocultural pressures have forced the policymakers to think afresh and on a global scale.

The various measures that are undertaken for this purpose is depicted on the food safety chain model as in Table 43.4 and briefly described as follows:

- **Efforts at global level:** Codex Alimentarius Food Hygiene Basic Texts published by Food and Agriculture Organization (FAO) of the United Nations, World Health Organization (WHO), Rome, 2001 has laid out standards of safe food material on a global level. The WHO and FAO coordinate food research and promote knowledge sharing between various countries. Ensuring enforcement of their guidelines is the limitation faced by the international bodies. However, the enfor-

Table 43.3: The impact of breach in food safety and impact of the efforts to maintain food safety on the community

Food safety chain	Level of Impact	Impact of breach in food safety	Socio-economic impact of maintenance of food safety
Production / Transport / Consumption	**Large scale, at community/national/international level**	• Outbreaks of diarrhea, vomiting • Withdrawal/destruction of entire shipments of food by regulatory authorities	• Expansion of food industry that caters for fast foods, MRE • Garbage disposal problem problem needed to dispose used packaging material
	Large scale at community level	• Food poisoning outbreaks • Wastage of degraded food • Pilfering during transport leads to contaminated food being consumed	• Diversion and consumption of resources for packaging • Consumption of fuel for transport of food stuffs
	Household level	• Acute: Gastroenteritis, Jaundice, Botulism, etc. • Choking: Sticky food, nuts, grapes, macro contaminants • Late: Cancers, precocious puberty, allergies, etc.	• Rising cost of packaged food aggravating food insecurity in developing countries

cement of these standards remains with the individual countries through their legislative powers.

- **Efforts at national level in India**
  - **Breastfeeding promotion:** The guidelines for feeding practices of infants has been laid out by the IAP and endorsed by the Government of India (GOI). The most important component of these guidelines has been exclusive breast-

feeding for the first six months of life. Breastfeeding hospital initiative (BFHI) has been an important step to educate and advocate both to the mothers as well as the health care professionals regarding exclusive breastfeeding right from birth. The GOI has also made provisions to benefit nursing mothers by provision of six months of maternity leave to cater for this period and an additional cumulative period of two

Table 43.4: Measures to ensure food safety chain at global, national and individual level

International	Chemical	Food safety chain	Mechanical	Lifestyle
• Codex Alimentarius * Food Hygiene	• Food Safety and Standards Act, 2006[#] • The Infant Milk Substitutes, Act, 1992 as Amended in 2003 (IMS Act)	Production	• Adherence to standards • Food research • Strict quality	• Knowledge • Attitudes • Behavior • Practices
• Packaging and transport research	• Food Safety and standards (packaging and labelling) regulations, 2011 • Maternity leave policy Sixth Pay Commission, GOI [@]	Transport	• Adherence to packing regulations • Packing technology research • Storage guidelines enforcement	• Avoid "All–in–one" containers for packing, heating and eating • Awareness regarding harmful packing material • Monitor storage
• Consumer education and awareness	• The Consumer Protection Act, 1986[$]	Consumption	• ***Appropriate food for appropriate age*** • Maintain hygiene and sanitation • Hand washing	• Supervised feeding of young infants • Avoid refrigerating and reheating of leftovers • Safe inert utensils • Avoid harmful plastics

* http://www.codexalimentarius.org/

www.fssai.gov.in/

@ http://finmin.nic.in/6cpc/

$ http://www.ncdrc.nic.in

years of child care leave. The Breast-feeding Promotion Network of India (BPNI) has laid down guidelines for working mothers to continue breast-feeding. This can be achieved by expressing their milk before going to work and also express milk at workplace to store it. Expressed breast milk (EBM) may be stored at room temperature for 6 hours and refrigerated for 24 hours. Another method would be to carry the baby to work and feed on demand. To support this endeavor the employer will need to provide facilities firstly to preserve EBM and secondly to establish crèches at work place. However, strong legislation is required to enforce every employer to provide these facilities.

The concept of breast milk banking is not yet established in India, but has begun earnestly in some countries. Efforts need to be taken to establish guidelines for breast milk banking to help mothers with lactation problems to provide EBM for their infants.[15]

– **FBI prevention:** The exact extent of FBI in developing countries, including India has not been fully understood. This may be due to the time lapse in reporting disease outbreaks, as the FBI appear to be non-epidemic in nature and are most often recognized neither by the public nor by the health authorities. The actual scenario of FBI can emerge only with proper emphasis on surveillance and establishment of a national FBI surveillance system. Despite these shortcomings, India has one of the most stringent policies on food safety.

Prior to 2006, a variety of acts and rules were in enforcement. In 2006, all the rules related to food safety were brought under the FSS Act, 2006 under the aegis of the FSSAI. In the light of increasing usage of packaged foods, FSSAI has also laid out specific standards for packing and labelling of foods in the form of the Food Safety and Standards (Packaging and Labelling) Regulations, 2011. Infant milk products in India follow the requirements of the FSSAI. Their production, supply and distribution are regulated by The Infant Milk Substitutes, Feeding Bottles and Infant Foods Act, 1992 as Amended in 2003 (IMS Act). However despite the regulations being in place, the occurrence of FBI cannot be prevented until the manufacturers adhere to these guidelines and the government does not strictly enforce them. Finally, the consumers have an important role in monitoring the standards of food and provide valuable feedback to the enforcement authorities.

• **Efforts and role of the consumer:** The most important link in the chain of food safety is the consumer. At an individual level the implementation of the various policies, both national and international can be observed, advocated and implemented. In this present era of rapid communication and heightened public opinion, requisite feedback may be provided to the concerned authorities with ease. However, this important link has one major weakness—*Lifestyle*…….

The changing lifestyle is the single largest driving force towards compromising food safety. Owing to socioeconomic pressures, the changes in lifestyles are seen in both the affluent class as well as the underprivileged; both rural as well as urban population. The various lifestyle changes that have occurred in contrast to the traditional habits are as described in Table 43.5. The rise of consumerism and commercial interests has largely replaced traditional practices with semi-rational, convenient and potentially harmful ones. The one practice that can be singled out is the use of formula feeds for infants—both PIF and commercial infant cereals. Additional driving force for its growth is the effect of advertisements on infant feeding practices. Despite the Infant Milk Substitution Act in place, it is noted in one study that subtle encouragement to formula feeding continues in the form of breastfeeding promotion material distributed by PIF manufacturers. Similarly marketing of commercial infant cereals is usually perceived as a mandatory requirement rather than an alternative by the mothers. Therefore the need of the hour is to "market" breastfeeding and traditional weaning practices actively rather than just advocate it.

Role of the family: The most important change in Indian society that has occurred in the last century has been a paradigm shift from joint family to nuclear and further "separated" families. As summarized in Table 43.5, the caregiver of a child has shifted from family elders to an unsupervised mother or more often a house maid or crèche with both working parents. As expected the commitment and experience too varies, thereby summarily increasing the risk of infection due to poor hygiene. The need for supervision is more so highlighted when the child begins to crawl and explore the surroundings. During this phase a child tends to put any small object in the mouth and carries the highest risk of choking. Thus, changing family dynamics now highlights the role of the father in sharing responsibility as a caregiver. The GOI has also appreciated this role by authorizing paternity leave for new fathers.

Table 43.5: The lifestyle changes with respect to feeding in both affluent and underprivileged population in comparison to traditional feeding practices

Feeding practice in infants	Traditional practice	Practice in underprivileged population	Practice in the affluent population
Primary feeding	• Breastfeeding	• Early initiation of diluted cow's milk or diluted PIF	• Early initiation of PIF
Mode of feeding milk **Sterilizing bottles**	• Bowl and spoon • Steel bottles/sippers • Boiling	• Plastic feeding bottles • Heating/inadequate boiling due to rising cost of fuel	• Plastic feeding bottles • Microwave sterilizing of plastic bottles
Weaning foods	• Homemade foods using traditional processing techniques	• Packaged cereals usually diluted • Street foods	• Packaged cereals • Precooked packed foods
Continued breastfeeding	• Till up to two years	• Abrupt cessation	• Early cessation • Primary PIF feeding
Caregiver and the risk of contamination of food	• Usual handling by primary caregiver under supervision of older experienced person at home– *minimal risk*	• Nuclear families with inexperienced mother handling more than one child • Elder sibling likely to have respiratory infection/diarrhea thus *increasing risk of infection to the young* infant	• Nuclear families with working mothers • Infant being taken care at a day care center or by household help • *High risk of infection[17]*

Conclusions

The progress and advances in food technology has thrown open innumerable choices regarding the availability, type, quality, quantity, appearance and taste of food for infants. Along with these choices, however the traditional food processing knowledge is slowly dying. Therefore, mankind is presently poised at a junction of forgoing centuries old knowledge for a convenient technology of extremely short duration; thus precariously balancing the safety of food on the current socioeconomic mantle. The chain of food safety has grown longer and complex, thus exposing several weak links. Therefore, despite several legislations being in place, it is still the responsibility of the society to ensure that the next generation at least has access to safe food if not a safe world.

44

Food Safety Issues During Pregnancy

Rakhi Kumari, Meenakshi Sharma, Puja Dudeja

It is 7 o' clock in the morning. Breakfast is ready. Smita has to rush!!. She packed the bag, water bottle and locked door in a hurry. Suddenly she screamed, "Oh God! I forgot to trim nails". This is not an extract from the life of a schoolgoing kid whose mother is scared of a diary note from the teacher for not trimming the nails. But my friend Smita who was strongly advised by the doctor to trim nails during pregnancy. She had warned her, "You should maintain good personal and kitchen hygiene to avoid any foodborne illness during pregnancy". It may prove problematic for you and your foetus. It is rightly said, 'A little precaution is worth a pound of cure'.

Pregnancy is an all together different experience for any woman. The joy of giving birth to a young one is difficult to be put in words. It can only be experienced. Pregnant mothers always wonder that what they should eat and what not. It is commonly said that a pregnant lady should take balanced diet so that mother and child both get proper nutrition for their body. It means that a she should be careful about the proper quantity and quality of any food she takes.

Diet during pregnancy has been a topic of discussion over the centuries, with science often taking a backseat to the old wives' tales surrounding the experience. There are a numbers of myths too regarding diet in pregnancy. These myths revolve around correlating a particular diet to the complexion of the unborn or intelligence or correlating food urges to the sex of unborn. For example, eating curd by the mother leads to a fair complexioned baby. Another common statement is that a pregnant lady has to 'eat for two' This is a myth as a pregnant lady does not need to 'eat for two'—even if she is expecting twins or triplets. A pregnant lady needs few extra calories than a normal female but definitely not double the baseline requirement.

Apart from the dietary content, it is important that whatever is consumed by a pregnant lady is safe. This issue remains neglected in various counseling sessions of antenatal care. It will be worthwhile here to discuss here the importance of safe food in pregnancy. Unsafe food can harm the foetus directly or indirectly (if the mother takes antibiotics).

Pregnant women are at increased risk for getting some FBI because of the hormonal changes that occur during pregnancy. While such changes are necessary for survival of the fetus, they also suppress the mother's immune system, thereby increasing the chance of infection from certain foodborne pathogens. Pregnant women are also more vulnerable to complications, so even a simple gastroenteritis can lead to dehydration and warrant hospitalization. Most of the FBI cause febrile illness along with other symptoms like diarrhea, vomiting. Any febrile illness in pregnancy may cause miscarriage or premature labour or other problems.

Pregnancy has its own symptoms of nausea, vomiting, etc. in first trimester. There is upper abdominal pain in the second trimester due to dyspepsia. Apart from this there may be false labour pains in third trimester. All these symptoms which are a part of normal pregnancy overlap with symptoms of a gastrointestinal tract infection. This makes it difficult for the health care provider to diagnose any FBI and hence delay in diagnosis and treatment. There are certain gastrointestinal infections which can directly harm the fetus, e.g. listeriosis and salmonellosis. Pregnant women are also at higher risk of travellers' diarrhea.

Many FBIes in pregnancy are self-limiting and relatively harmless, but some may be serious for the mother or fetus and warrant use of antibiotics. Some of these drugs are teratogenic and can affect the fetus. Many drugs are not recommended in pregnancy as they cross the placental barrier and reach the fetus. Pregnant women are considered a special risk group for FBI because of potential risks to the health of both mother and child.

Pregnancy is known to cause mood swings and cravings for different food items. Sometimes it is difficult to satisfy these food cravings in a short time interval by cooking at home. They are satisfied by eating the particular food from outside which also puts the pregnant lady at risk.

Some gastrointestinal infections which are of concern in pregnancy are as follows.

Infections that may Affect the Fetus

Listeria in Pregnancy

Listeriosis is a form of infection that may result when foods containing the bacteria *Listeria monocytogenes* are consumed. *L. monocy*togenes is widely distributed in nature and is found in soil, ground water, plants and animals. It is, however, easily destroyed by cooking. Infection from *L. monocytogenes* typically occurs in individuals with a weakened immune system, **including pregnant women**. There is an estimated 14-fold increase in the incidence of listeriosis among pregnant women compared to non-pregnant adults. Pregnant women make up 17 percent of all cases of listeriosis. Once in the bloodstream, Listeria bacteria can travel to any site, but seem to prefer the central nervous system and the placenta. The fetus is unusually prone to infection from *L. monocytogenes*, which can lead to a miscarriage, stillbirth, or infection of the neonate and health problems following birth.

Foods typically associated with listeriosis include refrigerated ready-to-eat perishable foods with a long shelf life that are eaten without further cooking. Examples of foods that may harbor this pathogen include unpasteurized milk, raw milk products, raw and any refrigerated ready-to-eat processed foods, such as burgers, sandwiches, etc. that have not been heated to proper temperatures before serving.

Salmonellosis in Pregnancy

According to the WHO, salmonella bacteria are one of the most common causes of FBI around the world. Most salmonella subspecies, including the two most common variants—*Salmonella enteritidis* and *Salmonella typhimurium*—produce "salmonellosis" which is a mild, self-limited gastroenteritis. However, pregnant women are more difficult to treat because fluoroquinolones—the antibiotic of choice in salmonella infection—are associated with birth defects. In rare cases, Salmonella can escape the intestine to enter the bloodstream. Bloodstream infection may itself be fatal and can produce longer-term complications when salmonella leaves the bloodstream to infect other areas of the body. Longer-term complications of salmonella infections include infection of the heart valves and lining of the heart (endocarditis), the bone (osteomyelitis), the kidneys (pyelonephritis), brain abcess and Reiter's syndrome, an autoimmune disease that produces chronic joint pain, eye irritation and urination problems. These complications appear to be **more common in pregnant women**, compared to other healthy adults.

This infection can occur through animal products like meat, poultry, dirty egg shell and products made from them, high risk foods exposed to warm temperatures like milk, fish, *mutton biryani*, sea food from canned foods that have been contaminated and are held without refrigeration once opened.

Salmonella infection crosses the placenta and may produce severe disease and death in the fetus, even when maternal symptoms are mild. In a 2004 report in the Scandinavian Journal of Infectious Disease, a pregnant woman admitted at 25 weeks gestation for *Salmonella gastroenteritis* underwent cesarean section for abnormal fetal heartbeat. Despite intensive medical intervention, the infant died four hours later from culture-proven salmonella bloodstream infection and infection-induced multi-system organ failure. Similarly, May 2008 issue of Archives of Obstetrics and Gynecology described a case of spontaneous abortion at 16 weeks gestation, 1 week after resolution of mild maternal salmonella infection.

Toxoplasmosis in Pregnancy

This infection is caused by the parasite *Toxoplasma gondii*, and can be passed to humans by water, dust, soil, or through eating contaminated foods. Most individuals do not experience recognizable symptoms, and will develop a protective resistance to the parasite. However, if a woman not previously exposed to *T. gondii* first acquires the parasite a few months before or during pregnancy, she may pass the organism to the fetus. This could result in stillbirth, early prenatal death, or serious health problems for the baby after birth such as eye or brain damage. Symptoms in the baby may not be visible at birth, but can appear months or even years later. Toxoplasmosis most often results from eating raw or undercooked meat or eating unwashed fruits and vegetables.

Campylobacteriosis in Pregnancy

Consuming food or water that contains the bacteria *Campylobacter jejuni* causes an infection called Campylobacteriosis. Although pregnant women are not at increased risk of becoming infected with *C. jejuni*, if they do get sick, the infection may spread to the placenta. Consequences of fetal infection include abortion, stillbirth or preterm delivery.

C. jejuni is most often found in raw (unpasteurized) milk and raw milk products, raw or undercooked meat and poultry, and raw shellfish.

Infections, which carry a higher risk of causing severe illness in the mother:

- ✓ *Escherichia coli* O157
- ✓ Shigellosis (bacillary dysentery)
- ✓ *Clostridium difficile* – (in case of a history of antibiotics or hospitalisation)
- ✓ Cholera
- ✓ Cryptosporidium

- • *Parasitic Infections Which may Require Specific Identification and Treatment*

  - ✓ *Giardia lamblia*
  - ✓ *Entamoeba* spp.

Ensuring Safe Food in Pregnancy

Safety of food in pregnancy will prevent FBI and hence keep the mother and fetus healthy. Food safety tips can be divided into two groups viz. outside home/during travel and at home. Food safety precautions which need to be taken care of outside home or during travel are:

- Carry water and food from home in case visiting a hospital with long waiting hours in the OPD/cinema hall/fair
- Avoid street food
- If eating outside prefer a place which is hygienic. Also prefer to eat hot foods than cold, for example, between samosa and sandwich select samosa.
- Avoid salads, curd, mayonnaise containing items, cream items

Food Safety at Home

Four golden principles which can be adopted in a home kitchen to ensure food safety are given in Table 44.1. Apart from this, whenever going outside home prefer to carry your own water bootle. If affordable a hand sanitizer can be added in the purse items. While working in kitchen ensure nails are short and clean and hair are tied. These precautions are applicable to all individuals but adherence is of special concern during pregnancy.

RAW FISH AND SHELLFISH

Seafood-related FBI is most commonly associated with the consumption of raw or under-cooked seafood. Pathogens associated with the consumption of seafood include noroviruses, Vibrionaceae and *Salmonella* (bacteria) species, and some helminthic and protozoan species. Pregnant women need not avoid raw fish if it is obtained from a reputable establishment, stored properly, and consumed soon after purchase. Women should limit their consumption of high mercury fish and shellfish, including fresh tuna and yellowtail although low mercury alternatives (e.g. salmon, crab, and shrimp) can be consumed more regularly.

Despite all precautions in case there is an episode of FBI during pregnancy then it needs to be given due attention to prevent any complications. In case of mild illness then home based care include replenishment of fluids through intake of Oral Rehydration Solution (ORS). Other fluids like, *lassi, nimboo*

Table 44.1: Precautions to be taken at home for safety of food	
Keep it cold	**Keep it clean**
• Put any food that needs to be kept cold in the fridge straight away • Do not eat food that's meant to be in the fridge if it is been left out for two hours or more	• Wash and dry hands thoroughly before starting to prepare or eat any food, even a snack • Keep kitchen equipment clean • Separate raw and cooked food and use different cutting boards and knives for each • Do not let raw meat juices drip onto other foods • Avoid eating food made by someone sick with something like diarrhea
Keep it hot	**Check the label**
• Cook foods until they are steaming hot • Reheat foods until they are steaming hot • Cook chicken thoroughly	• Do not eat food past the use-by date • Note the best before date • Follow storage and cooking instructions

pani, dal ka pani, khichdi, curd, yoghurt can be consumed. Avoid fruit juices and carbonated drinks.

In case the symptoms warrant hospitalization then maternal fetal monitoring is done along with IV fluids. Avoid self medication and drugs like antiemetics, antibiotics should be taken after consultation with the doctor only.

Case Study 1

One of our patients presented at 32 weeks period of gestation with history of diarrhea, 6–7 episodes of watery stools since 1 day. She told that she took *chicken biryani* one day prior to onset of illness. She stayed at home for a day thinking that it will resolve on its own. Next day she came to our OPD in view of non-resolution of diarrhea. She also had pain in abdomen on and off. On examination, her vitals were stable, was well hydrated and nails were clean and clipped. On per abdomen examination, fundal height corresponded to period of gestation, uterus was relaxed, i.e. no contractions with regular fetal heart rate of 146/min. Her fetal movement count was adequate. Ultrasonogaphy showed live fetus with good biophysical profile (adequate liquor).

She was advised to take khichidi, dalia, dahi, banana, etc. till diarrhea resolves. She was also advised to take plenty of fluids in any form like *naryal paani*, lemon water or oral rehydration solution, and lactobacilli sachet thrice a day, and to stop iron tablets for 1 week (iron itself can lead to or aggravate diarrhea) and to report to emergency if diarrhea does not resolve. She reported to the emergency 1 day later with persistent diarrhea. She was started on intravenous fluids ringer lactate. Strict input output was kept. Serum electrolytes and renal function tests were normal, Total Leucocyte Count was marginally raised 14000/cumm. Antibiotics were given and she was discharged after three days.

Case Study 2

Another pregnant lady came at 24 weeks of pregnancy with vomiting , 5–6 episodes per day. Vomitus contained food particles only and was nonprojectile. Her pregnancy induced vomiting has stopped by 13 weeks. She gave history of eating *golgappas* (water balls) a day before in a roadside stall. As per patient urine output was adequate. On examination she was hemodynamically stable with adequate urine output. She was slightly dehydrated. Her urine examination showed no urine albumin or sugar but ketones were 2+ due to dehydration.

She was kept under observation in the emergency. She was kept nil per oral till her vomiting stopped. Intravenous fluids and drugs were given. Strict fluid input and output monitoring was done. Serum electrolytes, renal and liver function tests were done. She recovered in 2 days and was discharged in a satisfactory condition with an advice to eat a healthy homemade diet.

The above two case studies illustrate a single fact that the cause of FBI was avoidable. Both the ladies consumed outside food which was probably contaminated and unsafe. They could have faced serious complications too. If due precautions had been taken in selection of food item/eating establishment even while eating outside food unnecessary medication and hospitalizations could have been avoided.

Conclusion

Food safety is a concern for all individuals but even more so for the pregnant woman and the fetus, as they might be more susceptible to some FBI with serious ramifications. As general guidelines to food safety, pregnant women should ensure that their food is obtained from reputable establishments; stored, handled, and cooked properly; and consumed in a timely manner.

45

Food Safety in HIV Positive Patients

Puja Dudeja, Sukhbir Singh

INTRODUCTION

HIV infection has been the biggest scourge of humankind in last 40 years. Until the advent of Anti Retroviral Therapy (ART) it was known that 'prevention is the only cure'. However, research is ongoing in this field to make Human Immunodeficiency Virus (HIV) positive people to HIV negative, i.e. free from HIV infection. The total number of People Living with HIV/AIDS (PLHIV) in India is estimated at 21 lakh (17.2–25.3 lakh) in 2011. Children (under 15 yrs) account for 7% (1.45 lakh) of all infections, while 86% are in the age group of 15–49 years. Of all HIV infections, 39% (8.16 lakh) are among women. The estimated number of PLHIV in India maintains a steady declining trend from 23.2 lakh in 2006 to 21 lakh in 2011.

Apart from hard-core clinic and lab based management of PLHIV there is a vast scope of supportive therapy. Nutrition is one such domain. Good nutrition has a definite scope in management of HIV cases. HIV causes immune impairment leading to malnutrition, which leads to further immune deficiency, and contributes to rapid progression of HIV infection to AIDS. A malnourished person after acquiring HIV is likely to progress faster to AIDS, because his body is weak to fight infection whereas a well-nourished person can fight the illness better. It has been proved that good nutrition increases resistance to infection and disease, improves energy, and thus makes a person stronger and more productive. On the other hand malnutrition adds fuel to the fire by accelerating the progress of HIV infection to AIDS.

HIV/AIDS is associated with biological and social factors that affect the individual's ability to consume, utilize, and acquire food. Once there is an infection with HIV, the patient's nutritional status declines further leading to immune depletion and HIV progression. During general management of HIV cases emphasis is given to the special nutritional requirements to improve immunity. However, the needs of food safety are generally ignored or overlooked. Major pathogens that may cause FBI in HIV positive patients are given in Table 45.1.

Food safety is important for all but more so for HIV positive people as these patients are more vulnerable to infections. This is because of the weakened immune system and reduced count of CD4 cells in the body. A properly functioning immune system helps to fight infections and other foreign agents from the body. Whereas in HIV positives this ability is markedly reduced leading to a more severe form of disease, a lengthier illness,

Table 45.1: Common causes of foodborne illness and their sources in HIV positive patients

Organism	Food items
Salmonella	Raw and undercooked poultry and eggs
Campylobacter	Untreated or contaminated water, unpasteurized ("raw") milk Raw or undercooked meat, poultry or shellfish
Cryptosporidium	Swallowing contaminated water, including that from recreational sources (e.g., swimming pool or lake), eating uncooked or contaminated food, placing a contaminated object in the mouth, soil, food, water, contaminated surfaces
Clostridium perfringens	Many outbreaks result from food left for long periods at room temperature and time and/or temperature abused foods. Meats, meat products, poultry, poultry products, and gravy
Listeria monocytogenes	Can grow slowly at refrigerator temperatures
	Improperly reheated foods, incompletely cooked poultry, pasteurized (raw) milk and soft cheeses made with unpasteurized (raw) milk, smoked seafood and salads, raw vegetables
Escherichia coli O157:H7	Unpasteurized milk and juices, like "fresh" apple cider, contaminated raw fruits and vegetables, and water. Person-to-person contact
Noroviruses (and other caliciviruses)	Shellfish and fecally-contaminated foods or water, Ready-to-eat foods touched by infected food workers; for example, salads, sandwiches, ice, fruits

hospitalization more often, or even death in event of a simple episode of FBI. HIV infection causes chronic progressive immunodeficiency through reducing CD4+ T-cell lymphocytes. This results in various opportunistic infections of the small intestine characteristically causing chronic disease with diarrhoea and malabsorption. About 70% of chronic weight loss in HIV is associated with gastrointestinal infection and diarrhea. Opportunistic parasitic gut infections cause severe diarrhea and profoundly compromised small bowel absorptive function and cause significant mortality in AIDS both in Western and developing countries.

Foodborne organisms causing diarrhea in people with HIV include nontyphoidal *Salmonella* and *Giardia* and, less commonly, *Shigella, Campylobacter, Microsporidium, Cryptosporidium, Isospora*, and *Cyclospora*. Patients with acquired immunodeficiency syndrome (AIDS) are also at greater risk of invasive listeriosis than the general population, although widespread CD4 counts below 200/µl are associated with toxoplasmosis encephalitis, and diarrhea caused by *Cryptosporidium*, which can be severe. People with advanced HIV are particularly susceptible to recurrent, invasive salmonellosis and occasionally to *Salmonella* meningitis. In view of above, it makes sense that due importance is given to food safety aspect in PLHIV in home as well in a hospital setting.

Eating at Home: Making Wise Food Choices

It is well said that from nutrition and food safety point of view home cooked food is better than eating outside. However, some foods can be the cause of illness even when eaten at home. For example, uncooked fresh fruits and vegetables, some animal products, such as unpasteurized (raw) milk; raw or undercooked eggs, raw meat, raw poultry and raw fish. The risks these foods impose also depend upon the origin or source of the food and how the food is processed, stored, and prepared. The risks associated with home cooked food can be lowered by reducing the risk as given in Table 45.2.

Food that appears completely fine can still contain pathogens. PLHIV should never taste a food to determine if it is safe to eat. During preparation of food for a person with HIV/AIDS, the basic four principles of—clean, separate, cook, and chill should be strictly followed. Full support of spouse, children and other family members is desirable in such endeavor.

Food Safety for PLHIV in Hospitals

The word hospital originates from the Latin word hospice where hospitality is rendered. Hospitals are not just curative centers but also promote healthful living. Hospitals must ensure that they do no harm to the patients. Nutrition/dietary/catering services are an important component of the overall care of the patient. Hospital kitchens cater to a large population group comprising patients, doctors, nurses, staff, etc. The dietetics department is generally involved with the composition, nutritional requirements and food exchanges in diet of all patients in the hospital. It is their responsibility to ensure safety of food at all stages viz. procurement, receiving, storage, cooking and transport. However, they must give due emphasis to safe handling of food meant for PLHIV. Nursing staff should be given special instructions to supervise feeding of these patients. They should also keep a check on food brought by relatives and visitors of such patients.

Apart from hospital kitchen there are various eating establishments (EE) in the hospital premises where the visitors, OPD cases, PLHIV visiting ART centers visit. The hospital administration along with the food safety authorities should ensure licensing/registration of these EEs. Regular inspections and training of the food handlers' employed in these EEs should be conducted to ensure provision of safe food.

Table 45.2: Lowering the food safety risk in home available food		
Type of food	*Avoid high risk form*	*Prefer low risk form*
Meat and poultry	Avoid in raw form or rarely done form	Cook well before consumption
Milk	Unpasteurized milk	Pasteurized milk. Boil before consumption
Eggs	Do not consume salad dressings with raw eggs like mayonnaise	Cook well before consumption
Sprouts	Avoid in raw form	Take cooked sprouts
Vegetables	Unwashed	Washed and cooked
Cheese	Soft cheese from unpasteurized milk	Hard cheese/chese from pasteurized milk
Fruits	Eat whole fruit	Peel fruits before eating

The Hospital Infection Control Committee (HICC) should strictly monitor and investigate occurrence of any nosocomial FBI in PLHIV .

Conclusion

HIV infection has assumed worldwide importance and continuous research is being done in many parts of the world regarding its treatment and vaccine development. Till the time there is a cure available the focus is on improving the quality of life of PLHIV through education, counselling, and nutritional support. Knowledge of essential components of nutrition and food safety will go a long way in improving quality of life and better survival in HIV-infected patients.

Section VIII

Newer Issues in Food Safety

46

Organic Farming

Mamta Bansal

With rapid increase in population it has become a necessity to use modern agricultural practices to meet the demand for food in the country. Before the advent of green revolution, in Indian agriculture which was based on the traditional knowledge and practices was not able to produce enough to feed the entire population. The green revolution fulfilled our aspirations by changing India from a food importing to a food exporting nation. The role of agricultural practices in economic development in an agrarian country like India is a predominant one. It provides food for more than 1 billion people and yields raw materials for agro-based industries.

Modernization of Indian agriculture began during the mid-sixties which also included the use of high yielding varieties of seeds, use of chemical fertilizers and pesticides, multiple cropping systems. But, after some time, it has been realized that these methodologies put severe pressure on natural resources like, land and water. This achievement was at the expense of ecology and environment and to the detriment of the well-being of the people. The negative effects of progress in agriculture on the environment manifested through soil erosion, water contamination and shortages, salination and soil contamination. The adverse effects of modern agricultural practices not only impact the farms but also harm the living and non living environment. Combination of heavy use of chemical fertilizers and pesticides by farmers has contaminated the farm produce. The continuous growth of modern agricultural technology along with the intensive use of natural resources, many of them of non renewable, has forced experts, scientists and environmentalists to conclude that these practices would not be sustained in future because the concern is now about the safety of food and damage to the environment.

Almost all pesticides are toxic in nature and pollute the environment leading to grave damage to ecology and human life itself. This indiscriminate use leaves toxic residues in foodgrains, fodder, vegetables, meat, milk, milk products, etc. besides in soil and water. The use of pesticides in India has increased from 24.32 thousand tonnes from 1970s to 44.58 thousand tonnes in 2001. Consumption of pesticides has almost doubled in past 30 years. In the same way the use of fertilizers in India has increased from 2.18 million tonnes in 1970s to 17.54 million tones in 2001. Consumption of fertilizers has increased almost seven times in past 30 years.

The necessity of having an alternative agriculture method which is conducive to evolution of a friendly eco-system while sustaining and increasing the crop productivity has been felt for quite some time. Policies have been designed to improve the environmental sustainability of agriculture. This includes bans on increasing use of pesticides, financial incentives and funding for agricultural improvement and technologies. In this perspective, organic farming is now being promoted as the best known alternative method for sustainable agriculture. Organic farming is based on the similar principles underlying our traditional agriculture (without the use of synthetic chemicals) but without any harm to the environment or to the human life itself.

Definition of Organic Agriculture

The term 'organic' was first used in 1940 by Northbourne in the book *Look to the Land*: 'the farm itself must have a biological completeness; it must be a living entity, it must be a unit which has within itself a balanced organic life'. Clearly, Northbourne was not simply referring to organic inputs such as compost, but rather to the concept of managing a farm as an integrated, whole system.

According to International food standards, Codex Alimentarius, Organic agriculture is a holistic production management system which promotes and enhances agroecosystem health, including biodiversity, biological cycles, and soil biological activity. It emphasises the use of management practices in preference to the use of off-farm inputs, taking into account that regional conditions require locally adapted systems. This is accomplished by using, where possible, agronomic, biological, and mechanical methods, as opposed to using synthetic materials, to fulfil any specific function within the system (FAO 1999).

Organic farming or agriculture is also defined as 'a system that is designed and maintained to produce agricultural products by the use of methods and substances that maintain the integrity of organic agricultural products without the use of extraneous synthetic additives or processing until they reach the consumers.

In organic agriculture, nutrients are returned to the soil in the form of manures and composts. These have to be cycled via the biological life of the soil before they become available to crops. In organic farming control of pests, disease and weeds is done through some practices like crop rotation, weeding, mulching, intercropping, maintenance of natural predator populations and biological controls. So, the organic agricultural practices are based on a maximum harmonious relationship with nature aiming at the nondestruction of the environment.

Origin of Organic Farming

Before the advent of pesticides and fertilizers, the organic agricultural practices were the only option for farmers. They worked within biological and ecological systems available. For example, the only source of fertilizer to replace nutrients from cropped fields was human and animal manure and leguminous plants. So, the origin of modern organic agriculture are intertwined with the birth of today's 'industry based' agriculture.

The research on organic agriculture started in the 19th century when it was discovered that it was the mineral salts contained in humus and manure that plants absorbed and not organic matter. The key founders of this theory were Sir Humphrey Davy and Justus von Liebig, who had published their ideas in *Elements of Agricultural Chemistry* (Davy 1813) and *Organic Chemistry in its Application to Agriculture and Physiology* (von Liebig 1840). According to the theory inorganic mineral fertilizers could increase the production and efficiency of crops and can replace manure.

The agricultural revolution began in the 1840s and with it came the first commercial production of inorganic fertilizers. In 1924 Rudolph Steiner, the founder of the philosophy of 'Anthroposophy' gave his agricultural lectures in 1924 which were the foundation of biodynamic agriculture. The first organic certification and labelling system, 'Demeter', was created in 1924 because of Steiner's actions.

Sir Albert Howard was also working in agriculture institute in India in 1920s. In his book *'The Waste Products of Agriculture'* (1931), he established the inextricable linkages between the health of the soil and the health of the plants and animals fed by that soil. This led him to research adapting oriental methods of composting to Indian conditions which resulted in the 'Indore process' of composting. The work and publications of some people like Howard, McCarrison and Steiner influenced the next wave of organic pioneers with the associations such as the Rodale Institute in the United USA, Soil and Health in New Zealand and the Soil Association in the UK. In the UK, Lady Eve Balfour in her book *The Living Soil* in 1943, compared organic and non-organic production over the long term. In 1947 J.I. Rodale from rural Pennsylvania, USA, realized the importance of restoring and protecting the natural health of the soil to preserve and improve human health and founded the Soil and Health Foundation that later become 'The Rodale Institute' and published literature on healthy soil, equals healthy food, equals healthy people.

The publication of *Silent Spring* by Rachel Carson (1962) was a key turning point for, and the start of, both the modern organic and environmental movements. *Silent Spring* opened the world's eyes to the damage that pesticides and other toxins were doing to the global environment. This change could well be considered a revolution and, at the least, a significant evolution of the organic movement.

Food Safety in Organic Agriculture

In this context food safety means to what extent do organic and non-organic foods contain undesirable components such as potentially harmful chemicals, drug residues and pathogens. While talking about organic farming and its relation to food safety, before reaching to any conclusion, the following areas need a discussion.

Pesticides in Organic Food

While many in agriculture believe that pesticides are necessary to produce and protect crops, it is universally agreed that consumer exposure to these toxins should be minimised on safety grounds. Due to the persistent nature of many pesticides, air, water and soils are inevitably contaminated by them. In organic farming, although the use of pesticides is restricted, but a small number of nonsynthetic pesticides can be used. These pesticides have been approved on the basis of their origin, environmental impact and potential to persist as residues. Some of them are copper ammonium carbonate; copper sulphate; copper oxychloride; sulphur; pyrethrum; soft soap and derris (rotenone). Some plant oils such as neem and microbial agents such as *Bacillus thuringiensis* (Bt) are also permitted. These pesticides are generally simpler substances than those used in non-organic agriculture. These tend to degrade quicker.

There are concerns about the safety of these compounds permitted in organic agriculture though, as would be expected given the prohibition of routine pesticide applications, organically grown food is usually found to have no residues.

Quality of Organically Grown Food

In recent years, people have become more concerned about safe food and quality. They feel that organic food is qualitatively better than non-organic. It is produced with the help of new technique where the use of fertilizers and pesticides are prohibited. About 59% people think that the term 'Organic food' means there is no chemical, additives or pesticides are present in food. Around 40% people think it is healthy and natural. But some 35% people think that organic food in very expensive, GM free and good for the environment. The people now feel that organic food is good for health (high nutritional value and minimal artificial chemical residues). It is also safe to environment (i.e. environmentally friendly production and processing). It has better taste and sometime it has become novelty and fashion to consume organic food. But the consumption of organic products is much limited by the public because of some other reasons like its high price and limited availability in the market. People are also not aware about the concept of organic food.

Not much research has been carried out to compare the nutritional quality of organically and non-organically grown foods. On the basis of the currently available data, the nutritional quality in organic grown food can be reviewed by its primary essential nutrients such as water, fibres, proteins, fats, carbohydrates, vitamins, dry matter and minerals or secondary metabolites' or 'phytonutrients' present in plants. There are some 5,000–10,000 secondary compounds in plants which are considered as the health-promoting and protective properties of food and thus necessary for health. Organically grown vegetables, tend to have 10–50 percent more nutrient level (primary and secondary or 'plant defence related' compounds) than conventionally cultivated vegetables. These metabolites are an important determinant of the nutritional value of fruits and vegetables in the diet, and hence expecting to be more health promoting than non-organic ones.

Benefits of Organic Farming

The potential benefits of the organic methods are as follows.

Healthy and Safe Foods

A study conducted in USA on the nutritional values of both organic and conventional foods found that consumption of the former is healthier. Apples, pears, potatoes, corn, wheat and baby foods were analyzed to find out 'bad' elements such as aluminum, cadimum, lead and mercury and also 'good' elements like boron, calcium, iron, magnesium sellenium and zinc. The organic food, in general, had more than 20 percent less of the bad elements and about 100 percent more of the good elements. There are other reports of positive health effects in humans resulting from the consumption of organically grown foods. For example, the exposures to pesticides either through diet or in occupation may influence sperm quality. The average sperm concentrations around the world, in the 50 years until 1990, fell by half from around 113 million/ml to around 66 million/ml and are still falling by around 2 percent a year.

Improvement in Soil Quality Increased Crop Productivity and Income

Soil quality is the foundation on which organic farming is based. Natural plant nutrients from green manures, farmyard manures, composts and plant residues build organic content in the soil. Efforts are directed to build and maintain the soil fertility through the farming practices. For example, multicropping, crop rotations, organic manures and minimum tillage are the various methods. The Central Institute for

Cotton Research, Nagpur conducted a study of economics of cotton cultivation in Yavatmal district of Maharashta. The cost of cultivation of cotton was lower in the organic farming than in the modern system which is based in the use of pesticides and other chemicals. Another study of 100 farmers of organic and conventional methods in five districts of Karnataka indicated that the cost of organic farming was lower by 80 percent than that of the conventional.

Low Incidence of Pests

The study of the effectiveness of organic cotton cultivation on pests at the farm of Central Institute for Cotton Research, Nagpur revealed that the mean monthly counts of eggs, larva and adults of were far lesser under organic farming than under the conventional method. Bio-control methods like the neem based pesticides to Ti-ichoderma are available in the country. Indigenous technological products such as Panchagavya (five products of cow origin) which was experimented at the University of Agricultural Sciences, Bangalore found to control effectively wilt disease in tomato.

Indirect Benefits

Several indirect benefits from organic farming are available to both the farmers and consumers. While the consumers get healthy foods with better palatability and taste and nutritive values, the farmers are indirectly benefited from healthy soils and farm production environment. Eco-tourism is increasingly becoming popular and organic farms have turned into such favourite spots in countries like Italy. Protection of the ecosystem, flora, fauna and increased biodiversity and the resulting benefits to all human and living things are great advantages of organic farming.

Organic Agriculture: International Perspective

The report from IFOAM suggests that of the US$23 billion global retail market for organic food and drink in 2002, North America accounted for US$11.75 billion, Europe US$10.5 billion, Japan US$350 million, Oceania US$200 million, Latin America US$100 million and the rest of Asia and the whole of Africa less than US$200 million. Of the roughly 24 million hectares managed worldwide for certified organic production in 2002, over 10 million were located in Australia, 5.8 million in Latin America, 5.5 million in Europe, 1.5 million in North America, 880,000 in Asia, and 320,000 in Africa. There has been a high level of adoption of organic agriculture in Central and South America in terms of certified land area and number of farms, with Argentina having the second highest amount of land under organic production in the world and Mexico having the greatest number of farms. By the end of the decade, the level of interest in organic agriculture and the volume of information compiled about organic methods had become sufficient to enable the highly successful publication of the landmark book *Organic Farming* by Nicolas Lampkin (1990).

According to recent data released in June 2012 by the IFOAM and the Research Institute of Organic Agriculture of Switzerland, there was a total of 37 million ha agricultural land that was organic in 2010. Table 46.1 presents the size of the organic food markets in some developed countries in 2000. The organic food market in the world has grown rapidly in the past decade.

The important organic products traded in the international market:

The important organic products traded in the international market are dried fruits and nuts, processed fruits and vegetables, cocoa, spices,

Table 46.1: Top countries in terms of area under organic farming, number of organic farms and World organic food market

Sr. No.	Country	Percentage of organic area to cultivated area	Number of farms	World organic food market (estimates) (in billion US $)
1.	Australia	2.31	1380	0.17
2.	Italy	7.94	56440	1.10
3.	USA	0.23	6949	8.00
4.	Germany	3.70	14703	2.50
5.	France	1.40	10364	1.25
6.	China	0.06	2910	0.12
7.	UK	3.96	3981	0.90

herbs, oil crops and derived products, sweeteners, dried leguminous products, meat, dairy products, alcoholic beverages, processed foods and fruit preparations, cotton, cut flowers, etc.

Organic Farming: At National Level

On the organic global map, India also does not figure in the list of the top 10 countries. India has only 0.03% land under organic cultivation, which becomes only 41,000 hectare area of total agricultural land.

The aggregate production of organic agriculture came to about 14,000 tonnes during 2002 and the exports amounted to 11,925 tonnes. The maximum export from India is tea, rice, pulses and vegetables around 3000, 2500 and 1800 tonnes respectively. Besides, cotton (1200 tonnes), wheat (1150 tonnes), spices (700 tonnes) and coffee around 550 tonnes are also exported. The little quantity of 250 and 100 tonnes of herbal products and oil seeds are also exported. Indian organic products are mainly exported to Europe (Netherlands, United Kingdom, Germany, Belgium, Sweden, Switzerland, France, Italy, Spain, etc.), USA, Canada, Saudi Arabia, UAE, Japan, Singapore, Australia and South Africa.

In India only ten states have clearly defined policies for organic farming. These are: Karnataka, Kerala, Andhra Pradesh, Maharashtra, Madhya Pradesh, Himachal Pradesh, Uttarakhand, Sikkim, Nagaland and Mizoram. Out of these Uttarakhand, Sikkim, Nagaland and Mizoram have declared their intention to go 100 percent organic. Uttarakhand is a third largest organic state with over 32,000 ha under organic cultivation and bringing 47,000 farmers under this tag. With the help of Uttarakhand Organic Commodity Board (UOCB), over 30 certified organic crop producer groups have come up here, producing wide range of organically safe food like amaranthus, basmati rice, maize, wheat, turmeric paddy, ginger, soyabean, rajma (kidney bean), finger millet, different type of pulses, medicines and aromatic plants.

Karnataka is a first state to announce an Organic Farming policy in 2004. State has set up the Jaivik Krishik Society that would facilitate farmers for certification and marketing. Here farmer also can sell their produce directly to consumers.

With the initiatives of Community Managed Sustainable Agriculture (CMSA), at Andhra Pradesh non-pesticide management (NPM) movements started. This state has freed 1.5 million ha and 1.5 million farmers from the tyranny of chemicals. By this movement, 124 villages declared pesticides-free, 26 villages deemed organic. The fundamental objective of CMSA is to provide healthy crops, healthy soil, healthy food and healthy life of public by ensuring food security locally. Table 46.2 shows the India's organic producers, with top organic state of India and the main crops grown by them.

Table 46.2: Top organic states of India and the main crops

States	Total certified area (in ha)	Main cultivated crop
Madhya Pradesh	2,866, 571.88	Cotton, oil seeds, cereals like maize and sorghum, pulses
Himachal Pradesh	631, 901.99	Fruits/vegetables, cereals like maize and sorghum, pulses, wheat
Rajasthan	217, 712.19	Cotton, oil seeds, cereals like maize and sorghum, spices
Maharashtra	177, 345.48	Cotton, oil seeds, fruits/vegetables, pulses
Uttar Pradesh	111, 644.83	Fruits/vegetables, wheat, rice, cereals like maize and sorghum
Uttarakhand	105, 465.98	Cereals like maize and sorghum, herbs and medicines, oil seeds, rice

Source: Ministry of Agriculture (Down to Earth)

Promotion of Organic Food

In order to promote organic products, the organic industry needs to address various issues including:

1. **Pricing:** Retail price premiums of organic produce remain considerably higher in most national markets than most consumers are willing to pay. This issue is really need to be addressed becaused it is a major barrier to organic food sales among less committed consumers. Retail organic foods have an average 70% price premium. Thus a household would need 70% more income to consume all organic products and continue spending the same proportion of its income on food. It will further increase the gap between rich and poor society. This raises a question—does only rich community have a authority to eat organic food which is assumed as a safe food. Rest which comes under middle and low class society will continue to eat unhealthy food. Is it a justice? Another question arises here after paying such a huge amount on organic items, how would we know, are we getting real organic or some fake item? In developing countries like? India, with in no time, we would start getting fake organic produce with same labels and same price. What should be done, if it is happening? There should be strict policies and law, who regulates it and impose huge penalties for the law breakers.

2. **Visibility:** Shopping habits are strongly influenced by the visibility. Thus, the appearance and visibility is the most important factor to change a habit of food purchasing. Although introducing organic foods certainly has increased their visibility, the appearance and layout of product displays. In most stores, the most influential area to display products is where buyers enter the retail outlet, and/or at eye level. Even where this is not possible, it is important to organize displays to identify organic products in some prominent manner.

3. **Labeling:** Inconsistent and inadequate labelling reduces consumer confidence and trust in the integrity of organic claims. National and international harmonisation or certified organic labels would be wel-

comed by most consumers if not by the certifying bodies who compete for farmers' business.

4. **Availability:** The supply and quality of organic products must be consistent enough that buyers are not tempted to substitute them for conventional products. There may be circumstances in which the organic industry is better served by a strategy of targeting a few key products than by attempting to provide a complete product range. Research has shown, for example, that a few fresh fruit and vegetables account for most of the expenditure, suggesting that the organic industry could have the greatest impact on its overall sales by targeting the top-selling items.

Organic Food: Certification

India has 22 certifying agencies accredited by Agriculture and Processed Food Products Export Development Authority (APEDA) also called third party certification (TPC). It is under the Ministry of Commerce, which set the norms for certification to help organic products find market abroad. These norms are based on the European Union Standards, known to be the toughest in the world. So, certificates in organic farming come at a huge cost and also require a huge amount of documentation. In developing countries like India, it poses a serious challenge, most of whom are small landholder farmers and illiterate. There is new alternative method developed to guarantee the organic integrity of products for small domestic producers, known as the participatory guarantee system or PGS. It is of low cost, involve minimal paper work and makes farmers responsible for their success and integrity of their group. PGS is governed by National Centre of Organic Farming (NCOF) which comes under the Ministry of Agriculture.

Conclusion

The Indian agriculture switched over to the modern system of production with the advent of the green revolution in the 1970s to fulfill the demand of food and grain. This method of farming had increased use of pesticides and fertilizers resulting in extensive damage to environment. We in India have to be concerned about this issue much more than any other nation of the world as agriculture is the source of livelihood of more than 6–7 million of our people and is the foundation of the economic development of the country. Sustainable agriculture based on technologies that combine increased production with improved environmental protection is a requirement of the day. Organic agriculture is best known alternative method to the modern system. Many countries have been able to convert 2–10 percent of their cultivated areas into organic farming. The demand for organic products is growing fast (at the rate of 20 percent per annum in the major developed countries).

India is lagging far behind in the adoption of organic farming. It has only 0.03% land under organic cultivation, i.e. only 41,000 hectare. The one achievement in this field is laying down of the National Standards for Organic Production (NSOP) and the approval of 4 accreditation agencies which are the government bodies. So, sensing the importance, the central and state governments have taken several initiatives to popularize organic farming in the country. There are valid and scientific evidence also which indicates that organically grown foods are significantly different in terms of their safety, nutritional content and nutritional value from those produced by non-organic farming. The concept of food quality should recognize wider aspects than mere external appearances, including authenticity, functionality, biological, nutritional value, sensual and ethical dimensions. The present day organic agricultural methods are far from fully developed. There is a lot of room for improvement especially in the complex area of soil microbiological activity, e.g. promotion of symbiotic relationships between microorganisms and crops. The government should also give financial support the farmers wishing to convert to an organic system. Funding should also be provided to research projects that wish to improve organic agricultural systems further.

47

Food Safety Aspects of Frozen Foods

Tavleen Kaur, Puja Dudeja, Amarjeet Singh

Urbanization has played a significant role in changing global food consumption patterns. This has lead to increasing penetration of frozen snacks from deep freezers in super-markets into home refrigerators. Young people, in general, are more open to experimentation in food tastes than the past generation. Consumers are attracted to frozen food owing to its vigorous marketing by corporate as 'convenience food' for working couples who are troubled by the demands of hectic and fast urban life. As a result retailers are under increased pressure to offer broader range of products. Now even small retailers have placed refrigerators in their shops for keeping stock of frozen products like peas and corns. Increasing prices of fresh fruits and vegetables in past few years have led to a reduction in price difference between fresh and frozen products. This has further increased a demand for frozen products which was earlier limited to ice creams only. This trend is likely to continue in coming years with more penetration of 'ready to eat' food in India.

Invention of Frozen Food

Freezing foods as means of preservation has a long history but modern frozen food industry came into existence over 70 years ago

in 1930. Clarence Birdseye—the king of frozen food left a permanent mark on the way the entire world consumes food. He invented and developed method of quick freezing food products without changing the original taste.

He began experimenting with freezing food in 1913 after observing the people of Arctic preserving fresh fish and meat by quickly freezing them. Faster the ice crystals are formed, smaller they are and cause less damage to the product. This observation was the basis of his invention. Finally in 1930 he introduced a line of quick-frozen vegetables, fruits, sea foods, and meat to the public for in Springfield, Massachusetts, under the trade name Birds Eye Frosted Foods.

Onset of World War II opened the doors for frozen foods and the industry quickly began to diversify and segment to boost profits. By the 1950s, frozen foods were the fastest growing sector of the food business. The distribution network caught up with the mass market's pace in the mid-1950s, as technology was developed to mechanically keep rail car contents frozen across long distances, linked together with strategically located centralized automated cold-storage warehouses.

By virtue of temperature control related imperatives frozen food option is more suitable for countries with temperate climate since ambient temperature is naturally low there. In tropical countries storage of frozen foods becomes a challenge because of higher ambient temperature and poor resources. Hence, cold chain and food safety is easily breached for frozen foods in these countries.

Increasing Consumer Demand

One of the primary factors affecting food consumption patterns is of course, the ability to purchase food. The last two decades have witnessed major increases in per capita income levels of households. 'Eating out concept' has become an essential part in the lives of Indians in urban areas due to rise in income and paucity of time. This has lead to increased demand for convenience foods.

Exposure to western lifestyles has led to change in mindset and preference of people regarding food. Redefinition of gender roles in households, with more women working has led to general loss in traditional values of cooking which was mostly confined to women. Household sizes have reduced with increasing number of one or two people houses; As a result, there has been an increase in demand for processed, ready-to-cook and ready-to-eat foods which are easy to serve and portion controlled. In such a scenario, the art

of cooking may be lost in the future generations because of availability of these convenience foods.

Younger consumers are more likely to try new foods and eat out more often. They have nontraditional food value. Most of them prefer convenient lifestyles. This causes their increased dependence on convenient foods. Frozen foods are gaining popularity because of their keeping quality, reduced waste generation and appeal to taste buds. Thus it is not just technology that fuelled the growth of such products but change in society also.

There are a number of advantages of frozen products over fresh. This results in an increasing demand as variety of products are available throughout the year. They are convenient, palatable, portable, longer shelf life, labour saving and downsizing kitchen inputs (time, skills, equipment, energy). Figure 47.1 shows conceptual framework on how various factors interrelate to influence acceptance and use of frozen foods.

Global Market for Frozen Foods

According to the report, "Frozen food market— Global industry analysis, size, share, growth, trends and forecast, 2013 – 2019", the global frozen food market is expected to reach $293.75 billion by 2019, up from $224.74 billion

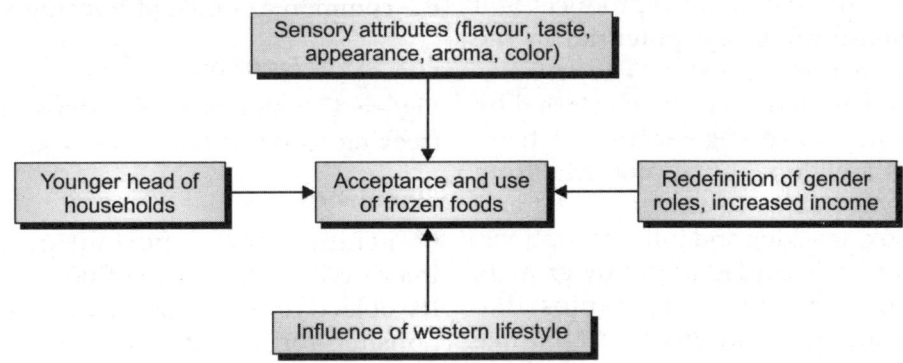

Fig. 47.1: Conceptual framework of increase in use of frozen foods

in 2012. The US was the largest market in 2012 and accounted for more than 80% share in the frozen food market followed by Japan and Germany. Brazil is the most attractive market for frozen food due to availability of raw materials in abundant quantity which makes frozen food products more accessible and affordable for consumers. The Brazil market is expected to grow at a compound annual growth rate (CAGR) of 4.7% from 2013 to 2019.

Frozen ready meals, including frozen pizza, desserts, snacks, entrees, etc. accounted for more than 30% of the total global market revenue in 2012. Expected CAGR is 3.9% from 2013 to 2019. The frozen fruit and vegetable market is expected to grow at a CAGR of 4.3% from 2013 to 2019.

The largest share in the global frozen food market in 2012 belonged to Europe (39.5%). The Asia Pacific market is another attractive market for frozen food due to a healthy growth rate and increasing consumer preference toward frozen food. Rest of the world is estimated to be the fastest growing region over the next six years, due to emergence of Brazil and Argentina as the new markets for frozen food. Expected CAGR is 4.3% from 2013 to 2019.

Frozen Food Market in India

India is a huge producer of food products. Still it has immense untapped potential in the frozen food export industry. The frozen/convenience food industry, which started by offering basic frozen vegetables and fries, today offers a wide range of products, from fruits and vegetables to frozen meats and ready-to-cook, snacking and full meal options. The segment has recorded a healthy growth, at a CAGR of 15–20%. Vegetables like drumsticks and peas and prepared food like *chapattis and paranthas* are nowadays available in frozen form in neat packets. The Indian frozen food market generated total revenues of $325.9 million in 2010, representing a CAGR of 16.6% for the period spanning 2006–2010.

Frozen meat products' sales proved the most lucrative for the Indian frozen food market in 2010, generating total revenues of $124.2 million, equivalent to 38.1% of the market's overall value. The performance of the market is forecast to decelerate, with an anticipated CAGR of 13.6% for the five-year period 2010–2015, which is expected to lead the market to a value of $617.5 million by the end of 2015. In terms of product categories, frozen vegetables and frozen snacks together make up a more than 65% share of the market and their collective volume share consumption for 2012 exceeded 85%.

According to trade estimates, the Indian frozen food market is expected to evolve at a CAGR of 45 percent to reach Rs 12,520 crore by 2014–2015. However, data released by India's Ministry of Food Processing indicate that the Indian frozen foods market is much smaller than even China's, which is also not considered a well-developed market. Although the frozen food sector in India is in the early stages of development, the growing 'out-of-home consumption' trend today points towards a successful innings.

To understand various losses suffered during generation of frozen foods let us take the common example of freezing vegetables.

Process of Freezing

Figure 47.2 shows all the steps involved in freezing foods.

Blanching

Blanching is an important procedure in freezing because of its influence on quality. Its objective is to inactivate enzymes responsible for alterations in sensory quality attributes and nutritional value during storage. After maturation, however, enzymes

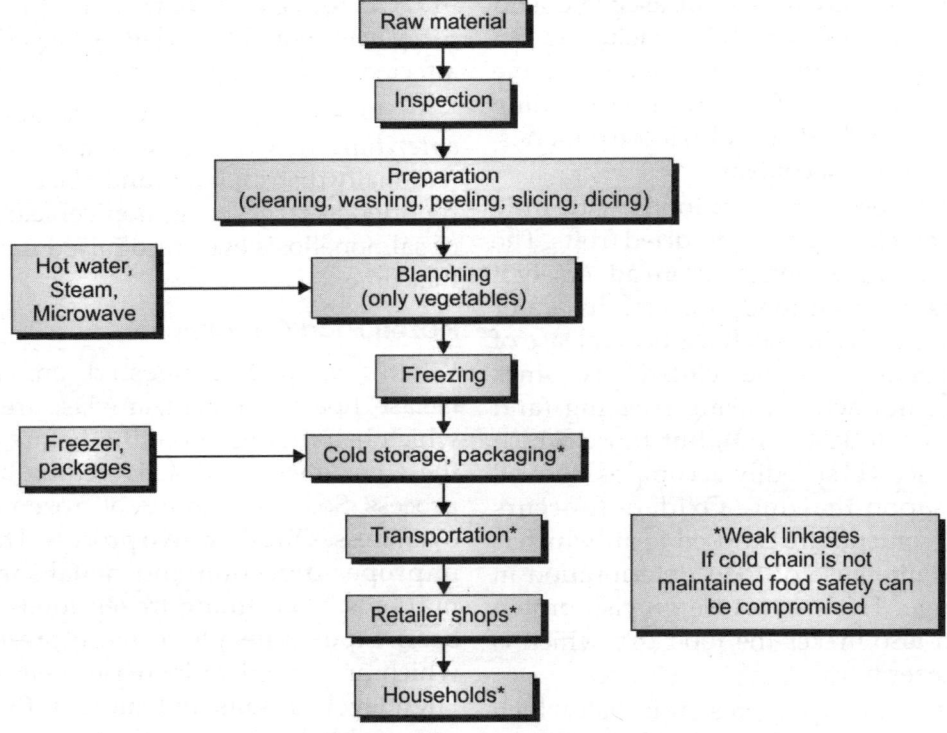

Fig. 47.2: Process of freezing

can cause loss in quality, flavour, colour, texture, and nutrients. If vegetables are not heated sufficiently, the enzymes will continue to be active during frozen storage. This may cause the vegetables to toughen or develop off-flavours and colours.

Blanching is the thermal treatment procedure in which product is heated by immersion in water at 85–100°C or steaming at 100°C or using microwave to inactivate enzymes for 1 to 10 minutes, depending on the size of vegetable pieces. Blanched vegetables should be promptly cooled down to control and minimize the degradation of soluble and heat-labile nutrients. The enzymes used as indicators of effectiveness of the blanching treatment are peroxidase, catalase, and more recently lipoxygenase. Peroxidase inactivation is commonly used in vegetable processing, since peroxidase is easily detected and is the most heat stable of these enzymes.

The heat produced during blanching causes irreversible alterations in cell structure which increase the permeability allowing blanch water to enter cells. Detrimental effects of blanching include destruction of nutrients, proteins, loss of soluble substances like vitamins, mineral salts, sugars. Color changes occur due to transformation of chlorophylls to pheophytins. These effects intensify with longer or higher blanching temperature.

Nutritional and Microbiological Effect of Blanching and Freezing

There is appreciable loss of ascorbic acid, B-complex vitamins, and folate during blanching and subsequent cooling. Peas blanched for 3 min lose 33% of initial ascorbic acid content, 20% riboflavin, 10% niacin and 5% thiamine, though these vitamins are stable during storage at –18°C or lower. Depending on blanching method peas also loose minerals, sugars and protein content.

Freezing does not affect foods with little water content like nuts, seeds, dried fruits. The more water a food contains more adversely it is affected as when food is frozen its water expands and cells bursts. It leads to release of self destructing enzymes called lysosomes which are not active during freezing (and some are even destroyed), but those which remain intact will speedily decompose the cell contents upon thawing. Oxidation occurs when cell contents are exposed to air which is an important cause of food deterioration in frozen stage. This may cause greyish brown spots and also makes the food dry which is called freezer burn.

Omega-3 fatty acids present in fish which have been found to reduce risk of heart attacks degrade over a period of time in frozen fish. Not all frozen fish can be labelled healthy like the frozen fish sticks which are deep fried and then frozen leads to gain in calories, sodium and saturated fat and loss of nutritional value

One common misconception is that these foods are free from microbial growth but there are certain species of microbes which although do not multiply but survive by going into hibernation or forming resistant cells and become active when food is thawed. Bacterial spores are resistant to freezing and thawing. Therefore these survive in most conditions. The survival rates of some vegetative bacteria in the same conditions are as follows: *Pseudomonas aeruginosa* 18%, *E. coli* 58%, *Staphylococcus aureus* 96%, and *Saccharomyces cerevisiae* 11%. Survival of *Salmonella* species and *E.coli* in frozen foods in of particular concern because of the low numbers that cause disease. Fish frozen at –22°C and stored at

–17.8°C for over a year had 10% of viable *S. typhymurium*. Toxins left by bacteria are not affected by freezing process.

In 1994 an outbreak of infection with *S. enteritidis* in US that was attributed to a nationally distributed brand of ice-cream was reported as largest common vehicle outbreak of salmonellosis ever recognised in the US at that time.

Frozen Food Consumption

When raw food is ingested, enzymes like lactase, lipase, protease, amylase are released which help in digestion. But before freezing, these enzymes are deactivated by blanching process. So, consumption of frozen products overstresses the digestive process. This causes improper digestion and malabsorption of nutrients. Consuming frozen foods on daily basis exposes people to more preservatives which are added to increase their shelf-life. High levels of salts and sugar in the form of syrup added to increase their shelf life makes the consumer more prone to health problems.

Storage and Safety of Frozen Food

Microbiological Aspect

Use of technology for freezing food properly at large scale takes a certain know-how. Understanding which foods freeze well and which ones can be finicky is just the first step. The choice of wrapping or container can make or break any attempt to keep frozen food at its best.

Development in kitchen technology like microwave oven to heat frozen foods has played strong role in frozen food evolution. Proper use of technology is important for keeping food product safe. An investigation into a case by Centre for Disease Control (CDC) where 44 people got food poisoning after eating frozen chicken-and-rice meals in 18 states in the US revealed that the outbreak of Salmonella bacteria—that caused food poisoning was owing to "improperly" microwaving food. If we think that frozen meals are ready to eat and just need to be reheated in microwaves, get alarmed. Microwave cooking is a critical control point to ensure raw and uncooked ingredients reach a sufficient temperature to render them safe from microbial hazards.

All foods can be safely frozen, but some foods should not be frozen for quality reasons (lettuce, tomatoes, cucumber, cream, etc.). Most frozen fruits maintain high quality for eight to 12 months. Unsweetened fruits lose quality faster than fruits packed in sugar or sugar syrups. Most vegetables will maintain high quality for 12 to 18 months at 0°F or lower. However, it is a good idea to use home-frozen vegetables before the next year's crop is ready for freezing.

Frozen food stored consistently at –18°C or lower will remain safe indefinitely. Most domestic freezers should operate at temperatures –18°C or lower. As a general rule, if the freezer cannot keep ice-cream solid, its temperature is above the recommended level. Temperature fluctuations during storage and handling promote ice crystal growth; the growth of ice crystals is another detrimental change which occurs during frozen storage therefore, temperatures should be kept as constant as possible.

Chemical Aspect

Important part of frozen food storage is the quality of the container the food is packed in. Air tight containers or freezer bags should be used for freezing as the air trapped within the bags causes food to deteriorate by development of rancid oxidative flavors through contact with the frozen product. This chemical change can be controlled by excluding oxygen through proper packaging. Quality of plastic bags also affects the safety of food products.

Grainy, brownish spots occur on the product because of moisture migration in frozen foods. This causes the tissue to become dry and tough and to develop off-flavors. This quality defect can be prevented by using heavyweight, moisture proof packaging during the freezing process. Textural difference is especially noticeable in frozen products normally consumed raw, as in the case of fruits. The water and dissolved solutes inside the rigid plant cell walls give texture to the fruit or vegetable tissue. In the process of freezing, when water in the cells freezes, an expansion occurs and ice crystals cause the cell walls to rupture. Consequently, the texture of the food is generally much softer after thawing when compared to fresh food.

Challenges

The manufacturers generally provide a deep freezer to the retailer for keeping and sale of frozen foods. However, in India particularly Northern part keeping a constant freezer temperature is a major concern in summers when power outage is expected to last several hours or days with lack of proper back up. A freezer full of food will usually keep about 2 days if the door is kept shut; a half-full freezer will last about a day. The freezing compartment in a refrigerator may not keep foods frozen as long. If the freezer is not full, keeping packages together helps retain the cold more effectively. Meat and poultry items should be separated from other foods so that if they begin to thaw, their juices would not drip onto other foods. We need to discard foods that have been warmer than 40°F for more than 2 hours. We should also discard any foods that have been contaminated by raw meat juices.

Even at the grocery shops a proper infrastructure is required for the storage with constant supply of electricity and deep freezers, temperature of which is maintained and regularly checked with an inbuilt thermometer as any temperature fluctuations affects the food quality and safety.

Inadequate Cold Chain and Transport Facilities in India

Frozen food is normally kept at −18°C and below. Any increase in the temperature of the environment may lead to deterioration in food quality. The cold chain in India is in nascent stages of development. The segment is largely dominated by fly-by-night suppliers and small businesses with poor networks. As the services are not integrated, it leads to wastage and damage to food due to frequent handling and transfer. In May 2012, according to centre for disease control and prevention 258 people were infected with Salmonella in the U.S. The CDC said that frozen yellow fish tuna imported from India was the likely source of outbreak as many of those sick people had eaten it in the week before they became ill.

Today, there are very few specialized distribution companies providing refrigerated transport and warehousing for perishable produce/processed food products. Prime reasons for the low adoption of cold chain facilities has been high costs and little knowledge of technical skills. India is a challenging market to operate in, considering that each operator must have a separate cold room and 24-hour uninterrupted supply of electricity. Unfortunately, most distributors may not have access to such facilities due to which many frozen products are damaged and rendered unsafe for human consumption.

The development of robust cold chain with efficiently designed infrastructure with the right network would contribute to an immediate reduction in waste, and improved product availability across the country.

Conclusion

What does the future hold for frozen foods? It is impossible to predict. Whether frozen foods constitute a boon or bane is still not clear. Maintenance of the cold chain is the weakest link in safety of frozen foods especially during transportation. Moreover frequent power cuts quite common in India both in homes and retail shops where frozen foods are stored can jeopardize food safety. Thawing of frozen food is also a tricky issue. For the sake of convenience microwave is used to thaw such foods. As food is defrosting in the microwave, the edges of the food may begin to warm or slightly cook while the inside of the food remains frozen. Uneven thawing is the major food safety concern, during thawing the microwave raises the temperature of food and if it enters the danger zone, bacteria begins to grow and multiply. Frozen food contains bacteria even though they are inactive. When a food's temperature rises, bacteria become active once again. It is also not safe to thaw food in the microwave and then put it in the refrigerator or a cooler to cook or grill later. The microwave a favorite kitchen appliance whether it is reheating, cooking or defrosting, can compromise safety of frozen foods.

All said and done frozen foods should not completely take over our kitchen at least not with an aim to make a fashion statement. Fresh foods have many advantages including extra nutritional value. Many factors affect the nutritional quality of fresh and frozen vegetables by the time they end up on our plate. We need to make sure that we make a right decision in choosing the food we eat.

48

Food Labeling, the Inevitable Fixture Required in Food Package: The Food Safety Concern

Neha Chanana, Puja Dudeja, Amarjeet Singh

Industrialization and urbanization have caused a shift not only in the kind of food that is consumed but also in its packaging. Because of the increase in the purchasing power of people especially. the Indian middle class, the consumption of pre-packaged or packaged food is on the rise. However, with increase in knowledge about etiology of FBI and non-communicable diseases (NCDs), there is also a growing awareness among people about attaining wellness through diet and nutritious food. In order to ensure that the food is safe and nutritious as well as in accordance with the dietary choices of the consumers, in modern society, the label on the food package holds special importance.

Food labeling is defined as any written, electronic or graphic panel on the package of food that contains a variety of information about its contents and the nutritional value.

From the manufacturer or Food Business Operator (FBO) point of view it as an instrument of marketing and product promotion among the discerning literate consumers. It is a way by which any manufacturer can introduce and promote his product among the target consumers. The food label also gives credibility to the manufacturer as well as the food product in the market. Several studies indicate that the label can reduce the infor-mation gap between producers and con-sumers. It also reduces search costs for consumers. Thus, a shift in trend has been seen in the packaging and labeling of food items.

CHANGING TRENDS IN FOOD LABELING

In the 20th century, till early 90s, majority most (95%) of the food articles available for consumption were sold loose and packaging or labeling of food products was not an issue. In the present times, more than 30% of the food available is packaged which either is factory packing or locally packaged. For example, wheat flour was available loose at the grocery shops or local *chakki* and was purchased by the consumers in their own drums or bags. In modern times, today, wheat flour available in grocery shops or super-markets is supplied by big factory flour mills in sealed and customized wheat *atta* packaging bags that have printed label giving information about the nutritional content. Other is the *atta* locally packaged and available at small grocery shops that bear a hand written label of just the item of food viz. wheat flour or *chakki ka atta*.

This shift and rising trend of packaging and labeling of food items has been seen more in cities. In rural areas of the country labeling is slow to pick up where food items are still sold loose. This increases the chances of these food items becoming adulterated or spoiled. This makes food items unsafe for consumption.

PRESENT CONTEXT OF LABELING AMONG CONSUMERS

India is still facing a problem of low literacy level of the masses. Moreover, majority people of the country cannot read or write English which unfortunately, is the language for labels of most products. In general, Indian consumers have a casual attitude towards labeling. Less than 5% of people read labels before buying food products. Majority of them just read the name of the food item, the brand and the Maximum Retail Price (MRP) on the label.

However, labeling is much more than identifying the brands or the MRP. In future labeling is sure to become an important tool to ensure food safety. The general and specific requirements for food labeling have been underlined under the Food Safety and Standards (Packaging and Labeling) Regulations, 2011.

GENERAL REQUIREMENTS FOR LABELING

Under Food Safety and Standards (Packaging and Labeling) Regulations, 2011 every pre-packaged food needs to bear label containing information as required for the interest of the consumer. An assessment of the following specifications need to be made on the food label before any purchase is made with regards to pre-packaged or packaged food articles:

- Language: The particulars of declaration under the regulation required are specified on the label in English or Hindi. However, depending on the local or regional situations other languages are also used.
- The label does not bear any false or misleading statement

- The label attached or pasted on the pre-packaged food is not separate from the container as this may lead to unscrupulous use of the label as well as the container.
- Contents on the label are clear, indelible and readily legible to the consumer.
- In case the container is covered by a wrapper, the label on the container or the necessary information on the wrapper is readily legible through the outer wrapper, i.e. the wrapper is such that it does not hide any relevant information.
- The label has a FSSAI logo and license number of the manufacturer
- Information on the label is not contradictory to the requirements laid under FSS Act, rules and regulations.
- The font size of the letters is appropriate and in colour contrast with the background of the label.
- For labeling of GM foods *refer* to Chapter 50 on GM crops.

Information Given on the Label

Before buying packaged food products in addition to the general labeling requirements, the following information needs to be examined on the label of every package of food:

- Name of the food: The label has the name of the food which includes the trade name or description of food contained in the package.
- List of ingredients: These are to be perused under the title *"Ingredients"* on the label in the following manner:

  i. The names of the ingredients used in the product are mentioned in descending order of their composition by weight or volume at the time of its manufacture.

ii. A specific name or category of ingredients is used in the 'list of ingredients'. For example, in case acidity regulator is used or emulsifier is used as an ingredient, either the name of the additive would be mentioned or the category along with the ISN number will be specified. Table 48.1 enlists the class titles that are usually used for ingredients falling in respective classes.

iii. In case of single ingredient foods, list of ingredients is not mentioned. For example, in case of packet of sugar, no list of ingredients may be mentioned. But the product name 'sugar' would be clearly mentioned.

iv. In cases where package of food is sold as a mixture or combination, the quantity of ingredients in percentage at the time of manufacture is mentioned

Table 48.1: Examples of class titles for ingredients

Classes	Class Titles
Edible vegetable oils/edible vegetable fat	Edible vegetable oil/edible vegetable fat or both hydrogenated or partially hydrogenated oil
Animal fat/oil other than milk fat	Give name of the source of fat. Pork fat, lard and beef fat or extracts is declared by specific names
Starches other than chemically modified starches	Starch
All species of fish where the fish constitutes an ingredient of another food and provided that the labeling and presentation of such food does not refer to a species of fish	Fish
All types of poultry meat where such meat constitutes an ingredient of another food and provided that the labeling and presentation of such a food does not refer to a specific type of poultry meat	Poultry meat
All types of cheese where cheese or mixture of cheeses constitutes an ingredient of another food and provided that the labeling and presentation of such food does not refer to a specific type of cheese	Cheese
All spices and condiments and their extracts	Spices and condiments or mixed spices/condiments as appropriate
All types of gum or preparations used in the manufacture of gum base for chewing gum	Gum base
Anhydrous dextrose and dextrose monohydrate	Dextrose or glucose
All types of caseinates	Caseinates
Press expeller or refined cocoa butter	Cocoa butter
All crystallized fruit	Crystallized fruit
All milk and milk products derived solely from milk	Milk solids
Cocoa bean, coconib, cocomass, cocoa press cakes, Cocoa powder (fine/dust)	Cocoa solids

on the label. If the ingredient has been used as a flavouring agent, the disclosure of such ingredient may not be there.

- **Nutritional information:** Nutritional facts or nutritional information is mentioned per 100 gm or 100 ml or per serving of the product. According to the regulations it is mentioned as:

  i. Energy value in kcal

  ii. Amount of proteins, carbohydrates (specified quantity of added sugar) and fat in grams (g)

  iii. Amount of any nutrient for which a nutritional or health claim is made

  iv. Vitamins and minerals are expressed in metric units

  v. When the nutritional information is given per serving, the amount in gram or milliliter is included for reference beside the serving measure

  vi. Declaration of trans fat: If a product claims regarding the quantity or type of fatty acids or the amount of cholesterol, the amount of the saturated fatty acids, monounsaturated fatty acids and polyunsaturated fatty acids are declared on the label in gram (g) and amount of cholesterol in milligram (mg). A food product is declared as 'trans fat free', if the quantity of trans fatty acids is less than 0.2 gm/serving. If the quantity of saturated fat does not exceed 0.1 gm/100 ml, the product is labeled as 'saturated fat free'. According to FDA, if there is less than 0.5 grams of fat in a serving the food is labeled as "Fat-Free". However, there is misbranding of food articles in the name of 97% Fat-Free food when the food actually contains more fat. A few related examples are as follows:

a. Many canned food available in supermarkets bear the tag of being 97% fat free. But looking at the given nutritional information, which says 145 calories per serving and the number of calories from fat is 70, it can be clearly computed that nearly 50% of calories in the canned food come from fat. But big brands selling such products get away easily with a little trick.

b. Another example is considering a food product X which is marketed to be 97% fat free. The product is labelled as: Serving size 56 g; Servings/container 4; Total weight 224 g. The nutritional information on the label is calories/serving 70; calories from fat 15; Total fat per serving is 1.5 g. The above information clearly indicates that the percentage of calories from fat is 21.4% (15/70*100) which is quiet high. This is far from the claim of food companies of the product being "97% fat free". However, the companies get away with this very smartly by highlighting the percentage of no fat by weight in the food item, i.e. is computed by considering total fat by weight in the package as 6 g (1.5 g * 4) and calculating its amount (percentage) with respect to the total weight of the product (6/224*100) which equals to 2.67% of fat by weight which the companies project as "97% fat free." The consumers are misled as they are given the percentage of the "weight" that is fat free rather than the percentage of "calories." Thus, instead of the going for the product by the big print in the front, always look for the calories per serving and what the companies refer to as serving size.

vii. According to the recent regulation, the label would contain the amount of trans fat content and total saturated fat content in the food product as follows:

Total trans fat content not more than...... percent by weight

Total saturated fat content not more than ...... percent by weight

viii. In case of foods such as raw agricultural commodities like, wheat, rice, cereals, spices, spice mixes, herbs, condiments, table salt, sugar, jaggery, or non-nutritive products, like, soluble tea, coffee, soluble coffee, coffee-chicory mixture, packaged drinking water, packaged mineral water, alcoholic beverages or fruit and vegetables, processed and pre-packaged assorted vegetables, fruits, vegetables and products that comprise of single ingredient, pickles, papad, or foods served for immediate consumption such as served in hospitals, hotels or by food services vendors or *halwais,* or food shipped in bulk which is not for sale in that form to consumers nutritional information is not given on the label as these food products are exempted from mentioning the nutritional facts or information. Table 48.2 highlight how nutritional information is given on the label.

- **Declaration of vegetarian and non-vegetarian food:** A brown circle enclosed in a square with brown outline indicates the product is non-vegetarian. Whereas a green coloured circle in a square with green outline suggests the food is vegetarian (Fig. 48.1).

Table 48.2: Pattern of nutritional information on the label

Nutritional elements	Amount
Energy	# kcal
Protein	# g
Carbohydrates sugars	# g
Fat	# g
Saturated fatty acids	# g
Polyunsaturated fatty acids	# g
Monounsaturated fatty acids	# g
Trans fatty acids	# g
Cholesterol	# mg

a. Non-vegetarian food symbol

b. Vegetarian food symbol

Fig. 48.1: Symbols for vegetarian and non-vegetarian foods

Moreover, the size of the logo depends on the size of the label of the product. The size of the sides of the square is double the diameter of the circle. Table 48.3 specifies the minimum size of the circle with respect to the area of principal display panel, i.e. the size of the label. The symbol is prominently displayed.

i. On the package having contrast background on principal display panel;

ii. Just close in proximity to the name or brand name of the product;

iii. On the labels, containers, pamphlets, leaflets, advertisements in any media

Sl. No.	Area of principal display panel	Minimum size of diameters in mm
1.	Up to 100 cm square.	3
2.	Above 100 cm square up to 500 cm square	4
3.	Above 500 cm square up to 2500 cm square	6
4.	Above 2500 cm Square	8

Table 48.3: Size of the logo

- Declaration regarding food additives:
  - i. For food additives, the class titles viz. *acidity regulator, acids, anticaking agent, antifoaming agent, antioxidant, bulking agent, colour, colour retention agent, emulsifier, emulsifying salt, firming agent, flour treatment agent, flavour enhancer, foaming agent, gelling agent, glazing agent, humectant, preservative, propellant, raising agent, stabilizer, sweetener, thickener,* are used together with the specific names or recognized international numerical identifications (INS no.).
  - ii. Addition of colours and/or flavours:
    - a. The extraneous colouring agents that are added to any food article are displayed on the label in capital letters just beneath the list of ingredients on the package of food as:

      CONTAINS PERMITTED NATURAL COLOUR(S)

      OR

      CONTAINS PERMITTED SYNTHETIC FOOD COLOUR(S)

      OR

      CONTAINS PERMITTED NATURAL AND SYNTHETIC FOOD COLOUR(S)

These statements have to be displayed along with the name or INS no. of the food colour.

- b. Extraneous addition of permitted flavouring agents: These are displayed, just beneath the list of ingredients on the label attached to any package of food to which flavouring agents have been added, as:

  Contains added flavour (specify type of flavouring agent viz. natural flavouring agent, synthetic flavours, natural identical flavouring substance, artificial flavouring agents)

- c. In case both colour and flavour are used in the product, one of the following combined statements in capital letters is displayed, just beneath the list of ingredients on the label attached to any package of food so coloured and flavoured as:

  CONTAINS PERMITTED NATURAL COLOUR(S) AND ADDED FLAVOUR(S)

  OR

  CONTAINS PERMITTED SYNTHETIC FOOD COLOUR (S) AND ADDED FLAVOUR(S)

  OR

  CONTAINS PERMITTED NATURAL AND SYNTHETIC FOOD COLOUR(S) AND ADDED FLAVOUR (S)

- **Name and address of the manufacturer:** These need to be checked on the label as under:
  - i. The name and complete address of the manufacturer and the manufacturing unit if these are located at different places and in case the manufacturer is not the packer or bottler, the name and

complete address of the packing or bottling unit has to be declared on every package of food.

ii. Where an article of food is manufactured or packed or bottled by a person or a company under the written authority of some other manufacturer or company, under his or its brand name, the label has to carry the name and complete address of the manufacturing or packing or bottling unit and also the name and complete address of the manufacturer or the company, for and on whose behalf it is manufactured or packed or bottled.

- **Net quantity:** The net quantity by weight or volume or number declared on the label needs to be seen before buying or consuming the packaged food as it gives an idea about the quantity that is to be purchased. For packing size and weight variations, legal metrology (packaged commodities) rules 2011 are followed. This rule states that a particular commodity needs to be packed in 1 kg or 500 g or 250 g and not packed in 680 g, etc.
- **Lot/Code/Batch identification:** A batch no./code no./lot no. on the label is a mark of identification by which the food is traced in the manufacture and identified in distribution. This number helps during various recall and withdrawal systems in the supply chain.
- **Date of manufacture or packing:** The date, month and year in which the commodity is manufactured, packed or pre-packed needs to be checked before purchase. The month and year of manufacture, packaging and pre-packing is given on the label if the "Best Before Date" of the product is more than 3 months. On the contrary, if the "Best Before Date" is less than 3 months the date, month and year of manufacture,

packaging and packing is given on the label.

- Best before and use by date: Best before date/use by date/expiry date mentioned on the label are important and need to be checked. The food usually becomes unsafe after this date, thus, keeping a check on these dates helps in reducing the chances of consuming unsafe food.

  i. The best before date is mentioned in capital letters in the following format:

  "BEST BEFORE .... MONTHS AND YEAR

  OR

  "BEST BEFORE.......... MONTHS FROM PACKAGING

  OR

  "BEST BEFORE............MONTHS FROM MANUFACTURE

  ii. In case of package or bottle containing sterilized or Ultra High Temperature treated milk, soya milk, flavoured milk, any package containing bread, dhokla, bhelpuri, pizza, doughnuts, khoa, paneer, or any uncanned package of fruits, vegetable, meat, fish or any other like commodity, the declaration is made as follows:

  "BEST BEFORE ....DATE/ MONTH/YEAR"

  OR

  "BEST BEFORE........DAYS FROM PACKAGING"

  OR

  "BEST BEFORE .... DAYS FROM MANUFACTURE"

  When the shelf life of the food article is in days, **the date** is mentioned in the "Best Before" pattern rather than the days from manufacture or date of manufacture or packing. For example, bread has a best before within 7 days, here the date is mentioned in the best before pattern.

iii. On packages of Aspartame, instead of best before date, use by date/recommended last consumption date/expiry date is given, which shall not be more than three years from the date of packing.

iv. In case of infant milk substitute and infant foods instead of best before date, use by date/recommended last consumption date/expiry date is given.

- **Instructions for use:** This caption on the label of pre-packaged food is essential as it ensures correct utilization of the food. The instructions include the directions like reconstitution or dilution of the product before use so that the food can be used as its intended usage and in quantity as desired or specified by the manufacturer.

- **Principal display panel/size of the label:** This needs to be in accordance with the size of the container so that the label is clearly legible to the consumer. A small label on a large container is not in concordance with the regulations. The size of the display panel is mentioned as follows:

i. Area of the principal display panel

a. Rectangular container: 40% of the product of height and width of the panel of such container.

b. In case of cylindrical or nearly cylindrical, round or nearly round, oval or nearly oval container, twenty percent of the product of the height and average circumference of such container; or

c. In the case of container of any other shape, twenty percent of the total surface area of the container except

where there is label, securely affixed to the container, such label will cover a surface area of not less than ten percent of the total surface area of the container.

- **Labeling of imported foods:** In case of buying food products imported from other countries, the following needs to be taken into consideration.

i. The country of origin of the food declared on the label of food imported into India.

ii. When a food undergoes processing in a second country which changes its nature, the country in which the processing is performed needs to be considered to be the country of origin for the purposes of labeling.

- **Labeling of irradiated foods:** When irradiated foods, i.e. the foods that are treated with ionizing radiation, are consumed the label indicating the treatment in close proximity to the name of the food needs to be checked. The pattern of declaration and logo is as follows:

PROCESSED BY IRRADIATION
METHOD DATE OF
IRRADIATION.................

LICENSE NO of Irradiation Unit..............
PURPOSE OF IRRADIATION.................

- **Labeling of proprietary food:** If the food product is manufactured having no specified standards and no guidelines are issued for adding ingredients and additives issued under FSS Act/Rules and

Regulations then such food product is called proprietary food. Approximately 370 articles of food ranging from different edible oils to cereals to bakery, milk products and others are standardized and their quality defined in Food Safety and Standards (Food Product Standards and Food Additives) Regulations, 2011. A check on the labeling of all the proprietary foods complying with FSS Act, Rules and Regulations needs to be done.

LABELING REQUIREMENTS SPECIFIC TO FOOD PRODUCTS

Approximately 50 food products have specific labeling requirements as are mentioned under the FSSA regulations, 2011. Table 48.4 speci-

Table 48.4: Specific labels for food products

Food product	Label
Infant milk substitute and infant food	"IMPORTANT NOTICE" (in capital letters) "MOTHER'S MILK IS BEST FOR YOUR BABY"
Every container of refined vegetable oil	Refined (name of the oil) Oil
Admixture of palmolein with groundnut oil	BLEND OF PALMOLEIN AND GROUNDNUT OIL Palmolein......per cent
Admixture of imported rape-seed oil with mustard oil	BLEND OF IMPORTED RAPE-SEED OIL AND MUSTARD OIL Imported rape-seed oil.....per cent
Vanaspati made from more than 30% rice bran oil	This package of vanaspati is made from more than 30% rice oil by weight
Food containing artificial sweetner	CONTAINS ARTIFICIAL SWEETENER AND FOR CALORIE CONSCIOUS
Package of pan masala	Chewing of Pan Masala is injurious to health
Package of supari	Chewing of Supari is injurious to health
Package of drinking water *The bottle needs to bear BIS marking*	PACKAGED DRINKING WATER CRUSH THE BOTTLE AFTER USE
Package of natural mineral water *The bottle needs to bear BIS marking*	NATURAL MINERAL WATER CRUSH THE BOTTLE AFTER USE
Package of low fat paneer/chhenna	LOW FAT PANEER/ CHHENNA
Package of fresh fruits coated with wax	Coated with wax (give the name of the wax)
Bread, biscuits, cakes containing oligofructose	Contains Oligofructose (dietary fiber) ——— gm/100 gm

fies some special labels for specific food products.

- Labeling of infant milk substitute and infant food: A label is affixed in a clear, conspicuous and in an easily readable manner with font size of letters not less than 5 millimeters as shown in Table 48.3.
- **Labeling of edible oils and fats:** The package, label or the advertisement of edible oils and fats expressions such as "Super-refined", "Extra-refined", "Micro-Refined", "Double-refined", Ultra-refined", "Anti-cholesterol", "Cholesterol Fighter", "Soothing to Heart", "Cholesterol Friendly", "Saturated Fat Free" or such other expressions need not be mentioned on the label. There is specific simple labeling of edible oils and fats.

LEGAL IMPLICATIONS

Under the FSSA, 2011 there are penalties with respect to faulty labeling in the form of misbranded food and misleading advertisements.

i. Penalty on the FBO for misbranded food is up to ₹ 3 lakhs
ii. Penalty on the FBO for misleading advertisement is up to ₹ 10 lakhs

FUTURE OF FOOD LABELING

Food labeling is an inevitable requirement of food packaging and it serves to be an essential component for ensuring food safety in the years to come as well. Along with awareness about labeling among consumers other suggested techniques that can make labeling more readable and accessible for the consumers are as follows:

- **Incorporating the concept of vaccine vial monitor (VVM) for perishable food items:** The concept of thermochromic labels used for polio vaccines can be used for perishable food items like meat, cheese and other dairy products in the form of Food Quality Monitor (FQM). The change in colour of the FQM would indicate the quality of the food item.

- **Use of scanning labels:** With use of scanning labels and scanners present in all smart phones, the details of labels consisting information on directions to use, how to use, the indicators suggesting the quality of food items, etc. will be readily detectable by the consumer.

CONCLUSION

The famous advertisements like *"Jago Grahak Jago"* have awakened the masses through print as well as electronic media, highlighting issues of (MRP) labeling and standardizing of various products including food items. However, a gap still exists in using food labeling as a reference among consumers for making their dietary choices. Still many times food products are purchased without referring to the "Best Before Date" or expiry date or loose food products such as oil, etc. are brought which results in consumers falling prey to foodborne illnesses. Moreover consumers are befooled by the big prints of "fat free" or "97% fat free" on the front of the food product. Examining food labels is an easy way to detect breach in food safety as no specific lab tests are required and is a self documentary evidence for identifying the infringe in the food supply chain. Thus, consumers need to become self conscious and incorporate food labeling in their daily purchases so as to keeping themselves as well as their families healthy.

49

Nutraceuticals

Nancy Sahini, Neha Channana, Amarjeet Singh

"Food is a fuel". The concept that food provides us energy needed to perform daily functions holds true since the time humans have come into existence. As we travel back in time say, 15,000 years back, primitive men were hunters-gatherers who use to hunt fauna and gather wild flora in the quest for something to eat that could provide them energy and keep them healthier. Moving ahead in time around 10,000 years ago, humans learnt the art of cultivating plants and herding animals. With this, humans began living in settled communities. Farming, started with cereals as the principal food for these established populations like rice and millets in the Far East and wheat and rye in Europe, gained advancements through the years with irrigation and other methods to serve even larger populations. With the rise of civilization, large organized societies appeared centered around villages and then in cities. This marked the establishment of industry and commerce.

The advancement in industrialization and commercialization has also caused a shift in the kind of food that is consumed. Traditionally, cereals like wheat and rice were consumed in unpolished form or were crushed to more consumable forms. In today's mechanized world, these are processed which makes them more palatable. However, it removes the nutrient rich outer layers to yield a fiber-free food. Moreover, the consumption of refined sugars in the form of bread, etc. has increased in comparison to cereals.

With industrialization people, especially women no longer are confined to household work are choosing for blue-color highly paid jobs which leave them with less or no time for cooking or eating at home. This has lead to people opting for ready cooked food. Because of consumption of ready to eat foods, fast foods, frozen foods coupled with unhealthy lifestyle, people have fallen prey to non-communicable diseases along with deficiency of essential nutrients.

The commercialization of ayurveda and yoga under the brand names of *Ramdev* and *Himalyas*, have highlighted the importance of nutritive value of food as a growing concern. People have realized that along with taste, health is also important. They are demanding for food products that are natural which they perceive to be undoubtedly healthy. The focus is dual treatment as well as prevention. With the fear of suffering from disease, consumers are also increasingly getting interested in disease prevention compounds contained in many foods. On the other hand the sufferers instead of taking capsules and tablets are also trying their hands on natural substitutes like herbal products in the hope of avoiding side effects and addictions caused by pharmaceuticals.

Pharmaceutical companies in order to maintain their monetary interest and also combating the demands of the consumers, although secondary, have come up with the concept of *'Medicalization in Nutrition'*. This idea involves the publicizing of natural products in the form of tablets, capsules or drinks. We are well aware of the famous probiotic drink *'yakult—piyo jaldi jiyo healthy'* which promises to keep your gut healthy or Dabur assuring their *chawanprash* composed of all natural ingredients like amla, etc. which will improve immunity and keep illness at bay especially during winters. These products are categorized into a new term, referred to as Nutraceuticals, hitting the research among pharmaceutical industry, biotechnologists, nutritionists, food companies and media.

Definition of Nutraceuticals

The term nutraceutical is coined from nutrition and pharmaceutical in 1989, by Dr. Stephen Defelice, founder and chairman of foundation for innovation in medicine. The term nutraceutical used in marketing industry has no regulatory definition. However, according to Defelice nutraceutical can be defined as, a food (or a part of food) that provides medical or health benefits, including the prevention and/or treatment of a disease". These products range from isolated nutrients, dietary supplements and diets to genetically engineered "designer" foods, herbal products and processed foods such as cereals, soups and beverages.

Approximately 40–50% proportion in cardiovascular disorders, 35–50% proportion in cancers, and 20% proportion in osteoporosis is attributable to dietary factors. Use of food as medicine for treatment and prevention of various disorders is not a recent development.

The term 'nutraceuticals' is newly heard but its concept dates back to around three thousand years. Hippocrates, the father of modern medicine, stated 'Let food be thy medicine and medicine be thy food' to predict the association between appropriate foods for health and their therapeutic benefits. The concept of nutraceuticals has evolved over the years. In early 1960s, the manufacturers in US began adding iodine to salt in an effort to prevent goiter representing one of the first attempt at creating a functional component through fortification of food product which substantiates the idea of therapeutic effects of food mentioned by Hippocrates.

The Indians, Egyptians, Chinese and Sumerians are few civilizations where foods have been effectively used as medicines to treat and prevent diseases. The so called 'home remedies' we take as was suggested by our grandmothers and the our foremothers which are today elaborated as traditional medicine along with Ayurveda mention the therapeutic role of food products in health and disease. For example, in Ayurveda, the concept of *dincharya* meaning daily routine mentions the foods to be included in our daily diet that would prevent and also treat diseases. With the rapid advances in science and technology, escalating healthcare costs, increasing purchasing power of the people especially the Indian middle class and growing awareness of attaining wellness through diet has fuelled the interest in nutracueticals in India. In England, Japan, United States and many other countries nutraceuticals have already become a part of the dietary pattern. Table 49.1 presents the origin and progress of nutraceuticals as we know them today.

Table 49.1: History of nutraceuticals

Year	Progress in nutraceuticals
1900s	Food manufacturers in USA added iodine to salt to prevent goiter (food fortification)
1954	Prevention of Food Adulteration Act, regulates packaged food
1989	Term nutraceuticals coined by Dr. Stepehen de Felice
1993	Foods for Specified Health Uses (FOSHU), legally permitted use of a few functional foods
1994	Dietary Supplements Health and Education Act—nutraceuticals expanded to include vitamins, minerals, herbs, amino acids, etc. Regulation on manufacturing and marketing of nutraceuticals in USA
1997	Food and Drug Administration Modernization Act (FDAMA). Widened options available to the manufacturers of nutraceuticals. Approves of their beneficial effects and protects public health by ensuring their safety and efficacy
2001	Food with Health Claims (FHC) and Foods with Nutrient Function Claims (FNFC) system was introduced
2002	European Parliament and Council approved a directive to harmonize rules for food-supplement labelling, and introduced specific rules on vitamins and minerals
2004	US Anabolic Steroid Control Act, manufacturers to list steroid hormones and their precursors on product packaging—however, they are still marketed as nutraceuticals and are regulated as foods despite drug-like actions
2005	EU was set to decide on the first set of laws to regulate health benefit claims for foods.

Nutraceuticals and other Related Terms

With advancement and research in the field of 'Medicalization in nutrition' new terms like functional foods and dietary supplements are gaining ears among the population. Nutraceuticals are often referred to as *"functional foods"*. However, there is a slight difference between the two.

When food is being cooked or prepared using "scientific intelligence" with or without the knowledge of how or why it is being used, then the food is called as "functional food." Thus, functional food provides vitamins, fats, proteins, carbohydrates necessary for health of an individual. Almost all functional foods are antianemic. When functional food aids in the prevention and/or treatment of disease(s)/disorder(s) other than anemia it is called a "nutraceutical". Fortified dairy products (milk) and citrus fruits (e.g. orange juice) and vegetables are examples of nutraceuticals.

With respect to dietary supplements, the Dietary Supplement Health and Education Act (DSHEA) has clearly defined dietary supplements as a product (other than tobacco) that is intended to supplement the diet that bears or contains one or more of the following dietary ingredients—a vitamin, a mineral, a herb or other botanicals, amino acids or a dietary substance for use by man to supplement the diet by increasing the total daily intake or a concentrate, metabolite, constituent, extract, or combinations of these ingredients. Its ingestion is in the form of pills, capsules, tablets or liquid form. It is not used as a conventional food or as the sole item of meat or diet. Moreover, it is labeled as "dietary supplement". It includes products such as an approved new drug, certified antibiotic, or licensed biologic that was marketed as dietary supplement or food before approval, certification or license.

Thus, nutraceuticals are slightly different from dietary supplements as the former not only supplements diet **but also prevent and/or treat disease/disorder**. Also they are represented for use as a conventional food or sole item of meal or diet.

Classification of Nutraceuticals

Nutraceuticals can be classified on the basis of their chemical constituents such as alkaloids, fatty acids, carotenoids, dietary fiber, etc. Other classifications are according to the type of food item and whether they are traditional or not as shown in Table 49.2.

i. Type of food item-dietary supplements, nutrients, dairy products, herbals

ii. Traditional/Non-traditional nutraceuticals: Traditional nutraceuticals are simply natural, whole foods with new

information about their potential health qualities. For example, lycopene in tomatoes, omega-3 fatty acids in salmon.

Non-traditional nutraceuticals are foods resulting from agricultural breeding or added nutrients and/or ingredients to boost their nutritional values. Examples include beta-carotene enriched rice, orange juice fortified with calcium and cereals with added vitamins and minerals.

Mode of Action

Research on how nutraceuticals find their way into the body and benefit the health of the individual in treatment as well as prevention of diseases is still in the pipeline. However, it has been stated by researchers that nutraceuticals function by increasing the supply of important building blocks of the body. These building blocks either act as buffering agents for relief by reducing signs of disease or directly provide health benefits to individuals.

Table 49.2: Classification of nutraceuticals

S.No.	Category	Details	Example
1.	**Based on food items**		
a.	Nutrients	Have established nutrient functions	Vitamins, minerals, amino acids
b.	Herbals	Herbs or botanical products as	Aloe vera, ginger, garlic extracts or concentrates
c.	Dietary supplements	Which are added to the food for a specific purpose and administered orally	Minimally refined grains, phytoestrogens (soya), dairy foods
2.	**Traditional/non-traditional**		
a.	Traditional	When the food is taken as whole food or without any change in it	Fruits, vegetables, dairy products.
b.	Non traditional	Outcome of agricultural breeding or addition of specific ingredients	Folic acid added to flour, calcium added to juices.

Potential Benefits of Nutraceuticals

Majority of the nutraceuticals available in recent times are potential nutraceuticals as there is still lack of sufficient data to demonstrate the benefit they assure to offer. However, nutraceuticals cover all therapeutic areas such as anti-arthritic, prevent and treat cold and cough, digestion, sleeping disorders, prevention of certain cancers, osteoporosis, blood pressure, cholesterol, depression and diabetes. Moreover, nutaceuticals can offer the following benefits:

i. Increase the health value of the diet.

ii. Prolong life.

iii. Avoid particular medical conditions like green tea prevents cold and cough by improving immunity, Lycopene, soyfoods and saponins from tomatoes, spinach are anti-cancerous agents.

iv. Fewer side effects as compared to the modern chemical compositions.

v. Provides food for populations with special needs (e.g. nutrient-dense foods for the elderly)

Table 49.3 gives an overview of the various nutraceuticals available in the market along with their functions.

Regulations Governing Nutraceuticals

The nutraceutical industy is expected to reach new heights in the coming years, however, an ambiguity regarding regulations governing nutraceuticals persists all over the world.

In United States, the FDA regulates it under the Dietary Supplement Health and Education Act (DSHEA), 1994. Under this act, FDA has laid down a framework for regulation of dietary supplements without considering nutraceuticals as a separate item. This act offers manufacturers of dietary supplements to give information about possible benefits of the product on labels. Thus, FDA needs to prove that a substance is unsafe rather than the manufacturer proving it to be safe in order

Table 49.3: Nutraceuticals available in the market and their functions

Brand name	Components	Function
Betatene	Carotenoids	Immune function
Xangold	Lutein esters	Eye health
Lipoec	a-lipoic acid	Potent antioxidant
Generol	Phytosterol	CHD reduction
Premium probiotics	Probiotics	Intestinal disorder
Soylife	Soyabean phytoestrogen	Bone health
Z-trim	Wheat	Zero calorie fat replacer
Linumlife	Lignan extract flax	Prostate health
Fenulife	Fenugreek galactomannon	Control blood sugar
Teamax	Green tea extract	Potent antioxidant
Marinol	w 3 FA, DHA, EPA	Heart health protection
Clarinol	CLA	Weight loss ingredient
Cholestaid	Saponin	Reduce cholesterol

to accomplish the quality of nutraceuticals available in the market.

In India, nutraceuticals has been defined in clause 22 of the Food Safety and Standards Act, 2006. However, issues which still need to be addressed include the packaging and labeling of nutraceuticals, restriction of advertisements and claims made by nutraceutical manufacturers, etc.

Nutraceutical Industry in India

Along with the growing health industry in India, there is an emerging trend in Fast Moving Healthcare Goods known as nutraceuticals which are ingredients with human health benefits beyond basic human nutrition. India is also one of the emerging markets for nutraceuticals with a CAGR (Compounded Annual Growth Rate) of 20%. The Indian nutraceutical industry has wide opportunity for growth as the country has seen a change in dietary and working patterns of its population which has led to an emerging trend of chronic diseases like diabetes, cardiovascular diseases, etc. Unfortunately these are not curable. An individual once gets affected is put to life-long treatment on medicines. The saying *"Prevention is better than cure"*, holds true for the rising trend of these non-communicable diseases. Therefore, in a drive to prevent these diseases people are resorting to natural products. The pharamaceutical companies like Amway, Glaxo Smith Kline are exploiting this need and have diversified into the production of nutraceuticals. In 2010, the Indian nutraceutical industry was estimated at US $2 Billion,

Table 49.4: Broad segments of Indian nutraceutical market

Dietary supplements (40%)	Functional foods and beverages (60%)
Vitamin and mineral supplement	Functional foods (fortified food items)
Herbal supplement Protein supplement	Functional beverages (energy drinks, fortified juices, sports drinks)
Chyawanprash	

roughly 1.5 percent of the global nutraceutical industry. Broad segments of Indian nutraceutical industry include dietary supplement (40%) and Functional food and beverage market (60%) (Table 49.4).

However, there are certain challenges which the nutraceutical industry still needs to overcome. It is still a product for the higher class because it is available at high price which makes it availability among only the urban residents which is further widening the urban rural divide. Its prices need to be lowered so that the rural population of the country can also benefit from these products. Moreover, its inclusion in the nutrional agenda of the country might help in overcoming the burden of nutritional deficiencies under which India is still lurching.

Therefore, nutraceuticals has advantages as well as disadvantages. The need of the hour is to focus on the health of the people of the country and not the personal wealth of the pharamaceutical industry.

50

Genetically Modified Crops

Neha Chanana, Puja Dudeja, Amarjeet Singh

Gene transfer between plants occurs naturally over an evolutionary time scale and plays a major role in dynamic changes to chromosomes during evolution. Traditionally, our farmers have been carrying out cross breeding of animals as well as plant/trees at the farm level. The introduction of foreign germplasm into crops has been done by traditional crop breeders by artificially overcoming fertility barriers. For example, a hybrid cereal was grown by crossing wheat and rye way back in 1875. This practice has now been used at a mass level in the form of Genetically Modified Crops (GM crops)

The idea of producing food with desirable qualities paved the way for the development of genetically modified crops worldwide. GM crops are those plants the DNA of which has been modified using genetic engineering techniques. The aim of producing such crops is to introduce a new trait to the plant which does not occur naturally in the species. For example, through GM technology we can change the resistance to certain pests, diseases, or environmental conditions, reduction of spoilage or improving the nutrient profile of the crop. Examples of GM technology in non-food crops include production of pharmaceutical agents, biofuels, and other industrially useful goods.

Milestones in the History of GM Plants

1982: First GM plant was produced using an antibiotic-resistant tobacco plant.

1986: First field trials of GM plants in France and the USA

1987: Plant Genetic Systems (Ghent, Belgium), the first company to develop genetically engineered (tobacco) plants with insect tolerance

1992: The People's Republic of China, the first country to allow commercialized transgenic plants, introducing a virus-resistant tobacco in

1994: First GM crop approved for sale in the U.S; FlavrSavr tomato, which had a longer shelf life, as it took longer to soften after ripening.

1995: Bt potato was approved safe by the Environmental Protection Agency, making it the first pesticide producing crop to be approved in the USA.

Since then, GM foods have been widely adopted. Worldwide between 1996 and 2012, the total surface area of land cultivated with GM crops had increased by a factor of 94, from 17,000 square kilometers (4,200,000 acres) to 1,600,000 km^2 (395 million acres). In 2012, GM crops were planted in 28 countries; 20 were developing countries and 8 were developed countries. 2012 was the first year in which developing countries grew a majority (52%) of the total GM harvest. 17.3 million farmers grew GM crops; around 90% were small-holding farmers in developing countries.

Benefits of GM Crops

GM crops appear to be promising, as they have been modified with traits intended to provide benefit to farmers, consumers, or industry. These traits include improved shelf life, disease resistance, stress resistance, herbicide resistance, pest resistance, production of useful goods such as biofuel or drugs, and ability to absorb toxins, for use in bioremediation of pollution. Following are the proposed benefits of GM crops:

- Crops are more productive and have a larger yield
- Could potentially offer more nutrition
- Inbuilt resistance to pests, weeds and disease
- More capable of thriving in regions with poor soil or adverse climates
- More environment friendly as they require less herbicides and pesticides
- Longer shelf life
- As more GM crops can be grown on relatively small parcels of land, they are an answer to feeding growing world populations

However, the costs both regulatory and research are high currently. To resolve this issue the majority of GM crops in agriculture consist of commodity crops, such as soybean, maize, cotton and rapeseed. To curb the cost the developing countries have been made as the soft target. Research and development have been targeted on crops that are locally important to them as insect-resistant cowpea for Africa and insect-resistant brinjal for India.

Controversies Over GM Crops

The availability of GM crops have also led to a dispute over their use and other goods derived from such crops as compared to conventional crops. Other uses of genetic engineering in food production have also been debated. The dispute involves consumers, biotechnology companies, governmental regulators, non-governmental organizations, and scientists. The key areas of controversy related to GM food are:

- Effect of GM crops on health and the environment
- Labeling of GM foods
- Role of government regulators
- Effect on pesticide resistance
- Impact of GM crops on farmer's interest
- Role of GM crops in feeding the world population
- GM crops as a tool for biological warfare

There is a broad scientific consensus that food derived from GM crops poses no greater risk than conventional food. No such reports of ill effects have been documented in the human population from GM food. However, it is anticipated that these crops in future might have proteins which are allergenic to humans. For example, in Brazil it was found that a GM soya bean containing Brazilian nut protein was allergenic to humans and was withdrawn from production.

Genetic engineering allows introducing animal products in plants. This may raise concerns for those with dietary restrictions, like vegetarian. GM technology is an unnatural way of producing food. There will always be unknown long term effects to the ecosystem and biodiversity when inducing unnatural ways to change the natural traits of crops. Altered genes in engineered food will multiply through generations, passing with it any unknown damaging trait to the future population. It will be one product of a kind that cannot be recalled when a malfunction is discovered. The harmful effects on the environment and human health are inevitable outcomes of GM crops.

Labeling of GM products in the marketplace is required in many countries. In 2006, the Ministry of Health and Family Welfare has proposed a mandatory labeling policy for all GM foods. The rule would require GM foods to bear a label stating that they have been subject to genetic modification after their approval for consumption by the safety authority. This would be required for all GM products, whether they are primary or processed food, food ingredients, or food additives derived from a GM food, even if there are no quantifiable traces of recombinant DNA in the food product (e.g., refined oils derived from GM products).

A gazette notification from the Ministry of Consumer Affairs makes it mandatory for packaged foods using GM products as ingredients to carry such labels from 1st January 1, 2013. Though it is a positive step but this move has tied our hands to ban GM foods in future. This has a direct implication to promote foreign agribusiness and GM food imports to the country. While labeling does give consumers a chance for avoiding genetically modified food in the market, but it is almost impractical in India where more than 90 percent of our food is unprocessed and available in open and non-packaged form. The permissible limit of GM ingredients in the food as proposed by Indian Council of Medical Research (ICMR) is slightly higher than the Eurpopean Union norms. While EU have fixed GM permissibility in food at 0.9 percent, the most stringent in the world, Japan has placed at one percent. ICMR has fixed up to a limit of 1 percent. However, the bottle neck is the non-availability of laboratory and infrastructure to test for GM products. For the imported food products the rule in our country says that it should have been cleared for marketing in country of origin. Thus the rule says that no verification tests are mandatory in India. But this is a questionable logic. In fact maximum harm contamination can be expected from this clause.

In India, the regulation of all activities related to GM foods is governed by "Rules for the Manufacture/Use/Import/Export and Storage of Hazardous Micro-organisms, Genetically Engineered Organisms or Cells, 1989" under the provisions of the Environment (Protection) Act, 1986 through the Ministry of Environment and Forests (MoEF). The rules, are primarily implemented by MoEF and the Department of Biotechnology (DBT), Ministry of Science and Technology through six competent authorities:

- Recombinant DNA Advisory Committee (RDAC)
- Review Committee on Genetic Manipulation (RCGM)
- Genetic Engineering Approval Committee (GEAC)
- Institutional Biosafety Committees (IBSC)
- State Biosafety Coordination Committees (SBCC)
- District Level Committees (DLC)

The Rules, of 1989 are very broad in scope and essentially capture all activities, products and processes related to or derived from biotechnology including foods derived from biotechnology, thereby making GEAC as the competent authority to approve or disapprove the release of GM foods in the market place.

The Food Safety and Standard Authority of India (FSSAI) has GM Food Safety Assesment Unit (GMFSAU) for safety assessment and approval process for GM foods that leverages existing regulatory capacity within the Government of India, notably within DBT, MoEF ICMR. The responsibilities of various agencies are given in Table 50.1.

There is a strong fear that herbicide-resistant and pesticide-resistant crops could give rise to super-weeds and super-pests in the long term use of GM crops. This would then need newer, stronger chemicals to destroy them.

GM crop producing companies patent their crops and also engineer crops in such a way that the harvested grain germs are incapable of developing. This harms the interest of impoverished third world farmers, since they are not able to save seeds for replanting. Thus they will have to buy expensive seeds from these companies every year. This is sheer monopolization which is undesirable in modern society. This will lead to exploitation of poor farmer. The new technology also interferes with traditional agricultural methods which may be more suited to local environments.

GM crops are not the answer to world hunger and health. Instead, we should focus on improving organic agricultural practices which are kinder to the earth and are healthier for humans. It is proposed that GM crops would help in ending of hunger from the world. The claim appears to be far from realistic. It is a known fact that world hunger is not caused by a shortage of food production, but by sheer mismanagement, and lack of access to food brought about by various social, financial and political causes.

Last but not the least presently the western world is leading in production and marketing of GM crops. Globalization of GM foods have its repercussions. There is a conflict between trade and human concerns at a global level. There is a possibility that the mutated crop may be used as a weapon for biological warfare in future to target the developing countries. So as of now, issue of GM crops is a mine field. In this context, the famous film maker Mahesh Bhatt also said, "Multinational seed companies would have had a free run in India had it not been for NGOs making it a big issue forcing the government to slow down its enthusiasm to promote GM crops. It is the responsibility of the government to protect farmers from the clutches of giant multinational seed companies. A crop failure is more than economic loss this failure deprives the farmer of their basic right to safe seeds. Modifying the DNA of the seeds with toxins is equivalent to replacing safe food with toxic food. Farmers and consumers do not need such toxins on their platter. Giant multinational seed companies have taken international patents and they are using these against the interest of the farmers".

Table 50.1: Responsibilities of governmental authorities as regards the regulation of GMOs

Activity	Responsible Authority
Contained research (laboratory and greenhouse)	RCGM (DBT)
Event selection trials/BRL 1 trials	RCGM and GEAC (MoEF)
Food safety asessment of GM foods (viable and processed)	FSSAI
Environmental risk assessment of GM organisms	GEAC
Approval for commercial release of GM foods (processed)	FSSAI
Approval for commercial release of GM foods (viable, i.e. LMOs)	GEAC
Approval for environmental (commercial) release of GM organisms	GEAC

51

Safety of Ready-to-Eat Foods

Puja Dudeja, Amarjeet Singh

Food is a significant part of any culture. Changes in lifestyle are closely associated with change in pattern of diet consumption. The change in eating style in modern Indian society has been reflected in form of more 'eating out' culture and increased consumption of convenience foods. These are commercially prepared and designed for ease of consumption. Although restaurants meals meet this definition, the term is seldom applied to them. Convenience foods include prepared foods such as ready-to-eat (RTE) foods, frozen foods prepared mixes such as *dosa, upma, and chutney mixes*, etc. For details *see* Chapter 47.

Indian cooking and lifestyle have undergone tremendous changes in the last 15 years. Many factors are responsible for this change. These are globalization, dual income, separate living of couples, role of media and pressure of marketing, etc. Due to lifestyle pressure, nowadays people prefer easy short way of cooking food rather spending too much time in elaborate cooking. Non-availability of raw materials to prepare *masala* and tedious process involved in doing so, has influenced people to choose such RTE products.

India has become the hub of many multinationals who catalyze, nurture and exploit rapid change in our lifestyles. Many people are migrating to cities for job and education. They find RTE products as a comfortable option to eat rather than depending on restaurants. Most of the dual income families want to avoid hassles of cooking because of lack of time. During weekends, they want to spend quality time with their family and go out to eat, whereas on weekdays the long working hours force them to go for buying such products. Other factors influencing increase use of these products is their easy availability and the wide variety. There has been a spectacular change as RTEs have become widely available in supermarkets. Not only this, these packages are available for most of the cuisines of the world to be eaten at home. These are ultimate processed foods with very high value addition, as they offer the convenience of "eating off the shelf", eliminating the kitchen drudgery associated with making a meal at home.

Various RTE foods are considered better over other food products as they do not contain any chemical preservatives and remain shelf-stable without refrigeration for at least one year. These changes are bringing a new revolution in processed food industry.

Bus do minute, Ready in two minutes, Taste bhi, Health bhi, Khushiyan bhi, Kal ki table book kar do, etc. are the common catch phrases associated with RTE foods available in the market. The working women of the Indian middle class prefer these as it saves time, labor and tediousness of cooking. They desire to spend less time in kitchen. Other factors are increase in eating out culture, weakening of family ties, the spread of television and its impact, the increasing difficulty and expenses involved in obtaining domestic help.

Thus, RTE foods, which have convenience as a key factor associated with them, are welcomed by all. Another section of society who is heavily dependent of these RTE products are college students staying away from homes or persons staying alone as the idea of cooking from scratch is unappetizing to them. There has also been an increase in travel and tours and these ready meals provide with an option of easy to handle and store the pre-cooked food during such excursion. They are handy and act as a speedier alternative to full cooking. They also provide free time for consumers to spend on other leisure activities.

Some examples of food companies marketing RTE are MTR, Tasty Bite, Mother's recipes, Satnam Overseas, Godrej, Al-Kabeer, etc. These companies manufacture food products from the northern, southern and western cuisine. There has been an overflow of these products in the supermarkets. Consumers have traditionally viewed the ready meals as less healthy than fresh foods. There have been concerns about the nutritional contents claimed and actually available in them. They are often high in saturated fats and salt content. In children over consumption of such foods can lead to obesity also. However, this view is beginning to change as a result of improvements in the quality of these foods supplemented by the promotional activities and massive advertisement campaigns by the manufacturers. Initially bread, jam, cheese, salted foods used to be the only available food is the category of RTE. With advances in food technology other kinds of foods were developed like such as candy, beverages, soft drinks, juices, processed meat and cheese, soups, pasta, potato chips, etc. RTE foods can be categorized as in Fig. 51.1.

Safety Aspects of RTE Foods

There are many food safety issues related to these products. All these products are available in packaged form. It is of utmost importance to take care of safety of RTE foods as they are no longer to be processed further. If these foods are contaminated at any stage from farm to fork they can be a cause of FBI. Following precautions needs to be taken in respect of RTE foods to ensure their safety:

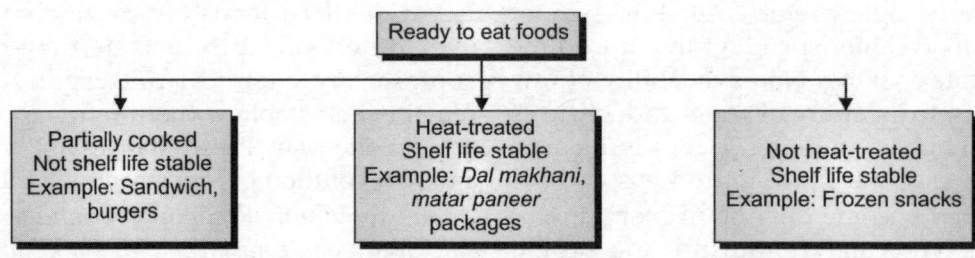

Fig. 51.1: Types of RTE foods

Raw Material

The microbial load of all the raw items to be used in preparation of RTE like raw fruits and vegetables, liquid milk, meat, eggs, flour, cereal grains, etc. should be within the acceptable limits at receiving as well as during their storage in the raw material store. In case perishable items are to be used like milk, meat then appropriate and adequate storage facilities must be ensured.

Manufacturing of RTE Foods

There are food-manufacturing units owned and run by reputed brands who manufacture RTE foods. These are located in remote areas and then the food items are transported to various places following safe transportation practices (*refer* Chapter 17 on safe transportation of food for details). There are large and small bakeries in cities and towns which also provide RTE foods. Most of the times the food safety practices are dismal. There is also a trend of making RTE food like sandwiches, burgers and hot dogs at homes in slums. Hygiene and sanitation of these areas is extremely poor. These foods are then covered with a cling foil (clean wrap) which gives a false sense of security to the consumer. There is no time and date seal on these unlabelled RTE foods (Figs 51.2 and 51.3).

Fig. 51.2: RTE foods like hot dogs, *bread* pakoras

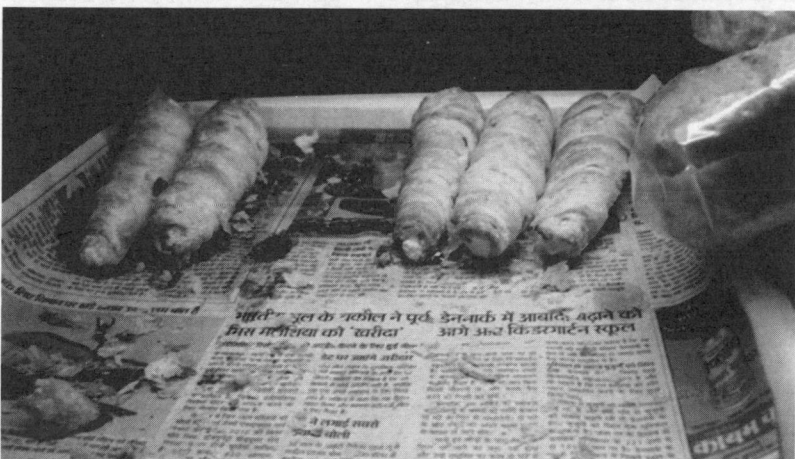

Fig. 51.3: Locally made cream rolls lying on a newspaper for sale

Their safety during their transportation form the place of manufacture to the retail shops is also questionable.

For the safety of RTE foods it is imperative that all pieces of food contact equipment viz. dough mixers, conveyors, rounders, dough dividers, racks, proofing equipment, oven, rollers, slicers, sifters, pasteurizer, homogenizer, retort, bottling unit, pulper, filtering screens, mixing vats, etc. should be clean, in good repair and free from evidence of rodent or insect activity. Time and temperature control of ovens, retort, heat exchangers and cooling area should be strictly adhered to ensure safety of food.

Before using any equipment it should be cleaned (in place, wherever possible). This will remove source of contamination. All vapor producing cooking equipment such as retort, ovens, grills, and fryers should be equipped with ventilation and an approved automatic extinguishing system to prevent unnecessary condensation in the working area. If this is not done micro-organisms may grow. Utensils like spoons, beaters, pans, bowls, trays, spatulas, etc. should be sanitized after every lot manufactured. For monitoring of food safety swabs of machine, working tables, utensils, food contact surfaces should be taken at regular intervals to ensure their microbial safety for food use. Antiseptic/disinfectant foot bath should be provided at the entrance of plant.

Packaging of RTE Foods

Packaging material (pouches, films, laminates, cans, glass/PET bottles, closures, jars, cardboard boxes) should be kept and stored under hygienic conditions in a room intended for that purpose. This aspect is generally ignored and packaging material is kept on the floor in unhygienic conditions. It is required that all packaging materials as bottles/ closures should be sanitized before use. These closures should be labeled for the product inside. In case these need to be stored before transport the room provided for storage should be having appropriate temperature and humidity conditions to prevent any spoilage. For the dispatch of all products First In First Out (FIFO) system should be applied. For details refer to chapter on good storage practices.

Detection of the foodborne pathogenic bacteria in RTE foods represent an unacceptable risk to health regardless of the number of bacteria present. The pathogens listed below should not be found in ready-to-eat food that has been adequately prepared.
Campylobacter spp., *Escherichia coli* O157, *Salmonella* spp. *Shigella* spp.
Vibrio cholera, Bacillus cereus, Bacillus spp, Clostridium *perfringens*
Listeria monocytogenes, Staphylococcus, Vibrio parahaemolyticus

According to Food Safety and Standards Regulations 2011 (FSSR), *Salmonella* contamination and other enteric infections are a problem with chocolate products. For these RTE foods critical raw materials such as skim milk powder, milk, eggs, cocoa, etc. should be adequately heat-treated, pasteurized, or handled in such a way that bacterial contamination is eliminated or minimized. Sanitation is a major problem, especially since many chocolate products are finished by hand-dipping; employee sanitation practices are, therefore, very important to prevent product contamination. These products are generally consumed by children, who are highly susceptible to enteric infections. Storage of cocoa beans, nuts and coconuts should be checked for insects, rodents, and mycotoxins. Samples collected for analysis of mycotoxins, unless otherwise directed, should consist of 30 individual portions of at least 125 g each.

Similarly for custard and cream-filled foods sanitation and good-quality raw materials are critical factors. Bacteria-sensitive materials, such as skim milk powder, milk and eggs, must have minimum bacteria levels, and must be stored, defrosted and handled in such a way that the addition or growth of bacteria is prevented. The products are not subjected to a heat treatment after filling; the filling operations must, therefore, be conducted in the most sanitary manner possible. Equipment sanitation, clean-up procedures and employee practices should be strictly as per the guidelines. The material should be handled and prepared as quickly as possible under the best sanitary conditions to minimize the number of bacteria present at the time of freezing. Frozen conditions must be adequately maintained during transportation of the frozen foods.

In a big set up raw materials should be subjected to a field examination. At the time of delivery, 100 pieces of the particular fruit, vegetable or other food arriving at the plant for preparation before freezing should be checked. All those pieces, which are unsatisfactory because of mould, decomposition, insect and rodent filth or foreign material should be sorted out. Report of unsatisfactory pieces as a percentage of the sample taken is made. After this another 100 pieces are examined after all sorting and grading have been completed, to determine the amount of unsatisfactory material being removed or being allowed to enter the process. This procedure can then be repeated at other times during the inspection to determine the overall quality of the food being prepared. The origin of raw food material being delivered to the plant should be reported.

Distribution of RTE Foods

Small EEs, tea shops, street vendors sell RTE foods in open markets. It is imperative that these vendors are educated about the health hazard associated with flies, birds, rodents and other vermin. The FBOs who sell such foods should observe basic hygienic measures to protect the consumer from environmental contamination and infections likely to be introduced during hawking. There should be a source of safe water available with them.

Fruits and vegetables on display, or their immediate container, should not be in contact with the ground. Unsheltered displays should be high enough above the ground surface to prevent contamination from any source. Dust and dirt on premises should be controlled to prevent contamination. Only a limited amount of perishable foods should be on display. If the market lasts all day, the bulk should be stored in a cold-store or room, or in an insulated container. Indications of spoilage as bad smell, unusual colour and changed consistency should be used to discard the spoiled items.

India has made lot of progress in agriculture and food sectors since independence in terms of growth in output, yields and processing. It has gone through a green revolution, a white revolution, a yellow revolution and a blue revolution. Now the time is to provide better food manufacturing units and its marketing infrastructure for Indian industries to serve good quality and safest processed and RTE foods. RTE market in India is expected to expand to reach ₹ 2,900 crore by 2015, according to an analysis done by Tata Strategic Management Group (TSMG). The key issue that remains is that how safe are food items which seem to be packaged with ultimate hygiene.

52

Food Safety Concerns in Home Delivery of Foods

Ishwarpreet Kaur, Puja Dudeja, Amarjeet Singh

Last weekend I got a phone call from my friend requesting me to lend an LPG cylinder. On enquiring, she told that her cooking gas has finished and the dinner was not ready yet. As I had none to spare, I jokingly suggested the idea of *"khushiyon ki home delivery"*! An advertisement tagline I just watched on TV of a food chain that promised speedy home delivery of food. She instantly liked the idea and asked for the phone numbers of the few eating establishments that provide home delivery of cooked meals.

Although this was a genuine case, ordering home delivered food has become a common practice in India. As advertised by suppliers nowadays 'hot sumptuous food' is just a phone call away. When there are unexpected guests at home or the lady of the house is not in a mood to cook or has come back home late in the evening, home delivery is a convenient option for her.

Home delivered food: The home delivery sector in the organized food services is crossing the annual turnover of more than ₹ 1000 crore in our country. It is a trend fast catching up with the youth and old alike. It is also grabbing a major share of the food service business. To meet present day consumer requirements both small and large restaurants are increasingly using the home delivery route which is expected to grow 30-40% over next few years. Double income nuclear family groups with increasing number of working women and people on the move are the main reasons that are changing the way customers dine today.

Classification of home delivered foods: Home delivered food is categorized on the basis of the volume of customers it caters to, i.e., individual, small or large group. Food is home delivered for individuals as in the case of tiffin service. This service is generally opted by students, paying guest (PG), bachelors, and elderly who live alone or unable to cook for themselves. Some food businesses deliver food, at home for small group of people, usually less than 50 like parties (kitty party, birthday party), get together, gatherings, etc. A few food businesses only provide home delivery of food for large number of people (greater than 50).

The various types of such services available in India are mentioned below:

a. Tiffin service is a common scenario where cooked food is delivered at place of work. Though the truth is that nothing can replace *"Ghar ka khana"*, there are people who move away from their homes because of work and this is one thing they miss more than anything else. And then there is that 9 am to 9 pm job which does not permit the person to even enter the kitchen after a long day. Eating out all the time is certainly not an option and the monotony of office canteen food gravitates one towards tiffin service. This provides home cooked food at the place of work or residence. These are often run as small business from homes, where food is transported to the customers in tiffins. It is a common sight to see a bicycle/scooters loaded with multiple tiffins being used for delivery of food before lunch time (Fig. 52.1).

A very well known specialized tiffin service of a different kind in Mumbai is known as "dabbawala". It is a very efficient tiffin/dabba supply chain. It has a Six Sigma performance rating, i.e. 99.99% of deliveries errorfree. According to quality standards the dabbawalla's are at par with various companies like Motorola, Honeywell and GE. The best (and the different) part with this service is that the dabawalla carries and delivers freshly made food from customer's own homes to them in lunch boxes (Fig. 52.2).

b. Food business—caterers deliver prepared food in bulk containers (casserole, tin, can, etc.) and serve at the desired venues like home.

c. Packed meals by branded restaurants are home delivered like Domino's, Pizza Hut, Subway, Yo China, etc. (Fig. 52.3). These are delivered through two wheelers in a variety of packaging materials, e.g. plastic containers, foil packs, cardboard, clay pots (Fig. 52.4).

Food safety issues in home delivered food and the ways to overcome them

There are a lot of food safety issues especially while considering Indian scenario.

a. **Quality of raw material:** Safety of food is taken for granted in institutes that cater to delivering food at door step. The FBOs know that clients ordering home delivered or tiffin services usually never visit the place where food is cooked and packed. Taking advantage of this, FBOs generally procure poor quality raw food material at lower prices to earn more profits. For example, buying rotten tomatoes to make gravies (these are commonly sold in market as "curry tomatoes"). This may lead to FBIs.

To avoid it, customers must ensure to order only from a licensed/registered food establishment while selecting the caterer.

Fig. 52.1: Tiffin service

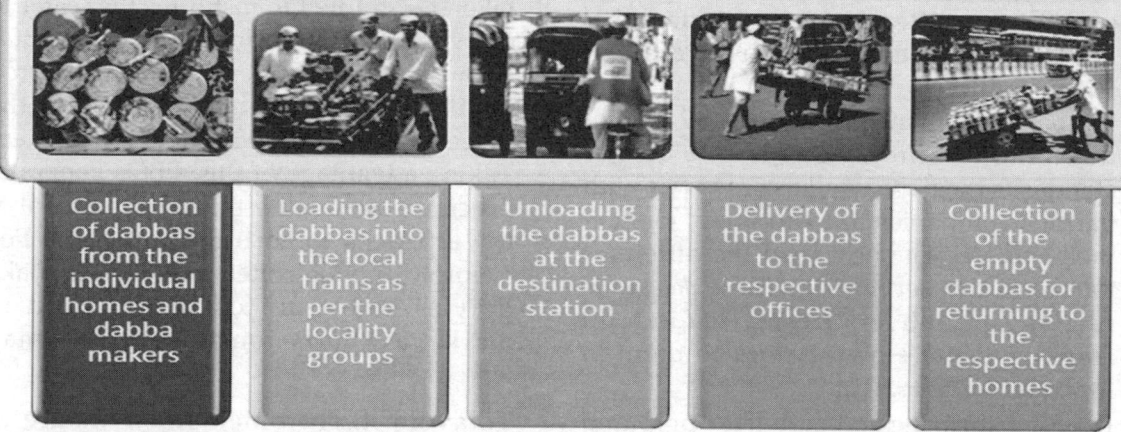

| Collection of dabbas from the individual homes and dabba makers | Loading the dabbas into the local trains as per the locality groups | Unloading the dabbas at the destination station | Delivery of the dabbas to the respective offices | Collection of the empty dabbas for returning to the respective homes |

Fig. 52.2: Dabbawalas and the process of carrying and delivering dabbas

Fig. 52.3: Brand restaurants home delivery bikes

Plastic containers

Clay pot

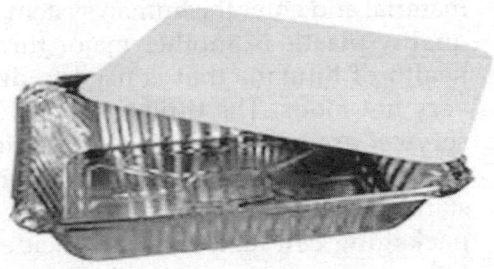

Aluminium box

Cardboard box

Fig. 52.4: Various food packaging materials

According to FSSA 2006, any individual in food business has to be registered with the food authority. In case his annual turnover is exceeding ₹ 12 lakhs then he should have the license from the authority.

Avoid ordering raw or uncooked food, particularly if you have young children, pregnant women or the elderly at home. It is imperative to check that hot food is delivered in hot boxes or insulated boxes and bags to keep the temperature above 60° C, and cold food is delivered in cooler or insulated boxes to keep the temperature below 5° C. One should order enough food for but not order in excess to avoid wastage. In case of any doubt regarding the quantity, it is advisable to discuss the matter with FBO for the serving size.

b. Premises of food preparation site: The small food outlets that provide home deliver food generally operate from very small dingy places. Such places do not have proper hygiene and sanitary conditions. These lack basic amenities for running food businesses. A few of these establishments do not even have safe water supply. Due to lack of water facility they do not even wash the vegetables and utensils properly. This lapse in quality standard can take its toll by compromising general public's health.

It is advisable that customers personally inspect the place from where food has to be ordered. They must also ensure the cleanliness of bike-box or hot case in which food is carried. The food case/package must not be contaminated. Always get food services from a hygienic and safe food business establishment.

c. **Temperature and time:** Electricity failure, poor cold storage and refrigeration facilities compromises food safety and quality in establishment running tiffin services and other food delivery services. The foods are left for much time in danger zone. This results in more likely spoilage of foods like chutneys, non-vegetarian foods, and foods with milk as main ingredients. All this can be potentially hazardous to the health of consumers.

Time of placing order is an important factor that can help to maintain food safety and quality. Order at a time such that the food to be delivered reaches at most **1 hour** before meal time and not earlier than that. This is to prevent food being kept under room temperature for too long. Food pathogens multiply rapidly between 5° C and 60° C, so no food should be kept in room temperature for more than **4 hours** from the time it is cooked at the restaurant's kitchen to the time it is consumed. These four hours include the time taken in packing and transport and set-up of food at the home table food. If the order for a high risk food which is to be served cold like salads, mayonnaise sandwiches, shakes, one must check that these are delivered chilled at below 5° C. Once delivered one should keep them chilled in refrigerator until ready to serve. In some countries (for example, Singapore) *"time stamp"* is mandatory on the delivered food package. This stamp informs the consumer the date and time that the food is cooked or prepared to a ready-to-eat state at the restaurant's kitchen, and when it should be consumed by. The "CONSUME BY" time should not be later than 4 hours from the time the food is cooked or

ready-to-eat if kept between 5° C and 60° C. If there is a party and the meal times are in different groups then, arrange for staggered delivery times with the caterer, instead of a single delivery time for all the packets.

d. **Packaging:** The quality of packaging material used is also often unsafe. Some use newspapers (which have poisonous ink) for wrapping purposes. The aluminum foil used is also of poor quality. The very thin and light foils generally stick to the food material and enter the human system. Poor quality plastic is another major threat to health of humans that is used to deliver very hot foods. The tiffins used for 'tiffin service' are not properly cleaned and at times have fungus in them. It is understandable thus, that lack of hygienic food packaging can lead to many incident of FBI.

Good quality packaging is the duty of FBO. They need to ensure that the packaging materials used for packed meals are food-grade, and can withstand the temperature at which the food is kept. In our country it is a common practice to use news papers, polythene bags to wrap food. This is done to reduce the cost of packaging. Such practices should be discouraged as they make safe food unsafe.

e. **Transport:** Food delivery on bike or bicycle is another critical point where the quality of food can be compromised. Generally there are few delivery boys and vehicles that cater to a large number of orders. Apart from this traffic jams, long distances and slow speed of vehicles further compromises the quality of food. The vehicles lack basic insulation boxes to maintain the right temperature of the food. And even if they have, they either are non-functional or are not cleaned regularly. This may also facilitate occurrence of FBI.

FBO must ensure that the transport vehicle is clean and paneled with stainless steel inside for easy cleaning. In case of both hot and cold foods delivery the chilled food items should be loaded into the delivery vehicle last. One delivery vehicle should be detailed for one single order and delivering food to multiple sites in a single trip should be avoided. This will result in prolonged storage of cooked food at incorrect temperature.

The new way of transporting food is through the use of drones. This may flourish in years to come. As of now it has just been introduced in food service sector. By using such high technology we can overall traffic and time constraints commonly faced in our country (Fig. 52.5).

Fig. 52.5: Delivering food package through use of drone

f. **Proper handling of delivered food:** In addition to above mentioned aspects; the hygienic handling of home delivered food is also crucial. If this is not adhered to FBI may result. To avoid it keep the hot food covered until most of the guests have arrived. Opening the covers too early will cause rapid heat loss and bring the food temperature into the danger zone of below 60°C. We often notice food warmers placed below the dishes in many parties. These can keep the food warm only to please the taste buds but do not extend the shelf life of the food.

In modern era, home delivery of food and tiffins has become an integral part of life in urban and metro areas. We cannot do away with such convenience food services. Therefore, both the FBOs and clients must take joint initiative and responsibility for ensuring safety of home delivered food. Surely a convenience food that ensures safety standards is like 'a cherry on the cake' for modern urban families.

53

Advertisement and Food Safety: Harms and Benefits

Nidhi Bhatnagar, Puja Dudeja, Amarjeet Singh

"Advertising is a non-moral force, like electricity, which not only illuminates but electrocutes. Its worth to civilization depends upon how it is used."

J. Walter Thompson

INTRODUCTION

Food safety attempts to ensure appropriate manufacture, storage, distribution, sale and import to ensure availability of safe and wholesome food for human consumption. Recently food safety has received the long awaited emphasis with establishment of FSSAI and also laying down of Food Safety and Standard Act (FSSA), 2006. Globalization forces and associated metamorphosis of Indian markets has deeply influenced our dietary pattern and preferences. Modern lifestyle and westernization has ushered in an era of new foods of young generation which have great acceptability with fewer health benefits. Ours is a growing economy which is often discredited with having a food market with surplus of misbranded and spurious products that may harm the consumers. Moreover, extensive Information Technology (IT) use has facilitated direct entry of markets in our homes. Advertisements are commonly employed through various mass media to promote products among consumers. An advertisement as defined by FSSA 2006 is any audio or visual publicity, representation or procurement made by means of any light, sound smoke, gas, print, electronic media, internet and website. It also includes any notice, circular, label, wrapper or other documents. Advanced information technology helps the manufacturer evolve innovative advertisements to reach to consumers. Message of food safety can be appropriately disseminated to the larger audience through advertisements. However, often these advertisements tend to make un-substantiated claims. Many a times these are likely to mislead consumers. Still quite often such advertisements are used as a device for marketing and hence profit making.

Reach of Advertising

Aggressive advertising and marketing of food products is done by the food industry. Indian markets are now-a-days flooded with food products from abroad as well as from local manufacturers. There was no standard agency till the establishment of FSSAI to approve safety of food products in India. Even now existence of such accreditation is not known to many. Knowledge and awareness of the same is essential as our markets are usually filled with spurious products. This is more so in rural areas where questioning of the quality of food purchased is rare. Food advertisements can very well be used as a medium for generating awareness on food safety.

Impact of television advertising is evident from the reach it has in the community. In urban slums of north India nearly 96.3% children have access to television. It has been reported that mean age of onset of TV viewing was 2.96 years and mean hours of TV viewing was 3.56 h. Another study conducted in South India found 49% children viewing television for more than two hours per day and 7% more than 4 hours per day. During weekends 71% children watch television for more than two hours and 24 percent for more than 4 hours daily. Many (44%) of the children were influenced by the advertisement of food items. A study conducted by the lead author in a government school of Chandigarh found that 33.4% children watched television for 3 hours or more and nearly similar percentage were able to recall more than eight advertisements they saw in television. This clearly supports the penetration of information technology in contemporary India and its potential utilization in propagating messages on food safety.

Impact of Television Advertising and Need of Regulations

Advertisements, the marketing tool of 21st century is beset with its own drawbacks. Food advertisements and labeling are two important sources of communication between manufacturers and consumers. Transmission of accurate information is necessary for decision making by the consumers regarding food purchase. In order to promote a food product amongst viewers, claims made are often in-complete, false or not substantiated by sound epidemiological studies. This may harm the consumer's health. For example, a common nutrient powder in India strongly claims improved growth of children who use this product. Its advertisement quotes a study which is methodologically incorrect and lacks scientific rigor. Age, socio-economic class were confounders with gross errors in the methodology of study. Moreover, the study was financially supported by manufacturers of the product which would have resulted in conflict of interest. Such flawed study cannot form the basis of strong recommendations given by the company in advertising products on television.

Similarly, two reputed manufacturer claimed that their products provided more stamina and made children smarter. In Chandigarh, a brand of Kulfi, locally produced was extensively advertised and well marketed in food joints and restaurants in 2013–14. This brand did not have FSSAI logo or license number either but the power of advertisement made it do well in the market. Another international corn flake brand's claim on the slimming qualities of product has also been questioned. A study done in Washington found that false claims, which are factually false or unsubstantiated, were rare, (1 in 10). However, most were potentially misleading, left out important information, exaggerated information, provided opinions, or made meaningless association with lifestyles.

Regulation of Food Advertising in India

Issue of advertising of food product is of great public concern. Advertising industry has recognized its sensitivity and has its own self-regulatory advertising guidelines. But till date the code is vague, compliance voluntary and enforcement not actively pursued. In most of the countries including India, self regulatory measures are adopted by food and beverage companies for regulating food advertising. An advertisement is called deceptive when it misleads people, alters the reality and affects buying behavior. According to Federal Trade Commission (USA) deception occurs under following situations:

1. There is misrepresentation, omission, or a practice that is likely to mislead.

2. The consumer is acting responsibly in given circumstances

3. The practice is material and consumer injury is possible because consumers are likely to have chosen differently if there is no deception. Deception exists when an advertisement is introduced into the perceptual process of the audience in such a way that the output of that perceptual process differs from the reality of the situation. It includes a misrepresentation, omission or a practice that is likely to mislead.

India has formulated regulations on these issues by the Information and Broadcasting Ministry and Ministry of Consumer Affairs. Most of the Self-Regulation Organizations around the world base their work on the codes prepared and published by the International Chamber of Commerce (ICC). 'Advertising Agencies Association of India', and the 'Advertising Standards Council of India' (ASCI), are business organizations. These can put moral pressure on advertisers and companies to withdraw objectionable advertisements. Their main objective is to promote responsible advertising, enhancing public confidence. ASCI's Code for Self-Regulation in advertising is now part of Ad code under Cable TV Act's Rules. The act and ASCI code is applicable to all states. Program and Advertising Codes prescribed under the Cable Television Network Rules, 1994 state that 'No advertisement which endangers the safety of children or creates in them any interest in unhealthy practices or shows them begging or in an undignified or indecent manner shall not be carried in the cable service.' ASCI as per the statistic provided by Customer Complaints Council reported nearly 58 claims of misleading advertisements in a month in health and personal care sector.

The FSSA, 2006 has clause 53 in Section 56 that clearly states that any person who publishes or is a party to the publication of an advertisement which—(a) Falsely describes any food or (b) is likely to mislead as to the nature or substance or quality of any food or gives false guarantee, shall be liable to a penalty which may extend to ten lakh rupees. Authorities entrusted for food safety will be responsible for enforcing the law.

Professional bodies like Indian Academy of Pediatrics (IAP) passed resolution to prevent sponsorship from companies that manufacture products covered under Infant Milk Substitutes Act 1992. The need to protect breastfeeding is becoming more urgent as the influence and sales ambitions of breastmilk substitute companies grow in emerging economies.

Advertisements for Food Safety

Advertisements should be the medium for disseminating mantra on food safety. Rising middle class and growing economy has made consumer more aware about their rights. Marketing of food safety can create awareness for safe food which will also help to ensure health and well being. Companies with FSSAI accreditation and good quality products should publicize the same. Sahara Q shops a newly opened chain of grocery stores in India assure the quality of the food products and well advertise the same. "Food Safe Families," was launched by US Department of Agriculture's Food Safety Inspection Service (FSIS), the Food and Drug Administration (FDA) and Centers for Disease Control (CDC) and Prevention, in cooperation with the Ad Council. This used videos, print ads and website to teach people about the risks of food poisoning and how they can reduce those risks by handling food properly at home. Safe food must be a rage in the food industry. This will check the sale of spurious products and ensure appropriate quality of products are put on sale by the manufacturers. Need for accreditation will surge among the manufacturers as demand for same will be raised by the consumers. Similarly, advertisements play an important role to allay consumer fears when product is linked with food safety concerns.

In times of today, advertisement is an important tool that propagators of food safety can capitalize. It can well act as a double edged sword, e.g. create awareness on food safety or result in much harm by misleading and false claims of the products advertised.

Section IX

Miscellaneous

54

Role of International Agencies in Ensuring Food Safety

Puja Dudeja, Ruchi Sharma, Amarjeet Singh

INTRODUCTION

Imagine you need to buy soap for yourself. A few years back the choices were restricted to Lux, Hamam and Lifebuoy only. However, in the present day world, there are many brands (included imported ones) available for everything we buy ranging from soap, cosmetics, jewellery, perfumes, cars, wines, and even food. It is the import and export of goods that has made the shelves of super-markets loaded with surplus of goods.

Trade is a boon to humankind and is important for the progress of any civilization. It provides countries with resources that they do not have. Everything cannot be grown everywhere. For example, coffee in England. The climate of England does not support production of coffee. So they do not produce it. But they import the same from Brazil. This helps them enjoy their mornings with a warm cup of coffee.

Trade in food between countries at times is also a necessity to ensure food security. In 1950s India experienced successive bad monsoons which led to a severe food crisis in 1955. At that time to prevent millions of people from starving in India, famous PL 480 wheat was imported through a deal with USA. This continued for more than twenty years .India received 50 million tons, or nearly 40 percent of all its food grains needs from USA from 1955 till 1971. However, the congress grass weed (*Parthenium*) one of 10 worst weeds in the world, traveled with it leading to a perennial problem we are still facing, viz-allergic problems in people all over the country.

In the ancient world, long distance trade started mainly for goods like spices, textiles (silk), precious metals and other luxury items. Later, Portuguese explorer and adventurer, Vasco da Gama established sea route from Europe to India, which was used for trading. In fact, East India Company came with the prime aim of trade only. In later half of 20th century the world experienced rapid globalization riding the tidal wave of advances in science and technology.

In today's globalized world, independent isolated existence of any country is unimaginable. Countries have to socialize for peaceful co-existence. Bilateral relationship between any two counties takes place at many fronts like political, commercial, economic and cultural. Such relations are aimed at growth of both the countries.

The word "globalization" simply means 'International Integration'. It is the process through which diverse world has been unified into a single society. It has contributed immensely to increase in cross border trade. Globalization has opened up gates for increase in international trade in various commodities like electronic goods, textiles, chemicals, food and even arms. Globalization also involves population migration, and increased mobility of goods, data and ideas between various countries. In 1947, 23 countries agreed to the General Agreement on Tariffs and Trade to rationalize trade among the nations. To facilitate such relations World Trade Organization (WTO) was created on January 1st, 1995.

The food trade has gained much momentum in recent past. It has vastly benefitted the consumers. They could get food from anywhere across the globe. Worldwide trade facilitates lower prices, year-round supplies, and a greater quality and variety of food. But, because of the associated hazards of FBI, food safety has emerged as shared concern for both developing and developed countries. Let us see some examples of food safety concern in the West. Till recently, 30 to 100 percent horsemeat was found in beef products of many top brands in 16 European Countries. In 2013, horse meat and traces of horse DNA were found in some food products where horse was not labeled as an ingredient, sparking the 2013 meat adulteration scandal across Europe. Irish food inspectors announced that they had found horsemeat in some burgers stocked by UK supermarket chains. But it was hushed up. Then in mid February 2013, up to 100% horsemeat was found in several ranges of prepared frozen food in Britain, France and Sweden. Even earlier, it was reported that Western fast food chains have used animal fats for cooking vegetarian dishes.

Another example is of the Guatemalan raspberry industry began exporting to the United States in the late 1980s, filling a market niche in the spring and fall when supplies were low. By 1996,

Guatemalan raspberry exports were increasing rapidly, up 113 percent from the previous season. That spring and early summer, the US Centers for Disease Control and Prevention (CDC) and Health Canada received reports of more than 1,465 cases of FBI from Cyclospora, a protozoan parasite. Although no one died, the large number of cases generated substantial adverse publicity. Initially, investigators linked the outbreak to California strawberries, but they finally decided that it was associated with Guatemalan raspberries. This case study reviews the efforts to resolve this food safety problem.

Therefore, safety is a shared responsibility as far as import and export of food products is concerned.

International trade allows for the rapid transfer of micro-organisms from one country to another. The increased time between processing and consumption of food leads to additional opportunities for contamination and time/temperature abuse, increasing the risk of FBI. Increasing trade also means that new and unfamiliar foodborne hazards can more easily reach consumers who have not developed immunities to those pathogens. The globalization forces have also led to many dramatic changes in the epidemiology of FBI. The emergences of newer pathogens have made the battle for food safety more difficult. For example, infection with *Listeria monocytogenes* after eating cantaloupe from Jensen Farms in Colorado lead to 32 deaths, *E. coli*, O104:H4, made worldwide headlines when an outbreak in Germany sickened approximately 4,000 people and killed 50, in 2011 after eating fresh sprouts. Emerging pathogens demand even greater food safety vigilance than what was required in previous generations because as pathogens are evolving and becoming more widespread, bacteria are becoming more resistant to treatments. Adding to the challenge, micro-organisms continue to adapt and evolve, often increasing their degree of virulence. For example, E. coli O157:H7 was first identified in 1982, but the bacterium has already been indicated as a cause for severe vomiting, bloody diarrhea, and even hemolytic uremic syndrome, which leads to kidney failure (CDC).

Another factor is introduction of newer methods of food processing, e.g. the method of *"infringement"* which has added many snack foods, breakfast cereals and confectionery items to our platter. The global fast food culture has also entered the land of "samosa and pav bhaji". With the advent of globalization, the fast food retail chains have penetrated Indian markets like KFC, Dominos, McDonalds, Subway. Multinational fast food joints take pride in declaring that the potatoes they use in making French fries are imported. They are competing with local fast food/street food, as well as with the elaborate meals of Indian homes by targeting the high end consumers. But their domination of our ethnic plates does have food safety concerns.

Alongside there has been progress in the food handling and packaging. In India, currently any imported good including food is considered as safe. Neatly packaged, branded and graded, foreign fruits and vegetables are the main attraction of the present day supermarkets. People in urban areas not only prefer to buy the exotic asparagus and melons but also the commonly grown apples, grapes and cauliflower, kiwis, strawberries, etc. But the beautifully packed and fresh looking fruits displayed on the stands in stylish markets are not always safe. In June 2011 India was placed on high alert against the deadly strain of Shiga toxin-producing E. coli, that has infected over 1,700 people across 12 European nations. The FSSAI

informed its officials in the five major ports and four airports which receive imports, to watch out for all food items, especially fruits and vegetables, coming in from Europe. All such items were first tested in FSSAI labs before being allowed into the country. WHO urged countries not to impose any trade restrictions in the face of this outbreak. However, Russia and Belgium clamped a ban on vegetables from Spain and Germany.

The lifestyle too has also changed in urban India. With globalization, eating trends have also changed. Women have started working outside home and are too busy to cook. They do not want to spend time and energy in household work and are against exclusive confinement of women in the kitchen. There is not enough time to shop for groceries or and to eat with one's family at home. The double income group families find it affordable to go out and eat. The fast pace of life has also brought the concept of fast food culture in our society. However along with advantage of availability of fast food and exotic foods there is a genuine concern about food safety.

The relative importance of different risks to food safety varies with climate, diets, income levels, and public infrastructure. Some food safety risks are greater in developing countries, where poor sanitation and inadequate drinking water pose greater risks to implications for developing country food producers and processors.

The *Morinaga Milk Arsenic Poisoning* incident in Japan where milk powder contaminated with Arsenic and *Chinese Melamine Scandal* emphasize the fact that food safety hazards do not spare anyone even if it is the feed of the young children.

In postindustrial era of modern society, service economy has dominated the job sector in developed countries viz. Banking, insurance, stock market, etc. In most cases, menial jobs of industrial production are outsourced to developing countries. For example, ship dismantling industry is present mainly in four countries of the world Bangladesh, India, Pakistan, and China. It has always been viewed as a polluting industry that has adverse effects particularly on the workers. Ship breaking activity is associated with dirty jobs, numerous deadly accidents, insecure labor, environmental injustice, and violation of human rights. Similarly, rich countries are also in a position to import raw food stuff from developing countries. So, there is dominance by the rich developed countries over the poor developing ones in importing rather than producing raw food.

Thus, there is a mutual dependence between rich and poor countries where rich countries get food like grapes, mango and rice at low cost while the poor ones get foreign currency which boosts their economy. However, because of their much better evolved society and exalted standard of living the developed countries impose higher and stringent standards to any stuff they import more so in case of food products. The reason is simple. Food is a potential source of disease (both infectious and non-infectious). So to protect its people against any hazard for such food imports developed countries enforce strict and stringent standards.

These are mediated through various laws/checks/monitoring systems. Also among exporters, there is a tough competition in. So, even in low income developing countries focus is on maintaining high standard of food products to meet international standard.

At international level food safety is guarded by both private and public sector. The private sector is more careful as in case of a controversy it suffers a reputation loss. These can even be blacklisted for food trade. They are pioneers in food safety, for example,

Nestle have developed their own stringent standard for ensuring food safety in international trade. Food safety in private sector is ensured through self-regulation, vertical integration with an outside agency to look after food safety along food supply chain or third party certification like ISO certifications. There are various international agencies that are committed to ensure safety of food across borders. A brief description of some of the international agencies in food safety is given in this chapter.

After the two world wars, which left agony, deaths and pain for the humankind there were international efforts to create peace in the world. The birth of United Nations Organization (UNO) took place to maintain peace between countries. The aim was to resolve conflicts between nations in case the need arose. So far this has worked well and prevented the eventuality of another world war. On similar lines, when the counties engage in trade they are bound to have controversies and disagreement on certain issues. This applies to food products also. There is often a need to arbitrate on any emerging controversy related to international food trade.

There are certain issues related to food trade and food safety, which sour relations between different countries. In such situations, there are specific obstacles to joint problem solving approach. This may be due to disagreement on existence or severity of a food safety risk. Many issues need to be resolved like who will bear the responsibility, who will pay for the cost and who will manage the response. Trade disputes over food safety, may require public intervention. Such food safety issues become Technical Barriers to Trade (TBT). For example, the issue of export of *King of fruits*; Alphanso mangoes from India to European Union and Britain.

Case History (May 2014)

European Union has temporarily banned the import of Alphanso mangoes, the king of fruits, and four vegetables from India from 1st May, 2014 as 207 consignments of fruits and vegetables from India imported into the EU in 2013 were found to be contaminated by pests such as fruit flies and other quarantine pests. The potential introduction of new pests could pose a threat to EU agriculture and production. This implies a ban on 16 million mangoes from India and a huge set back to economy worth 6 million pounds. The Indian traders firmly believe that the decision has been taken in haste.

Great Haste makes Great Waste

There have been many such examples in the past (Export of beef in case of Bovine spongiform encephalopathy). However, there are no perfect solutions to such disagreements. There are only proposed mechanisms to address such issues. There do exist law, code of ethics, agreements and accountability for such situations in international trade.

World Trade Organization (WTO) conceptualized agreement on Sanitary and Phytosanitary Measures (SPS) to addresses food safety and animal and plant standards for traded goods. The SPS Agreement confirms the right of WTO member countries to apply measures to protect human, animal and plant life and health. The Agreement covers all relevant laws, regulations testing, inspection, certification and approval procedures and packaging and labeling requirements directly related to food safety. Member States are asked to apply only those measures for protection that are based on scientific principles, only to the extent necessary, and not in a manner which may constitute a disguised restriction on international trade.

The TBT Agreement requires that technical regulations on traditional quality factors, fraudulent practices, packaging, labeling, etc. imposed by countries will not be more restrictive on imported products than they are on products produced domestically. It also encourages use of international standards.

The WTO also encourages its members to use standards set by the Codex Alimentarius Commission (discussed as separate chapter), or to set up their own. Codex work has created worldwide awareness of food safety, quality and consumer protection issues, and has achieved international consensus on how to deal with them scientifically, through a risk-based approach. These standards can even be higher than internationally agreed ones. But they must be based on scientific evidence. Also these must not discriminate between countries. Thus the agreement tries to balance health protection with trade openness.

There are many international bodies which play an important role in setting standards for food safety is given in Fig. 54.1.

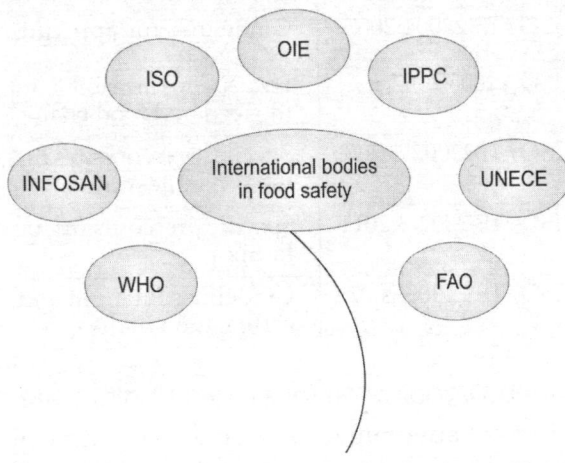

Fig. 54.1: International bodies in food safety

International Organization for Standardization (ISO)

ISO is derived from the Greek *isos*, meaning equal. ISO is an independent, non-governmental organization and has members from 162 countries. The standards developed by ISO are voluntary and there is no legal authority for implementation of these standards. The structure consists of a Central Secretariat in Geneva, General Assembly, ISO council and Technical Management board. To improve food safety ISO's food safety management standards help organizations identify and control food safety hazards. Food products repeatedly cross national boundaries and these standards help to ensure the safety of the global food supply chain. There are more than 17000 ISO standards. The ISO 22000 family contains a number of standards focusing on different aspects of food safety management. The ISO 22000 standards are given in Table 54.1.

Table 54.1: ISO 22000 standards related to food safety

ISO 22000:2005	Overall guidelines for food safety man agement
ISO/TS 22004:2005	Guidelines for applying ISO 22000
ISO 22005:2007	Focuses on traceability in the feed and food chain
ISO/TS22002-1:2009	Specific prerequisites for food manufacturing
ISO/TS 2002-3:2011	Specific prerequisites for farming
ISO/TS 22003:2007	Guidelines for audit and certification bodies

World Organization for Animal Health (OIE)

Humans are omnivorous. To ensure safety of animal feed there was felt need to fight animal diseases at global level. This led to the creation of the **Office International des Epizooties**

(OIE) through the international Agreement, which was signed on January 25th 1924. In May 2003 this office became the World Organisation for Animal Health but kept its historical acronym OIE. The OIE is the intergovernmental organisation responsible for improving animal health worldwide. It is the sole international reference organization for animal health and is recognised as a reference organization with 178 member countries in its organization.

As a result of increase in consumers demand worldwide for safe food, the OIE is dedicated to reduce foodborne risks to human health due to hazards arising from animal production. It works towards providing safety of food of animal origin. Its main focus is on eliminating potential hazards existing in animals prior to their slaughter or before primary processing of their products (meat, milk, eggs, etc.) That could be a source of risk for consumers. Department of Animal husbandry, Ministry of Agriculture represents India in this organization.

International Plant Protection Convention

International Plant Protection Convention (IPPC) is a treaty which aims to secure coordinated and effective action to prevent and to control the introduction and spread of pests of plants and plant products. Plant Protection Advisor (PPA) to government of India from Ministry of Agriculture represents India in this treaty. It covers both direct and indirect damage to plants by weeds. The provisions extends to cover conveyances, storage covers, storage places, soil and other objects or materials capable of harbouring or spreading pests.

United Nations Economic Commision for Europe (UNECE)

This is an organization of European community with 56 member states. The focus in respect of food safety is with establishing safety standards for motor vehicles and transport of dangerous goods.

Food and Agriculture Organisation (FAO)

FAO is a United Nations body with 193 member countries working to reduce hunger, food insecurity and malnutrition. Apart from this, FAO works with governmental authorities, local industry and other relevant stakeholders to ensure that the food available to the customers on domestic markets is safe and of the expected quality. It tends to improve systems of food safety and quality management, based on scientific principles that lead to reduced FBI. Its Endeavour is to support fair and transparent food trade. FAO has a food safety and quality programe which provides independent scientific advice on food safety and nutrition which serves as the basis for international food standards. This organization also works towards capacity building and food safety management programmes in many countries, including the management of food safety emergencies. It also supports processes for the development of food safety policy frameworks and facilitates global access to information for development of food safety/quality networks.

The Food Safety and Quality Programme coordinates activities in collaboration with other concerned technical divisions is their respective regional Offices. FAO also works with a wide range of national stakeholders, according to the particular problem being addressed, to ensure that holistic and feasible approaches are taken.

WHO is the body of United Nations which deals with health issues. It provides leadership in global health matters. It formulates the research agenda, provides technical support and sets norms and standards. WHO and its Member States are committed to food safety.

They plan and take multisectoral and multidisciplinary actions to promote the safety of food at local, national and international levels. It works towards protecting the health the consumers by providing public health leadership, technical assistance and cooperation, normative frameworks, science-based policy guidance and by consolidating health-related data. There is a department of Food safety and Zooneses in WHO which broadly carries out the tasks related to food safety in collaboration with other international agenies within and outside WHO. The International Health Regulations (IHR) which came in vogue in 2007 covers some food safety events which can constitute public health emergency of international concern. In such situations there is an existing plan to coordinate and collaborate among established networks in the area of food safety and foodborne zoonoses. These are namely International Food Safety Authorities Network (INFOSAN), the Global Early Warning System (GLEWS) for Major Animal Diseases, including Zoonoses -and the network of National IHR Focal Points.

International Food Safety Authorities Network (INFOSAN)

This is a joint initiative of WHO and FAO of United Nations which includes 181 member states. Each member state has a designated INFOSAN Emergency Contact Point for communication between national food safety authorities and INFOSAN secretariat. Each country is asked to identity Focal Points in other ministries or relevant agencies to receive INFOSAN communications.

This network aims to

- Promote the rapid exchange of information during food safety related events
- Share information on important food safety related issues of global interest

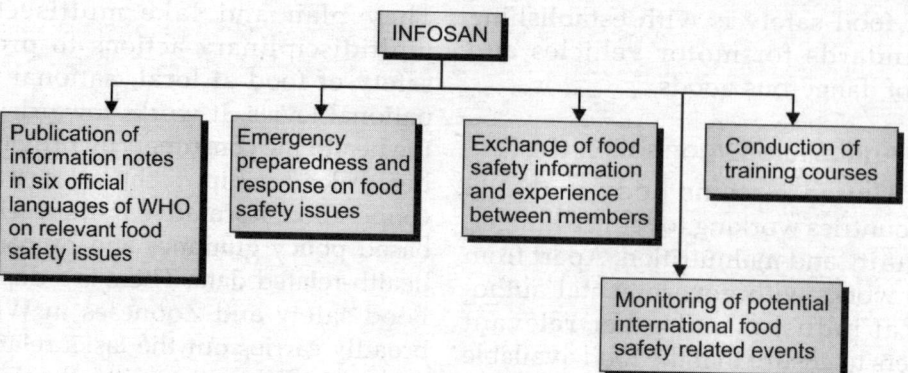

Fig. 54.2: Activities of INFOSAN in respect of food safety issues

- Promote partnership and collaboration between countries
- Help countries strengthen their capacity to manage food safety risks

Other activities of INFOSAN are shown in Fig. 54.2.

Case History: INFOSAN

Recently, it was reported through INFOSAN that several batches of milk whey protein concentrate (WPC) (an ingredient used in the manufacture of food products such as powdered infant formula, juice, dairy beverages, yoghurt and sports beverages) were suspected to be contaminated with *Clostridium botulinum*. The affected batches had been distributed to many countries worldwide.

The main concern was that WPC is an ingredient in other products that are then secondarily distributed, making it increasingly difficult to track and trace products through the food chain.

In this particular case, there was a great concern because WPC is also used in infant formula

C. botulinum is a bacterium which under certain conditions form heat-resistant spores that can germinate and produce toxins. Ingestion of these pre-formed toxins can cause a serious disease, botulism. Infants are of particular concern, because unlike adults, ingestion of *C. botulinum* spores can result in colonization in the gut, germination and release of toxins.

At this stage, the affected batches and products potentially containing them needed to be identified and recalled from the market.

WHO, through INFOSAN, assisted member states with the tracing of products which contained affected WPC as an ingredient. Information on product names and production batches was shared, allowing national authorities around the globe to respond appropriately to swiftly identify affected products to remove them from the market. So far, no reports of cases of botulism linked to this event have been reported through INFOSAN and the Secretariat is continuing to follow the issue closely with affected Member States.

It is rightly said that a stitch in time saves nine

55

Health Risks due to Pesticide Contamination of Food: A Growing Scourge in India

Mamta Bansal

INTRODUCTION

Safety of food and water is a basic requirement to ensure and maintain health of the masses. Their contamination leads to many health problems in human beings. There are various chemical and biological hazards which make food injurious to health. These hazards can arise from different ways. For example, it may arise from use of pesticides and other chemicals in agricultural practices, poor hygiene, lack of preventive controls in food processing operations, inappropriate storage and handling of food. Exposure to high levels of pesticides and other chemicals is assumed to be very risky to health by causing cancer, mutations and other problems. A study conducted by Department of Agriculture and the Indian Agricultural Research Institute proved that, the fruits, vegetables, poultry and even milk carry high pesticide residues which are much above the maximum residue limits (MRLs) set by the Prevention of Food Adulteration Act of 1954. The samples of popular milk brand collected from Ahmedabad, also had the highest traces of pesticides chlorpyriphos, a known carcinogen which also cause neural disorders.

According to a report, food items from 13 states in 20 laboratories across the country between 2008 and 2009 were tested for pesticide residue. The results revealed that the samples had residues of pesticides that are either banned in the country or are recommended for restricted use. For instance, in tomatoes, the traces of DDT were 108 times the recommended MRL. Residues of banned pesticides like aldrin, chlordane, chlorfenyinfos and heptachlor were found in samples of other vegetables, apple, rice, wheat, milk and butter. Most of these samples were from Uttar Pradesh. Other example from Assam shows that the tea samples had 4.280 ppm of pesticides fenpropathrin but its recommended MRL is at 2 ppm. All these studies confirm the risk of use of pesticides in basic day to day eatables. Still, there is no holistic approach to identify and prevent sources of contaminants.

Why there is a Need to use Pesticides in Agricultural Practices?

In modern agricultural practices, the uses of pesticides have become an important part of food production. The pests destroy 30% ($20 billion worth) of crops each year in the US and food costs could increase up to 50%. The use of pesticides in agricultural production can lower food costs by preventing direct loss of a product due to pests. Their usage also increases food's value cosmetically. Food's safety is also improved by reducing harmful organisms and through improved storage life. Without pesticides, food would be more expensive, require more labor and more extensive management. Our food supply's quality and the storage life would be lessened if pesticides are not used some food as they may contain harmful organisms.

However, pesticides use has its own risks. They can affect human health in three ways:

a. Pesticide residue in food

b. Exposure of person involved in spray and use of pesticide in agriculture

c. Suicidal or accidental consumption of available pesticide

Risks of Pesticides and their Residues in Food

Many of the pesticides are proven carcinogenic and mutagenic. In case of exposure to pesticides, some of the commonest effects are in the form of irritation to the skin, eyes and throat and sneezing and coughing while prolonged inhalation may cause dizziness, vomiting, burning sensation in the stomach, diarrhea and muscle twitching. A number of epidemiologic studies have found a significant association between cancer like brain tumours, acute lymphocytic leukaemia, non-Hodgkin lymphoma. Long term effects can be genotoxic, endocrine disruption, neurological disorder and immunotoxic in nature. Exposure of either mother or father to pesticide before conception, or exposure of the mother during pregnancy, has been associated with an increased risk of fetal death, spontaneous abortion and early childhood cancer. There is increasing evidence that *in utero* exposure increases the risk of growth retardation: a small-for gestational age baby, low birth weight, reduced length and small head circumference.

All India Coordinated Research Project on Pesticide Residues (AICRPPR) analyzed 4,111 samples of fruit and vegetables from different states between 1986 and 1996. According to their analysis, about 55% of samples were found contaminated with one or more pesticides. The most contaminated samples were pigeon pea, cow pea, snake gourd and cauliflower. The most common pesticides found were monocrotophos, methyl parathion and DDVP. Food samples from Uttar Pradesh and Kerala found MRL being surpassed by as high as 43 percent and 53 percent respectively. It was also found that the fruits, vegetables and milk in India were most contaminated. Recently the European Pesticides Residues Committee tested 4,000 fruits and veggies and found that a number of them contained more than the legal amount of pesticides. The levels of pesticides varied considerably, and imported fruits and vegetables were found to have a highest quantity. Apples, peas, yams, tomatoes, grapes, Chinese cabbage, cucumber, melon, oranges, spinach, potatoes, and pears were all found to have illegal levels of pesticides. Many of the foods contained multiple pesticides. It is not possible to test for all of the pesticides in use. In one of the survey conducted by AICRPPR on branded baby food and milk found it to be highly contaminated (Table 55.1).

Another study was carried out by University of California, Davis and University of California, Los Angeles, on 11 foodborne toxins in the US. This study was done using consumption of 44 key food products and the concentration of contamination in them. These toxins in food affect children the most as their brains and other organs are still developing (Table 55.2).

Health Risks due to Pesticides

The signs and symptoms of pesticide poisoning are:

Table 55.1: Pesticides residues found in baby milk powder by AICRPPR

Baby milk brand	Himachal Pradesh			Hyderabad		Kerala	West Bengal	Bangalore
HCH	DDT	HCH	DDT	HCH		HCH	HCH	
Brand I	3.74	1.47	0.57	0.22	0.25	0.52	0.22
Brand II	1.12	0.83	1.06	0.32	0.24	0.49	0.01
Brand III	1.88	0.34	0.41	0.04	0.35	0.14	0.08
Brand IV	2.86	0.46	0.45	0.02	0.24	0.69	0.07
Brand V	3.03	–	0.38	0.05	0.16	0.27	0.02
Average	2.52	0.78	0.58	0.13	0.25	0.42	0.08
Excess*	252.8	78.0	58.1	13.3	25.1	42.6	8.3

*Number of times higher than EU body food norms (0.01 mg/kg for all pesticides)

Table 55.2: Children at risk due to presence of toxins in food

Toxin	Mean daily intake (mg/kg body weight/ day) × 10–5	Reference dose*(mg/ kg body weight/day) × 10–5	Cancer benchmark	Percent Participants above Reference dose	Percent Participants above Cancer benchmark	Important sources
Acrylamide	180	20	Not specified	97.1	Not specified	Chips, cookies, French fries
Arsenic Lead Dieldrin	19.8 13.6 0.417	30 0 5	0.0667 Not specified 0.00625	18.36 100 0	100 Not specified 100	Fish Fish Meat, diary products, fish
DDT	0.000339	0	0.294	100	100	Meat, diary products, fish
Dioxins	0.000101	0.00023	0.0000001	2.42	100	Meat, diary products, fish

* Safe limit set by the US Environmental Protection Agency, (*Source*: Environmental Health, Nov. 9, 2012)

- Dermal and ocular irritation (or allergic response)
- Upper and lower respiratory tract irritation
- Allergic responses and asthma
- Gastrointestinal symptoms: usually vomiting, diarrhoea and abdominal pain
- Neurological symptoms: excitatory signs in the case of exposure to organochlorines, lethargy and coma; also polyneuritis

The specific syndromes of pesticide poisoning are:

- Cholinergic crisis (organophosphorus pesticides)
- Bleeding (warfarin-based rodenticides)
- Caustic lesions and pulmonary fibrosis (paraquat insecticides)

Different Ways of Exposure to Pesticides and their Evidence

There are some major groups of pesticides which are of specific concern to human in term of exposure. For example, anticholinesterase insecticides which belong to class organophosphates and carbamates and certain groups of fungicides are known for endocrine disruptors. The exposure of these pesticides and their residue pose higher risk to certain groups in the population, i.e. pregnant women and young children because of developing brain and endocrine systems of fetus and the children. The well known example is from district of Kerala where hundreds of victims are leading a life of perpetual misery due to undiagnosed illnesses, attributed to the over-exposure to the pesticide endosulfan. These people are still awaiting a news from the State government which has agreed to implement the relief and rehabilitation packages suggested by the National Human Rights Commission (NHRC).

According to Ministry of Agriculture, in India the entry of pesticides into food products due is mainly due to the following reasons:

- Indiscriminate use of chemical pesticides
- Non-observance of prescribed waiting periods
- Use of sub-standard pesticides
- Wrong advice and supply of pesticides to the farmers by pesticide dealers
- Effluents from pesticides manufacturing units
- Wrong disposal of left over pesticides and cleaning of plant protection equipment
- Use of pre-marketing pesticides
- Treatment of fruits and vegetables by pesticides for increased production

Determination and Calculation of Exposure to Pesticides

The entry of pesticides and other chemical residues in human body can also be through a number of ways, but primarily through eating. Therefore, knowing human dietary patterns is extremely crucial to determining exposure. It is calculated individually for an individual pesticide. The Food Standards Agency insisted that by knowing the levels of pesticides on foods and the level of exposure, it becomes easy to quantify the harm due to pesticides. To assess the risks it is important to determine the residue levels of pesticides which are allowed on a food commodity, it is called MRL in a food commodity. Enforcement of MRL levels enables us to minimize the pesticides exposure. If pesticide residue in a food commodity exceeds its MRL, then the food is legally considered adulterated and penalties can be imposed.

For the calculation of exposure of pesticides in general population, it is important to identify the food we eat daily, preferably at the national level and multiply the two quantities to know at the pesticide intake a person can be legally exposed to. For instance, Indian MRL for pesticide monocrothophos in rice is 0.025 mg per kg. Indian diet for rice is 209 gm per day. Thus, exposure to this pesticide through rice is 0.005 mg per day. This quantity which arrived at through multiplication is called theoretical maximum daily intake (TMDI). TMDI is the first step to measure the exposure but it can be further refined by studying the actual residues found in the food which is called estimated daily intake (EDI). Exposure of pesticides can also be estimated by measuring its intake in cooked food, which is called Total Diet Study (TDS). According to 1999 All India Coordinated Research Project on Pesticide Residues (AICRPPR) report, only 2% of food commodities worldwide were found to be above MRL. But in India this figure was as high as 20 percent.

A Brief Description of ADI and NOAEL Levels in Food

Acceptable Daily Intake (ADI) is that amount of a pesticides we can ingest daily over a life time without any damage to health. It is expressed in relation to bodyweight (bw). The safety levels for adults and children may vary. So, it should be calculated separately according to the bodyweight and senstivity. At a particular dose where a selected pesticide can not cause harm is called No Observable Adverse Effect Level (NOAEL). At the point where first sign of adverse effects appear, is called Lowest Observable Adverse Effect Level (LOAEL). Both these measures indicate the risk of long term effect or also called chronic toxicity. But it is also important to measure short term risk or acute toxicity of a pesticide.

For this purpose, USEPA established some safety limits to calculate the risk for acute toxicity which is called the Acute Reference Dose (ARfD). It is defined as the maximum residues that can be safely consumed at a meal or in a day. It is calculated by knowing the LD50 dose of a pesticide. It is called lethal dose (LD) that provides the potent quantity of a pesticide that can kill 50 percent of test animals either through ingestion or through contact with skin. To determine the safety it is important to identify both exposure and acceptable daily intake (ADI) of pesticides and its residue.

Managing Risks and its Monitoring: To What End?

After assessing the risk, the next important step is to manage it properly. To manage risks the important and crucial tools are ADI and ARfD. But for the consistent effectiveness, they need to be, constantly updated as the science improves; and calculated on the basis of the latest, most credible data.

For this purpose information received from pesticides company and from member nations are used by JMPR and USEPA. For example, JMPR's ADI for malathion is 0.3 mg/kg bw/day. Therefore, a 60 kg adult could safely consume 0.3 mg/day × 60 kg, or 18 mg of malathion each day. For a child of 10 kg, the amount would be 0.3 mg/day × 10 kg, or 3 mg of malathion each day.

Pesticide residue are monitored by AICRPPR under the Indian Agricultural Research Institute (IARI). But its mandate is to research and not enforce standards. There are still no pesticide residue standards for these products under Food Safety and Standards Act (FSSA) 2006. The report's publication was also an exception as usually, AICRPPR residue monitoring data is treated as a secret. Regulatory framework for pesticides in the country is still not well established. They are supposed to enforce MRLs but do not monitor residues. The monitoring agencies calculate the contamination, but cannot regulate the poisonous presence of pesticides in food.

International Scenario to Combat Health Risks from Pesticides

US has assigned clear responsibilities to two nodal agencies: the US Environment Protection Agency (USEPA) and the US Food and Drug Administration (USFDA). USEPA is the standard setting agencies and entrusted with registering a pesticides for use. Before registration of any pesticides, it follows all the steps to establish ADI and sets MRLs for residues on food commodities. This agency makes sure that the exposure of pesticides is well below ADI. USFDA is an enforcing agency. It ensures MRLs are adhered to. This agency collects raw or processed samples of food and analyses them for pesticide residue. Both are the federal agencies which are responsible for evaluating, setting and enforcing safe levels of pesticide residues in food for human consumption. Pesticides cannot be registered and cannot be sold in the US unless the safe level (tolerance) is set. These tolerances are set after rigorous field tests, which involve the maximum usage of that pesticide. Also, scientists must determine that no observable effect is found in sensitive laboratory animals.

EPA scientists calculate the safe daily intake of any particular pesticide for humans. The EPA determines how much of that pesticide's residue consumers are exposed to and what the maximum possible exposure could be. To figure that out, they suppose that a certain crop is treated with the highest legal rates of a pesticide and that consumers eat that crop every day for a lifetime. If the maximum possible exposure to a chemical is less than the legal residue level, the EPA grants the tolerance. The researchers extrapolate the data received from animal toxicity on humans, keeping in mind that human are more sensitive than animals. They adjust it downward usually by a factor of 100: a division factor of 10 is used as humans are sensitive, and further division factor of 10 is used to allow for difference between sensitivity among humans. Nowadays, there is another concern to make this calculation for infants and children. To solve this problem, scientists and health activists want a further safety factor of 10 for children. If this happens, the toxicity data will be adjusted downward by a factor of 1000.

Pesticide Regulation in India

Two legislations regulate pesticides in India—the Insecticide Act, 1968 (IA) under the Union Ministry of Agriculture; and Food Safety and Standards Act 2006 (FSSA), under the Union Ministry of Health and Family Welfare. The former's provisions are enforced by the Central Insecticide Board (CIB) and Registration Committee (RC). The over 25-member strong CIB, headed by the Director General of Health Services, meets once in six months to advise on matters related to administering the Insecticide Act. The RC is headed by the agriculture commissioner and meets once every month to register pesticides for use in India and for export. It is supposed to do so after satisfying itself about a pesticide's efficacy and safety to human beings, animals and environment and relevant data to this end are collected from companies. But the major problem lies here that they neither fix ADI of a pesticide to be registered, nor set MRLs on food commodities.

In India, the pesticides regulations are governed under the following Acts/Rules:
1. Pesticides Management Act 2008
2. Food Safety and Standards Act 2006
3. The Environment (Protection) Act 1986
4. The Factories Act 1948
5. Bureau of Indian Standards Act
6. Air (Prevention and Control of Pollution) Act 1981

7. Water (Prevention and Control of Pollution) Act 1974

8. Hazardous Waste (Management and Handling) Rules 1989

In western countries, the agency registering the pesticide establishes ADI, sets MRLs and then ensures cumulative exposure is within the safety levels. However, in India, a pesticide is registered without any of these mandatory safety regulations. In fact, there is no legislative provision to link pesticide registration to setting MRLs. IA mandates registration, but FSSA mandates MRLs. Such legislative blindness has ensured that, of the 180 pesticides currently registered, MRLs have been set only for 71. In other words, more than 60% of pesticides currently registered have no MRLs.

In 2003, an effort was made to ensure that registration of a pesticide by the CIB. For this purpose, the Ministry of Health fixed the MRL for the pesticide in different foods. But the fixation of MRL further delayed registration. MRLs are set on the basis of recommendations made by the Pesticide Residues Sub-Committee of the Central Committee of Food Standards (CCFS), Union Ministry of Health. The CCFS meets once or twice a year, and standards are set on the basis of information supplied by government research institutions and companies.

FSSA is oblivious of ADI. The CCFS has no mandate to establish it. Thus, when CCFS develops MRLs, it does not cross-check exposure levels against ADI. If any product meets the MRLs there is no guarantee of safety, for there is no way to find the safety threshold in the absence of ADI. Moreover, there is no communication between the two sets of nodal regulatory agencies created under IA and FSSA. There is a mismatch between the pesticides the CIB recommends for use on a food commodity, and those for which MRLs have been set under FSSA for the same commodity. For example, for sugar cane, the CIB recommends 13 pesticides to be used. However, under FSSA, MRLs for only 2 of the recommended pesticides have been established. Similarly in rice, 56 percent of recommended pesticides have no MRLs; in wheat, 43 percent have no MRLs; in mango, 44 percent. In coffee, 80 percent of recommended pesticides have no MRLs.

Current Regulation of Pesticides

Government under the scheme titled "Strengthening and Modernisation of Pest Management Approach in India" has adopted Integrated Pest Management (IPM). It is an eco-friendly approach for pest management that involves cultural, mechanical and biological methods and need-based use of chemical pesticides with preference to use of biopesticides, biocontrol agents and indigenous innovation potential. The human resource development in IPM is done by imparting training to agriculture/horticulture extension officers and farmers at grass root level through organising Farmers' Field Schools. In collaboration with State Department of Agriculture/Horticulture/ICAR Institutions/State Agriculture Universities the IPM package of practices for pest/disease management in 77 major crops have been developed which have been circulated to all States/UTs.

Although the effects of pesticides are generally target-specific yet, in actual practice, they are not always selective for intended target species as they also tend to damage the nontarget species of plants and animals including man. In most cases pesticides are sprayed on the crops by uneducated farmers who cannot even read the warnings related to their safe use. Farmers and other field workers directly come in contact with the pesticides during their use in agricultural practices. It is suggested in a monitoring report of the Department of Agriculture and the Indian Agricultural Research Institute that the country's regulators have failed to check the flow of pesticides into the food chain, which may cause mutagenic damage also.

At last it is also important to mention here that many of the foods we eat contain numbers and levels of cancer-causing agents that may or may not come from pesticides. The threat posed by them may be higher than the levels posed by pesticides. More research on biology and non-chemical control of pests is needed, so that in the future non-chemical treatments can be an option for farmers and consumers alike. There would be development of such methods which are able to determine more precisely the real risks of pesticide residues as compared to natural toxins. In states, food inspectors are appointed and can take legal action for violation of the FSSA Act. They are supposed to keep proper track of pesticide residues in food commodities. For this purpose, they send samples to state-run laboratories and there are seventy such laboratories in our country. From time to time, these laboratories may find contamination and then send to Central Food Laboratories established under the FSSA. Four such laboratories exist to verify whether contamination exists or not. This process is so tedious that enforcing standards is almost non-existent.

Organic Farming: An Alternate to Modern Agriculture

The need for an appropriate method in agriculture which should be suitable to environment and the human being is being felt. Organic farming is then recognized as the best known alternative method for sustainable agriculture free from pesticide use. Organic farming is based on the similar principles underlying our traditional agriculture (without the use of synthetic chemicals) but without any harm to the environment and also to the human life itself. Organic farming or agriculture is also defined as 'a system that is designed and maintained to produce agricultural products by the use of methods and substances that maintain the integrity of organic agricultural products without the use of extraneous synthetic additives or processing until they reach the consumers. The potential benefits of the organic methods are healthy and safe foods, improved soil quality, increased crop productivity and low incidence of pests.

Some Recommendations

1. The approval of pesticides and authorization of similar compounds used on crops should consider all sources of exposure.

2. Scientific and systematic framework should be established to decide when it is appropriate to carry out combined risk assessments of exposures to more than one pesticide. Analysis of exposure to more than one pesticide will require changes in the methods used for risk assessment, including, in some cases, the use of probabilistic exposure assessment.

3. The methods be developed to provide valid and cost-effective biomarkers or other robust indicators of population exposure and systemic (body) burdens of mixtures of pesticides and its residues.

4. Alternative methods of crop production may be implemented in large scale. The government is promoting research on the use of alternative and safe pesticides, for example neem seeds. A country programme entitled "Development and production of neem products as environment friendly pesticides" is being undertaken by the Department of Chemicals and Petro-

chemicals with the financial assistance of United Nations Development Programme (UNDP) which will lead to the following:

a. Preservation of ecosystem

b. Conservation of biological diversity

c. Improved water quality both ground and surface water

d. Environmentally-sustainable economic development

e. Propagation of receptive models for development, promotion and conservation of biologically diverse bio-pesticides as an alternative to persistent organic pollutants (POPs) pesticides.

56

Food Safety in Indian Weddings

Suninder Kaur, Puja Dudeja, Amarjeet Singh

INTRODUCTION

Marriage is more or less a universal pheno-menon. It involves a ceremony where two persons are socially declared as *married*. Although wedding rituals differ across cultures, ethnic groups, countries and reli-gions, one thing remains the same, the ceremony involves a large gathering of people which include both family and friends, relatives and colleagues, neighbours and well wishers from the bride as well as groom's side. Weddings are an occasion for excitement and enjoyment. The wedding reception involves blessing of the newly weds by the invited guests. Over the millennia, the tenor and duration of marriage ceremonies have changed. However, feasting is a common feature, which has persisted. People remember marriage for the kind of food served to the guests. Food remains the centre point of all weddings. This is true for both modern as well as traditional weddings. In most weddings food is cooked and served to large number of people. Hence any breach in food safety can lead to large number of cases of food poisoning.

Cases of FBI have been reported after having lavish wedding feasts across the globe, both in developing and developed countries. There have been food poisoning cases reported from almost every part of India due to consumption of food during weddings. Newspapers frequently report articles relating to food safety and people getting ill after consumption in large scale feasts. A review of recent cases of FBI identified an outbreak in March, 2010, in Vadodara, India, where as many as 150 persons suffered from food poisoning after eating a wedding feast including *dal, rice, laddu* and various vegetables. In May, 2010, wedding food also hospitalized at least one hundred people in Northern India.

For various reasons, a lot of weddings in India take place during May and June, the hottest months of the year. Food spoils very quickly, and it is common for many wedding guests to become ill. Additionally, during the summer season meals are often prepared and served outdoors, increasing the risk of FBI. On 2nd May 2014, twenty-four people got ill after a wedding feast in Ullai Village in Bhilwara district of Rajasthan in India. In another incident at least 100 people were admitted to Rao Tula Ram Hospital in New Delhi with food poisoning symptoms, after they consumed food at a wedding.

Stories of people falling ill are numerous. However, many go under-reported. Some victims resort to home remedies without realizing the symptoms of food poisoning

458

such as dirrhoea, vomiting, etc. Ensuring food safety can help reduce such episodes.

Changing Times

With the passage of time, there have been variations in ways in which the wedding ceremonies are planned. In older times, the entire village used to help the bride's family on her wedding. Neighbours and relatives would help in cooking food. Local *halwais* would be called for preparing sweets. This trend has however changed. Now-a-days wedding functions are organised in large banquet halls and caterers are hired who are given the entire responsibility from procuring, transport, storage, cooking and serving. Over the years, the marriage ceremony, though a religious affair, accompanied with chanting of vows of rights, duties, prayers, hymns, etc. has come to focus more in entertaining the guests through foods and drinks. The more lavish the party the more it is applauded for its hospitality which includes quality and the variety of items served. In earlier times, since the people were involved themselves personally in food preparation, there were less chances of any flaw or food safety error. Also there was more involvement of the host in overall management. Now, hiring catering professionals has made the task convenient for the host but at the cost of risk of compromised food safety.

Socioeconomic Status and Wedding Styles

Affordability and expectations, both influence the kind of wedding preparations and cere-monies taking place including the snacks and foods involved.

Low Income Group

Among the low income group people, marriage of girls comes as a burden as the tradition and customs involved in performing ceremonies are a big financial strain on the hosts. Food remains the centre point of welcoming and honoring the guests. Hence, though quantity is arranged, there is neither the financial means to assure safe drinking water and food nor the facilities or knowledge to understand the importance of food hygiene and safety. Hence many a times it has been seen that there are illnesses, food poisoning and diarrhea among the guests in weddings.

Middle Income Group

The middle class families do try and take care of food safety and hygiene. The friends, relatives and neighbours and even at times the colleagues of the brides parents, in middle income groups of people, participate and help the brides parents in planning and managing the wedding ceremonies. Usually a dep-endable senior relative is assigned to supervise the catering arrangement. He literally sits there and keeps a watch over the *halwais*. The main factors which remain the reason for poor adherence to hygiene and safety standards of food in the community are ignorance and limited financial resources to reach up to the expectations of the groom and his relatives.

High Income Group

The rich and the affluent do spent a lot of money in entertaining the guest with a number of grandeur functions and lavish feasting. However, the entire program is outsourced and the hosts participate in the function as guests only. They also reach the spot/venue being unaware of the pre-parations. They do not bother how and where are the dishes being prepared and whether trained cooks and food handlers are being involved in maintaining hygiene and food safety standards or not. Many-a-times it has been seen that the same oil is used in frying *pooris and pakoras* and non veg-food items. However, drinking water used is distilled water bottles or sealed plastic glasses which ensures water safety.

In spite-of-all this, it has also been observed in some lavish parties that cut fruits are displayed, which are cut and kept much earlier than the actual feast begins, and sometimes uncovered fruit items attracts all kinds of insects which is unhygienic to consume thereafter. Many a times, the breakfast, lunch or dinner takes place much later than the expected time, as in India, usually the grooms side people reach much later than the scheduled time. By that time all the food items become stale, including the cut fruits and salad and the slow heat on which it is kept all the time makes possible for bacteria to breed. Lots of time the curd or yogurt served becomes sour as well, since it is kept for long hours.

Other common practices which lead to cross contamination during food preparation, e.g; using unwashed knives for cutting vegetables after using them for peeling raw vegetables. The cut vegetables are placed usually on gunny bags. Another problem is lack of availability of clean utensils from the tent houses which are hired. Since there is no continuous water supply often two tubs are placed one with soap water and another one with clean water for washing utensils. This can also compromise food safety through use of unclean utensils. Food remains the main attraction of all weddings. Not only quantity and variety, even serving styles, and decoration of food items is focused upon more in wedding. However food safety is not ensured with the same intensity and minuteness.

Role and Importance of Food in an Indian Wedding

The ceremonies accompanying the Indian wedding are incomplete without a massive feast accompanying the ceremonies. Depending upon the status and lifestyle of the bride's and groom's side, the feast now-a-days includes not only Indian dishes but also Chinese Thai and Continental cuisine as well, especially in weddings among the upper class families. The Indian cuisine also consists of an assortment of dishes ranging from North to South, i.e. Punjabi food to Gujrati snacks and South Indian recipes as well. Snacks also cover a wide range of vegetarian and non-vegetarian items and at times includes sea food as well. Desserts have also hit a wide range of items in modern weddings. In earlier times, the relatives and women folk of the household would gather a few days in advance. The women would cook for the relatives in the wedding. Now-a-days this is usually skipped. Food/catering is outsourced. So safety has emerged as an issue. Earlier when the women-folk would look after the preparation of food items, food safety was never a concern. All cleanliness and hygiene related issues were adhered to as a standard pattern of working and serving.

Large quantities of food items to cater to the huge gathering is prevalent in large scale among high profile weddings. Food safety is a major issue, but often is not adequately addressed. Such a scenario becomes a threat to the lives of people relishing and enjoying meals in weddings. Food safety is taken for granted at the host's end. It is assumed that hiring a reputed and expensive caterer will imply safe food. However, a good host should understand that in case of an adverse event, only he will be ultimately blamed for poor organization/management. Any wedding involves mass cooking and eating. Foods at high risk are also served during such gatherings like salads, salad dressings, curd, etc. Since it is difficult to reheat the entire food, they are kept in food warmers for a long time before actual consumption. This makes all food potentially hazardous.

Factors Leading to FBI and their Control

Some of the common factors leading to FBI in a wedding feast are:

a. Improper holding temperatures

b. Contaminated equipment, utensils for cooking and storage

c. Unsafe water

d. Poor storage conditions of cooked food

First and foremost is to ensure cleanliness at all costs. Unless utensils, cooking surface and hands are not washed well by the food handlers illness causing bacteria will prevail in and around the kitchen. Most weddings are held in open places where *shamianas*/large size tents are put up. The caterer makes a local/temporary kitchen arrangement and there is not much storage facility and hand washing facilities. The host should ensure that there is availability of clean running water in required quantities for washing fruits and vegetables as well as utensils. All fruit and vegetables need to be washed even if we need to peel them and remove the outer covering. This is important because illness causing bacteria can spread from the outside to the inside of these food items and can make food unsafe. Food is usually cooked much in advance. In wedding functions food is served over a long duration of time. Hence, temperature control is important especially in summers.

In Indian weddings, the main style of food preparation is by deep frying, be it *pooris* or fried vegetables or a variety of *pakoras* served as snacks with *chutney*. Deep frying helps kill bacteria and other illness causing germs. This in turn proves to be not just a quick method of cooking but also a good method of ensuring food safety. Similarly, preference of tea/coffee or cold drinks of a large number of guests over water also ensures food safety especially among low and middle class families where mineral water is not served. Many Indian weddings avoid non-vegetarian dishes in wedding functions as a customary practise linking it with religious values and norms. This also accounts for food safety as non vegetarian food has a higher possibility of food poisoning among people in comparison with vegetables, if food safety guidelines are not adhered to properly.

If food is reheated well up to 75°C then likelihood of its spreading disease is remote. Other causes of food poisoning are unhygienic and/or unclean surroundings. If covers for big sized vessels used for mass cooking are not available then chances of insects falling in open food vessels are high.

Precautions to be Taken While Organizing Food for Wedding

A few key points which should be kept in mind prior to arranging wedding feasts are:

- Caterer selection: Hiring caterers has become a common norm among almost every wedding. However it should be ensured that licensed/registered caterers are only hired.
- Hygiene: Ensure hygiene and cleanliness at the place where food is being cooked. It has been observed that a many a times, family members and associates of the hosts keep a check on the food items lest they get pilfered, but safety measures take a back seat.
- Quality of raw material: The raw material purchased should be of good quality, which includes both spices used in cooking as well as raw vegetables and fruits.
- Careful menu planning: Avoid high risk items in the list of menu items. Prefer fried food in place of boiled or steamed food or even raw eatables. Avoid non-vegetarian items in the menu.
- Hygiene of people working in wedding parties: The chefs, helpers and serving stewards should be ensured.

- Gap between time of preparation and time of serving: There should be minimum gap between preparation and serving time. The hosts play a big role in ensure this as they need to inform the chefs and caterers the correct time of food serving wherein the guests would arrive and eating would begin.

- Employ a family member to supervise: Usually this is prevalent, more commonly among middle class families, however, the family members are not clear on what they need to supervise and they mainly focus on guests being served well and food containers not staying empty of long. Though this is essential but equally important is too have hygienically clean and well cooked food. Family members should supervise the cooking and serving part.

- Caters should ensure that plates and other utensils are not wiped with a dirty cloth after washing.

Precautions While Attending Wedding

The guests should also ensure their safety by selecting food items which are safe to consume in mass celebrations and feasts. Some of the basic safe choice of food items are:

- Avoid cut fruits, salads, curds, salad dressings
- Avoid chutneys with snacks
- Avoid raw ice in drinks

The guests can select the right foods in these wedding ceremonies and avoid getting ill. Foods heated well, or below freezing point are more safe to consume than items kept warm on low flame for a long time. Uncovered Food items are also a source of illness. Mineral water should be prefered, else tea, coffee and soft drinks (without ice) are a better option. Avoid the consumption of meats, egg and sea food especially if they are not cooked at high temperatures. Eggs should be cooked well as else germs can propagate.

Post-wedding Food Safety

It is common to see that often large amount of food is left in such gatherings after the wedding is over. To prevent wastage of food the left over food is either taken to host's residence, distributed to friends/relatives or given to the caterer for further distribution to workers. Consumption of this food in subsequent days which has already been in temperature danger zone, can lead to FBI.

Conclusion

Food safety is a wedding is a matter that affects each one of us either as a host or a guest. A food safety model in wedding feast would ensure serving of safe water for drinking, for washing raw material and utensils in food preparation, using good quality food items, ensuring minimum time lapse between preparation and serving of food and avoiding high risk foods. One should select a menu which is both nutritious as well as safe for consumption. More preference should be given to items of instant cooking rather than preparing in advance. Monitoring food safety both by the caterer as well as the host should be done. Finances should not be compromised if they jeopardize food safety. The selection of caterer should be done with due care based on his reputation and availability of license/registration. Hygiene of food handlers should also be checked before assigning them various tasks.

People who organize large scale weddings catering to massive gathering do ensure quantity and assortment of food items to please the guests. But what usually lacks is the due attention to food safety. Such feasts should be planned and organized with due care to food safety aspects for the well being of our society.

57

Bihar's Midday Meal Tragedy: A Case Study

Neha Chanana, Shalini Dwivedi, Jaideep, Puja Dudeja

It was 16th July 2013, like a usual day, children (aged 5–12 years) of Dharmashati Gandawan village (District Chapra) of Bihar went to the nearby primary school. The clock struck 12:30 pm, the school bell rang… *its lunch time*!! The children put away their bags and books to pick up plates for the free lunch they are served everyday at school, popularly known as the 'midday meal'. Laughing and chatting they sat on the verandah floor and got ready for the meal—rice, soyabean-potato curry-to be served to them. The meal is served. This was the usual routine followed at school every day. However, what happened next was not only unusual but also tragic!!

Soon after having the meal, many children started vomiting few fainted and several collapsed. 23 children died.

The newspaper for the next few days was full of updates about this tragic incident with headlines such as:

"Midday meal horror in Bihar"………. "Poison served on the platter"…….."Poison theory floats as Bihar midday meal kills 23 kids"….. and many more……..

The death of 23 innocent children in Bihar in July 2013 was indeed a shock. Majority of them were aged below 12 years. Midday meal (MDM) served to primary and upper primary school children was supposed to provide energy and nutrition. Instead it turned fatal. This midday meal tradegy of Bihar was a huge setback for the largest feeding and free lunch programme of the country, the MDM Programme.

In fact, there have been events highlighting poor standards of food hygiene in the past all over the country wherein dead frog, lizard, cockroach, rat, pests, snakes were detected in the food served to children. On various occasions children have been hospitalized on account of food poisoning. However, because of this incident in Chhapra district, this free lunch programme has suffered a huge setback.

In this case, the autopsy reports indicated that children died because of insecticide poisoning. The forensic report confirmed the presence of 'monocrotophos', an organo-phosphorus compound in the samples of oil from the container that was used for cooking. Food remains on the platter and mixture of rice with vegetables on the aluminum utensils which were used for cooking and serving food also had remains of the insecticide. Organophosphorus compounds are usually used as insecticides in cultivation of rice and wheat. The reports also suggested that the amount of insecticide in the food items was 5 times more than what is usually used as a control against insects in the fields.

This 'Bihar MDM horror' is indeed one of the biggest tragedies in the 'MDM Scheme' since its inception in 1995.

MDM Scheme Over the Years

The MDM Programme was first introduced in 1925 by the Madras Municipal Corporation for disadvantaged kids so as to enroll more students in schools. Table 57.1 shows the major milestones of MDM Programme.

Prior to 2004, the concept was to distribute dry rations and biscuits. Dry ration included free supply of food grains @ 100 grams per child per school day, and subsidy for transportation of food grains up to a maximum of ₹ 50 per quintal. However, experience showed that most of the times the food was shared in the family. Also it had low nutritional value. Thus, with the renaming of the programme to MDM Scheme in 2004, the idea shifted to cooked meal. In September 2006, the norm to offer 450 calories and 12 grams of proteins to every child was incorporated under the scheme The main objective of the scheme was to improve nutritional status of children by offering them cooked meal. Encouraging poor kids to attend school, improve enrollment in schools, reduce absenteeism, increase retention and reduce gender gap in education were several other intentions that drove towards the implementation of this scheme. In 2008, it was extended to all schools and learning centres and since then is supported by the *Sarva Shisksha Abhiyan.*

Table 57.1: Milestones of MDM Programme

Year	Milestone
Pre-Independence	
1925	MDM programme introduced in Madras Corporation by British India
1930	A MDM Programme was introduced in the Union Territory of Puducherry by the French administration
After Independence	
August 15, 1995	National Programme of Nutritional Support to Primary Education **(NP-NSPE)** was launched as a Centrally Sponsored Scheme under the Ministry of Education
1980	Three States viz. Gujarat, Kerala and Tamil Nadu and the UT of Pondicherry had universalized a cooked MDM Programme with their own resources for children
1990–1991	12* states implemented some form of the programe with their own resources
1997– 98	NP-NSPE was introduced in all blocks of the country
2001	State Government were directed to provide cooked MDM to the students in schools
July 2004	National MDM Scheme was launched so as to ensure that a child is not hungry and provided free cooked meal at school
2006	Programme extended to upper primary schools in government and aided schools.

*Goa, Gujarat, Kerala, Madhya Pradesh, Maharashtra, Meghalaya, Mizoram, Nagaland, Sikkim, Tamil Nadu, Tripura and Uttar Pradesh. In another three States, namely Karnataka, Orissa and West Bengal

The children studying in primary and upper primary schools are enrolled under this programme. On the other hand, pre-school children (2–6 years) are enrolled under the Integrated Child Development Services (ICDS) programme at the Anganwadi centres (AWC). Initially, *khichdi* was served as a means for supplementary nutrition by the Anganwadi worker (AWW) at the AWC. *Khichdi* provided 300 kcal and 8–10 g protein per serving as was mandated under scheme. However, with the amount of supplementary nutrition being revised to 500 kcal and 12–15 g of proteins, the meal served to these pre-school children has undergone a change and includes morning snack and lunch on a weekly prescribed menu for the states and is no longer limited to just *khichdi*.

Financial Assistance under MDM Programme

With regards to financial assistance, the centre pays fully (100%) for the supply of food grains under this scheme (100 grams for primary student per day and 150 grams for upper primary). The centre and state share cooking cost in 75:25 ratio; Centre gives cooking cost @ ₹2.89 for primary students, @ ₹4.33 for upper primary out of total of ₹3.11 and ₹4.65 respectively. These norms vary from state to state. The centre provides 100% assistance for monitoring of scheme, cook's salary, utensils and construction of kitchen – cum- sheds and transportation of food grains to the school.

The movement of funds is from centre to states in two installments 60% by June and remaining 40% by October through e transfer to state accounts in the RBI. State Finance Departments issue sanction orders to release funds to each district. District authorities order the district Treasury to pay District Education Officer (DEO). The DEO sanctions money to Block Education Officer (BEO) through treasury. The BEO releases money through sanction orders to accounts of respective schools. The whole process takes four to six months and till then the schools operate the scheme on loan or credit. Cooks are paid a salary only when money reaches the school.

In this scheme raw materials are cooked and meal is served to students studying in a particular school. The raw materials from the field passes through various stages after which it reaches the beneficiaries in cooked form—the farm to fork chain. As discussed in other chapters that how food, if not handled properly in the supply chain, is vulnerable to become unsafe, the food supplied under the MDM Scheme (MDMS) that feeds approx. 12 crore children on schooldays is no exception. Although the scheme has succeeded in enrolment and reducing dropout rates all over the country, the quality of food served as midday meal in primary and upper primary schools is often compromised.

But how and why does it happen? What went wrong with the food items at the "Dharmashati Gandawan" primary school in Bihar that the poor school children had to bear the burden of compromised quality of food with their lives?

The scheme which was conceptualized to serve the dual purpose of health and education has been associated with many ills not only in Bihar but all over the country. Let us see how rules meant to ensure regular and hygienic feeding are jeopardized along the chain with examples from the Bihar's MDM tragedy.

Supply of Food Grains

Norm: The food grains, like finances, are also supplied from the centre to the schools. Food grains, usually rice, are supplied by the Food Corporation of India (FCI) to the state food corporations. The state MDM directorate makes quarterly allotments to the districts on the basis of availability of food grains. Depending upon the requirements each district lifts food grains from the regional godown of the FCI. These grains are now transported to the block godowns and onwards to the individual school through PDS shops.

Loopholes: Many a times, the food grains are not picked up from the FCI godowns as they are purchased locally. The gunny bags full of these food grains are left unattended for months which increase its possibility of being infested by pests and insects, thus making the food grains unsafe for consumption. Moreover, food grains are pilfered on their way to schools. Government officials, transport contractors, NGOs and even school authorities are involved in the loot. Moreover, to cover up this pillage the school authorities also do not weigh the grains when it is received.

Bihar MDM tragedy: Since the forensic reports suggest the problem was with the curry that was served in the meal, the contamination of food grains can be ruled out in the Bihar's MDM incident. Also the reports suggested that children who ate just rice did not suffer from any complaints of vomiting, etc. which suggest the rice used was safe for consumption.

Procurement of Raw Materials Like Vegetables, Cooking Oil and Other Ingredients

Norm: These are brought from the local market. A quality control committee at the school is expected to monitor these purchases. All purchases are to be made though an inventory.

Loopholes: The quality control committee is ineffective in majority of the schools across the country. Also there exists a nexus between the school authorities and the local business men who consider mid day meal programme as a profit making scheme and use inferior quality ingredients for more money. Sometimes adulterated ingredients are procured

purposely by the school authorities so as to have their share of profit in the purchase.

Bihar MDM tragedy: At the Dharmashati Gandawan primary school cooking oil contaminated/adulterated with insecticides was used. It was because of the insecticide that children died. It was reported that the ingredients were procured from the nearby grocery shop which was owned by the headmistress's husband that too without any purchase inventory. Moreover, since all the ingredients bought were loose, so it was not possible for the cooks to check the expiry date of the ingredients especially the cooking oil. Even the cook reported that the oil was foul-smelling and even turned black on heating. However, this report was also ignored. All these events-buying of ingredients from the husband's shop without any purchase inventory, report of the foul smell of the cooking oil being ignored, use of contaminated oil for cooking probably indicated how the head-mistress along with her husband was involved in profit making by using inferior adulterated ingredients.

Storage and Preparation of Food

Norm: Official guidelines insist on *pukka* kitchens which should be clean and hygienic. The kitchens in schools are a mandate even under the Right to Education Act. Moreover, a separate shed or room for storage of raw materials needs to be maintained. The ingredients are to be stored in containers to protect them from moisture, pests, etc.

Loopholes: Many schools across the country are one room structures so the meal is usually prepared and cooked in unhygienic tempo-rary sheds or even in the open. This makes the meal vulnerable to be contaminated by rats, dead lizards, insects and even iron chips. In majority of schools there is no proper place for storage and raw materials are even stored in toilets.

To overcome this problem of space constraint there is a provision of outsourcing the supply of cooked food from NGOs. These NGOs prepare food in centralized kitchens and provide cooked meals to a cluster of schools. But unfortunately the hygiene and sanitary conditions for cooking are also compromised in these kitchens as well. Moreover, cooked hot meals are transported under poor hygienic conditions. But the pest of corruption has not spared our NGOs also. There is disregard of norms in selection of these NGOs. They are involved in cutting costs and making money.

Bihar MDM tragedy: Reports suggest that in Bihar only 56% of schools had stores for keeping raw materials and storage cans for storing these ingredients. The rest 44% neither had a storage place or containers for ingredients. The Dharmashati Ganadwan primary school was one of these. Good storage practices were ignored due to absence of any storage shed or room, thus, the ration was stored in the headmistress house. Moreover, there were no cans for storage of cooking oil, so the cooking oil was stored in a container meant for pesticides as the later was purchased by the headmistress husband and kept in the same room with the food ingredients. This way the pesticides got mixed with the food which caused death of the beneficiaries.

It is not clear yet that whether the oil was adulterated when it was purchased or it got mixed with the insecticide when it was stored in the old container which was used for keeping insecticides.

Checks and Tasting

Norm: The food before being served needs to be tasted by three adults including a teacher before being given to the students.

Loopholes: This norm is often given a miss since the bad taste of MDM has become the present norm.

Bihar MDM tragedy: In this disastrous event, the food was not tasted by any adult and was directly served to the students. Moreover, complain about the bad taste of food by the students was also ignored by the headmistress.

With regards to redressal under the case, the headmistress and her husband were charged with sections under murder and were put in jail.

All the norms under the MDM Scheme were violated to a large extent at the Dharmashati Gandawan primary school. Figure 57.1 shows the loopholes in the supply chain that caused the disaster.

Monitoring and Feedback Mechanism under MDM Scheme in Bihar: The Underlying Loophole for the Tragedy

Monitoring is the backbone for successful implementation of any programme. The policy makers have been very particular about this issue and therefore, a separate fund is always allocated for monitoring of any programme. The centre had given funds for monitoring of MDM Programme as well. The state which is entrusted with the responsibility of implementing the monitoring mechanism was not active in Bihar. It had utilized less than 50% of the allocated fund. Moreover, the mandate of quarterly meeting of state committees and monthly meetings at district and block levels, to get a feedback on the issues related to MDM Scheme, were given a miss in the state. Only 3 meetings of the state committee took place in the last five years as against the standard 20 meetings which should have taken place.

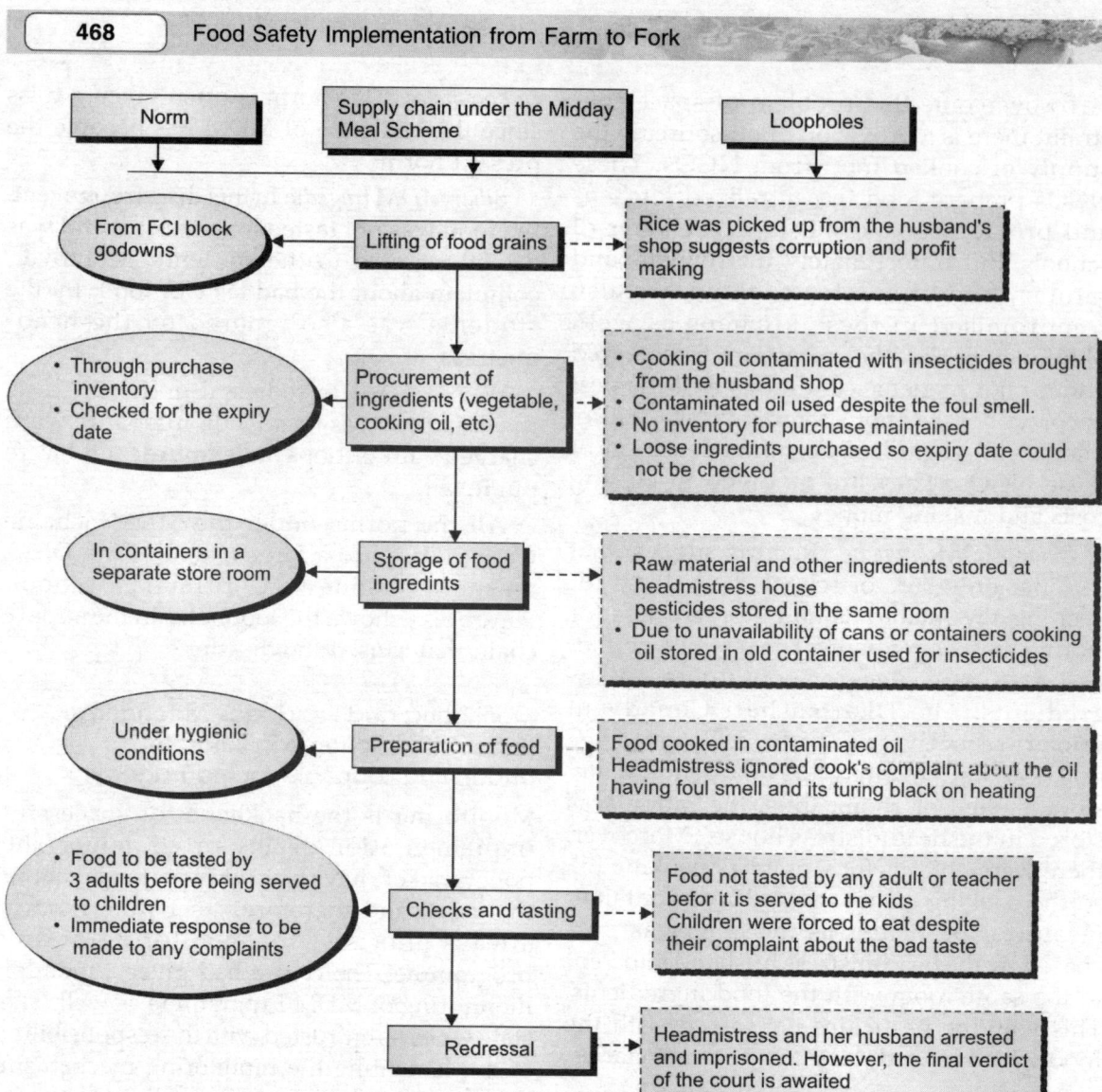

Fig.57.1: Norms and loopholes in the MDM supply chain

Thus, a poor monitoring mechanism was one of the basis which led to the tragedy. Often repeated warning signals were pointed out by various monitoring agencies, one being the lack of trained cooks cum helpers in the state, which were ignored. This accomplished a perfect recipe for a disaster in the form of Bihar's MDM tragedy.

Consequences Following the Tragedy

The MDM disaster has alerted the government and also the children as well as their parents.

On children and their parents: Following the incident a fear has surrounded children and their parents across the state and country on the quality of food being served under the programme. Parents had requested schools headmasters not to serve their children MDM. This tragedy had burdened the impressionable minds as well. However, as it is said with darkness comes light, the same holds true in case of MDM programme. The incident caused children their lives but the ones who survived and heard about the incident are now at least aware that food needs to be tasted by the school authorities before it is served on their plates. This awareness among the school children of Sundarpur nursery school forced the principal and teachers to taste the food before it was served to them.

Government's action: As a step to improve the mechanism underlying the preparation of MDM and mending the loopholes, the centre following the disaster has issued a toll free number for all complaints related to midday meal.

It has also planned to expedite its integrated voice response system (IVRS) for daily monitoring of MDM scheme. Under this system the service providers will have the responsibility of real-time monitoring of the scheme. Approximately, seven service providers will be engaged nationally and each will be assigned 1.5 lakh schools. Responses from the teachers on a standard set of questions like whether meal was served or nor, if yes was it tasted, has the salary of cooks arrived in time etc. will be collected by these service providers through an out bound dialing solution wherein calls are placed to all the teachers from a virtual number using Primary Rate Interface (PRI) lines. Along with question from the teachers, data of number of children who availed MDM would be keyed in by the teachers and displayed on web the same day. This system intends to make individual schools responsible and accountable for the scheme.

Moreover, the government and policy makers need to take strict measures to improve the system under the MDM Programme. The involvement of the community can be harnessed for better quality and quantity of food under the MDMS.

Suggested Measures to Improve the MDM Supply Chain

Listed below are few suggested measures to improve the quality and quantity of food served in the MDM Scheme.

i. With regards to supply of food grains under the programme, lessons can be learnt from successful models. The supply crunch in the delivery system can be overcome as is done by Gujrat and Tamil Nadu, where food grain delivery system for MDM Scheme has been delinked from PDS structure and a separate system of delivery of food grains and condiments to the school has been started.

ii. Training the school catering staff on healthy and hygienic cooking and eating habits will surely cut down on the problems that instigate unsafe food while preparation.

iii. Community solutions to MDM monitoring will to a certain extent ensure safe food along the supply chain. Instating a school meal review panel with community as a key stakeholder to monitor the MDMS at school level on a regular basis can be a step in this direction.

iv. Setting up a grievance redressal mechanism through which the community can approach the authorities for any lacunae after the monitoring of the scheme will help in making the monitoring system transparent.

The tragic death of innocent kids because of eating unsafe midday meal clearly suggests that food which is served as a medium for providing nutrition and energy can cause serious ill effects if not procured, prepared, stored, cooked and served properly. The political parties play the blame game in every case of food poisoning caused by the midday meal. The officials working under the programme are involved in profit making and living an easy life. However, sufferers are the poor children who far from understanding the essence of corruption and politics involved in the programme.

Thus, rather than policy shift, problem analysis of the current loopholes needs to be done so as to improve the system underlying the MDM Scheme. Effective measures that suit the local context need to be thought of.

58

Food Safety on Religious Occasions in India

Ruchi Sharma, Suninder Kaur, Amarjeet Singh

There is not a single week when we do not come across a large number of outbreaks of FBI in India. Some happen after wedding parties, some of them in schools and many of them after religious occasions. India is well known all over the world as a country of cultural and traditional festivals as it has many cultures and religions. India presents a cultural assortment of number of religions with their own festivals and celebrations but the four major religions followed in India are Hinduism, Islam, Christianity and Sikhism. Bright colors, brightly lit religious places, illuminated houses, sweets, feasting, traditional dresses, dances and unwavering enthusiasm are the characteristics of all the festival holidays in India. It is fact that no other country holds so many festivals of antiquity as India does. Hardly a day passes without a festival taking place somewhere in India. They range from small one-day village or temple celebrations to weeklong religious functions. Most of these festivals have religious outlook and some food/*prasada*/feasting is often associated with all these festive occasions.

Feasting at Religious Occasions

Celebration and feasting is a part and parcel of almost every culture especially on religious occasions. Festivals have been around for as long as religion has been around and the etymology of 'feast' is very similar to that of 'festival'. This practice of feasting and fasting, lasts till today. Every religion and culture in the world, be it the Muslims during Ramadan or the Hindus during *navratras*, have festivals and religious observances which circle around fasting and feasting. Therefore, religious occasions/festivals have a particular 'cuisine' or tradition of cookery, associated with their culture. For example, hot cross bun is traditionally eaten on Good Friday to break the fast among Christians. In Judaism religion, apples and honey are eaten on Rosh Hashanah, to symbolize a sweet new year. *Pongal*, a popular rice dish in South Indian cuisine is a Tamil dish associated with many Hindu rituals and feast celebrated at the end of the harvest season (*pongal festival*). It signifies thanking the 'sun god' for agricultural abundance. Apart from Pongal Day celebrations, cooking pongal is a traditional practice at Hindu temples during any Temple Festival in Tamil Nadu. The community will convene to cook pongal rice, distribute it to those present. Even non-vegetarian dishes are a part of some religious occasions. **Bakr-Id** is Muslim festival of great rejoicing. It is celebrated by sacrificing animals and distributing meat among the poor. Similarly, Lamb is considered to be the traditional meat for Easter due to its religious connections. Many different religions throughout the ages have ritually sacrificed lambs in honor of their respective Gods. Turkey is cooked and served to the family on Christmas, the birthday of Lord Jesus Christ.

The food item served on religious occasions is at times considered 'Holy' and 'sacred', also called *"Prasada"* in India. In India, food plays an important role in rituals and worship, and the food offered to the Gods is called prasada. The Sanskrit word *"prasada"* means "mercy," or the divine grace of God. Here part of food is served to God and distributed among masses. *Prasada* may be liquid in form of *charanamrita*, in powder form made from wheat, etc. and in solid form of *boondi* or *ladoos*. For example, in Hindu culture, *Panchagavya* or *panchakavyam* is a concoction prepared by mixing five products of cow. The three direct constituents are cow dung, urine, and milk; the two derived products are curd and ghee. It is also served as a *Prasada* in temples.

In Hindus, a specific food is offered to a particular God as *prasada* according to tradition as token of respect. Lord Ganesha is offered *"modaks"* and Lord Shiva "raw diluted milk". Similarly, *boondi* and *ladoos* are presented to Lord Hanumana. Yellow food/rice are offered to Mata Saraswati.

Feasting in Indian Religious Ceremonies

There are many Indian religious festivals where feasting is a compulsory event. Hindus, Muslims and the Sikh festivals in India include *'langars'* or feasts cooked and served on a large scale. The famous festival of *Makar Sankranti* is celebrated among Hindus in January and a number of prayers are recited during the day. A special sweet made of sesame seeds symbolizing life is distributed to relatives and friends and children. *Deepawali* is another festival where sweets are prepared in abundance and distributed to family and friends. Sweet water is served to masses on *nirjala ekadashi* and lassi on martyram day of Shri Guru Arjan Dev ji. *Karva Chauth*, a Hindu festival wherein wife fasts the entire day for the long life of her husband and enjoys a heavy dinner when the moon is sighted. In the modern times, dining in restaurants has become the norm on such occasions. Muslims also fast from dawn to dusk during Ramadan and dine out when they break the fast. The festival of Eid-ul-Fitr is celebrated by the Muslims at the end of the month of Ramadan to mark a successful completion of the period of fasting. The early part of the day is spent offering prayers at a Mosque followed by hearty meals at home or with relatives.

Therefore, food is an essential part of the religious festivals. Food is cooked and served to the masses in these festivals, which swell up in thousands at times. And when we talk of food on these occasions, food safety becomes an obvious issue. Food quantity and preparation is ensured on these occasions but what gets neglected at times is the safety of the food prepared. Food safety at religious festivals is a sensitive issue because here sentiments of community are at stake.

Food Safety in Religious Festivals

Food offered to Gods/Godesses and distributed among masses is often not tasted after cooking. Hence even if it is spoiled while cooking, nobody would know that and it is distributed as such. Health and hygiene of food handlers is often not rigidly adhered to on these occasions. Other issues linked with food safety during religious festivals are given in succeeding paragraphs

Adulteration: In view of the adulteration being rife in India, food safety becomes an important issue in religious feasts. People dilute products with water, add cheap oil, use adulterated milk, etc. Food is also cooked

Table 58.1: Some important religious Indian festivals and foods/*parsada* associated with them

Religion with their festive occasions	Associated foods/Parshad
Hindu	
Makar sankranti/Pongal	Pongal-made of rice and jaggery
Mahashivaratri	Milk and fruits
Navratras	Kuttu ka atta/chana-puri-halwa
Nirjal ekadashi	Channe and sweet water
Onam	Ata prepared from rice flour and molasses
Krishna Janamashtmi	Fruits/ Milk/ Butter with crystal sugar
Ganesh Chaturthi	Coconut, jaggery, modakas
Durga Puja	Sweets and fish dishes
Dussehra	Sweets-Jalebi, etc.
Deepavali	Sweets
Chhath puja	Kheer, puri and bananas
Muslim	
Bakr-Id	Goat
Ramzan	Iftar dinner-kebabs, biryani, etc.
Eid-ul Fitr	Saviyaan made of dates/raisins, etc.
Muharram	Meat, rice, kebabs and puddings
Sikhs	
Guru Nanak Birthday Jayanti	Langar-Pulses, vegetable, roti and kheer
Guru Govind Singh's Birthday	Langar-Pulses, vegetable, roti and kheer
Guru Teg Bahadur's martyrdom day	Langar
Baisakhi	Sweet yellow rice
Christmas	
Good Friday	Hot cross buns
Easter	Egg
Christmas	Cake and turkey

much before it is served which may lead to its spoilage.

Parsada and almost all food on religious occasions is usually made in substandard 'Ghee' and 'refined oil'. They are threat to health of devotees especially ones suffering from heart ailments. Though, by and large most religious ceremonies are accompanied by vegetarian dishes, some religious festivals like Bakr-Id accounts for consumption of non-vegetarian food. Where non-vegetarian food is served to masses, adequate cleaning and cooking is often compromised.

Intentional contamination of food: At times food may be contaminated by miscreants to harm religious sentiments of the people and create violence.

Transportation: *Prasad/langar* if prepared at different site is often transported for distribution in large drums to the venue. Food may get spoiled during transportation on the way. It may be due to contamination from utensils/rodents or lack of temperature control.

Storage: Food being served is kept in open and devotees offer free service to distribute it. It may be spoiled intentionally or due to lack of hygiene among the distributors. At times, people tend to take these cooked food items home, after having eaten enough, to use it during later meals and sometimes for several days. As this is free food for all, collecting it in polythene bags or boxes and storing it for days violates food safety measures and spoils/adulterates the food making it unfit for consumption.

Bhandaras/mass serving: Often families and communities organize *bhandaras/langars* (Fig. 58.1) to serve the community on religious occasions. Quality of food during this mass distribution of food remains unchecked. People sometimes distribute home cooked food at religious places an example being black grams being distributed on Saturdays in the temples. Such food distributed through group of people or community has no quality control or check.

Packaging: Packed food can be a source of toxin to the consumer. If *prasada* is packed its shelf life also becomes an important issue of concern. Packed *ladoos/prasada* often donot have expiry date printed on them. Even if they do have one, people do not throw then because of religious sentiments. During packaging more focus is on the cosmetic appearance of the food, artificial means are used to make it look fresh and plumpier hence, bypassing food safety.

Fruits: Officially banned chemicals such as calcium carbide (banned as per FSSAI) is often used to ripen fruits artificially. Bananas and other fruits offered to God or served as prasada are often ripened artificially to cope up with their increasing demand during festive season. Similarly, coconut given in temples as prasada is many a times rotten. Fruits are often dipped, polished or sprayed with artificial colors to make them appear fresh for sale.

Hygiene: Sweetshop keepers generally prepare sweets in compromised hygiene, e.g.,- without wearing gloves. In festival season, to cope up with increase in demand of sweets, they are often prepared in bulk and months in advance and kept in unhygienic conditions.

Similarly, *prasada* is served with bare hands without appropriate hygiene measures. Recently, on *Janamasthmi* eve a priest in a temple was distributing butter and crystal sugar *(mishri)* from bowl with bare hands. People were consuming it without bothering for hygiene.

Fig. 58.1: Navaratri bhandara: Prasad distribution

Sweet water *"Chabeel"* served on *nirjala ekadashi* is kept in open with flies fluttering around. Ice put into it is often kept in open and utensils are not washed properly because of lack of running water supply. Often same plastic glasses are used for serving without being properly washed. This sweet water then becomes culprit for causing various water borne illnesses.

Water from *sarovars* and holy rivers is taken away by people for drinking purpose with view that it heals various ailments because of religious sentiments. This water is drunk as such without boiling or purifying.

Vendors/street food: During festival season, especially during *Dussehra, Navratra* and *Durga Puja*, local fairs or *"melas"* are organized by local communities. It is not unusual to see people eating outside from street vendors on these occasions. Open food and cut fruits are enjoyed lavishly by the people. This food being served is poor in hygiene and content.

Exploitation of religious sentiments of the community: *Prasada* is considered pure and sacred. Hence, due to religious sentiments people often do not hesitate in consuming spoiled *prasada*/food. They reheat and consume it keeping their health at stake. Right from childhood, we are taught that Prasada is a sacred gift, we should respect it and consume it even if it has spilled on floor/ spoiled or contaminated with rodents and flies.

Some Incidents of FBI on Religious Occasions

Where large scale production of *prasada* is being done like on religious occasions, utmost care should be taken, as a single negligence or careless attitude can lead to food poisoning or related ailments to not a few but a large number of people, as has been seen many a times in such festivals

On September 29, 2011 as many as 219 people had become ill in Pune after consuming *varai* also known as *bhagar* during Navaratri festival. A few of them took days to fully recover. The laboratory report later revealed that *varai (samo)* flour was adulterated with *kodra* seeds (a wild variety of millet) and glucosites.

In May, 2014 over 400 people turned ill and one died after eating *prasada* at a temple festival at Solapur, outside Mumbai. Illnesses begun 2–3 hours after food was served to the crowd at the temple.

In April, 2014 as many as 350 people were treated for food poisoning at a local public hospital after consuming contaminated *panakam* at Srirama Navami Festival, held at Damarcheral Mandal Headquarters in Andhra Pradesh. More people were sickened, but they either did not seek medical attention, or did so at private hospitals.

On August 6, 2011 over 400 people were hospitalized in Delhi and neighbouring cities in Uttar Pradesh with symptoms of food poisoning after consuming adulterated *'Kuttu Ka Atta'* during fasting for Navaratri.

Advantages of Food Served on Religious Occasions

Food considered *'Parsada'* prepared in large amounts in the religious festivals is cooked by the people of the community themselves, not by any catereres, and hence, religious people who care to maintain the purity of food, linking it to God and religion, take utmost care in food preparation. Hands are washed, cooking vessels are washed properly, even heads are covered and shoes/chappals removed from the feet before preparing food. An attempt is also made to sing hymns, religious/holy songs or recite prayers while cooking, and food is prepared with poise and dedication as a service to God and his fellow beings.

In Hindus, mostly vegetarian food is prepared during religious ceremonies and hence no poultry products, meat, seafood, etc. is used. Even alcohol is strictly forbidden in these festivals, hence there is more focus on 'purity, health and hygiene'.

Usually no junk food is served and food with nutritional value is usually cooked and served. Many a times fruits are given intact as *prasada*, these are good and healthy substitute of *ladoos/halwa* which may have issues of adulteration and hygiene.

Food served is generally prepared fresh, served and consumed.

Guidelines to Prepare Prasada and Distribute among Masses

Before we can offer any food to God, however, we must include guidelines for preparation and distribution of '*prasada*' during religious functions. We must sensitize authorities at religious places about sticking to proper and safe ways of preparing and distributing *prasada*. There are guidelines as to what precautions one has to take while preparing. It is important to maintain a high standard of cleanliness while preparing, cooking, and offering the food. The kitchen, utensils and foods used should be clean. Food handlers should also be clean and have bathed before beginning to cook.

Strict vigil should be kept on quality of food during preparation, transportation and storage. It should be made sure that persons involved in food handling are healthy and clean.

It should be tried to keep *prasada* in dry form such as crystal sugar and packed in small packing to ensure quality. Its expiry should be denoted on the packing.

Educate masses not to store in abundance and eat spoiled food.

Quality control and strict vigilance on the sweet shops during festival occasion is utmost importance. Strict guidelines should be issued to the sweet shops to adhere to permissible food standards to check adulteration.

Quality check on open food sold through street vendors. Licensing them could be one of the possible options.

It should be compulsory for temples/mosques and authorities of other religious places to keep a record of all food items purchased, including the name of the grocery shop and telephone number. Authorities of religious places need to keep a record of devotees who donate food items to temples/mosque and other places.

There is need to put check on mass distribution of food by unknown people without prior permission from concerned religious organizations.

A dashboard urging devotees not to accept *prasada* from unknown people should be displayed at a conspicuous location. *Prasada* should be prepared in clean surroundings, covered adequately and consumed without much delay. Stale *prasada* should not be distributed among devotees.

Certain SOPs for food safety on religious festivals needs to be developed. Guidelines should be set to regulate food preparation, storage and consumption. There is also need to develop proper measures to dispose off leftover food so that it does not hurt religious sentiments off the community.

As it is difficult to find fault in food served on religious occasions there is a responsibility of each one of us to protect our fellow beings, i.e. the public at large from health risks which we can do by adhering to food safety methods, hygiene and also by spreading awareness to others about safety measures and the harmful effects by violating them.

Conclusion

The scale of the compromised food safety during religious occasions cannot be measured accurately, in a country with many religions and religious rituals and festivals. To curb the problems arising from non-adherence to food safety regulations, authorities (management of religious bodies) should conduct surprise checking and food raids and take strict measures against those violating food safety guidelines. Raids are especially important during religious festivals or weddings and other such occasions where a large gathering consumes food, as contaminated food can result in hundreds of people taken ill and in hospitals.

Enforcing India's food safety laws is a challenge, as the laws are very good but their correct implementation is quite weak. To implement food safety sternly means an attack on the livelihood of the local vendors, who are already living on meager wage. Adding to it is the problem in the system where the sample of food is not collected by the same person whoes appointed to perform the tests, so one can never be sure if the sample provided to him/her is of best quality representative.

The only silver lining in the cloud is that there is growing awareness among Indians now, of the issue at hand and Indians are becoming more food safety conscious.

59

Advice for Travelers with Respect to Food Safety in India

Sukhpal Kaur, Neha Chanana

"Food allows an individual to experience the 'Other' on a sensory level, and not just an intellectual one"—Long (1998)

"Incredible India—truly India" initiative by Ministry of Tourism, Govt. of India with Amitabh Bachhan as its brand ambassador has created a distinctive identity of the country's tourism across the globe. It has appropriately showcased India's bounteous heritage including beautiful architecture, its mesmerizing and scenic landscapes, rich traditions and culture of *"aththi devo bhava"*. The campaign has created a colourful image of India in the minds global citizens, which has led to an increase in the tourist traffic in India both domestic as well as international. The tourism of the country is so popular around the globe that 'Conde Nast Traveler', one of the ace travel magazines, profiled India as the most preferred tourist destination in recent times.

According to a report by World Economic forum in 2009, India ranked 11th in the Asia-Pacific region and 65th globally out of 140 economies on travel and tourism competitiveness index. The country had received 6.58 million international tourists in 2012; 15.81% of which travelled from the USA. The other top sources of international travelers were from the UK at 11.98% and Bangladesh at 7.40% of foreign tourist arrivals. Domestic tourism is about 158 times the amount of international tourism accounting for 1036.35 million travelers in 2012. Domestic travel has been steadily increasing at an average of about 13% per year since 2000, when 220.11 million travelers were recorded.

Apart from the scenic beauty, historical and cultural heritage tourists are also attracted to the variety of cuisines available in the country. The connection of food and tourism has also been highlighted by travel shows hosted on various channels like *'Zayika India ka'* on NDTV or the famous show *'Highway on my plate'*. These shows have indeed brought to light the delicious food available across destinations in India. All the more majority of times the food showcased is none other than the mouth-watering street food with the famous *gol gappas, bhel puri, tikki* and many more such varieties. Be it the *'chatori gali'* in Delhi or the famous *'chaat wali gali'* in Agra, street food is not only tried but enjoyed by domestic as well as international tourists. However, one aspect that withholds tourists from enjoying this yummy food, available only in India, is it being unsafe. Many times tourists fall sick merely because of eating out in India.

Often international tourists suffer from *'Delhi Belly'* while travelling to India or

'Karachi's Krouch' in Pakistan. These terms are slangs for what is known as traveler's diarrhea. This is also common among Indians travelling within the country visiting various places of interest. It has been reported that 30 to 70% of travelers suffer like this, depending on the destination and season of travel. Bacterial pathogens account for 80–90% of such diarrhea. Intestinal viral infestations usually account for 5–8% of illnesses, although improved diagnostics and their increased use for recognizing norovirus infections in travelers may change the percentages in the future. Protozoal pathogens are slower to manifest symptoms and collectively account for approximately 10% of diagnoses in long duration travelers. Food poisoning may also occur. It involves the ingestion of preformed toxins in food. Vomiting and diarrhea may both be present, but these symptoms usually resolve spontaneously within 12 hours.

Risk factors for these problems primarily include ingestion of contaminated food and water. Moreover, the reason for diarrhea among travelers is that their immunity is low with respect to particular pathogenic organisms prevalent in these places, i.e. the microbial gut of the travelers is not used to such kind of food and water.

Existence of such situation even in 21st century is mainly because the laws of food safety are not strict in India especially with food available in restaurants at bus stands and railway stations, *dhaabas* and street vendors. Majority of times tourist enjoy eating at these places.

Moreover, the food served in trains is also not considered safe. Complaints have been filed regarding substandard food served even superfast trains like *Shatabdi* and *Rajdhani*. When food served on so called luxury trains is unsafe we can very well imagine the plight of food served in other trains. The pantries of majority of the trains are dirty and food is prepared in unhygienic conditions. At times the remains of rats and cockroaches have been found in food served in these trains. Thus, while travelling, tourists are susceptible to illnesses mainly because of consumption of unsafe food and water. It has also been reported that water across the country is generally unsafe for consumption and one-fifth (21%) of communicable diseases in India are related to unsafe water.

Thus, it is essential to take measures both by the government and at the individual level by tourists so as to reduce and avoid the risks involved due to food.

Measures to Reduce Illness due to Contaminated Food during Travel

A. Actions Taken by the Government

Lip smacking street food available in India is enjoyed by domestic as well as international tourists. In view of the potential food safety risk involved, the government has planned to set up designated street food zones in Delhi. In this initiative, the National Association of Street Vendors of India (NASVI) will train vendors in the safe zones to follow basic hygiene practices such as wearing aprons, caps, gloves and also taught on safe handling of food. After the training, the vendors will receive a safety stamp for their outlets. Moreover, through focused inspections, these food zones will help reduce the incidence of foodborne diseases.

The government needs to implement certain other measures to ensure food safety.

For example:

- Regular monitoring of pantries and food served in trains.
- Implement strict laws and levy heavy fine and punishment on the catering agencies involved with serving of food on trains as

well as eating joints at railway stations and bus stands.

B. Travelers' Role: Precautions to be Taken while on Tour

Apart from the government's effort on training food handlers, travelers need to take precautions at their end so as to reduce the chances of problems they can suffer from consuming food and water while travelling.

a. Precautions related to food

- Selection of the eating joint/restaurant: Less occupied restaurants need to be avoided. Fewer turnovers indicates the food available is not likely to be safe. Fewer costumers suggest the food served is not fresh. While travelling one must look for popular places, particularly those patronized by families. Busy restaurants typically serve fresh, clean and safe food. Also, avoid eating at dingy restaurants. Make sure the place is clean, well lighted and well kept.
- Eating food that is cooked fresh and served hot reduces the chances of it being unsafe and contaminated.
- Moreover, eateries those have been freshly fried or boiled is safe to eat. For example, the freshly prepared fried *'puri and freshly boiled and cooked sabzi'* is better than *'kulcha chana'* or say the mouth watering famous *'aaloo ki tikki'* is safer than *'gol gappas'*. Chances of fried food being contaminated, unless the oil is unsafe, are less.
- Hard boiled eggs should be preferred. Half cooked eggs should be avoided.
- Sandwiches with lots of raw vegetables are likely to be unsafe for consumption as the mostly the vegetables are not fresh and also not washed properly before they are used as dressing in sandwiches. Moreover, the bread used may not be fresh and not within the 'best before' period.

- India is majorly famous for its diverse vegetarian cuisines. So eating non-vegetarian food should be avoided. If at all it is to be tasted, prefer a restaurant that local people may recommend or from famous outlets like *Tunde kebab*i of Lucknow or *Karims* in *Chandni Chowk*, Delhi.
- Travelers should opt for pasteurized dairy products like milk or curd. Pasteurization kills bacteria and makes these products safe for consumption.
- Street vendors without access to running water need to be avoided. Moreover, places where all utensils are washed in one bucket, which is common with street vendors, should be avoided.
- *Fruits and salads*
  - Cut fruits and salads need to be avoided since chances of these being contaminated are high. Unpeeled and washed fruits should be preferred, as they are the safest.
  - Salads, leafy fresh vegetables, and thin skinned fruits such as apples, berries, cherries, etc. should be avoided. Green-leafey vegetables (e.g. green salads) can contain dangerous microorganisms which are difficult to remove.
  - Thick skinned fruits such as bananas, mangoes, etc. that can be peeled by the person who eats them are usually safe.
  - Fruits and vegetables with damaged skin should also be avoided because toxic chemicals may be formed in these.

b. Precautions related to drinks (water and other beverages)

The most common source of diarrhea while travelling is drinking water, including locally made ice or the ice which is

prepared with non-purified water. In many parts of the country, where water treatment, sanitation, and hygiene are inadequate, tap water may contain disease-causing contaminants, including viruses, bacteria, and parasites. This makes tap water in those places unsafe for drinking, preparing food and beverages, making ice and cooking.

- While travelling boiled, filtered or purified water, mineral water or sodas should be used for drinking. Tap water needs to be avoided at all cost unless it has been treated or made safe for drinking.

- In restaurants, bottled drinking water-packaged drinking water or mineral water should be preferred. Packaged drinking water is treated and made healthy for drinking, while mineral water has been obtained naturally at its underground source and hygienically bottled. Both are safe to drink, although mineral water is better as it is chemical free. However, if the source is dubious even these are also not necessarily safe and may be contaminated. Quiet often used bottles are refilled and sold with fake seals. Therefore, extra care needs to be taken while selecting for bottled water.

  - Bottled water of reputed brands should be preferred.

  - Seal of the bottle needs to be checked. A coloured plastic seal wrapped on the lid is ideal.

  - The manufacturing and expiry date on the container needs to be checked before buying. Shelf life of a packaged water jar is one month from date of manufacture.

  - The label should mention quality of water.

  - Water in the bottle should be checked for any suspended particles by holding the bottle in front of a bright light. The bottle should be rejected in case of presence of suspended particles.

  - Sometimes, water drops/mist outside the chilled cans and bottles may be contaminated. This should be wiped clean and dried before opening the bottle/can.

  - Chemical disinfectants in the form of aqua tablets, chlorine tablets or iodine tablets or packets of powder may be used as water treatment method by travelers so as to ensure consumption of safe water.

- Ice made from bottled or disinfected water should be preferred; the ice prepared from well or tap water needs to be avoided. Thus, care needs to be taken before drinking iced tea or iced coffee as the ice used may be made from contaminated water.

- Beverages made with boiled water and served steaming hot (such as tea and coffee) are generally safe to drink. Even pasteurized or canned milk of reputed brand names is also safe for consumption.

- Carbonated drinks of reputed brands are also safe to drink. The cap of the bottle should be checked before it is opened. Spurious drinks may have caps of a different brand. If possible, drinking straws (check quality) should be used so as to avoid drinking straight from the bottle as its (bottle) top can be dirty and not clean.

- Reconstituted juices or drinks need to be avoided as these are usually prepared from tap or well water which is usually contaminated and unsafe for consumption.

- Packaged juices that are branded can be used. However, their expiry date should always be checked before these are consumed.

c. **Miscellaneous precautions**

- Nuts and other shelled foods are usually a good choice for energy food.
- Coffee and tea are generally harmless. But it's best to take hot drinks black (thus avoiding potentially contaminated milk). Cream from sealed containers, if pasteurized or dry milk is usually safe.
- Condiments such as mayonnaise, ketchup and salad dressings are safest in sealed packages.
- Try to maintain a well-balanced diet. Make sure your diet includes all the nutrients as far as possible. Stay hydrated by drinking lots of (safe) water.
- Keep sachets of Oral Rehydration Solution (ORS) while travelling.
- Carry antidiarrheal medicines (Tinidaizole, Cotrimaxazole, Furazolidone)
- In case of foodborne illness, consultation with a general physician is necessary.
- Public toilets may not have soaps, or at times not even have water. It is always better to carry a small bar of soap, soap strips, wipes, or instant hand sanitizers solution and toilet papers with you.

Consumption is an integral aspect of the tourist experience, with the tourist consuming

General do's and don'ts	
Do's	**Don'ts**
Eat	*Don'ts Eat*
• Food that is cooked and served hot	• Food served at room temperature
• Properly cooked eggs	• Raw or soft-cooked eggs
• Fruits and vegetables you have washed in clean water or peeled yourself	• Raw or undercooked meat or fish
	• Unwashed or unpeeled raw fruits and vegetables
• Pasteurized dairy products	• Chutney, local sauce
	• Salads
	• Flavoured ice
	• Unpasteurized dairy products
Drink	**Don'ts Drink**
• Water, sodas, or sports drinks that are bottled and sealed (carbonated is safer)	• Tap or well water
• Water that has been disinfected (boiled, filtered, treated)	• Fountain drinks
• Ice made with bottled or disinfected water	• Ice made with tap or well water
• Hot coffee or tea	• Drinks made with tap or well water (such as reconstituted juice)
• Pasteurized milk	• Unpasteurized milk

not only the sights and sounds, but also the taste of a place. Local food is a fundamental component of a destination's attributes, adding to the range of attractions and the overall tourist experience. However, the food should be safe. Thus, consuming safe food and following simple precautions will help in making the trip and journey safe and enjoyable.

60

Poisonous Plants—An Ignored Problem in Food Safety

Satinder Pal Singh, Sumeet Kaur, Dalbir Singh

INTRODUCTION

The plant world is indispensible for life to exist on earth. The different plants have different characteristics, some of which are useful to humans and some are not. Since ages, man has tried to harness the medicinal properties of plants and utilize it to provide relief in sickness and boost health. Many plants of significance to humans also have certain features that produce detrimental effects on health and can even cause death. Such flora poses a significant public health hazard especially for children as most of these plants are brightly colored and attract children. Though a large number of such plants exist in India, the relevant data regarding the morbidity and mortality due to consumption of such plants is lacking except a few sporadic cases as reported in news. These poisonous plants may also prove lethal for animals. For example, a report in Times of India newspaper dated 23 January 2013 reported that 51 spotted deer died in Kanha National Park, Madhya Pradesh after consuming Lantana camara, a poisonous plant found in forested area of that locality. So there is a need to make the public aware of such plants that are widely available and thus frequently encountered in India.

Accurate identification is essential for diagnostic and treatment purposes as most of these plants or their products produce a variety of clinical features in humans and pose a diagnostic dilemma for inexperienced physicians. This leads to failure on the part physicians to diagnose the condition well within time to begin remedial measures. This not only brings agony to the patient and his relatives but also puts an additional burden on the already stressed health sector in India. The authors have made a generous attempt to highlight this problem especially among the commonly occurring poisonous plants. The authors also have suggested suitable diagnostic clues to their early detection to reduce the burden of disease on society.

Following are some of the plants that are of clinical significance to humans

1. Abrus precatorius
2. Ricinus communis
3. Calotropis
4. Croton tiglium
5. Ergot
6. Capsicum
7. Plumbago
8. Semecarpus anacardium
9. Argimone maxicana
10. Eucalyptus globules

1. Abrus Precatorius

This plant is widely available throughout India and is also known by other names like *rati, gunchi or Indian liquorice*. The plant is woody and slender (Fig. 60.1a). All of its parts are poisonous. The seeds are oval in shape, about 0.8 cm in length and 0.6 cm in breadth. The average weight of a seed is about 110 mg and they have any taste or smell. The seeds are red in color with a black spot at one pole (Fig. 60.1b). They are heat labile and some-times consumed as food after proper boiling. When ingested, the seed coat does not permit the toxin to come in direct contact with the gastrointestinal lining. The seeds pass out as such without causing any effect. It is only after crushing/chewing the seeds before ingestion that results in toxicity.

Active Agent

It is known as abrin and resembles in action with the venom of a viper. It has antigenic pro-perties. Then needles of this paste are made and dried under sunlight. These needles are kept between the fingers and the enemy is slapped so that the needle is injected into the body. The seeds are also powdered and employed to produce conjunctivitis by malingerers. The seeds are sometimes used as an abortifacient and also as arrow poison.

2. Ricinus Communis

This plant is also commonly available throu-ghout our country. It is also known by the other names like *castor oil plant* or *arandi*. The plant is also cultivated for castor oil (Fig. 60.2a). All parts of the plant are pois-onous but seeds are most poisonous. Two types of seeds are produced by the plant-small and big. The small seeds are about 1.2 cm long and 0.8 cm broad (Fig. 60.2b).

Fig. 60.1a: The plant (*Abrus precatorius*)

Fig. 60.1b: The seeds of the plant

Fig. 60.2a: The plant (*Ricinus communis*)

Fig. 60.2b: The seeds of the plant

Active Agent

It is called as Ricin that causes agglutination of red blood cells (RBCs). Ricin acts by blocking protein synthesis.

Clinical Features

Raw seeds when ingested, produce burning in throat, salivation, vomiting, bloody diarrhea, abdominal pain, muscleaches, convulsions, collapse and death.

Fatal dose: About 5 to 10 seeds; 20 to 30 mg of ricin

Fatal period: About 2 to several days

Treatment

(a) Gastric lavage (b) Activated charcoal (c) Demulcents d) Symptomatic

Significance

The children may get poisoned by accidentally ingesting its seeds. The seeds may be powdered and applied to eyes to simulate conjunctivitis. The seeds are also used in villages to commit suicide.

3. Calotropis

It is a wild plant but commonly grows at roadsides throughout the country (Fig. 60.3).

The plant exists in two forms—Calotropis giganta (*akand*) that has purple colored flowers and Calotropis procera (*madar*) having white flowers. The leaves and stem of the plant produces a thick milky juice when cut.

Fig. 60.3: The plant (Calotropis)

Active Agent(s)

Calotropin, Uscharin, calactin, and calotoxin.

Clinical Features

When the juice of the plant is applied to the skin, it produces redness with occasional blister formation. If applied to the eyes, it produces conjunctivitis. If it is ingested, it tastes bitter and produces burning pain in mouth, throat and stomach, nausea, vomiting and diarrhea. Tetanic convulsions and dilated pupils can result. Failure of circulatory system may result in death.

Fatal dose: Uncertain

Fatal period: Within a few hours

Treatment

Stomach wash, demulcents and symptomatic.

Significance

The leaves, flowers, roots and juice have traditionally been in use in Indian medicine. The juice is used as an abortifacient. Accidental poisoning can result from medicines prescribed by quacks. It is sometimes used as arrow poison. It is also used to produce artificial bruise to bring a false charge against an enemy.

4. Croton Tiglium

It is also known as *Jamalghota* or *Nepala*. Of all the parts of the plant (Fig. 60.4a), seeds contain the maximum concentration of active agents. The seeds dark brown in color, oval shaped with longitudinal lines (Fig. 60.4b). The oil is extracted from the seeds.

Fig. 60.4a: The plant (*Croton tiglium*)

Fig. 60.4b: Seeds of the plant

Active Agent(s)

Crotin, a toxalbumin and crotonside.

Clinical Features

Burning pain in gastrointestinal tract associated with salivation, vomiting, purging, vertigo, prostration, collapse and death.

Fatal dose: About 4 to 5 seeds; 1 to 2 ml of croton oil

Fatal period: A few hours to 3 days

Treatment

Stomach wash, demulcents and symptomatic.

Significance

The croton oil may be accidentally consumed in place of some other substance. The oil and roots are sometimes used as abortifacient. The oil is also used as arrow poison.

5. Ergot

It has vasoconstrictor properties and is obtained from the dried sclerotium of fungus Claviceps purpura that grows on rye (Fig. 60.5), barley, oats and wheat.

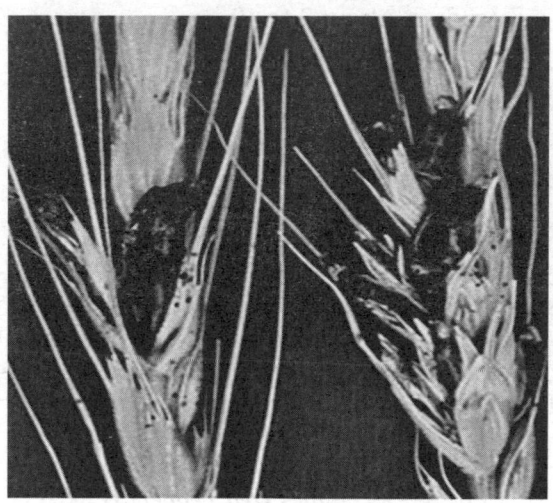

Fig. 60.5: Claviceps growing on rye

Active Agent(s)

Ergotoxin, ergotamine and ergometrine.

Clinical Features

It produces both acute and chronic poisoning.

Acute Poisoning

The entire gastrointestinal tract is irritated and features such as nausea, vomiting, diarrhea, smooth muscle contraction. The respiratory system is depressed with tightness of chest, labored breathing and exhaustion. Tingling and numbness are felt in hands and feet. The victim may feel dizzy with diminished vision and bleeding from nose can occur in case of a large dose.

Chronic Poisoning

In chronic poisoning, features are divided into two main groups: (a) Convulsive features, (b) Gangrenous features. *Convulsive* features include numbness, tingling and twitching of muscles and sometimes contractures. In the *gangrenous* type, the feet feel like burning, inflamed, swollen with alternating hot and cold sensations. The affected limb shows gangrenous ulceration and sloughing. The fingers and toes may develop dry gangrene.

Fatal dose: About 2 to 10 gm

Fatal period: A few days

Treatment

Stomach wash, activated charcoal, purgatives, vasodilators, diazepam for convulsions.

Significance

Accidental poisoning occurs due to consumption of food articles prepared from affected grains. Ergot is used in the treatment of migraine or prolonged uterine hemorrhage. Ergot is a known abortifacient.

6. Capsicum

Also known as *chillies (red pepper)* (Fig. 60.6) and is known for its pungent taste and odour.

They are commonly used in Indian kitchen. It is not fatal.

Fig. 60.6: Capsicum

Active agent(s)

Capsaicin and capsicin

Clinical Features

If large dose is ingested, it acts as an irritant and produces difficulty in swallowing and pain in stomach and esophagus. When applied to the eyes, severe irritation results with profuse lacrimation, burning pain and redness. It also irritates the skin.

Treatment

If ingested, ice cubes should be provided for sucking or ice cold water for drinking. Bulky foods and demulcents also bring relief. If applied to skin or eyes, wash with a large quantity of water. Lignocaine ointment may be used if the reaction is severe.

Significance

The children may accidentally touch the eyes with hands contaminated with powdered capsicum. Therefore, it needs to be kept away from reach of children. leading to severe irritation. The powdered form is sometimes thrown into the eyes of an enemy with malafide intention.

7. Plumbago

All parts of this plant are poisonous but roots have the highest concentration of poison. Two forms of the plant exist, *Plumbago rosea* (*lal chitra*) (Fig. 60.7a) and *Plumbgo zeylanica* (*chitra*)(Fig. 60.7b).

Fig. 60.7a: *Plumbago rosea*

Fig. 60.7b: *Plumbago zeylanica*

Active Agent
Plumbagin

Clinical Features

If the roots are ingested, there is burning sensation from mouth to stomach along with thirst, vomiting, diarrhea, collapse and death.

Fatal dose: Uncertain

Fatal period: Within a few hours

Treatment
Stomach wash, demulcents and symptomatic

Significance

Roots have been used in producing abortion, either by direct ingestion after application in the form of paste to the cervix. Rarely a paste made from roots has been employed to irritate the skin and simulate a bruise to falsely implicate an enemy. The juice turns black after exposure to air.

8. Semecarpus anacardium

It is also known by other names like *Bhalia* or *marking nut* (Fig. 60.8a). The seeds are heart shaped and blackish in color (Fig. 60.8b). The seeds yield an oily juice that is capable of irritating the skin.

Fig. 60.8a: The plant (*Semecarpus anacardium*)

Fig. 60.8b: The seeds of the plant

Active Agent(s)

Semicarpol and bhilawinol

Clinical Features

When the juice is applied to the skin, it causes irritation, painful blistering and eczematous eruption of the skin. Itching of skin followed by ulceration is present. When ingested, the juice produces blistering of the throat. Dysponea, cyanosis, hypotension and loss of reflexes are some of the other associated features. Delirium and coma are followed by death.

Fatal dose: 5 to 10 gm

Fatal period: About 12 to 24 hours

Treatment

Wash the affected part of the skin with plenty of water and soap or antiseptic. Gastric lavage is done and demulcents are given.

Significance

The most common manner of poisoning is accidental especially in the form of medicines administered by quacks. The juice is sometimes introduced into the vagina as a punishment for infidelity. The juice is also employed in the production of artificial bruise.

Table 60.1: Differences between a bruise and an artificial bruise produced by Semecarpus anacardium

Feature	True Bruise	Artificial bruise
Margins	Diffuse	Clear and sharp
Swelling	Present	Absent in some cases
Shape	Regular	Irregular
Blisters	Absent	Present
Extravasation in tissues	Present	Absent

9. Argimone Mexicana

All parts of this plant (Fig. 60.9a) and especially the seeds (Fig. 60.9b) are poisonous.

Fig. 60.9a: The plant (*Argimone mexicana*)

Fig. 60.9b: The seeds of the plant

Active Agent(s)

Berberine, protopine, sangunarine

Clinical Features

The features result from consumption of edible oil adulterated or contaminated with the argimone oil. It results in circulatory system failure, epidemic dropsy and neuropathy.

Treatment

The consumption of the offending oil should be stopped. Symptomatic treatment with good nursing care and physiotherapy form the mainstay of treatment in dropsy cases.

10. Eucalyptus Globules (Fig. 60.10)

Fig. 60.10: Eucalyptus globules

Active Agent

Cinehole

Clinical Features

Include cyanosis, dysponea, excitement convulsions and death due to failure of circulatory system.

Treatment

Maintain circulation and protect respiration, rest of the treatment is symptomatic.

Poisoning in adults may be accidental, suicidal or homicidal. In children, it is mainly accidental as they are easily attracted by the appearance of such plants. Due to lack of knowledge about the toxicity of the plant, they might ingest parts of such plants and it brings a trouble not only for the child but also for his family. As most of these plants are wild and grow on the roadside, the parents should regularly keep a vigil on such plants on the way to school of their children. The teachers of schools should also be trained about the identification of these plants and immediate remedial measures in a case of such poisoning. Sincere efforts should be made to remove these plants from school premises and playgrounds. Awareness campaigns should be organized to educate the community about the identification and appropriate immediate remedial measures to be taken in such poisoning cases.

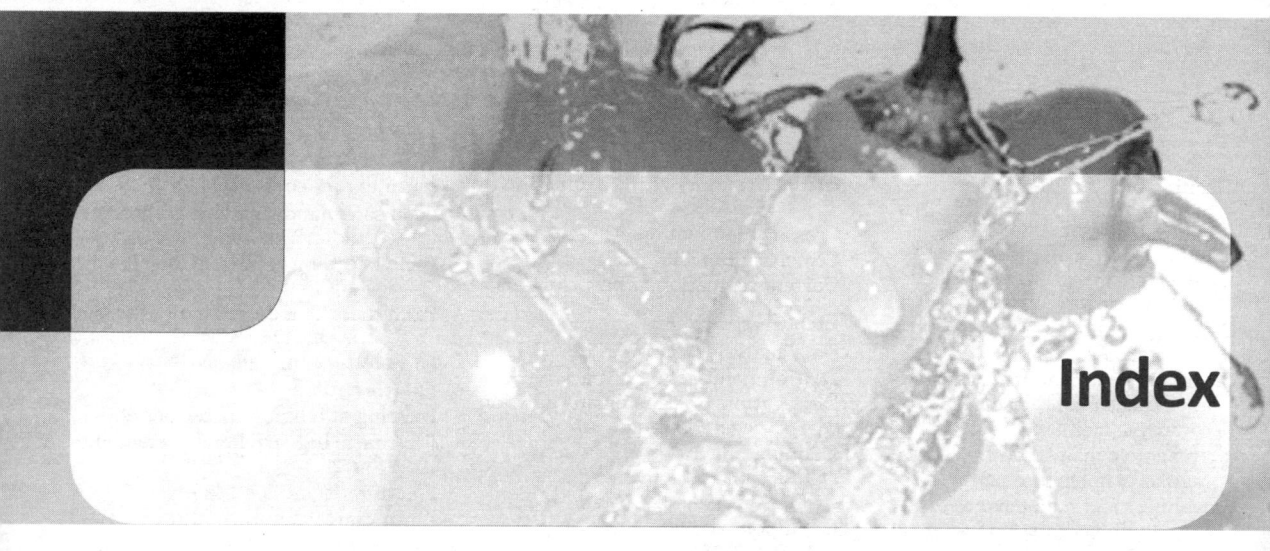

Index